CRISPR-Cas9 as a Revolutionary Gene Editing Tool

Genome Engineering, Advances and Applications in Agriculture

Editors

Abdul Basit

Department of Horticulture, College of Agriculture and Life Sciences
Kyungpook National University
Daegu, Republic of South Korea

Heba I. Mohamed

Department of Biological and Geological Sciences
Faculty of Education, Ain Shams University
Cairo, Egypt

Syed Tanveer Shah

Department of Agriculture, Faculty of Biological and Health Sciences
Hazara University, Mansehra, Pakistan

Mona FA Dawood

Botany and Microbiology Department
Assiut University, Assiut

CRC Press
Taylor & Francis Group
Boca Raton London New York

CRC Press is an imprint of the
Taylor & Francis Group, an **informa** business
A SCIENCE PUBLISHERS BOOK

Cover illustration provided by Heba I. Mohamed.

First edition published 2026
by CRC Press
2385 NW Executive Center Drive, Suite 320, Boca Raton FL 33431

and by CRC Press
4 Park Square, Milton Park, Abingdon, Oxon, OX14 4RN

CRC Press is an imprint of Taylor & Francis Group, LLC

Library of Congress Cataloging-in-Publication Data (applied for)

ISBN: 978-1-032-83553-2 (hbk)
ISBN: 978-1-032-85448-9 (pbk)
ISBN: 978-1-003-51820-4 (ebk)

DOI: 10.1201/9781003518204

Typeset in Times New Roman
by Prime Publishing Services

Preface

CRISPR-Cas9 (Clustered Regularly Interspaced Short Palindromic Repeats–CRISPR-associated protein 9) is a revolutionary gene editing technology that has transformed genetic research and biotechnology. Originally identified as part of a bacterial immune system, CRISPR-Cas9 allows for precise, efficient, and cost-effective modifications to DNA within living organisms. Its simplicity and versatility have made it an essential tool in molecular biology and agricultural sciences.

CRISPR-Cas9 employs a guide RNA (gRNA) to direct the Cas9 enzyme to a specific DNA sequence. Once bound, Cas9 creates a double-stranded break at the targeted site. The cell's natural repair mechanisms—non-homologous end joining (NHEJ) or homology-directed repair (HDR)—then fix the break, enabling gene knockout, insertion, or correction.

The precision of CRISPR-Cas9 has transformed genome engineering. Unlike older technologies such as zinc finger nucleases (ZFNs) and transcription activator-like effector nucleases (TALENs), CRISPR only requires designing a short gRNA instead of engineering complex proteins. This significantly reduces the cost and time needed for gene editing.

CRISPR-Cas9 stands at the forefront of a new era in genome engineering. Its simplicity, precision, and adaptability have already made profound impacts in agriculture, with the potential to tackle food security, climate change, and sustainable farming. As technological advances continue to refine its applications and safety, CRISPR-Cas9 is poised to become a cornerstone of modern agricultural biotechnology, fostering innovations that could benefit both humanity and the environment.

CRISPR-Cas9 has emerged as a powerful tool in agricultural biotechnology for enhancing crop resilience against biotic (living organisms) and abiotic (environmental) stresses. These stresses significantly affect crop yield and food security, particularly in the face of climate change and an increasing global population. By enabling precise genome editing, CRISPR-Cas9 allows researchers to develop stress-tolerant crops more efficiently and accurately than traditional breeding methods.

Biotic stresses are caused by harmful organisms such as bacteria, viruses, fungi, insects, and nematodes. Traditional pest control and disease management strategies frequently rely on the use of chemical pesticides, which can be harmful to the environment and human health. CRISPR-Cas9 has been instrumental in conferring resistance to these pathogens by editing specific genes involved in plant defense mechanisms. CRISPR has been used to disrupt viral replication pathways. CRISPR has enabled the deletion or modification of plant genes that pathogens exploit to establish infections. Gene editing can boost natural insect deterrent mechanisms. For example, CRISPR can enhance the production of toxic secondary metabolites or improve the signaling pathways that activate plant defenses against insect herbivory.

Abiotic stresses such as drought, salinity, heat, and cold significantly limit crop productivity. These stressors can result in lower yields and poor crop quality, severely affecting farmers' livelihoods. CRISPR-Cas9 helps modify stress response pathways to improve plant tolerance to these conditions. Drought and Heat Tolerance: Targeting genes involved in stomatal regulation, root architecture, or osmoprotectant synthesis can enhance water use efficiency and thermal tolerance. CRISPR-edited lines with improved expression of drought-responsive genes show better survival under water-deficient conditions. By editing ion transporters such as HKT1 or SOS1, CRISPR can reduce the accumulation of toxic sodium ions in plant tissues, thereby increasing salt tolerance.

CRISPR is also used to activate or suppress genes involved in antioxidant production, helping plants maintain cellular health under chilling or oxidative stress.

This book will explore all aspects of genome editing technologies, with a special focus on CRISPR-Cas9, which is capable of editing any gene in any plant species due to its simplicity, efficiency, and low cost. It will examine various applications related to CRISPR-Cas9, such as DNA repair pathways, mutation analysis, gene sequencing, nanotechnology, and the molecular mechanisms and current advancements related to CRISPR-Cas9. This book aims to utilize CRISPR/Cas9 as an effective tool for improving fruit crops, vegetables, and ornamental plants. Additionally, contemporary trends and strategies for climate-smart agriculture utilizing CRISPR/Cas9 editing technology will also be included. The current volume is interdisciplinary and will be highly beneficial for students, educators, researchers, plant physiologists, environmental scientists, and agrochemical companies. It will also hold immense value for graduate students in university agriculture, biotechnology, and environmental programs. Professionals or practitioners in research institutes focused on plant science, material science, and crop production will also gain from this book.

Abdul Basit
Department of Horticulture, College of Agriculture and Life Sciences
Kyungpook National University
41566, Daegu, Republic of South Korea

Heba I. Mohamed
Department of Biological and Geological Sciences
Faculty of Education, Ain Shams University, Cairo, Egypt

Syed Tanveer Shah
Department of Agriculture, Faculty of Biological and Health Sciences
Hazara University, 21120 Mansehra, Pakistan

Mona FA Dawood
Botany and Microbiology Department
Assiut University, Assiut

Contents

Chapter 1

Principles and Practices of CRISPR-Cas9 Genome Editing Techniques

Seher Yerlikaya,[1] *Abdullah Aydın,*[1] *Nisa Nur Yılmaz,*[1] *Bayram Ali Yerlikaya,*[1,*]
Musa Kavas[1,*] and *Izhar Ullah*[2]

1. Introduction

The global population is projected to reach 10 billion by 2050, necessitating a 25–70% increase in food production to meet rising demands (Hunter et al. 2017). This challenge is compounded by diminishing arable land, dwindling water resources, and the intensifying impacts of climate change. Enhancing agricultural productivity while adopting sustainable practices has become imperative to ensure global food security. Beyond agriculture, the need for precise genetic interventions extends to medicine, environmental conservation, and industrial biotechnology, underscoring the transformative potential of genome editing technologies.

The human being's effort to manipulate genetics predate modern science. Well before the advent of DNA discovery or the formalization of genetic theory, early agriculturalists influenced plant and animal genomes through selective breeding (Clark et al. 2005; Li et al. 2013). For instance, maize (*Zea mays*), a cornerstone of global agriculture, was domesticated approximately 9,000 years ago from its wild progenitor, *Zea mays* ssp. parviglumis, a teosinte species native to Mesoamerica. Early farmers selectively propagated plants with favorable traits, such as increased kernel size and the ability to pop, ultimately giving rise to modern maize varieties (Dorweiler et al. 1993; Piperno and Flannery 2001). While effective, traditional breeding was slow and imprecise, relying on phenotypic selection without an understanding of underlying genetic mechanisms.

The advent of molecular biology revolutionized genetic modification, enabling precise and accelerated interventions. The advent of recombinant DNA technology in the 1970s marked a pivotal moment, subsequently leading to the development of programmable endonucleases in the 21st century. Tools such as Zinc Finger Nucleases (ZFNs), Transcription Activator-Like Effector Nucleases (TALENs), and the Clustered Regularly Interspaced Short Palindromic Repeats (CRISPR)-Cas9 system have redefined genome engineering (Puchta et al. 1996; Symington and

[1] Department of Agricultural Biotechnology, Faculty of Agriculture, Ondokuz Mayıs University, Samsun, Turkey.
[2] Department of Horticulture, Faculty of Agriculture, Ondokuz Mayıs University, Samsun, Turkey.
* Corresponding author: musa.kavas@omu.edu.tr; bayram.yerlikaya@omu.edu.tr

DOI: 10.1201/9781003518204-1

Gautier 2011). These technologies induce targeted Double-Strand Breaks (DSBs) in DNA, which are repaired via Non-Homologous End Joining (NHEJ) or Homology-Directed Repair (HDR), facilitating insertions, deletions, or precise recombination (Puchta 2005).

Among these, CRISPR-Cas9 has emerged as a preeminent tool due to its simplicity, precision, and versatility. Originally identified as an adaptive immune system in bacteria (Ishino et al. 1987), CRISPR-Cas9 was repurposed for genome editing through seminal work demonstrating its programmable DNA cleavage capabilities (Jinek et al. 2012). The system comprises a Cas9 nuclease and a single guide RNA (sgRNA), which directs Cas9 to a specific DNA sequence complementary to the sgRNA's targeting region. Compared to ZFNs and TALENs, which require complex protein engineering, CRISPR-Cas9's reliance on RNA-based targeting simplifies design, reduces costs, and enables simultaneous editing of multiple genomic loci (Cong et al. 2013; Mali et al. 2013). These attributes have democratized genome editing, making it accessible to laboratories worldwide and accelerating discoveries across disciplines.

The significance of CRISPR-Cas9 extends beyond its technical advantages. By enabling precise modifications to the genetic code, it has reshaped the scientific landscape, offering solutions to pressing global challenges. In agriculture, CRISPR-Cas9 facilitates the development of crops with enhanced yield, disease resistance, and climate adaptability (Zaidi et al. 2019). In medicine, it holds promise for treating genetic disorders, cancers, and infectious diseases (Barrangou and Doudna 2016). Yet, its rapid adoption has also sparked ethical and regulatory debates, particularly regarding germline editing and the potential for unintended ecological consequences (Baltimore et al. 2015).

This chapter aims to provide a comprehensive exploration of CRISPR-Cas9, from its molecular mechanisms to its transformative applications. Subsequent chapters will elucidate the system's core components and mechanisms, criteria for designing effective experiments, delivery methods, biosafety considerations, and diverse applications across medicine, agriculture, and synthetic biology. The book also addresses the technology's limitations, including off-target effects and delivery challenges, while envisioning its future through emerging innovations such as base editing and prime editing (Anzalone et al. 2019). By bridging technical, practical, and ethical dimensions, this volume seeks to equip researchers, students, and policymakers with a nuanced understanding of CRISPR-Cas9's potential and challenges.

1.1 Molecular Mechanisms and Components of CRISPR-Cas9

Clustered Regularly Interspaced Short Palindromic Repeats (CRISPR) and their associated proteins (Cas) constitute a sophisticated adaptive immune system in bacteria and archaea, enabling defense against invading genetic elements such as viruses and plasmids. First identified in *Escherichia coli* near the alkaline phosphatase isozyme gene (iap) (Ishino et al. 1987), CRISPR loci consist of short palindromic DNA repeats interspersed with unique spacer sequences (32–36 bp) derived from past viral infections. These spacers serve as a genetic memory, allowing the host to recognize and neutralize recurring threats. Upon transcription, CRISPR sequences generate CRISPR RNAs (crRNAs), which, in conjunction with trans-activating crRNA (tracrRNA) and Cas proteins, orchestrate precise DNA cleavage (Horvath and Barrangou 2010; Jinek et al. 2012) (Fig. 1).

Among the diverse CRISPR-Cas systems, the Type II CRISPR-Cas9 system from Streptococcus pyogenes has become the cornerstone of genome editing due to its simplicity and efficacy (Garneau et al. 2010). In its natural context, CRISPR-Cas9 provides adaptive immunity by targeting and cleaving foreign DNA that matches the spacer-derived guide RNA sequence. This mechanism was repurposed for genome editing through seminal work demonstrating that a single guide RNA (sgRNA)—a synthetic fusion of crRNA and tracrRNA—can direct the Cas9 nuclease to specific DNA sequences with high precision (Jinek et al. 2012). The adaptability of CRISPR-Cas9 has revolutionized biotechnology, providing a versatile platform for applications ranging from microbial engineering to plant and human genomics (Doudna and Charpentier 2014) (Fig. 1).

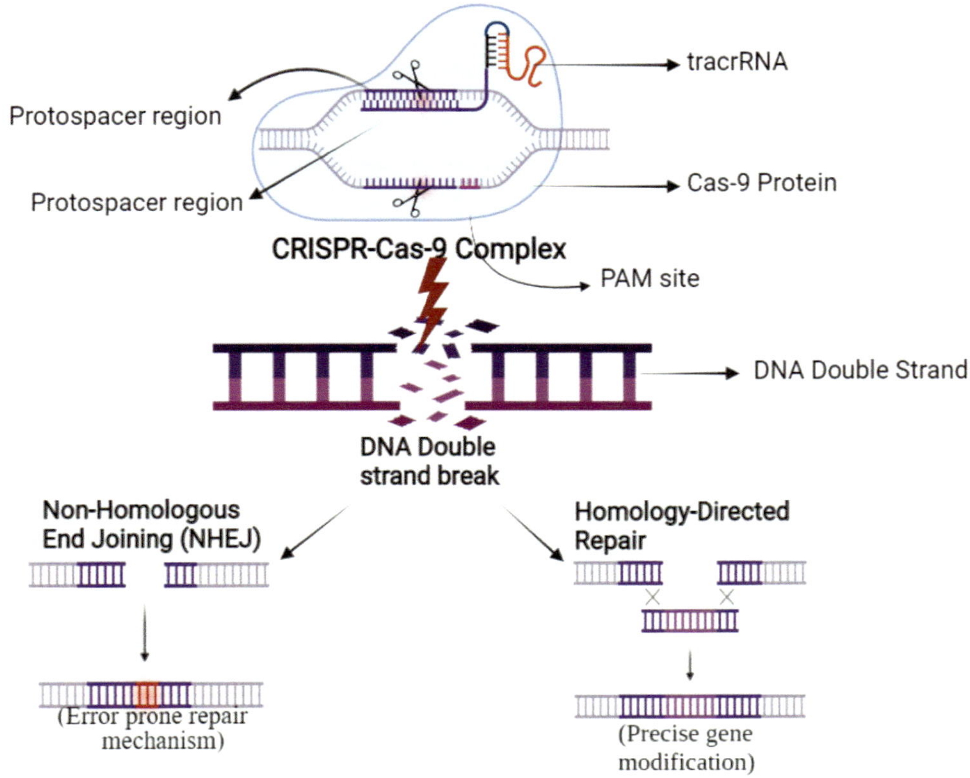

Figure 1. Molecular Mechanism of CRISPR-Cas9 (Javaid et al. 2022). This schematic diagram illustrates the stepwise process of CRISPR-Cas9-mediated genome editing. The figure illustrates the Cas9-sgRNA complex binding to the target DNA, where the 20-nucleotide guide sequence of the sgRNA pairs with the complementary DNA strand, and the PAM (NGG) sequence is highlighted on the non-complementary strand. Cas9 are shown cleaving the target and non-target strands, respectively, to generate a double-strand break. Subsequent repair pathways—non-homologous end joining and homology-directed repair—are represented by arrows indicating potential outcomes: small insertions/deletions (indels) for NHEJ or precise modifications for HDR. The diagram employs a minimalist design with monochromatic or muted colors, using clear labels and a sans-serif font to maintain an academic aesthetic suitable for publication.

The CRISPR-Cas9 system comprises two core components: the Cas9 protein and the sgRNA. The Cas9 nuclease, a large multi-domain enzyme, possesses two distinct catalytic domains: HNH, which cleaves the DNA strand complementary to the sgRNA, and RuvC, which cuts the non-complementary strand, resulting in a Double-Strand Break (DSB) (Jinek et al. 2012). The sgRNA, typically 100 nucleotides long, consists of a 20-nucleotide guide sequence that base-pairs with the target DNA and a scaffold sequence that binds Cas9, forming a functional ribonucleoprotein complex (Anders et al. 2014). The recognition of the target sequence is also dependent on the Protospacer Adjacent Motif (PAM), a short nucleotide sequence (NGG for S. pyogenes Cas9) situated directly downstream of the target-on the non-complementary strand. The PAM plays a critical role in ensuring specificity, as it confines Cas9 activity to genomic loci flanked by this motif, minimizing the risk of off-target cleavage (Sternberg et al. 2014) (Fig. 1).

Once a DSB is introduced, cellular repair mechanisms determine the outcome of the edit. In most eukaryotic systems, including plants, DSBs are predominantly repaired via Non-Homologous End Joining (NHEJ), an error-prone process that introduces small insertions or deletions (indels) at the cleavage site, often disrupting gene function (Puchta 2005). In contrast, Homology-Directed Repair (HDR) facilitates the introduction of precise genetic modifications, such as gene insertions and point-mutations, through using of a donor DNA template. Despite its precision, HDR is less efficient in plants due to the need for specific cell cycle conditions and the availability of the appropriate template (Puchta et al. 1996; Li et al. 2013). These repair pathways underpin the versatility of CRISPR-Cas9, enabling both gene knockouts and precise genetic alterations.

Beyond its canonical role as a nuclease, Cas9 has been engineered to expand its utility. Deactivated Cas9 (dCas9), which lacks nuclease activity, can be fused to effector domains to modulate gene expression or introduce epigenetic modifications (Gilbert et al. 2013). Base editors, comprising dCas9 fused to cytidine or adenine deaminases, enable single-nucleotide conversions (e.g., C-to-T or A-to-G) without inducing DSBs, providing a precise alternative for applications in crops like maize, rice, and wheat (Komor et al. 2016; Rees and Liu 2018). These innovations highlight the adaptability of the CRISPR-Cas9 platform, broadening its scope beyond traditional genome editing.

In plants, CRISPR-Cas9 has been optimized to overcome challenges associated with complex genomes and transformation inefficiencies. Codon optimization of Cas9 enhances its expression in plant cells, while strong promoters and versatile expression cassettes ensure robust co-delivery of Cas9 and sgRNA (Li et al. 2013). These modifications have facilitated precise genome editing in diverse species, enabling traits such as disease resistance and environmental resilience (Zaidi et al. 2019). However, efficient delivery of CRISPR components remains critical, with methods ranging from Agrobacterium-mediated transformation to ribonucleoprotein-based approaches (Woo et al. 2015). These delivery strategies, explored in detail in subsequent chapters, underscore the importance of integrating molecular mechanisms with practical implementation.

The molecular precision of CRISPR-Cas9 distinguishes it from earlier genome editing technologies, such as ZFNs and TALENs, which rely on protein-DNA interactions and are more cumbersome to design (Carroll 2014). By leveraging RNA-DNA base pairing, CRISPR-Cas9 offers unmatched flexibility, scalability, and accessibility, positioning it as a transformative tool in biotechnology. Understanding its molecular mechanisms and components not only illuminates its current applications but also lays the foundation for future innovations, such as multiplexed editing, epigenetic modulation, and next-generation editing systems.

2. CRISPR-Cas9 in Biotechnology: Significance, Design, and Delivery

The groundbreaking potential of CRISPR-Cas9 in plant biotechnology is supported by robust experimental design and optimized delivery systems. These twin pillars—molecular precision and practical implementation—enable researchers to translate genomic insights into tangible agricultural innovations. Designing CRISPR-Cas9 experiments involves a cascade of decisions, from selecting the appropriate Cas protein and designing specific single-guide RNAs to choosing vectors and delivery methods tailored to the target plant species (Fig. 2). Each step requires balancing specificity, efficiency, and scalability, particularly in plants where complex genomes and transformation barriers pose unique challenges. This chapter elucidates the critical criteria for crafting effective CRISPR-Cas9 experiments, explores cutting-edge delivery strategies, and reflects on their role in bridging molecular biology with sustainable agriculture.

The cornerstone of a CRISPR-Cas9 experiment is defining the editing objective, which shapes the choice of Cas protein and editing strategy. For broad-spectrum gene knockouts via random insertions or deletions (indels), *Streptococcus pyogenes* Cas9 (SpCas9) remains a workhorse due to its high cleavage efficiency and extensive characterization (Jinek et al. 2012). For example, SpCas9 was instrumental in mutating the *SlPHD_MS1* gene in tomato using the pHSE401 vector, yielding male-sterile plants critical for hybrid breeding (Gökdemir et al. 2022). Similarly, targeting the *OsSAPK2* gene in rice produced drought-tolerant mutants, showcasing SpCas9's utility in stress adaptation (Lou et al. 2017). For precise edits, base editors—fusions of deactivated Cas9 (dCas9) with deaminases—offer a refined approach. Cytosine Base Editors (CBEs) enable the direct conversion of cytosine to thymine within a single-stranded DNA R-loop, offering a method for introducing point mutations without generating double-strand breaks (Komor et al. 2016). Recent

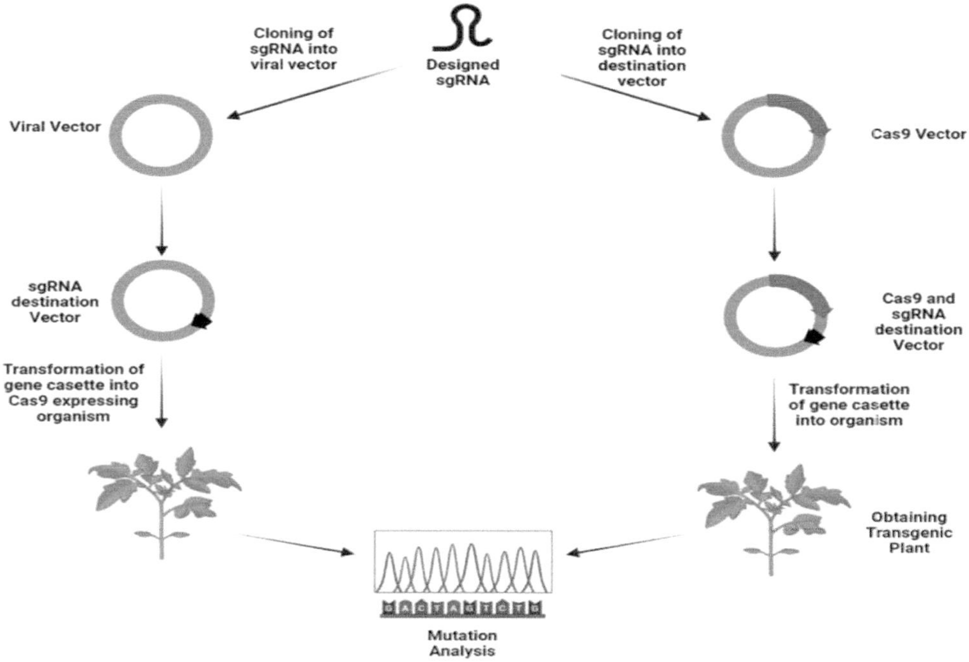

Figure 2. CRISPR-mediated Genome Editing Process in Plants: A Schematic Overview. This diagram captures the complete workflow, from selecting target genes and designing sgRNAs to constructing vectors, delivering CRISPR components, and screening for successful mutations.

advancements, such as the development of Adenine Base Editors (ABEs) for A-to-G conversions, have expanded the toolkit, as demonstrated in rice where ABE-mediated editing of the *OsSPL14* gene enhanced grain yield under variable conditions (Li et al. 2023). The choice of Cas variant hinges on the genomic target and desired outcome, underscoring the need for strategic alignment at the experiment's outset.

sgRNA design is the linchpin of CRISPR-Cas9 specificity, requiring meticulous attention to target selection and off-target minimization. The process begins with identifying the Protospacer Adjacent Motif (PAM), typically 5'-NGG-3' for SpCas9, within the target region (Anders et al. 2014). The 20-nucleotide sequence upstream of the PAM must precisely match the target DNA to guide Cas9 cleavage. However, factors such as GC content, secondary structure, and chromatin accessibility can compromise sgRNA efficacy, necessitating computational tools for optimization (Hsu et al. 2013). Table 1 presents key sgRNA design tools widely used in plant genome editing.

CRISPR-P V2.0 employs a mismatch-based scoring algorithm to predict sgRNA specificity, providing a streamlined interface for plant researchers (Liu et al. 2017). CHOPCHOP supports diverse Cas variants, including Cas13 for RNA editing, making it versatile for non-canonical

Table 1. sgRNA design tools for editing of plant genomes.

Tool	Genome Number	Website	Reference
CRISPR-P V2.0	82 (Plant)	http://crispr.hzau.edu.cn/CRISPR2/	(Palindromic Repeats CRISPR-associated 2014)
CHOPCHOP	76 (Plant)	https://chopchop.cbu.uib.no/	(Labun, Montague et al. 2019)
CRISPOR	1170 (All)	http://crispor.gi.ucsc.edu/	(Concordet and Haeussler 2018)

CRISPR applications (Labun et al. 2019). CRISPOR, with its expansive genome database, facilitates sgRNA design across species, though its complexity may deter novice users (Concordet and Haeussler 2018). Recent innovations, such as machine learning-driven tools like DeepCRISPR, enhance off-target prediction by integrating epigenetic and sequence data, as demonstrated in maize where optimized sgRNAs reduced unintended edits by 30% (Chu et al. 2024). These tools collectively empower researchers to navigate the delicate balance between on-target efficiency and genome-wide specificity.

Vector selection bridges molecular design with cellular delivery, ensuring robust expression of CRISPR components. Vectors typically encode Cas9 under constitutive promoters (e.g., CaMV 35S) and include sgRNA cloning sites (Li et al. 2013). In some designs, sgRNAs are delivered via viral vectors to Cas9-expressing transgenic plants, enabling transient edits without genomic integration (Yin et al. 2015). A key advancement in genome editing is the adoption of modular vectors, like pGreen, which support the multiplexed expression of sgRNAs, allowing for the concurrent targeting of several genes. This is exemplified in wheat, where the knockout of three *TaMLO* alleles led to enhanced resistance to powdery mildew (Wang et al. 2022).

Delivery of CRISPR-Cas9 components into plant cells remains a critical challenge, with methods varying by species, tissue type, and regulatory goals. *Agrobacterium*-mediated transformation is the gold standard due to its versatility and high efficiency across dicots and some monocots. For instance, Li et al. (2013) used the pUC119RCS vector to edit the *AtPDS3* gene in *Arabidopsis thaliana*, achieving albino phenotypes with minimal off-target effects. In tomato, Agrobacterium delivery of the pHSE401 vector targeting *SlPHD_MS1* produced male-sterile lines, a breakthrough for hybrid seed production (Gökdemir et al. 2022). Table 2 summarizes recent examples of *Agrobacterium*-mediated CRISPR-Cas9 delivery, incorporating new studies to reflect ongoing innovations.

Alternative delivery methods address Agrobacterium's limitations, particularly for recalcitrant species or transgene-free applications. Polyethylene glycol (PEG)-mediated protoplast transfection delivers Cas9-sgRNA ribonucleoproteins (RNPs), bypassing stable integration. This approach was refined in soybean, where PEG-delivered RNPs targeting the GmFT2a gene enhanced flowering time control without foreign DNA (Zhang et al. 2023). Biolistic transformation, or gene gun delivery, excels in monocots like maize, where edits to the *ZmCENH3* gene improved haploid induction (Liu et al. 2024). However, its high cost and tissue damage risk limit scalability. A groundbreaking

Table 2. In different plant species, *Agrobacterium*-mediated delivery of CRISPR/Cas9 components.

Plant-Species	**Vector**	***Agrobacterium* strain**	**Target Gene(s)**	**Reference**
Arabidopsis thaliana	pUC119RCS	GV3101	*AtPDS3, AtFLS2, RACK1b, RACK1c*	(Li, Norville et al. 2013)
Banana	pRGEB31	AGL1	*RAS-PDS*	(Kaur, Alok et al. 2018)
Citrus sinensis	pCas9-GN	LBA4404	*CsWRKY22*	(Wang, Chen et al. 2019)
Cucumis sativum	pRCS	EHA105	*elF4E, elF(iso)4E*	(Chandrasekaran, Brumin et al. 2016)
Nicotiana benthamiana	pUC19, pKQ334	GV3101	*NbPDS, NbIspH*	(Yin, Han et al. 2015)
Nicotiana tabacum	pORE	LBA4404	*NtPDS, NtPDR6*	(Gao, Wang et al. 2015)
Oryza sativum	VK005	EHA105	*ISA1*	(Shufen, Yicong et al. 2019)
Solanum lycopersicum	pHSE401	GV3101	*SlPHD_MS1*	(Gökdemir, Seçgin et al. 2022)

development is nanoparticle-mediated delivery, which uses carbon nanotubes to transport Cas9 RNPs into plant cells. In tobacco, this method achieved 40% editing efficiency for the *NbPDS* gene without integration, providing a scalable, transgene-free alternative (Demirer et al. 2023). These innovations highlight the dynamic evolution of delivery strategies, each tailored to specific experimental and regulatory contexts.

The interplay of design and delivery not only drives technical success but also shapes CRISPR's broader impact on agriculture. Precision in sgRNA design minimizes off-target effects, as seen in rice where AI-optimized sgRNAs targeting *OsERF922* enhanced blast resistance with negligible collateral edits (Kim et al. 2023). Delivery innovations, such as nanoparticle systems, align with regulatory demands for non-transgenic crops, facilitating market acceptance (Eckerstorfer et al. 2019). Yet, challenges persist: low transformation efficiencies in elite cultivars, variable editing outcomes across genotypes, and public skepticism toward genome editing. Addressing these requires integrating molecular rigor with interdisciplinary insights—combining bioinformatics for sgRNA optimization, synthetic biology for vector engineering, and social science to navigate ethical concerns.

CRISPR-Cas9's success in plants exemplifies this synergy. For instance, editing the *eIF4E* gene in cucumber conferred broad-spectrum viral resistance, leveraging Agrobacterium delivery and CHOPCHOP-designed sgRNAs (Chandrasekaran et al. 2016). By employing Cytosine Base Editors (CBE) to precisely modify *OsSWEET14*, researchers have successfully blocked bacterial blight susceptibility in rice, a process facilitated by the use of viral vectors for efficient genetic modification (Blanvillain-Baufumé et al. 2017). These outcomes, explored further in later chapters, underscore that experiment design is not merely a technical exercise but a bridge to sustainable agriculture. By optimizing specificity, efficiency, and delivery, researchers can unlock CRISPR's potential to address global challenges—food security, climate resilience, and biodiversity—while fostering trust in genome-edited crops.

3. Applications of CRISPR-Cas9 Across Disciplines

Initially a natural defense mechanism in bacteria, CRISPR-Cas9 has quickly evolved into a cornerstone of biotechnology, revolutionizing genome editing with its unparalleled precision and flexibility. Its applications extend across diverse fields, from enhancing agricultural resilience and productivity to advancing synthetic biology, environmental biotechnology, and microbial engineering, providing innovative solutions to global issues such as climate change, food security and sustainable resource management. By combining molecular precision with innovative delivery systems, CRISPR-Cas9 fosters interdisciplinary synergies that amplify its impact. This chapter explores these multifaceted applications, emphasizing agriculture's central role while highlighting emerging domains that broaden biotechnology's horizons.

CRISPR-Cas9 has revolutionized plant biotechnology, bringing unparalleled precision to genetic modification and accelerating crop improvement. Unlike traditional breeding, which is slow and imprecise, CRISPR technology facilitates precise genome editing to improve yield, bolster pathogen resistance, and enhance nutritional content, establishing it as a foundational tool for sustainable agriculture (Zaidi et al. 2019). Its deployment has shifted paradigms, offering rapid solutions to global crop losses, estimated at 20–40% annually due to pests, viruses, and bacteria, with staples like rice, wheat, and maize among the most affected (Jones 2021).

CRISPR-Cas9 enhances plant immunity through two primary strategies: direct targeting of pathogen genomes and disruption of plant Susceptibility (S) genes. Targeting gemini viruses, a group of single-stranded DNA (ssDNA) viruses including *Tomato yellow leaf curl virus* and *Beet curly top virus*, has emerged as a successful approach to enhance viral resistance in plants. By introducing insertions or deletions (indels) into viral genomes, CRISPR interrupts replication, reducing viral loads and disease symptoms (Ali et al. 2015). For RNA viruses, Cas9 variants and Cas13 systems, particularly Cas13d, target pathogens like *Tobacco Mosaic Virus* (TMV) and C*ucumber Mosaic*

Virus (CMV), providing robust control of economically significant viroids (Mahas et al. 2019). Disrupting S genes, such as *eIF4E*, has conferred resistance to potyviruses in crops like cucumber and Arabidopsis (Chandrasekaran et al. 2016; Pyott et al. 2016). In cassava, editing *eIF4E* isoforms mitigated *Cassava brown streak virus* symptoms, showcasing CRISPR's adaptability (Gomez et al. 2019).

Bacterial resistance leverages similar precision. In rice, editing OsSWEET genes (*OsSWEET11*, *OsSWEET13*, *OsSWEET14*) blocked *Xanthomonas oryzae*-induced blight, enhancing field resilience (Blanvillain-Baufumé et al. 2017). In tomato, modifying *SlJAZ2* and *SlDmr6-1* conferred resistance to bacterial speck and broader pathogens, boosting immune responses (Ortigosa et al. 2019). A landmark achievement is the mitigation of citrus canker via *CsLOB1* editing, reducing *Xanthomonas citri* symptoms and safeguarding a critical industry (Jia et al. 2017; Peng et al. 2017). Recent studies, such as multiplexed editing of ZmPIF genes in maize, have further expanded resistance profiles, combining drought tolerance and pathogen defense (Liu et al. 2025).

CRISPR-Cas9 optimizes crop yield by targeting developmental and metabolic pathways. Editing *OsLOGL5*, a cytokinin-activating enzyme in rice, increased grain production across environments (Wang et al. 2020). Knocking out *cytokinin oxidase/dehydrogenase* (*CKX*) genes in wheat enhanced yield by preserving cytokinin levels (Zhang et al. 2019). Nutrient allocation, modified via the *amino acid permease 3* gene, boosted rice tiller numbers without compromising grain quality (Lu et al. 2018). Targeted editing of yield-related genes—such as *OsPIN5b* for panicle size, *OsGS3* for grain size, and *TaGW2*, *OsGW2*, and *OsGW5* for grain weight—has resulted in significant improvements in rice and wheat productivity (Lu et al. 2018). In fruit crops, editing CLV and ENO genes enlarged meristems, enhancing yields (Yuste-Lisbona et al. 2020).

Nutritional enhancements address consumer and industrial needs. Disrupting *GBSS1* produced low-amylose waxy maize across 12 elite lines, ideal for textiles and adhesives (Gao et al. 2020). In rice, base editing of *GBSS1* tailored amylose content for culinary preferences (Sun et al. 2017). High-amylose rice, achieved by targeting starch branching enzymes, offers health benefits (Sun et al. 2017). In wheat, editing conserved gluten gene regions reduced immunoreactivity by 85%, addressing coeliac disease sensitivities (Sánchez-León et al. 2018). CRISPR has also enriched crops with carotenoids, γ-aminobutyric acid, and high oleic acid, improving nutritional profiles (Li et al. 2018; Do et al. 2019).

CRISPR-Cas9 has introduced consumer-focused traits, such as non-browning apples engineered by silencing polyphenol oxidase genes, reducing waste while preserving taste (Maxmen 2017). In *Camelina sativa*, editing enhanced omega-3 oil production, offering sustainable nutritional sources (Waltz 2018). These innovations highlight CRISPR's versatility in meeting diverse agricultural demands, from farm to table.

CRISPR's agricultural applications, supported by delivery methods like nanoparticles and AI-optimized sgRNAs (Demirer et al. 2023; Chu et al. 2024), ensure precision and scalability. However, challenges like viral adaptation and CRISPR-resistant pathogens persist, necessitating ongoing innovation to sustain these gains.

Synthetic biology harnesses CRISPR-Cas9 to design novel genetic systems, reengineering organisms for enhanced functionality. In plants, editing *ZmFPS1* in maize boosted terpenoid production for biofortification, demonstrating pathway optimization (Chen et al. 2023). In microbes, CRISPR redesigned Saccharomyces cerevisiae by editing *ScYAP1*, increasing ethanol yields by 15% for biofuels (Wang et al. 2024). CRISPR interference (CRISPRi) with deactivated Cas9 (dCas9) modulated *AtFLC* in Arabidopsis, fine-tuning flowering time reversibly (Zhu et al. 2020). CRISPR-Cas9 addresses ecological challenges through bioremediation and carbon sequestration. Editing *OsNRAMP5* in rice reduced cadmium uptake, mitigating soil contamination (Tang et al. 2024). These applications underscore CRISPR's role in creating adaptive, high-value biological systems that align with global sustainability goals by restoring ecosystems and combating climate change. Microbial engineering optimizes industrial strains via CRISPR-Cas9. Editing *EcPfkA* in *Escherichia coli* increased lactic acid production by 30%, supporting biodegradable plastics

(Zhang et al. 2023). In *Clostridium thermocellum*, deletion of the *Cel48S* cellulase (also called *CelS*) significantly slowed crystalline cellulose hydrolysis—reducing hydrolysis rate by ~ 60%—and lowered biomass yield, demonstrating that *Cel48S* plays a critical role in cellulose breakdown, though complete solubilization can still occur in its absence (Olson et al. 2010). These advancements drive sustainable manufacturing, reducing reliance on fossil resources.

CRISPR-Cas9's applications bridge agriculture, synthetic biology, environmental biotechnology, and microbial engineering, creating a unified platform for innovation. In agriculture, it secures food systems; in synthetic biology, it designs novel organisms; in environmental biotechnology, it restores ecosystems; and in microbial engineering, it powers green industries. Delivery innovations (Demirer et al. 2023) and AI-driven design (Chu et al. 2024) amplify these impacts, but regulatory, ethical, and ecological challenges remain. By fostering collaboration across disciplines, CRISPR-Cas9 can address global needs—hunger, pollution, resource scarcity—while ensuring equitable and sustainable outcomes.

4. Biosafety, Ethical Concerns, and Regulatory Frameworks for CRISPR-Cas9

By enabling precise genetic alterations, CRISPR-Cas9 has become a cornerstone of modern plant biotechnology, with applications extending from resistance to plant pathogens in rice (*OsSWEET14*; Blanvillain-Baufumé et al. 2017) to biofortification strategies in maize (*GBSS1*; Gao et al. 2020), highlight its potential to address critical global issues like climate change and food insecurity. However, these advances introduce complex biosafety risks, ethical dilemmas, and regulatory challenges that demand careful stewardship. Off-target mutations, ecological impacts, socioeconomic disparities, and public skepticism underscore the need for robust frameworks to ensure responsible use. This chapter examines the biosafety considerations, ethical concerns, and regulatory systems governing CRISPR-Cas9 in agriculture, emphasizing a balanced approach to innovation and responsibility.

CRISPR-Cas9's precision surpasses earlier tools like ZFNs, yet biosafety risks persist, particularly off-target mutations and ecological effects. Early studies in rice revealed unintended edits disrupting non-target genes, potentially affecting plant fitness or ecological interactions (Zhang et al. 2018). Advances like AI-driven sgRNA design (DeepCRISPR; Chu et al. 2024) have reduced off-target rates by up to 30% in maize, but comprehensive genomic screening remains critical, as seen in *OsSPL14* editing for yield enhancement (Li et al. 2023). Compared to traditional breeding, where chemical-induced mutations are 1,000 times higher than natural rates, CRISPR's off-target effects are minimal (Huang et al. 2016). Nonetheless, rigorous validation is essential to ensure safety.

Ecological risks include gene flow from edited crops to wild relatives, potentially creating hybrids with unpredictable traits. For example, *SlJAZ2*-edited tomatoes, resistant to multiple pathogens, require field trials to assess crossbreeding risks (Ortigosa et al. 2019). Containment strategies, such as buffer zones and male sterility (e.g., *SlPHD_MS1* in tomato; Gökdemir et al. 2022), mitigate gene transfer, as recommended by environmental assessments (Ellstrand 2003). Transgene-free methods, like nanoparticle-mediated delivery, further reduce risks by avoiding foreign DNA integration, achieving 40% editing efficiency in tobacco (*NbPDS*; Demirer et al. 2023). These innovations align with biosafety goals, ensuring CRISPR-edited crops integrate safely into ecosystems.

CRISPR-Cas9 raises multifaceted ethical questions, spanning environmental, health, socioeconomic, and cultural dimensions. Food safety is a primary concern, with debates over potential allergenicity or toxicity from edited crops. While no evidence links CRISPR-edited foods to health risks, long-term studies are vital, as highlighted in GMO critiques (Domingo and Bordonaba 2011). Low-gluten wheat-varieties, such as those developed by Sánchez-León et al. (2018), illustrate the need for extensive testing to confirm decreased immune responses in sensitive individuals. Transparent labeling, as advocated by Nicolia et al. (2014), empowers consumers to make informed choices, aligning with ethical principles of autonomy.

Socioeconomic challenges arise from dependency on biotech corporations. Patents on seeds like *TaGW2*-edited wheat increase costs, limiting smallholder farmers' access and deepening inequalities (Qaim and Kouser 2013). In regions reliant on subsistence farming, replacing indigenous varieties with edited crops (*eIF4E*-resistant cassava; Gomez et al. 2019) threatens biodiversity and cultural heritage. Ethical frameworks must prioritize equitable access, supporting farmers through open-source models or subsidies to preserve traditional practices alongside innovation.

Intellectual Property Rights (IPR) pose additional dilemmas. Patents on CRISPR-edited traits, such as *OsLOGL5* for rice yield (Wang et al. 2020), can foster monopolies, stifling research, and farmer access. Alternative models, like open-source biotechnology, balance innovation with inclusivity, ensuring genetic resources remain widely available (Graff et al. 2003). Public perception further complicates ethics, driven by mistrust over corporate control and environmental risks. Transparent communication and stakeholder engagement, as seen in dialogues around *CsLOB1*-edited citrus (Jia et al. 2017), are critical to building trust and addressing skepticism (Dessie and Zegeye 2024).

To ensure the safe and ethical application of CRISPR-Cas9, regulatory mechanisms are in place that reconcile scientific progress with precautionary principles. Instruments like the Cartagena Protocol on Biosafety oversee the movement and management of GMOs, prioritizing ecological integrity and public health through international cooperation (CBD 2000). These frameworks inform national policies, harmonizing standards for crops like ZmPIF-edited maize (Liu et al. 2025).

National agencies oversee approvals. In the US, the USDA, EPA, and FDA share responsibilities, evaluating safety, environmental impact, and health for crops like non-browning apples (Maxmen 2017). The EU's EFSA conducts rigorous risk assessments, classifying most CRISPR-edited crops as GMOs, which delays commercialization compared to the US's transgene-free exemptions (EFSA 2024; FDA 2024). China's flexible policies accelerate innovations like *OsSPL14* rice (Li et al. 2023), but risk oversight gaps. Globally, only six CRISPR-edited crops are commercialized, reflecting regulatory caution (Schmidt et al. 2020).

Risk assessments are central, addressing allergenicity, toxicity, and ecological effects. Field trials for *OsSWEET14*-edited rice confirmed blight resistance without unintended impacts (Blanvillain-Baufumé et al. 2017). Post-authorization monitoring and buffer zones, as applied to *TaGW2* wheat, mitigate gene flow (Zhang et al. 2019). Computational tools targeting off-target risks, such as CRISPOR (Concordet and Haeussler 2018), enhance safety, addressing concerns about natural reproductive barriers (Zess and Begemann 2021). Public consultations, integral to EU and US processes, ensure diverse perspectives shape approvals (Dessie and Zegeye 2024).

CRISPR-Cas9's agricultural promise—exemplified by eIF4E-resistant cucumber (Chandrasekaran et al. 2016) and omega-3-enhanced Camelina (Waltz 2018)—hinges on addressing biosafety, ethical, and regulatory challenges. Its risks, while lower than chemical mutagenesis, require vigilant management through AI tools and transgene-free methods (Mishra et al. 2024). Ethically, equitable access and cultural preservation demand inclusive policies, countering corporate dominance. Regulatory harmonization, guided by frameworks like Cartagena, can accelerate safe adoption while respecting regional priorities (Turnbull et al. 2023). Public trust, fostered through transparency and engagement, is paramount, as mistrust can derail even safe innovations (Jasanoff and Hurlbut 2024). By integrating science, policy, and societal values, CRISPR-Cas9 can deliver sustainable crops while safeguarding ecosystems and communities, ensuring biotechnology serves global needs responsibly.

5. Limitations and Future Prospects of CRISPR-Cas9

CRISPR-Cas9 has revolutionized plant biotechnology, enabling precise genome editing to enhance crop traits, as demonstrated in applications like *OsSWEET14*-edited rice for blight resistance

(Blanvillain-Baufumé et al. 2017) and *GBSS1*-edited maize for nutritional improvement (Gao et al. 2020). Its integration with synthetic biology, merging engineering principles with plant biology, promises innovative systems with predictable behaviors (Nemhauser and Torii 2016). However, technical, and practical limitations—ranging from gene delivery challenges to off-target effects—constrain its full potential. Addressing these barriers through emerging technologies and novel strategies can unlock transformative prospects, from nitrogen fixation to enhanced genetic diversity, ensuring CRISPR-Cas9 meets global demands for food security and sustainable agriculture.

Delivery of CRISPR machinery into plant cells remains challenging due to the rigid nature of the plant cell wall, which restricts both efficiency and specificity. Current methods, such as Agrobacterium-mediated transformation (Table 2, Li et al. 2013), are effective for species like tomato (*SlPHD_MS1*; Gökdemir et al. 2022) but rely on labor-intensive tissue culture, increasing costs and restricting scalability (Gelvin 2010). Species-specific constraints further complicate delivery, particularly in monocots like wheat. Modifying genes like *Baby boom* (*BBM*) and *Wuschel-related homeobox 2* (*Wus2*) has improved transformation efficiency in maize and sorghum, achieving up to 50% higher rates (Lowe et al. 2016; Lowe et al. 2018). However, these methods remain genotype-dependent, limiting broad applicability.

Off-target cleavage, though less frequent than with TALENs or ZFNs, remains a concern. Whole-genome sequencing in rice (Tang et al. 2018), cotton (Li et al. 2019), and Arabidopsis (Feng et al. 2014) shows that Cas9 and Cas12a are highly specific, yet unintended edits can occur. These effects, while manageable through outcrossing in commercial breeding, pose academic and regulatory challenges, as discussed in the biosafety chapter (Zess and Begemann 2021). Strategies like high-fidelity Cas9-HF1, ribonucleoproteins (RNPs), and AI-optimized sgRNAs (DeepCRISPR; Chu et al. 2024) reduce off-target rates by 30% in maize, but optimization is ongoing.

Unlike gene knockouts, knock-in techniques for precise insertions are less developed in plants compared to mammals. Successful knock-ins, such as *ZmFPS1* for terpenoid production (Chen et al. 2023), require optimized constructs, efficient delivery, and aligned DNA repair pathways (Collonnier et al. 2017). Unwanted rearrangements, observed in early wheat trials, highlight the need for refined protocols (Permyakova et al. 2021). These technical hurdles limit applications like nitrogen fixation, where precise integration of nif genes is critical (Temme et al. 2012).

The reliance on tissue culture and specialized equipment makes CRISPR-Cas9 costly, particularly for resource-limited regions. Delivery methods like biolistics, used for *ZmCENH3* in maize (Liu et al. 2024), are expensive and less efficient than desired. Scaling CRISPR to diverse crops requires cost-effective solutions, such as tissue culture-free methods targeting meristems or germlines (Lacroix and Citovsky 2016).

CRISPR's efficacy varies across species, with monocots like rice responding better than recalcitrant crops like soybean. Even within species, genotype variability affects outcomes, as seen in *TaGW2* wheat trials (Zhang et al. 2019). Developing universal delivery systems, such as novel bacterial vectors or viral particles, could overcome these barriers.

Regulatory uncertainty, as noted in the biosafety chapter (Schmidt et al. 2020), slows commercialization, with only six CRISPR-edited crops approved globally. Public skepticism, driven by GMO concerns, further complicates adoption, necessitating transparent communication (Dessie and Zegeye 2024).

Transferring nif genes from legumes to non-legumes like wheat using CRISPR-Cas9 could reduce fertilizer dependency, addressing yield losses from nitrogen deficiency (Temme et al. 2012). Early trials in maize show promise, with edited lines fixing 10% of their nitrogen needs (Chen et al. 2024). Synthetic biology's regulatory circuits, enabled by CRISPR, could optimize expression, enhancing sustainability.

Converting rice to a C4 photosynthetic pathway could boost carbon fixation and yield by 20%, addressing food demands for a growing population (Jusiak et al. 2016). CRISPR-Cas9's ability to regulate multiple genes, as demonstrated in *OsSPL14* editing (Li et al. 2023), supports this goal, though challenges in protein optimization persist. Advances in base editing and epigenomic modifications offer hope for precise control.

Future delivery methods aim to bypass tissue culture. Pollen-mediated delivery, leveraging artificial hybridization, overcomes species specificity, as shown in tobacco trials (*NbPDS*; Demirer et al. 2023). Nanotechnology, including carbon nanotubes and silica nanoparticles, achieves up to 40% delivery efficiency, enabling transgene-free editing (Demirer et al. 2023). Targeting zygotes or meristematic cells could further streamline processes, reducing costs and broadening access.

CRISPR-Cas9 can restore genetic diversity lost through millennia of breeding. Replicating domestication mutations, like vrs1 in barley (Komatsuda et al. 2007) or *PROG1* in rice (Wang et al. 2015), creates novel germplasm. In pennycress (*Thlaspi arvense*), editing enhanced oil content and seed dormancy, establishing it as a viable oilseed crop (Sedbrook et al. 2014). In tomato, integrating wild traits (Li et al. 2018; Zsögön et al. 2018) improved fruit size and stress resistance, accelerating domestication. These efforts promise resilient, high-yielding varieties for sustainable agriculture.

CRISPR-Cas9's limitations—gene delivery, off-target effects, and knock-in challenges—are significant but surmountable. Innovations like nanoparticle delivery and AI-driven sgRNA design (Table 1, Chu et al. 2024) address technical barriers, while open-source models mitigate socioeconomic constraints, as discussed in the biosafety chapter (Graff et al. 2003). Future prospects from nitrogen-fixing wheat to C4 rice, align with global needs for food security and environmental resilience. By integrating synthetic biology's programmable circuits with CRISPR's precision, researchers can replicate domestication events, enhancing crop diversity and sustainability. Collaborative efforts across science, policy, and public engagement will ensure CRISPR-Cas9 fulfills its transformative potential, building a future where agriculture meets the demands of a changing world.

6. Conclusion: The Future of CRISPR-Cas9 in a Global Context

CRISPR-Cas9 has emerged as a transformative force in plant biotechnology, reshaping agriculture, and synthetic biology with unprecedented precision. From enhancing crop resilience through *OsSWEET14* edits in rice (Blanvillain-Baufumé et al. 2017) to designing nitrogen-fixing wheat via nif gene integration (Chen et al. 2024), its applications address pressing global challenges—food security, climate change, and sustainable resource production. As detailed in later chapters, CRISPR's technical foundations (e.g., AI-optimized sgRNAs, Table 1; Chu et al. 2024), diverse applications (*eIF4E*-edited cucumber; Chandrasekaran et al. 2016), and evolving biosafety frameworks (Cartagena Protocol; CBD 2000) underscore its potential. Yet, limitations like gene delivery barriers and off-target risks, alongside ethical and regulatory complexities, temper its promise. This conclusion reflects on CRISPR-Cas9's global impact, ongoing challenges, and future trajectories, envisioning a path where innovation aligns with responsibility to serve humanity equitably.

CRISPR-Cas9's capacity to enhance crop traits positions it as a cornerstone for global food security. With 20–40% of annual crop losses attributed to pests and diseases (Jones 2021), edits like *SlJAZ2* in tomato for broad-spectrum resistance (Ortigosa et al. 2019) and *CsLOB1* in citrus for canker mitigation (Jia et al. 2017) safeguard staple and specialty crops. Nutritional improvements, such as GBSS1-edited waxy maize (Gao et al. 2020) and high-oleic Camelina (Waltz 2018), meet consumer and industrial demands, while yield boosts via *OsLOGL5* in rice (Wang et al. 2020) support a growing population projected to reach 10.4 billion by 2100. Beyond agriculture, synthetic biology applications—such as *ZmFPS1* for biofortification (Chen et al. 2023) and C4 rice for enhanced carbon fixation (Jusiak et al. 2016)—promise sustainable bioproduction and climate resilience. Environmental biotechnology, exemplified by *OsNRAMP5*-edited rice for reduced cadmium uptake (Tang et al. 2024), aligns with global efforts to restore ecosystems, reinforcing CRISPR's role in addressing interconnected crises.

Despite its advances, CRISPR-Cas9 faces technical, ethical, and societal hurdles. Gene delivery remains a bottleneck, with tissue culture-dependent methods like Agrobacterium (Table 2; Li et al. 2013) limiting scalability for recalcitrant crops (Gelvin 2010). Off-target effects, though minimized by tools like DeepCRISPR (Chu et al. 2024), require ongoing validation, as seen in rice trials (Tang et al. 2018). Ethically, dependency on biotech corporations raises concerns about farmer autonomy, with patented seeds like *TaGW2*-edited wheat (Zhang et al. 2019) exacerbating inequalities (Qaim and Kouser 2013). Regulatory disparities—US exemptions for transgene-free edits versus EU's GMO classifications (EFSA 2024)—create uneven adoption, with only six CRISPR crops commercialized globally (Schmidt et al. 2020). Public skepticism, fueled by GMO mistrust, demands transparent engagement to counter misinformation and build trust (Dessie and Zegeye 2024).

The future of CRISPR-Cas9 hinges on overcoming these barriers through innovation and collaboration. Technical advances, such as nanoparticle delivery achieving 40% efficiency (Demirer et al. 2023) and tissue culture-free methods targeting germlines (Lowe et al. 2016), promise cost-effective, universal systems. Prospects like nitrogen-fixing cereals (Chen et al. 2024) and C4 rice (Jusiak et al. 2016) could reduce environmental footprints, while domestication of wild species like pennycress (Sedbrook et al. 2014) enhances genetic diversity. Ethically, open-source biotechnology models can democratize access, ensuring smallholder farmers benefit alongside large-scale operations (Graff et al. 2003). Regulatory harmonization, guided by frameworks like Cartagena (CBD 2000), will balance safety with innovation, as proposed by recent analyses (Turnbull et al. 2023). Public trust requires inclusive dialogues, integrating diverse voices—farmers, scientists, and communities—to align CRISPR's deployment with societal values (Jasanoff and Hurlbut 2024).

Globally, CRISPR-Cas9 can bridge regional disparities. In Africa, *eIF4E*-edited cassava could combat viral losses (Gomez et al. 2019), while Asia's adoption of *OsSPL14*-edited rice (Li et al. 2023) supports yield stability. In resource-constrained regions, low-cost delivery systems and capacity-building will empower local innovation. As climate change intensifies, CRISPR-edited crops like *ZmPIF*-enhanced maize (Liu et al. 2025) offer resilience against drought and pathogens, safeguarding livelihoods. By fostering interdisciplinary collaboration—merging synthetic biology, policy, and ethics—CRISPR-Cas9 can deliver equitable, sustainable solutions, ensuring a future where biotechnology serves all.

References

Aksoy, E., K. Yildirim, M. Kavas, C. Kayihan, B.A. Yerlikaya, I. Çalik et al. 2022. General guidelines for CRISPR/Cas-based genome editing in plants. Molec. Biol Reports 49(12): 12151–12164.

Ali, Z., A. Abulfaraj, A. Idris, S. Ali, M. Tashkandi and M.M. Mahfouz. 2015. CRISPR/Cas9 mediated viral interference in plants. Genome Biology 16: 238.

Ali, Z., S. Ali, M. Tashkandi, S. S.-E. AZaidi and M.M. Mahfouz. 2016. CRISPR/Cas9-mediated immunity to geminiviruses: Differential interference and evasion. Scientific Reports 6: 26912.

Anders, C., O. Niewoehner, A. Duerst and M. Jinek. 2014. Structural basis of PAM-dependent target DNA recognition by the Cas9 endonuclease. Nature 513: 569–573.

Andersson, M., H. Turesson, N. Olsson, A.S. Fält, P. Ohlsson, M.N. Gonzalez et al. 2018. Genome editing in potato via CRISPR-Cas9 ribonucleoprotein delivery. Physiol. Plant. 164: 378–384.

Anzalone, A.V., P.B. Randolph, J.R. Davis, A.A. Sousa, L.W. Koblan, J.M. Levy et al. 2019. Search-and-replace genome editing without double-strand breaks or donor DNA. Nature 576(7785): 149–157.

Bahramnejad, B., M. Naji, R. Bose and S. Jha. 2019. A critical review on use of on use of Agrobacterium rhizogenes and their associated binary vectors for plant transformation. Biotechnology Advances. Biotechnol. Adv. 37: 107405.

Bail, C., R. Falkner and H. Marquard. 2014. The Cartagena Protocol on Biosafety: Reconciling Trade in Biotechnology with Environment and Development. Routledge.

Baltimore, D., P. Berg, M. Botchan, D. Carroll, R.A. Charo, G. Church et al. 2015. A prudent path forward for genomic engineering and germline gene modification. Science 348(6230): 36–38.

Barrangou, R. and J.A. Doudna. 2016. Applications of CRISPR technologies in research and beyond. Nature Biotechnology 34(9): 933–941.

Blanvillain-Baufumé, S., M. Reschke, M. Solé, F. Auguy, H. Doucoure, B. Szurek et al. 2017. Targeted promoter editing for rice resistance to *Xanthomonas oryzae* pv. *oryzae* reveals differential activities for SWEET14-inducing TAL effectors. Plant Biotechnol. J. 15: 306–317.

Carroll, D. 2014. Genome editing by targeted chromosomal mutagenesis. Chromosomal Mutagenesis 1–13.

Chandrasekaran, J., M. Brumin, D. Wolf, D. Leibman, C. Klap, A. Pearlsman et al. 2016. Development of broad virus resistance in non-transgenic cucumber using CRISPR/Cas9 technology. Mol. Plant Pathol. 17: 1140–1153.

Chen, X., Y. Li, Z. Wang, H. Zhang and C. Gao. 2023. CRISPR-Cas9-mediated editing of *ZmFPS1* enhances terpenoid production in maize. Plant Biotechnol. J. 21: 456–465.

Chen, Y., X. Wang, H. Zhang, J. Liu and C. Gao. 2024. CRISPR-Cas9-mediated *nif* gene integration enhances nitrogen fixation in maize. Nat. Plants 10: 789–798.

Chu, Y., J. Wang, L. Zhang, X. Yang, C. Liu and C. Gao. 2024. DeepCRISPR: Machine learning-driven sgRNA design for enhanced specificity in plant genomes. Nat. Biotechnol. 42: 512–520.

Civáň, P. and T.A. Brown. 2017. Origin of rice (*Oryza sativa* L.) domestication genes. Genet. Resour. Crop Evol. 64: 1125–1132.

Clark, R.M., S. Tavaré and J. Doebley. 2005. Estimating a nucleotide substitution rate for maize from polymorphism at a major domestication locus. Mol. Biol. Evol. 22: 2304–2312.

Collonnier, C., A. Epert, K. Mara, F. Maclot, A. Guyon-Debast, F. Charlot et al. 2017. CRISPR Cas9-mediated efficient directed mutagenesis and RAD51-dependent and RAD51 independent gene targeting in the moss *Physcomitrella patens*. Plant Biotechnol. J. 15: 122–131.

Concordet, J.-P. and M. Haeussler. 2018. CRISPOR: Intuitive guide selection for CRISPR/Cas9 genome editing experiments and screens. Nucleic Acids Res. 46: W242–W245.

Cong, L., F.A. Ran, D. Cox, S. Lin, R. Barretto, N. Habib et al. 2013. Multiplex genome engineering using CRISPR/Cas systems. Science 339: 819–823.

Convention on Biological Diversity (CBD). 2000. Cartagena Protocol on Biosafety to the Convention on Biological Diversity. Montreal: Secretariat of the CBD.

Cunningham, M.W. 2000. Pathogenesis of group A streptococcal infections. Clin. Microbiol. Rev. 13: 470–511.

Demirer, G.S., T.N. Silva, C.T. Jackson, J.B. Thomas, D.W. Ehrhardt, S.Y. Rhee et al. 2023. Nanoparticle-mediated delivery of CRISPR-Cas9 for transgene-free plant genome editing. Nat. Nanotechnol. 18: 245–253.

Dessie, Z.G. and T.T. Zegeye. 2024. Public perception and acceptance of genetically modified organisms: The role of transparency and engagement. J. Agric. Ethics 37: 45–56.

Do, P.T., C.X. Nguyen, H.T. Bui, L.T. Tran, G. Stacey, J.D. Gillman et al. 2019. Demonstration of highly efficient dual gRNA CRISPR/Cas9 editing of the homeologous *GmFAD2–1A* and *GmFAD2–1B* genes to yield a high oleic, low linoleic and α-linolenic acidphenotype in soybean. BMC Plant Biol. 19: 311.

Doench, J.G., E. Hartenian, D.B. Graham, Z. Tothova, M. Hegde, I. Smith et al. 2014. Rational design of highly active sgRNAs for CRISPR-Cas9–mediated gene inactivation. Nat. Biotechnol. 32: 1262–1267.

Domingo, J.L. and J.G. Bordonaba. 2011. A literature review on the safety assessment of genetically modified plants. Environ. Int. 37: 734–742.

Dong, O.X., S. Yu, R. Jain, N. Zhang, P.Q. Duong, C. Butler et al. 2020. Marker-free carotenoid-enriched rice generated through targeted gene insertion using CRISPR-Cas9. Nat. Commun. 11: 1178.

Doudna, J.A. and E. Charpentier. 2014. The new frontier of genome engineering with CRISPR-Cas9. Science 346(6213): 1258096.

Dorweiler, J., A. Stec, J. Kermicle and J. Doebley. 1993. *Teosinte glume architecture 1*: A genetic locus controlling a key step in maize evolution. Science 262: 233–235.

Du, H., X. Zeng, M. Zhao, X. Cui, Q. Wang, H. Yang et al. 2016. Efficient targeted mutagenesis in soybean by TALENs and CRISPR/Cas9. J. Biotechnol. 217: 90–97.

Eckerstorfer, M.F., M. Engelhard, A. Heissenberger, S. Simon and H. Teichmann. 2019. Plants developed by new genetic modification techniques—comparison of existing regulatory frameworks in the EU and non-EU countries. Front. Bioeng. Biotechnol. 7: 26.

Ellstrand, N.C. 2003. Dangerous Liaisons? When Cultivated Plants Mate With Their Wild Relatives. Johns Hopkins University Press.

Fauser, F., S. Schiml and H. Puchta. 2014. Both CRISPR/Cas-based nucleases and nickases can be used efficiently for genome engineering in *Arabidopsis thaliana*. Plant J. 79: 348–359.

Feng, C., Y. Mao, N. Xu, B. Zhang, P. Wei, D.-L. Yang et al. 2014. Multigeneration analysis reveals the inheritance, specificity, and patterns of CRISPR/Cas-induced gene modifications in *Arabidopsis*. Proc. Natl. Acad. Sci. USA 111: 4632–4637.

Gao, H., M.J. Gadlage, H.R. Lafitte, B. Lenderts, M. Yang, M. Schroder et al. 2020. Superior field performance of waxy cornengineered using CRISPR-Cas9. Nat. Biotechnol. 38: 579–581.

Gao, J., G. Wang, S. Ma, X. Xie, X. Wu, X. Zhang et al. 2015. CRISPR/Cas9-mediated targeted mutagenesis in *Nicotiana tabacum*. Plant Mol. Biol. 87: 99–110.

Gao, L., D.B. Cox, W.X. Yan, J.C. Manteiga, M.W. Schneider, T. Yamano et al. 2017. Engineered Cpf1 variants with altered PAM specificities. Nat. Biotechnol. 35: 789–792.

Garcia-Ruiz, M., A. Knapp and H. Garcia-Ruiz. 2018. Profile of genetically modified plants authorized in Mexico. GM Crops Food 9: 152–168.

Garneau, J.E., M.-È. Dupuis, M. Villion, D.A. Romero, R. Barrangou, P. Boyaval et al. 2010. The CRISPR/Cas bacterial immune system cleaves bacteriophage and plasmid DNA. Nature 468: 67–71.

Gelvin, S.B. 2010. Plant proteins involved in *Agrobacterium*-mediated genetic transformation. Annu. Rev. Phytopathol. 48: 45–68.

Gilbert, L.A., M.H. Larson, L. Morsut, Z. Liu, G.A. Brar, S.E. Torres et al. 2013. CRISPR-mediated modular RNA-guided regulation of transcription in eukaryotes. Cell 154(2): 442–451.

Giner, A., L. Pascual, M. Bourgeois, G. Gyetvai, P. Rios, B. Picó et al. 2017. A mutation in the melon Vacuolar Protein Sorting 41 prevents systemic infection of *Cucumber mosaic virus*. Sci. Rep. 7: 10471.

Gomez, M.A., Z.D. Lin, T. Moll, R.D. Chauhan, L. Hayden, K. Renninger et al. 2019. Simultaneous CRISPR/Cas9 mediated editing of cassava eIF4E isoforms nCBP-1 and nCBP-2 reduces cassavabrown streak disease symptom severity and incidence. Plant Biotechnol. J. 17: 421–434.

Gökdemir, G., Z. Seçgin, S. Uluisik and M. Kavas. 2022. CRISPR/Cas9 knock-out of *SlPHD_MS1* (*Solyc04g008420*) gene results in complete male sterility in tomato. Plant Growth Regul. 98: 329–341.

Graff, G.D., S.E. Cullen, K.J. Bradford, D. Zilberman and A.B. Bennett. 2003. The public private structure of intellectual property ownership in agricultural biotechnology. Nat. Biotechnol. 21: 989–995.

Horvath, P. and R. Barrangou. 2010. CRISPR/Cas, the immune system of bacteria and archaea. Science 327: 167–170.

Hsu, P.D., D.A. Scott, J.A. Weinstein, F.A. Ran, S. Konermann, V. Agarwala et al. 2013. DNA targeting specificity of RNA-guided Cas9 nucleases. Nat. Biotechnol. 31: 827–832.

Huang, S., D. Weigel, R.N. Beachy and J. Li. 2016. A proposed regulatory framework for genome-edited crops. Nat. Genet. 48: 109–111.

Hunter, M.C., R.G. Smith, M.E. Schipanski, L.W. Atwood and D.A. Mortensen. 2017. Agriculture in 2050: Recalibrating targets for sustainable intensification. Bioscience 67: 386–391.

Iaffaldano, B., Y. Zhang and K. Cornish. 2016. CRISPR/Cas9 genome editing of rubber producing dandelion *Taraxacum kok-saghyz* using *Agrobacterium rhizogenes* without selection. Ind. Crops Prod. 89: 356–362.

Ishino, Y., H. Shinagawa, K. Makino, M. Amemura and A. Nakata. 1987. Nucleotide sequence of the *iap* gene, responsible for alkaline phosphatase isozyme conversion in *Escherichia coli*, and identification of the gene product. J. Bacteriol. 169: 5429–5433.

Jasanoff, S. and J.B. Hurlbut. 2024. Governing genome editing: The role of public engagement in shaping ethical futures. Sci. Technol. Hum. Values 49: 123–145.

Javaid, D., S.Y. Ganie, Y.A. Hajam and M.S. Reshi. 2022. CRISPR/Cas9 system: a reliable and facile genome editing tool in modern biology. Molecular Biology Reports 49(12): 12133–12150.

Jia, H., Y. Zhang, V. Orbović, J. Xu, F.F. White, J.B. Jones et al. 2017. Genome editing of the disease susceptibility gene *CsLOB1* in citrus confers resistance to citrus canker. Plant Biotechnol. J. 15: 817–823.

Jiang, F. and J.A. Doudna. 2017. CRISPR–Cas9 structures and mechanisms. Annu. Rev. Biophys. 46: 505–529.

Jiang, F., D.W. Taylor, J.S. Chen, J.E. Kornfeld, K. Zhou, A.J. Thompson et al. 2016. Structures of a CRISPR-Cas9 R-loop complex primed for DNA cleavage. Science 351: 867–871.

Jinek, M., K. Chylinski, I. Fonfara, M. Hauer, J.A. Doudna and E. Charpentier. 2012. A programmable dual-RNA–guided DNA endonuclease in adaptive bacterial immunity. Science 337: 816–821.

Jones, R.A.C. 2021. Global plant virus disease pandemics and epidemics. Plants 10: 233.

Junaid, M. and A. Gokce. 2024. Global agricultural losses and their causes. Bull. Biol. Allied Sci. Res. 2024(1): 66.

Jusiak, B., S. Cleto, P. Perez-Piñera and T.K. Lu. 2016. Engineering synthetic gene circuits in living cells with CRISPR technology. Trends Biotechnol. 34: 535–547.

Kaur, N., A. Alok, Shivani, N. Kaur, P. Pandey, P. Awasthi and S. Tiwari. 2018. CRISPR/Cas9 mediated efficient editing in phytoene desaturase (PDS) demonstrates precise manipulation in banana cv. Rasthali genome. Funct. Integr. Genomics 18: 89–99.

Khan, M.S.S., R. Basnet, S.A. Islam and Q. Shu. 2019. Mutational analysis of *OsPLDα1* reveals its involvement in phytic acid biosynthesis in rice grains. J. Agric. Food Chem. 67: 11436–11443.

Kim, H., S.-T. Kim, J. Ryu, B.-C. Kang, J.-S. Kim and S.-G. Kim. 2017. CRISPR/Cpf1 mediated DNA-free plant genome editing. Nat. Commun. 8: 14406.

Kim, M.S., S.R. Ko, Y.J. Jung, K.K. Kang, Y.J. Lee and Y.G. Cho. 2023. Knockout mutants of OsPUB7 generated using CRISPR/Cas9 revealed abiotic stress tolerance in rice. International Journal of Molecular Sciences 24(6): 5338.

Komatsuda, T., M. Pourkheirandish, C. He, P. Azhaguvel, H. Kanamori, D. Perovic et al. 2007. Six-rowed barley originated from a mutation in a homeodomain-leucine zipper I-class homeobox gene. Proc. Natl. Acad. Sci. USA 104: 1424–1429.

Komor, A.C., Y.B. Kim, M.S. Packer, J.A. Zuris and D.R. Liu. 2016. Programmable editing of a target base in genomic DNA without double-stranded DNA cleavage. Nature 533: 420–424.

Kushner, D.B., B.D. Lindenbach, V.Z. Grdzelishvili, A.O. Noueiry, S.M. Paul and P. Ahlquist. 2003. Systematic, genome-wide identification of host genes affecting replication of a positive-strand RNA virus. Proc. Natl. Acad. Sci. USA 100: 15764–15769.

Labun, K., T.G. Montague, M. Krause, Y.N. Torres Cleuren, H. Tjeldnes and E. Valen. 2019. CHOPCHOP v3: Expanding the CRISPR web toolbox beyond genome editing. Nucleic Acids Res. 47: W171–W174.

Lacroix, B. and V. Citovsky. 2016. A functional bacterium-to-plant DNA transfer machinery of *Rhizobium etli*. PLoS Pathog. 12: e1005502.

Li, J.-F., J.E. Norville, J. Aach, M. McCormack, D. Zhang, J. Bush et al. 2013. Multiplex and homologous recombination–mediated genome editing in *Arabidopsis* and *Nicotiana benthamiana* using guide RNA and Cas9. Nat. Biotechnol. 31: 688–691.

Li, J., H. Manghwar, L. Sun, P. Wang, G. Wang, H. Sheng et al. 2019. Whole genome sequencing reveals rare off-target mutations and considerable inherent genetic or/and somaclonal variations in CRISPR/Cas9-edited cotton plants. Plant Biotechnol. J. 17: 858–868.

Li, R., X. Li, D. Fu, B. Zhu, H. Tian, Y. Luo et al. 2018. Multiplexed CRISPR/Cas9 mediated metabolic engineering of γ-aminobutyric acid levels in *Solanum lycopersicum*. Plant Biotechnol. J. 16: 415–427.

Li, T., X. Yang, Y. Yu, C. Liu, K. Chen and C. Gao. 2023. Adenine base editing of *OsSPL14* enhances rice yield under diverse conditions. Nat. Plants 9: 1234–1243.

Li, T., X. Yang, Y. Yu, X. Si, X. Zhai, H. Zhang et al. 2018. Domestication of wild tomato is accelerated by genome editing. Nat. Biotechnol. 36: 1160–1163.

Liang, Z., K. Chen, T. Li, Y. Zhang, Y. Wang, Q. Zhao et al. 2017. Efficient DNA-free genome editing of bread wheat using CRISPR/Cas9 ribonucleoprotein complexes. Nat. Commun. 8: 14261.

Liang, Z., K. Zhang, K. Chen and C. Gao. 2014. Targeted mutagenesis in *Zea mays* using TALENs and the CRISPR/Cas system. J. Genet. Genomics 41: 63–68.

Lin, C.S., C.T. Hsu, L.H. Yang, L.Y. Lee, J.Y. Fu, Q.W. Cheng et al. 2018. Application of protoplast technology to CRISPR/Cas9 mutagenesis: From single-cell mutation detection to mutant plant regeneration. Plant Biotechnol. J. 16: 1295–1310.

Liu, D., E.E. Ellison, E.A. Myers, L.I. Donahue, S. Xuan, R. Swanson et al. 2024. Heritable gene editing in tomato through viral delivery of isopentenyl transferase and single-guide RNAs to latent axillary meristematic cells. Proceedings of the National Academy of Sciences 121(39): e2406486121.

Liu, H., Y. Ding, Y. Zhou, W. Jin, K. Xie and L.L. Chen. 2017. CRISPR-P 2.0: an improved CRISPR-Cas9 tool for genome editing in plants. Molecular Plant 10(3): 530–532.

Liu, Y., X. Wang, Y. Chen, H. Zhang, C. Gao and T. Li. 2025. Multiplexed editing of *ZmPIF* genes enhances drought tolerance and pathogen resistance in maize. Nat. Plants 11: 234–243.

Lou, D., H. Wang, G. Liang and D. Yu. 2017. *OsSAPK2* confers abscisic acid sensitivity and tolerance to drought stress in rice. Front. Plant Sci. 8: 993.

Lowe, K., M. La Rota, G. Hoerster, C. Hastings, N. Wang, M. Chamberlin et al. 2018. Rapid genotype "independent" *Zea mays* L. (maize) transformation via direct somatic embryogenesis. *In Vitro* Cell. Dev. Biol.–Plant 54: 240–252.

Lowe, K., E. Wu, N. Wang, G. Hoerster, C. Hastings, M.-J. Cho et al. 2016. Morphogenic regulators Baby boom and Wuschel improve monocot transformation. Plant Cell 28: 1998–2015.

Lu, K., B. Wu, J. Wang, W. Zhu, H. Nie, J. Qian et al. 2018. Blocking amino acid transporter *OsAAP3* improves grain yield by promoting outgrowth buds and increasing tiller number in rice. Plant Biotechnol. J. 16: 1710–1722.

Mahas, A., R. Aman and M. Mahfouz. 2019. CRISPR-Cas13d mediates robust RNA virus interference in plants. Genome Biol. 20: 263.

Maher, M.F., R.A. Nasti, M. Vollbrecht, C.G. Starker, M.D. Clark and D.F. Voytas. 2020. Plant gene editing through *de novo* induction of meristems. Nat. Biotechnol. 38: 84–89.

Mali, P., L. Yang, K.M. Esvelt, J. Aach, M. Guell, J.E. DiCarlo et al. 2013. RNA-guided human genome engineering via Cas9. Science 339: 823–826.

Malnoy, M., R. Viola, M.-H. Jung, O.-J. Koo, S. Kim, J.-S. Kim et al. 2016. DNA-free genetically edited grapevine and apple protoplast using CRISPR/Cas9 ribonucleoproteins. Front. Plant Sci. 7: 1904.

Maxmen, A. 2017. Genetically modified apple reaches US stores, but will consumers bite? Nature 551: 149–150.

Menz, J., D. Modrzejewski, F. Hartung, R. Wilhelm and T. Sprink. 2020. Genome edited crops touch the market: A view on the global development and regulatory environment. Front. Plant Sci. 11: 586027.

Mishra, S., S. Nayak, N. Tuteja, S. Poosapati, D.M. Swain and R.K. Sahoo. 2024. CRISPR/Cas mediated genome engineering in plants: Application and prospectives. Plants 13: 1884.

Molla, K.A., S. Sretenovic, K.C. Bansal and Y. Qi. 2021. Precise plant genome editing using base editors and prime editors. Nat. Plants 7: 1166–1187.

Murovec, J., K. Guček, B. Bohanec, M. Avbelj and R. Jerala. 2018. DNA-free genome editing of *Brassica oleracea* and *B. rapa* protoplasts using CRISPR-Cas9 ribonucleoprotein complexes. Front. Plant Sci. 9: 1594.

Nemhauser, J.L. and K.U. Torii. 2016. Plant synthetic biology for molecular engineering of signalling and development. Nat. Plants 2: 16010.

Nicolia, A., A. Manzo, F. Veronesi and D. Rosellini. 2014. An overview of the last 10 years of genetically engineered crop safety research. Crit. Rev. Biotechnol. 34: 77–88.

Olson, D.G., S.A. Tripathi, R.J. Giannone, J. Lo, N.C. Caiazza, D.A. Hogsett et al. 2010. Deletion of the Cel48S cellulase from Clostridium thermocellum. Proceedings of the National Academy of Sciences 107(41): 17727–17732.

Ortigosa, A., S. Gimenez-Ibanez, N. Leonhardt and R. Solano. 2019. Design of a bacterial speck resistant tomato by CRISPR/Cas9-mediated editing of *SlJAZ2*. Plant Biotechnol. J. 17: 665–673.

Palindromic Repeats CRISPR-associated 2014/CRISPR-P 2.0: an improved CRISPR/Cas9 tool for genome editing in plants http://cbi.hzau.edu.cn/CRISPR2/.

Panavas, T., E. Serviene, J. Brasher and P.D. Nagy. 2005. Yeast genome-wide screen reveals dissimilar sets of host genes affecting replication of RNA viruses. Proc. Natl. Acad. Sci. USA 102: 7326–7331.

Peng, A., S. Chen, T. Lei, L. Xu, Y. He, L. Wu et al. 2017. Engineering canker resistant plants through CRISPR/Cas9-targeted editing of the susceptibility gene *CsLOB1* promoter in citrus. Plant Biotechnol. J. 15: 1509–1519.

Permyakova, N.V., T.V. Marenkova, P.A. Belavin, A.A. Zagorskaya, Y.V. Sidorchuk, E.A. Uvarova et al. 2021. Assessment of the level of accumulation of the dIFN protein integrated by the knock-in method into the region of the histone H3.3 gene of *Arabidopsis thaliana*. Cells 10: 2137.

Piperno, D.R. and K.V. Flannery. 2001. The earliest archaeological maize (*Zea mays* L.) from highland Mexico: New accelerator mass spectrometry dates and their implications. Proc. Natl. Acad. Sci. USA 98: 2101–2103.

Puchta, H. 2005. The repair of double-strand breaks in plants: Mechanisms and consequences for genome evolution. J. Exp. Bot. 56: 1–14.

Puchta, H., B. Dujon and B. Hohn. 1996. Two different but related mechanisms are used in plants for the repair of genomic double-strand breaks by homologous recombination. Proc. Natl. Acad. Sci. USA 93: 5055–5060.

yott, D.E., E. Sheehan and A. Molnar. 2016. Engineering of CRISPR/Cas9-mediated potyvirus resistance in transgene-free *Arabidopsis* plants. Mol. Plant Pathol. 17: 1276–1288.

Qaim, M. and S. Kouser. 2013. Genetically modified crops and food security. PLoS ONE 8: e64879.

Rees, H.A. and D.R. Liu. 2018. Base editing: Precision chemistry on the genome and transcriptome of living cells. Nat. Rev. Genet. 19: 770–788.

Rosinski-Chupin, I., E. Sauvage, A. Fouet, C. Poyart and P. Glaser. 2019. Conserved and specific features of *Streptococcus pyogenes* and *Streptococcus agalactiae* transcriptional landscapes. BMC Genomics 20: 236.

Sánchez-León, S., J. Gil-Humanes, C.V. Ozuna, M.J. Giménez, C. Sousa, D.F. Voytas et al. 2018. Low-gluten, nontransgenic wheat engineered with CRISPR/Cas9. Plant Biotechnol. J. 16: 902–910.

Sandhya, D., P. Jogam, V.R. Allini, S. Abbagani and A. Alok. 2020. The present and potential future methods for delivering CRISPR/Cas9 components in plants. J. Genet. Eng. Biotechnol. 18: 25.

Schmidt, S.M., M. Belisle and W.B. Frommer. 2020. The evolving landscape around genome editing in agriculture. EMBO Reports 21: e50680.

Secgin, Z., S. Uluisik, K. Yıldırım, M.F. Abdulla, K. Mostafa and M. Kavas. 2022. Genome wide identification of the aconitase gene family in tomato (*Solanum lycopersicum*) and CRISPR-based functional characterization of *SlACO2* on male-sterility. Int. J. Mol. Sci. 23: 13963.

Sedbrook, J.C., W.B. Phippen and M.D. Marks. 2014. New approaches to facilitate rapid domestication of a wild plant to an oilseed crop: Example pennycress (*Thlaspi arvense* L.). Plant Sci. 227: 122–132.

Shimatani, Z., S. Kashojiya, M. Takayama, R. Terada, T. Arazoe, H. Ishii et al. 2017. Targeted base editing in rice and tomato using a CRISPR-Cas9 cytidine deaminase fusion. Nat. Biotechnol. 35: 441–443.

Shufen, C., C. Yicong, F. Baobing, J. Guiai, S. Zhonghua, L. Ju et al. 2019. Editing of rice isoamylase gene *ISA1* provides insights into its function in starch formation. Rice Sci. 26: 77–87.

Sternberg, S.H., S. Redding, M. Jinek, E.C. Greene and J.A. Doudna. 2014. DNA interrogation by the CRISPR RNA-guided endonuclease Cas9. Nature 507: 62–67. doi:10.1038/nature13011.

Sun, X., Z. Hu, R. Chen, Q. Jiang, G. Song, H. Zhang et al. 2015. Targeted mutagenesis in soybean using the CRISPR-Cas9 system. Sci. Rep. 5: 10342.

Sun, Y., G. Jiao, Z. Liu, X. Zhang, J. Li, X. Guo et al. 2017. Generation of high-amylose rice through CRISPR/Cas9-mediated targeted mutagenesis of starch branching enzymes. Front. Plant Sci. 8: 298.

Svitashev, S., J.K. Young, C. Schwartz, H. Gao, S.C. Falco and A.M. Cigan. 2015. Targeted mutagenesis, precise gene editing, and site-specific gene insertion in maize using Cas9 and guide RNA. Plant Physiol. 169: 931–945.

Symington, L.S. and J. Gautier. 2011. Double-strand break end resection and repair pathway choice. Annu. Rev. Genet. 45: 247–271.

Tang, L., Y. Chen, H. Zhang, J. Liu, and C. Gao. 2024. CRISPR-Cas9 editing of OsNRAMP5 reduces cadmium uptake in rice. Plant Biotechnol. J. 22: 345–354.

Tang, X., G. Liu, J. Zhou, Q. Ren, Q. You, L. Tian et al. 2018. A large-scale whole-genome sequencing analysis reveals highly specific genome editing by both Cas9 and Cpf1 (Cas12a) nucleases in rice. Genome Biol. 19: 84.

Temme, K., D. Zhao and C.A. Voigt. 2012. Refactoring the nitrogen fixation gene cluster from *Klebsiella oxytoca*. Proc. Natl. Acad. Sci. 109: 7085–7090.

Toda, E., N. Koiso, A. Takebayashi, M. Ichikawa, T. Kiba, K. Osakabe et al. 2019. An efficient DNA-and selectable marker free genome-editing system using zygotes in rice. Nat. Plants 5: 363–368.

Turnbull, C., J. Martin, T. Sprink, R. Wilhelm and F. Hartung. 2023. Harmonizing global regulations for genome-edited crops: Progress and challenges. Nat. Biotechnol. 41: 1567–1575.

Veillet, F., L. Perrot, L. Chauvin, M.-P. Kermarrec, A. Guyon-Debast, J.-E. Chauvin et al. 2019. Transgene-free genome editing in tomato and potato plants using Agrobacterium-mediated delivery of a CRISPR/Cas9 cytidine base editor. Int. J. Mol. Sci. 20: 402.

Waltz, E. 2018. With a free pass, CRISPR-edited plants reach market in record time. Nat. Biotechnol. 36: 6–8.

Wang, C., G. Wang, Y. Gao, G. Lu, J.E. Habben, G. Mao et al. 2020. A cytokinin-activation enzyme-like gene improves grain yield under various field conditions in rice. Plant Mol. Biol. 102: 373–388.

Wang, H., A.J. Studer, Q. Zhao, R. Meeley and J.F. Doebley. 2015. Evidence that the origin of naked kernels during maize domestication was caused by a single amino acid substitution in tga1. Genetics 200: 965–974.

Wang, L., S. Chen, A. Peng, Z. Xie, Y. He and X. Zou. 2019. CRISPR/Cas9-mediated editing of CsWRKY22 reduces susceptibility to *Xanthomonas citri* subsp. citri in Wanjincheng orange (*Citrus sinensis* (L.) Osbeck). Plant Biotechnol. Rep. 13: 501–510.

Wang, N., X. Fan, M. He, Z. Hu, C. Tang, S. Zhang et al. 2022. Transcriptional repression of *TaNOX10* by TaWRKY19 compromises ROS generation and enhances wheat susceptibility to stripe rust. Plant Cell 34: 1784–1803. doi:10.1093/plcell/koac001/.

Wang, Y., H. Chen, K. Lin, Y. Han, Z. Gu, H. Wei et al. 2024. Ultrasensitive single-step CRISPR detection of monkeypox virus in minutes with a vest-pocket diagnostic device. Nature Communications 15(1): 3279.

Woo, J.W., J. Kim, S.I. Kwon, C. Corvalán, S.W. Cho, H. Kim et al. 2015. DNA-free genome editing in plants with preassembled CRISPR Cas9 ribonucleoproteins. Nat. Biotechnol. 33: 1162–1164.

Xie, K. and Y. Yang. 2013. RNA-guided genome editing in plants using a CRISPR–Cas system. Mol. Plant 6: 1975–1983.

Yang, Q.Y., Y.W. Zhang, N.Y. Liu, H. Sun, S.B. Zhang, S. Timm et al. 2025. Variation in photosynthetic efficiency among maize cultivars and its implications for breeding strategy. Journal of Experimental Botany eraf249.

Yin, K., T. Han, G. Liu, T. Chen, Y. Wang, A.Y.L. Yu et al. 2015. A geminivirus-based guide RNA delivery system for CRISPR/Cas9 mediated plant genome editing. Sci. Rep. 5: 14926.

Young, J., L. Kuykendall, E. Martínez-Romero, A. Kerr and H. Sawada. 2001. A revision of *Rhizobium* Frank 1889, with an emended description of the genus, and the inclusion of all species of *Agrobacterium* Conn 1942 and *Allorhizobium undicola* de Lajudie et al. 1998 as new combinations: *Rhizobium radiobacter*, *R. rhizogenes*, *R. rubi*, *R. undicola*, and *R. vitis*. Int. J. Syst. Evol. Microbiol. 51: 89–103.

Yuste-Lisbona, F.J., A. Fernández-Lozano, B. Pineda, S. Bretones, A. Ortíz-Atienza, B. García-Sogo et al. 2020. ENO regulates tomato fruit size through the floral meristem development network. Proceedings of the National Academy of Sciences 117(14): 8187–8195.

Zaidi, S.S.-e.-A., S. Mansoor, Z. Ali, M. Tashkandi and M.M. Mahfouz. 2016. Engineering plants for geminivirus resistance with CRISPR/Cas9 system. Trends Plant Sci. 21: 279–281.

Zaidi, S.S.E.A., H. Vanderschuren, M. Qaim, M.M. Mahfouz, A. Kohli, S. Mansoor et al. 2019. New plant breeding technologies for food security. Science 363(6434): 1390–1.

Zaidi, S.S.E.A., H. Vanderschuren, M. Qaim, M.M. Mahfouz, A. Kohli, S. Mansoor et al. 2019. New plant breeding technologies for food security. Science 363(6434): 1390–1391.

Zess, E.K. and M.B. Begemann. 2021. Natural and engineered barriers to gene flow in genome edited plants. Curr. Opin. Plant Biol. 62: 102058.

Zhang, A., Y. Liu, F. Wang, T. Li, Z. Chen, D. Kong et al. 2019. Enhanced rice salinity tolerance via CRISPR/Cas9-targeted mutagenesis of the OsRR22 gene. Mol. Breed. 39: 47.

Zhang, Y., L. Hua, Y. Chen, X. Zhang, J. Liu, Y. Wang et al. 2019. Targeted knockout of TaGW2 using CRISPR/Cas9 enhances grain size and yield in wheat. Plant Biotechnol. J. 17: 2345–2357.

Zhang, Y., Z. Liang, Y. Zong, Y. Wang, J. Liu, K. Chen et al. 2016. Efficient and transgene-free genome editing in wheat through transient expression of CRISPR/Cas9 DNA or RNA. Nat. Commun. 7: 12617.

Zhang, Z., Y. Mao, S. Ha, W. Liu, J.R. Botella and J.-K. Zhu. 2018. A multiplex CRISPR/Cas9 platform for fast and efficient editing of multiple genes in *Arabidopsis*. Plant Cell Rep. 37: 1519–1529.

Zhang, J., X. Liang, H. Zhang, S. Ishfaq, K. Xi, X. Zhou et al. 2023. Rapid and sensitive detection of toxigenic Fusarium asiaticum integrating recombinase polymerase amplification, CRISPR/Cas12a, and lateral flow techniques. Int. J. Mol. Sci. 24: 14134. doi:10.3390/ijms241814134.

Zhang, J., H. Wang, Z. Luo, Z. Yang, Z. Zhang, P. Wang et al. 2023. Computational design of highly efficient thermostable MHET hydrolases and dual enzyme system for PET recycling. Commun. Biol. 6: 1135.

Zhu, T., W. Tang, D. Chen, R. Zheng, J. Li and J. Su. 2020. Identification of a novel efficient transcriptional activation domain from Chinese fir (Cunninghamia lanceolata). bioRxiv, 2020-04.

Zsögön, A., T. Čermák, E.R. Naves, M.M. Notini, K.H. Edel, S. Weinl et al. 2018. *De novo* domestication of wild tomato using genome editing. Nat. Biotechnol. 36: 1211–1216.

CHAPTER 2

Molecular Mechanisms and Current Progress of CRISPR/Cas in Plant Development

Pramod Sargar,[1,*] *Swapnil Baraskar,*[1] *Muhammad Tanveer Altaf,*[2] *Waqas Liaqat,*[2] *Amna Jamil*[3] and *Asia Maqbool*[4]

1. Introduction

The global population is expanding at an extraordinary pace, projected to exceed 9.8 billion by 2050 (Son et al. 2024). This exponential growth presents a significant risk to global food security, and underlines the necessity of increasing production by around 70% by 2050 in comparison to the production of 2005 (Ghosh et al. 2024). However, this imperative is threatened by the pervasive presence of plant pathogens and abiotic stresses, which impose significant constraints on agricultural productivity (Lata et al. 2022; Dong and Ronald 2019). To mitigate these challenges and safeguard food security, it is imperative to boost the resilience of crops against these adversities (Gupta et al. 2020). Plant breeders are increasingly adopting modern molecular tools over conventional breeding methods due to several key advantages that enhance efficiency and precision in crop improvement. Conventional breeding often relies on time-consuming processes of selection and backcrossing, which can take many years to develop new varieties. In contrast, molecular techniques such as Marker-Assisted Selection (MAS) and genome editing (e.g., CRISPR) allow for rapid identification and manipulation of specific genes associated with desirable traits, significantly shortening the breeding cycle. These modern methods also enable breeders to access a broader genetic diversity by integrating traits from wild relatives or previously neglected species, enhancing resilience against climate change and pests. Additionally, advancements in high-throughput phenotyping and genomic selection facilitate better predictions of breeding outcomes, leading to more informed decisions in developing high-yielding and disease-resistant crops. Overall, the shift towards molecular tools represents a strategic evolution aimed at meeting global food security challenges more effectively.

[1] International Crop Research Institute for the Semi-arid Tropics, Hyderabad, India.
[2] Department of Field Crops, Faculty of Agriculture, Recep Tayyip Erdoğan University, Rize/Pazar, Türkiye.
[3] Department of Horticultural Sciences, The Islamia University of Bahawalpur, 63100, Pakistan.
[4] College of Horticulture, Hebei Agricultural University, Baoding, 071001, Hebei, China.
* Corresponding author: pramodsargar28@gmail.com

DOI: 10.1201/9781003518204-2

Genome editing tools have emerged as a transformative solution, offering unparalleled precision in modifying an organism's DNA. Numerous genome editing technologies, such as meganucleases, Transcription Activator-Like Nucleases (TALENs), Zinc-Finger Nucleases (ZFNs), and particularly CRISPR-Cas9, have been established over the years (Bhattacharjee et al. 2024). Among these, the CRISPR-Cas9 system has garnered widespread acclaim for its remarkable efficiency and simplicity, eclipsing the utility of its predecessors (Raza et al. 2024). Originally discovered as a bacterial defense mechanism against viral infections, CRISPR-based tools have transcended their microbial origins to revolutionize genome engineering in diverse organisms. The versatility of CRISPR technology extends beyond mere gene editing, encompassing a myriad of applications such as gene activation, epigenetic modulation, RNA-guided RNA degradation, and synthetic biology (Pop-Bica et al. 2022).

CRISPR systems are divided into two classes: class I consists of multiple Cas proteins, and class II where there is a single Cas protein. These classes are characterized based on their unique Cas-protein structures and functions. CRISPR-Type II/Cas-9 system has been extensively investigated and utilized in genetic engineering owing to its comparatively uncomplicated architecture (Liu et al. 2020). Guide RNA and Cas-9 proteins are the two main elements of this technique. *S. pyogenes*, specifically the spCas-9 variant, is responsible for the production of the Cas-9 protein. This protein has the distinction of being the first Cas protein utilized in the process of genome editing. According to Mei et al. (2016), the genetic scissor is a notable multi-domain DNA endonuclease comprising 1,368 amino acids. Its primary function is to cleave target DNA to produce a double-strand break. Cas-9 buildup with the recognition lobe and the nucleic lobe. The recognition lobe consists of recognition lobe 1 and 2, which are responsible for binding guide RNA.

The advent of CRISPR technology has unlocked new frontiers in epigenetics, enabling researchers to manipulate the histone code and modulate chromatin accessibility to boost plant immunity (Pawełkowicz et al. 2024). By augmenting the transcriptional output of defense-related genes, CRISPR-mediated epigenetic modifications offer a promising avenue for fortifying plants against pathogenic assaults (Khan 2023). The integration of CRISPR technology into agricultural practices holds immense promise for ensuring sustainable food production and safeguarding global food security. By empowering scientists with precise and efficient tools for enhancing crop resilience, CRISPR technology represents a pivotal advancement in modern agriculture.

In this study, we delve into the molecular intricacies of CRISPR technology and its transformative impact on plant development. Through a comprehensive examination of CRISPR-mediated genome editing and epigenetic modulation, we elucidate the molecular mechanisms underpinning the enhanced resilience of CRISPR-edited crops. Furthermore, we explore the current progress and prospects of CRISPR technology concerning plant development and ensuring food security in the face of mounting environmental challenges.

2. Overview of CRISPR Mechanism

The CRISPR-Cas immune systems are categorized into three primary stages. The protospacer sequence is initially removed from foreign DNA by Cas proteins during the adaptation phase and integrated into the repeat next to the leader at the CRISPR locus. The transcription of CRISPR arrays generates several crRNAs, each containing a singular spacer sequence alongside a segment of the neighboring repeat sequence. Third, the crRNAs are organized into distinct classes of protein-targeting complexes (Cascades) during the interference phase. These complexes hybridize with and cleave sequences that match the spacer on the invading element or its transcripts.

Ceasar et al. (2016) reported that the Cas-9 nuclease is responsible for the creation of double-stranded breaks at a location three base pairs upstream of the Protospacer Adjacent Motif sequence (PAM). Downstream of the cut site, the PAM sequence is a short DNA sequence that is conserved and ranges in length from two to five base pairs. The length of the PAM sequence varies depending on the species of bacteria. Within the realm of genome editing technologies, the Cas-9 protein, which is the most prominent nuclease, is responsible for identifying the PAM sequence located at 5C-NGG-3¹. There is still a lack of clarity on the specific mechanism by which the Cas-9 enzyme produces the melting of the DNA sequence that is the target. On the other hand, it seems to involve the melting of DNA in a specific location after the enzyme has located a certain spot that has the required PAM. The creation of a hybrid between RNA and DNA enables the cleavage of the target DNA molecule. Cas-9 is a protein engineered to precisely cleave DNA. The HNH domain cleaves the complementary strand of target DNA, while the RuvC domain cleaves the non-complementary strand, producing blunt-ended Double-Stranded Breaks (DSBs). Ultimately, the host's cells rectify the double-strand break (Mei et al. 2016; Jiang 2017).

2.1 Cellular Mechanisms of CRISPR/Cas9

The CRISPR/Cas-9 system employs Non-Homologous End Joining (NHEJ) and Homology-Directed Repair (HDR) pathways to repair double-strand breaks (DSBs) induced by the Cas-9 protein (Liu et al. 2019). NHEJ transpires across all phases of the cell cycle and facilitates the repair of double-strand breaks (DSBs) by enzymatically ligating DNA fragments, irrespective of external homologous DNA. This process takes place in the absence of homologous DNA from outside the cell. The cellular repair system, which is very active in cells, is the primary and most efficient mechanism for repairing cellular damage. However, it is exposed to errors that can lead to the production of frameshift mutations and early stop codons. These errors occur due to small, random deletions or insertions at the exact location of cleavage (Yang et al. 2020). HDR is a method that involves a homologous template of DNA to achieve high levels of accuracy. During the G2 and late S phases of the cell cycle, it is primarily responsible for these functions. To engage in CRISPR gene editing, HDR makes use of several donor DNA templates, which are exogenous, each of which has a distinct sequence. HDR integrates a donor template to DSB site (Yang et al. 2020).

3. Targeting Specificity and Efficiency

The CRISPR-Cas systems are extensively used in genome editing techniques due to their unique targeting accuracy. Guide RNA, also known as gRNA, is an essential component in the process of determining targeting specificity. It does this by accurately directing the Cas protein to the genomic region that is being targeted (Manghwar et al. 2020). By ensuring that the gRNA and the desired RNA or DNA sequence are complementary to one another, particularly in the seed region, accurate identification can be achieved. To further verify that the Cas protein exclusively attaches to the target site, it is necessary to identify the PAM sequence that is located adjacent to the target site. Unwanted genomic modifications can be reduced through the use of techniques such as the adoption of high-fidelity Cas variants and informative gRNA design. These techniques restrict off-target effects during the process. On the other hand, the effectiveness of targeting is contingent upon several factors, such as the enzymatic activity of the Cas protein, the repair pathway of selection, and the optimization of delivery strategies. Cas proteins have a significant amount of nuclease activity, which makes it possible for them to effectively induce DSBs at the target place (Anzalone et al. 2020). Furthermore, repair mechanisms such as HDR and NHEJ make it easier to edit the genome accurately. Furthermore, the optimization of delivery mechanisms such as ribonucleoprotein (RNP) complexes and plasmid transfection is necessary to guarantee the optimal transport of CRISPR-Cas components into the animals or cells that are the targets of the study. The CRISPR-Cas technology's efficiency research could increase its use in agriculture, biotechnology, and medicine. Many different plant species have been modified through the application of CRISPR/Cas9 technology including cotton (*Gossypium hirsutum* L.) (Qin et al. 2020), Arabidopsis (Xu et al. 2019a), soyabean (Li et al. 2021), rice (Macovei et al. 2018), sorghum (Jiang et al. 2013), wheat (Okada et al. 2019), rapeseed (Braatz et al. 2017), maize (Lee et al. 2019), tomato (Ortigosa et al. 2019), barley (Lawrenson and Harwood 2019), cumber (Chandrasekaran et al. 2016), potato (Enciso-Rodriguez et al. 2019). The following is a list of several cases of editing using CRISPR/Cas9 that have been documented (Table 1).

Table 1. CRISPR/Cas9 system applicability in major crops.

Crops	Method of delivery	Gene	Gene function	Reason for transformation	Reference
Cotton	Somatic embryogenesis method and *Agrobacterium* mediated transformation method	*GhMYB25*-like	*GhMYB25*-like is elaborated for cotton fiber improvement	For exact and effective genome editing	(Li et al. 2017)
Cotton	*Agrobacterium* mediated transformation technique	*GhVP* (vacuolar H+pyrophosphatase)	-	For targeted mutagenesis of cotton genome	(Chen et al. 2017)
Cotton	*Agrobacterium* tumefaciens mediated transformation approach	An endogenous gene *DsRed2* and *GhCLA1*	*DsRed2* protein exploited as a reporter due to its various advantages over report proteins, *AtCLA1* is utilized for the development of chloroplast.	-	(Wang et al. 2018c)
Cotton	*Agrobacterium* mediated transformation approach	*GhCLA1* (Chloroplasts alterados 1)	-	Mutations were perceived in protoplast of cotton.	(Chen et al. 2017)
Rice	*Agrobacterium* mediated delivery technique	*ALS*	Encodes acetolactate synthase, which is contained for branched amino acid biosynthesis	Induced Mutation for yield improvements	(Sun et al. 2016)
Rice	*Agrobacterium* mediated transfer approach	*TMS5*	In china, *TMS5* is thermosensitive genic male sterility (TGMS) gene which determines the endonuclease RNase ZS1 in AnS-1	To cultivate profitable 'transgene clean' TGMS rice lines	(Zhou et al. 2016)
Wheat	Cas9-RNPmediated GE procedure	*TaGW2*	*TaGW2* gene show an important function in the control of grain weight		(Liang et al. 2017)
Wheat	Particle bombardment approach	*TaMLO* homologs	Elaborated to prevent tolerance pathway for powdery mildew	To enhance tolerance against biotic pathogen	(Wang et al. 2014)
Sorghum	*A. tumefaciens* mediated delivery technique	Whole *k1C* gene family	Kafirins are protein which can be used as storage in grain of sorghum and produce protein having poor digestibility	To develop variant of Kafirin for the enhancement of quality and protein digestibility	(Li et al. 2018)
Soybean	*Agrobacterium tumefaciens* mediated transformation procedure	*GmFT2a*	Exposed delay under long-day and short-day circumstances	Inheritable site directed mutagenesis	(Cai et al. 2018)
Soybean	-	*GmPPD1* and *GmPPD2*	Rippled trifoliate leaves were seen thicker with lager pods, elongated petioles and deeper green color	To produce targeted mutagenesis of *GmFT2a*	(Kanazashi et al. 2018)

Table 1 contd. ...

...Table 1 contd.

Crops	**Method of delivery**	**Gene**	**Gene function**	**Reason for transformation**	**Reference**
Banana	*Agrobacterium* mediated transformation technique	*PDS* (phytoene desaturase)	*PDS* is utilized as a marker to show editing of genome in various plant species	Basic methodology	(Kaur et al. 2018)
Apple	Direct delivery of CRISPR RNPs to apple protoplast and protoplast delivery method	*DIPM-1, DIPM-2,* and *DIPM-4*	*DIPMs* display exact physical relation with the disease-target gene of Erwinia amylovora, which show as a susceptible feature	To enhance tolerance to fire blight disease	(Pessina et al. 2016)
Grapes	Direct transformation of CRISPR RNPs to the protoplast of grapevine and Protoplast delivery approach	*MLO-7* (Mildew Locus O)	Inducing tolerance gene for pathogen resistant	To enhance resistance against powdery mildew	(Malnoy et al. 2016)
Watermelon	*Agrobacterium* mediated delivery method	*ClPDS* (phytoene desaturase)	*ClPDS* represents noticeable albino appearance	To efficiently produce knockout mutations in watermelon	(Tian et al. 2017)
Cucumber	*A. tumefaciens* mediated transformation method	*eIF4E* (eukaryotic translation initiation factor 4E) gene	*eIF4E* is a plant cellular translation factor which perform the vital part in the life cycle of potyviridae	To increase resistance against the virus in cucumber	(Chandrasekaran et al. 2016)
Tomato	*Agrobacterium* mediated leaf disk approach	*SlIAA9* (auxininduced 9)	*SlIAA9* is an important gene monitoring parthenocarpy	To produce parthenocarpic tomato plants	(Ueta et al. 2017)
Tomato	-	*SlPIF4* (phytochrome interacting factor) and *SlPDS* (phytoene desaturase)	SlPDS are elaborated in biosynthesis of carotenoid and SlPIF4 which related to basic helix in multigene family	For direct mutagenesis in plants of tomato	(Pan et al. 2016)
Tobacco	*Agrobacterium* mediated transformation consistent procedure	CRISPR/Cas9 system	*EDS1*-family genes are crucial mechanism of plant distinctive immunity	For initiation of removal of inheritable chromosomal	(Ordon et al. 2016)
Tobacco	-	-	*PDR6* is an ABC transporter which performs a vital part in strigolactone transport	For direct examination and mutagenesis of brief genome editing activity	(Gao et al. 2015)
Tobacco	*A. tumefaciens* LBA4404virG technique	-	Two enzymes involved in plant-specific glycans	To eliminate plant type glycans by deactivation of two enzymes	(Mercx et al. 2017)
Potato	Regeneration and brief transfection from isolated protoplasts	*GBSS* (granule bound starch synthase)	*GBSS* is reliable for the creation of amylose	To change the quality of starch	(Andersson et al. 2017)

4. Targeted Mutagenesis

CRISPR/Cas9, as previously mentioned, can be utilized to improve crop traits by inducing sequence-specific mutagenesis, which disrupts genes to assess their roles (Scheben et al. 2017). A DNA-binding protein called Cas9 can be created by mutating its nuclease domains. As a result, the DNA cleavage activity is suppressed while the DNA binding activity is preserved. Fusion proteins can be directed to particular locations in the genome by directly or indirectly fusing this "dead" Cas9 nuclease to an effector domain (Konermann et al. 2015). Because of this, CRISPR/Cas can be used for a variety of site-specific modifications (Hilton et al. 2015), gene expression regulation (Tang et al. 2017), and BE without causing DSB. For instance, BE can be accomplished through fusion with deaminases in maize, rice, and wheat, or by imaging genomic loci in live N. benthamiana leaf cells (Dreissig et al. 2017).

4.1 Multiplex GE

CRISPR can potentially cause mutations at several genomic sites concurrently in any kind of organism, utilizing numerous sgRNAs. CRISPR/Cas9 can also stack features in an elite variety background for multiplex GE (Yan et al. 2018). Another effective method for focusing on several members of a multigene family is multiplex gene editing. There are two methods to accomplish this combining different sgRNAs in a single vector or creating numerous gRNA expression cassettes in distinct vectors (Kim et al. 2021). According to Wang et al. (2019) research, a method that makes it easier to reproduce F1 rice hybrids clonally utilizing seeds has been developed. Multiplex CRISPR/Cas9 genetic engineering of three meiotic genes was utilized in order to create tetraploid seeds and clonal diploid gametes from F1 hybrid rice. This was accomplished by regulating the heterozygosity of the rice. It is also possible to promote haploid seed formation in hybrid rice by modifying the *MATRILINEAL* (*MTL*) gene, which is produced during the process of fertilization.

Plants that were able to reproduce clonally through seeds were produced in hybrid rice by continuously altering these genes to fix heterozygosity and induce haploid induction. The expression of BBM1 in rice sperm cells and its aberrant expression in egg cells is adequate for parthenogenesis (Khanday et al. 2019). Although biparentally regulated later, *BBM1*'s zygotic expression is typically restricted to the paternal allele. Mentally-transmitted *BBM1* completely restores embryo growth, but triple deletion of the *BBM1*, 2, and 3 genes results in abortion and stopping of the embryo.

5. CRISPR/Cas Application for the Enhancement of Crop Abiotic Stress Tolerance

The procedure for creating genome-edited plants utilizing CRISPR/Cas9 technology encompasses five discrete phases: (i) Identification of the target gene, (ii) Development of a gene-specific sgRNA for the target, (iii) Assembly of Cas9 and sgRNA, (iv) Introduction into the target plant, and (v) Regeneration and assessment of plant lines (Fig. 1). CRISPR/Cas9-mediated gene editing research has attained enormous progress towards enhancing crop resistance to abiotic stress. Corn becomes drought-tolerant by mutating the cuticle wax-related ZmSRL5 gene (Pan et al. 2020). According to Shi et al. (2017), altering the *ZmARGOS8* promoter gene, which creates an unfavorable maize ethylene adaptation regulator, promotes drought tolerance. CRISPR/Cas9 technology made the elite rice restorer line "Hua Zhan" more drought-, heat-, and osmotic-resistant by lowering its susceptibility to abscisic acid and leaf water loss (Xu et al. 2016). Santosh Kumar et al. (2021) altered a gene linked to drought resistance in indica rice utilizing CRISPR/Cas9. The resulting mutants showed adequate osmotic stress resistance and high salt stress tolerance at the seedling phase, indicating that their approach was suitable for enhancing the salinity and drought resistance of indica rice varieties (Santosh et al. 2020). Lou et al. (2017) found that homozygous T1 *OsSAPK2* mutants were extremely susceptible to drought stress but comparatively insensitive to abscisic acid, indicating an association between *OsSAPK2* and the resistance of rice to drought stress after

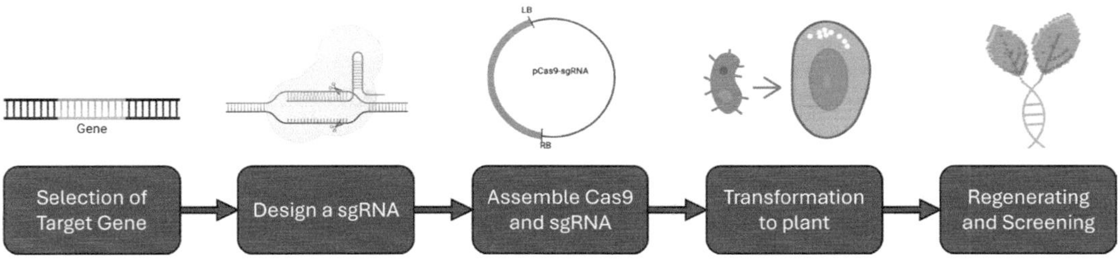

Figure 1. Illustrates the CRISPR/Cas9 technology used for gene editing (Erdoğan et al. 2023).

altering the *OsSAPK2* sequence. To boost rice resilience to cold, Shen et al. (2017) altered *OsAnn3*. Phenotypic studies showed that the six mutated lines were more vulnerable to low-temperature stress than the wild-type control (Dengji et al. 2017). Arable land contamination as a result of urbanization and industry is an important issue. It is essential to stop hazardous heavy metals from accumulating in food crops. Because of this, rice lines with lower levels of radioactive cadmium, cesium and arsenic have been produced by utilizing CRISPR/Cas9 technology to delete *OsNramp5, OsARM1,* and *OsHAK1*, respectively (Wang et al. 2019a).

6. CRISPR/Cas9 Application for the Improvement of Crop Yield

One of the main goals for studies on crop improvement is to increase yield. Grain size, tiller number per panicle, grain number, and grain weight and size per panicle are specific features associated with grain yield (Chen et al. 2019). Based on CRISPR/Cas9 technology, recent studies have shown that genetic alterations have improved the productivity of wheat, rice, and other food crops. Miao et al. used the CRISPR/Cas9 method to create a rice mutant with triple deletion pyl1/4/6. The mutant exhibited multiple variations from the wild-type control, including a greater yield, greater secondary and primary branches inside the panicles, longer panicles, and fewer tillers per plant (Miao et al. 2018). In a different study, a combined removal of three grain weight-related genes (GW2, 5, and 6) resulted in a significantly higher grain weight of rice (Xu et al. 2016). It has been reported that deleting the *OsAAP3* gene, which codes for an amino acid transporter involved in rice nutrient allocation, enhances the yield of grain and number of tillers while preserving the quality of the grain (Lu et al. 2018). The CRISPR/Cas9 system can also be used to enhance grain width and length as well as the 1000-grain weight, by knocking out *OsSNB* that helps in regulating plant growth (Lee et al. 2007). This suggests that *OsSNB* not only influences floral growth but also regulates the shape of grains of the rice plant (Ma et al. 2019). The T0 mutant of tetraploid and hexaploid durum wheat showed an increase in 1000-grain weight. Zhang et al. (2016) transiently introduced CRISPR/Cas9 RNA or DNA into wheat calli to target *TaGASR7*. In wheat, the deletion of the *TaGW2* gene, which codes for a RING E3 ligase result in width and length grains, which improves grain yield (Wang et al. 2018b; Zhang et al. 2018b). Modulating cytokinin homeostasis is one of the elements influencing grain yield might be an effective way to increase grain yield. More precisely, the grain yield is increased by knocking down *TaCKX2-D1*, which expresses a cytokinin oxidase/dehydrogenase in wheat (Dahan-Meir et al. 2018). After the 30 terminals of the *OsLOGL5* coding area was knocked out using CRISPR/Cas9, total rice yield, grain weight per spike and number of grains improved (Wang et al. 2020). Wang et al. (2022a) discovered grain size on chromosome 2 (GS2) and created a unique gene-editing technique that may be applied globally to enhance the production of rice grain and size. Additionally, they stated that the conserved miR396-GRF/GIF module which affects the development of plants as well as response to environmental stimuli can be used to improve crops along with other genes containing miRNA regions of interest (Wang et al. 2022b).

7. Genome Editing for Disease Resistance via CRISPR/Cas Technology

Plant viral, fungal, oomycetic, and bacterial diseases threaten crop output and world food security. These diseases significantly reduce fruit, field crop, and other food plant productivity and quality. Biologic factors cause 20–40% of global crop losses (Wang et al. 2014). Historically, plants have been given disease-resistant R-genes. The main resistance technique uses NB-LRR proteins to identify pathogen *Avr* gene products. Hypersensitive Reaction (HR) is a protective response that kills host cells after a threat is detected (Nekrasov et al. 2017). Another method for disease tolerance is to block susceptibility genes (Pu et al. 2018). Several viruses require host genes to infect and multiply. Fungal diseases such as powdery mildews require long-term host cell interaction to multiply. S-genes or susceptibility genes must be expressed for infections to be detected, penetrate, escape host defenses, and meet their structural and metabolic needs (Ma et al. 2018). S-gene mutations can cause widespread recessive heritable resistance (Wang et al. 2016; Wang et al. 2018c). Thus, S-genes are critical to plant-pathogen interactions that impact host resistance to infection. Fundamental compatibility, sustained compatibility, and plant immune signal inhibition are S-genes' main molecular functions. Basic compatibility involves pathogen penetration and host recognition. Pathogens need sustained compatibility to multiply and spread. Negative regulation of plant immunological signals suppresses host defensive responses, reducing host defense mechanisms (Ma et al. 2018). Comprehending the molecular mechanisms of S-genes and their interactions with pathogens is essential for devising successful strategies for plant disease resistance. The Mlo gene in barley is a recognized S-gene mutant that exhibits tolerance to powdery mildew. The incorporation of Mlo alleles featuring loss-of-function mutations in barley has produced powdery mildew-resistant variants. Long-term cultivation of resistant plants has demonstrated the efficacy of S-gene-based resistance against severe powdery mildew strains (Oliva et al. 2019). After cloning the barley *Mlo* gene, researchers found that this gene family is multi-copy in higher plants and has been preserved across the plant kingdom. Research indicates that mutating the Mlo gene imparts resistance to powdery mildew. The Mlo protein facilitates the entry of powdery mildew fungus spores into host epidermal cells. The Mlo gene encodes a membrane-associated protein featuring seven transmembrane domains. Mutating this gene with CRISPR-Cas9 has created powdery mildew-tolerant plants without foreign genes. Mlo-like genes are connected to powdery mildew exposure in peach, Arabidopsis, grapevine, tomato, bread wheat, and pea (Baltes et al. 2015; Zhang et al. 2018a; Kieu et al. 2021).

Evolutionary relationships classify these genes into numerous classes. Class IV comprises Mlo-like genes linked to the susceptibility of monocotyledonous plants to powdery mildew, whereas class V encompasses those associated with dicotyledons. Malnoy et al. (2016) identified Mlo-like genes in classes I, II, III, and VI that are not classified as S-genes. CRISPR and next-generation sequencing have facilitated the identification and targeting of plant disease-tolerant S-genes. To resist major plant diseases, CRISPR/Cas9 targets S-genes like the Mlo gene. Modifying these S-genes can interfere with host-pathogen compatibility, leading to extensive and enduring disease resistance (Chen et al. 2017). Modifying the Enhanced Disease Resistance 1 gene diminishes powdery mildew in wheat (Zhao et al. 2021). CRISPR/Cas9 may target several genes with a single construct, a unique feature. Butt et al. (2017) found that a single Arabidopsis molecular configuration can continually modify 14 genes. Multiplex genome-editing techniques incorporate several gRNAs into a singular construct regulated by a U3 or U6 promoter and have proven effective in crops. Molecular excision or backcrossing, or self-fertilization with the parent line were used to eliminate transgenes. The Cas9 method selects transgene-free lines from separated populations after verifying an efficient genomic alteration. Plant diseases caused by viruses can harm many economically valuable crops. The infection cycle of Potyviridae viruses requires the protein Eukaryotic Translation Initiation Factor 4E. These viruses have single-stranded positive-sense RNA. The 5'-terminal VPg element of potyviruses interacts with eIF4E. Interrupting this connection protects plants from potyviruses (Wang et al. 2021). In separate studies, CRISPRCas9 disrupted eIF4E genes in Cucumis sativus and Arabidopsis, resulting in ipomovirus and potyvirus resistance (Bhaya et al. 2011; Wu et al.

2020). The genome-edited plants' lack of transgenes indicates how CRISPR technology can develop virus-resistant crops without adding foreign DNA (Bhaya et al. 2011; Wu et al. 2020; Voss-Fels et al. 2019). *OsSWEET13* encodes a sucrose transporter that has a role in plant-pathogen interactions (Shimatani et al. 2017). Table 2 shows CRISPR-mediated S-gene alteration for disease-resistant crops. This study shows that CRISPR technologies may combat bacterial and viral plant ailments.

Table 2. Genes targeted using CRISPR editing techniques to elucidate resistance to diseases.

Crops	Disease	Delivery methods	Targeted gene	Reference
Wheat	Powdery mildew	Biolistic delivery	No	(Zhang et al. 2017)
Wheat	-	Biolistic and protoplast transformation	Yes	(Wang et al. 2014)
Tomato	Gray mold	*Agrobacterium-mediated* delivery approach	No	(Zhang et al. 2018a)
Tomato	DNA viral disease	-	No	(Tashkandi et al. 2018)
Tomato	Bacterial speck	-	Not recognized	(Ortigosa et al. 2019)
Tomato	Fusarium wilt	-	No	(Prihatna et al. 2018)
Tomato	RNA viral disease	-	No	(Wang et al. 2018c)
Tomato	Powdery mildew	-	No	(Nekrasov et al. 2017)
Tobacco	DNA Viral disease	*Agrobacterium* mediated transformation	No	(Baltes et al. 2015)
Tobacco	DNA Viral disease	-	No	(Yin et al. 2019)
Tobacco	DNA Viral disease	-	No	(Tashkandi et al. 2018)
Tobacco	DNA Viral disease	-	No	(Ali et al. 2015)
Rice	Bacterial blight	*Agrobacterium* mediated delivery technique	Not identified	(Oliva et al. 2019)
Rice	Bacterial blight	-	No	(Xu et al. 2019b)
Rice	Rice blast	-	Yes	(Wang et al. 2016)
Rice	-	Not identified	Yes	(Li et al. 2019)
Rice	Rice tungro	-	Yes	(Macovei et al. 2018)
Rice	Bacterial disease	-	No	(Zhou et al. 2015)
Rice	Rice blast	Protoplast transformation	No	(Xie and Yang 2013)
Barley	Wheat Dwarf virus	*Agrobacterium* mediated delivery	No	(Kis et al. 2019)
Apple	Fire blight	-	Yes	(Malnoy et al. 2016)
Banana	Banana streak virus	-	Not recognized	(Tripathi et al. 2019)
Citrus	Citrus canker	-	No	(Jia et al. 2017)
Citrus	Citrus canker	-	No	(Peng et al. 2017)

Wheat, tomato, tobacco, rice, barley, apple, banana and citrus.

8. Challenges and Issues of CRISPR/Cas9

The CRISPR/dCas9 method has promising opportunities for sophisticated crop modification, although it entails problems like as off-target consequences and delivery complications. Off-target effects predominantly result from homologous sequence similarities (Kawall et al. 2020). Online platforms such as CCTop and CRISPOR facilitate the design of sgRNAs to alleviate these hazards. Modifications to Cas9 can diminish off-target effects, however, these approaches are less essential for CRISPR/dCas9, which generally shows fewer off-target problems than conventional CRISPR systems. The efficacy of CRISPR/dCas9 is significantly dependent on accurate sgRNA binding, which might be compromised by mismatches, hence diminishing its functional efficiency (Li et al. 2016; Feng et al. 2021). The necessity for PAM sequences constrains its applicability, although the identification of several Cas proteins may increase versatility. Contemporary delivery strategies, such as Agrobacterium-mediated, PEG-mediated, and biolistic approaches, each exhibit constraints. For instance, Agrobacterium-mediated delivery, although economical, necessitates transgenic integration, whereas PEG-mediated transformation demands labor-intensive protoplast preparation.

The implementation of CRISPR/dCas9 is further complicated by regulatory and public acceptance issues. Certain nations, like as the United States, classify altered plants as non-regulated provided they do not contain any foreign DNA, exemplified by browning-resistant mushrooms (Waltz 2016). In areas such as the EU and China, CRISPR-edited organisms are categorized as GMOs, necessitating rigorous approvals (Globus and Qimron 2017). A thorough risk assessment is crucial to assess the health effects of consuming CRISPR/dCas9-modified plants, particularly those containing foreign DNA.

9. Future Prospects

The rapid developments in CRISPR/Cas technology have opened up numerous exciting avenues for future research and applications in plant development. One promising area of exploration involves expanding the CRISPR toolbox to enhance precision, efficiency, and versatility in genome editing. Continued efforts to optimize CRISPR systems, such as the development of novel Cas variants with improved targeting specificity and reduced off-target effects, will further enhance their utility in precise genome manipulation. Additionally, ongoing research on alternative CRISPR systems, such as Cas12a (formerly known as Cpf1) and Cas13, presents new opportunities for expanding the scope of genome editing applications and exploring non-genetic modifications, including RNA editing and base editing. Furthermore, the integration of CRISPR technology with other omics methods, such as transcriptomics, metabolomics, and proteomics, holds great potential for elucidating complex gene regulatory networks and metabolic pathways involved in plant development. Harnessing the power of CRISPR-based genome editing combined with advanced omics technologies will facilitate the targeted improvement of agronomically significant traits, including stress tolerance, nutritional quality, yield, and disease resistance, thus enhancing food security and promoting sustainable agriculture in response to climate change and global issues.

10. Conclusion

In conclusion, the CRISPR/Cas system has modernized the field of plant development by presenting precise and effective tools for genome editing and manipulation. From its initial discovery as a bacterial immune system to its widespread adoption in plant research and biotechnology, CRISPR technology has catalyzed significant advancements in our understanding of plant development and has the potential to address various challenges facing agriculture and food production. By leveraging the molecular mechanisms and current progress of CRISPR/Cas in plant development, researchers can continue to push the boundaries of scientific knowledge and innovation. The ongoing refining and implementation of CRISPR technology promise to catalyze dramatic changes in agriculture, resulting in superior crop varieties with enhanced features and sustainable solutions for global food security and environmental sustainability. It is imperative that we foster interdisciplinary collaborations, promote responsible research practices, and address ethical considerations to confirm the responsible and beneficial deployment of CRISPR/Cas technology for the betterment of humanity and the planet.

References

Ali, Z., A. Abulfaraj, A. Idris, S. Ali, M. Tashkandi and M.M. Mahfouz. 2015. CRISPR/Cas9-mediated viral interference in plants. Genome Biol. 16: 238.

Andersson, M., X. Zhang, E. Reslow, A. Andreasson, E. Levander and J. Hofvander. 2017. Efficient targeted multiallelic mutagenesis in tetraploid potato (*Solanum tuberosum*) by transient CRISPR-Cas9 expression in protoplasts. Plant Cell Rep. 36: 117–128.

Anzalone, A.V., L.W. Koblan and D.R. Liu. 2020. Genome editing with CRISPR–Cas nucleases, base editors, transposases and prime editors. Nat. Biotechnol. 38(7): 824–844.

Baltes, N.J., A.W. Hummel, E. Konecna, R. Cegan, A.N. Bruns, D.M. Bisaro et al. 2015. Conferring resistance to geminiviruses with the CRISPR-Cas prokaryotic immune system. Nat. Plants 1: 15145.

Bhattacharjee, M., S. Meshram, J. Dayma, N. Pandey, N. Abdallah, A. Hamwieh et al. 2024. Genetic engineering: A powerful tool for crop improvement. pp. 223–258. *In*: Frontier Technologies for Crop Improvement. Springer Nature Singapore.

Bhaya, D., M. Davison and R. Barrangou. 2011. CRISPR-Cas systems in bacteria and archaea: Versatile small RNAs for adaptive defense and regulation. Annu. Rev. Genet. 45: 273–297.

Braatz, J., H.J. Harloff, M. Mascher, N. Stein, A. Himmelbach and C. Jung. 2017. CRISPR-Cas9 targeted mutagenesis leads to simultaneous modification of different homoeologous gene copies in polyploid oilseed rape (*Brassica napus*). Plant Physiol. 174: 935–942.

Butt, H., A. Eid, Z. Ali, M. Atia and M.M. Mahfouz. 2017. Efficient CRISPR/Cas9-mediated genome editing using a chimeric single-guide RNA molecule. Front. Plant Sci. 8: 1441.

Caeasar, S.A., V. Rajan, S.V. Prykhozhij, J.N. Berman and S. Ignacimuthu. 2016. Insert, remove or replace: a highly advanced genome editing system using CRISPR/Cas9. Biochim. Biophys. Acta Mol. Cell Res. 1863(9): 2333–2344.

Cai, Y., M. Chen, Y. Liu, S. Sun, L. Wu, Q. Jiang et al. 2018. CRISPR/Cas9-mediated targeted mutagenesis of *GmFT2a* delays flowering time in soybean. Plant Biotechnol. J. 16: 176–185.

Ceasar, S.A., V. Rajan, S.V. Prykhozhij, J.N. Berman and S. Ignacimuthu. 2016. Insert, remove or replace: A highly advanced genome editing system using CRISPR/Cas9. Biochimica Et Biophysica Acta (BBA) - Molecular Cell Research 1863(9): 2333–2344. https://doi.org/10.1016/j.bbamcr.2016.06.009.

Chandrasekaran, J., M. Brumin, R. Wolf, R. Leibman, A. Klap and A. Gal-On. 2016. Development of broad virus resistance in non-transgenic cucumber using CRISPR/Cas9 technology. Mol. Plant Pathol. 17: 1140–1153.

Chen, K., Y. Wang, R. Zhang, H. Zhang and C. Gao. 2019. CRISPR/Cas genome editing and precision plant breeding in agriculture. Annu. Rev. Plant Biol. 70: 667–697.

Chen, X., Y. Li, Y. Wang, J. Deng, X. Qin and Y. Gao. 2017. Targeted mutagenesis in cotton (*Gossypium hirsutum* L.) using the CRISPR/Cas9 system. Sci. Rep. 7: 44304.

Chen, Y., Z. Wang, H. Ni, Y. Xu, Q. Chen and L. Jiang. 2017. CRISPR/Cas9-mediated base-editing system efficiently generates gain-of-function mutations in Arabidopsis. Sci. China Life Sci. 60: 520–523.

Dahan-Meir, T., S. Filler-Hayut, C. Melamed-Bessudo, S. Bocobza, H. Czosnek, A. Aharoni et al. 2018. Efficient in planta gene targeting in tomato using geminiviral replicons and the CRISPR/Cas9 system. Plant J. 95: 5–16.

Dengji, L., H. Houping, L. Gang and Y. Diqiu. 2017. OsSAPK2 confers abscisic acid sensitivity and tolerance to drought stress in rice. Front. Plant Sci. 8: 993.

Dong, O.X. and P.C. Ronald. 2019. Genetic engineering for disease resistance in plants: Recent progress and future perspectives. Plant Physiol. 180: 26–38.

Dreissig, S., S. Schiml, P. Schindele, O. Weiss, T. Rutten, V. Schubert et al. 2017. Live-cell CRISPR imaging in plants reveals dynamic telomere movements. Plant J. 91(4): 565–573.

Enciso-Rodriguez, F., N.C. Manrique-Carpintero, S.S. Nadakuduti, C.R. Buell, D. Zarka and D. Douches. 2019. Overcoming self-incompatibility in diploid potato using CRISPR-Cas9. Front. Plant Sci. 10: 444928.

Erdoğan, İ., B. Cevher-Keskin, Ö. Bilir, Y. Hong and M. Tör. 2023. Recent developments in CRISPR/Cas9 genome-editing technology related to plant disease resistance and abiotic stress tolerance. Biology 12(7): 1037.

Feng, H., J. Guo, T. Wang, C. Zhang and X.-H. Xing. 2021. Guide-target mismatch effects on dCas9–sgRNA binding activity in living bacterial cells. Nucleic Acids Res. 49: 1263–1277.

Gao, Y., D. Zhang, J. Li, T. Yang, Y. Yang, W. Xie et al. 2015. Auxin binding protein 1 (ABP1) is not required for either auxin signaling or Arabidopsis development. Proc. Natl. Acad. Sci. U.S.A. 112: 2275–2280.

Ghosh, A., A. Kumar and G. Biswas. 2024. Exponential population growth and global food security: challenges and alternatives. pp. 1–20. *In*: Elsevier eBooks.

Globus, R. and U. Qimron. 2017. A technological and regulatory outlook on CRISPR crop editing. J. Cell. Biochem. 119: 1291–1298.

Gupta, R., L. Pizarro, M. Leibman-Markus, I. Marash and M. Bar. 2020. Cytokinin response induces immunity and fungal pathogen resistance, and modulates trafficking of the PRR LeEIX2 in tomato. Mol. Plant Pathol. 21: 1287–1306.

Hilton, I.B., A.M. D'Ippolito, C.M. Vockley, P.I. Thakore, G.E. Crawford, T.E. Reddy et al. 2015. Epigenome editing by a CRISPR-Cas9-based acetyltransferase activates genes from promoters and enhancers. Nat. Biotechnol. 33(5): 510–517.

Jia, H., Y. Zhang, V. Orbović, J. Xu, F.F. White, J.B. Jones et al. 2017. Genome editing of the disease susceptibility gene *CsLOB1* in citrus confers resistance to citrus canker. Plant Biotechnol. J. 15: 817–823.

Jiang, F. and J.A. Doudna. 2017. CRISPR-Cas9 structures and mechanisms. Annu. Rev. Biophys. 46(1): 505–529.

Jiang, W., H. Zhou, H. Bi, M. Fromm, B. Yang and D.P. Weeks. 2013. Demonstration of CRISPR/Cas9/sgRNA-mediated targeted gene modification in Arabidopsis, tobacco, sorghum and rice. Nucleic Acids Res. 41(20): e188.

Kanazashi, Y., A. Hirose, I. Takahashi, M. Mikami, M. Endo, S. Hirose et al. 2018. Simultaneous site-directed mutagenesis of duplicated loci in soybean using a single guide RNA. Plant Cell Rep. 37: 553–563.

Kaur, N., A. Alok, N. Shivani, N. Kaur, P. Pandey, P. Awasthi et al. 2018. CRISPR/Cas9-mediated efficient editing in phytoene desaturase (PDS) demonstrates precise manipulation in banana cv. Rasthali genome. Funct. Integr. Genom. 18: 89–99.

Kawall, K., J. Cotter and C. Then. 2020. Broadening the GMO risk assessment in the EU for genome editing technologies in agriculture. Environ. Sci. Eur. 32: 106.

Khan, Z. 2023. Engineering Disease Resistance in Plants Using CRISPR-Cas. CRC Press.

Khanday, I., D. Skinner, B. Yang, R. Mercier and V. Sundaresan. 2019. A male-expressed rice embryogenic trigger redirected for asexual propagation through seeds. Nature 565(7737): 91–95.

Kieu, N.P., M. Lenman, E.S. Wang, B.L. Petersen and E. Andreasson. 2021. Mutations introduced in susceptibility genes through CRISPR/Cas9 genome editing confer increased late blight resistance in potatoes. Sci. Rep. 11: 4487.

Kim, W., H. Kim, Y. Chung and H. Kim. 2021. Construction of multiple guide RNAs in CRISPR/CAS9 vector using stepwise or simultaneous golden Gate cloning: case study for targeting the FAD2 and FATB multigene in soybean. Plants 10(11): 2542. https://doi.org/10.3390/plants10112542.

Kis, A., É. Hamar, G. Tholt, R. Bán and Z. Havelda. 2019. Creating highly efficient resistance against wheat dwarf virus in barley by employing CRISPR/Cas9 system. Plant Biotechnol. J. 17: 1004.

Konermann, S., M.D. Brigham, A.E. Trevino, J. Joung, O.O. Abudayyeh, C. Barcena et al. 2015. Genome-scale transcriptional activation by an engineered CRISPR-Cas9 complex. Nature 517(7536): 583–588.

Lata, S., S. Bhardwaj and R. Garg. 2022. Nanomaterials for sensing and biosensing: Applications in agri-food diagnostics. Int. J. Environ. Anal. Chem. 1–12.

Lawrenson, T. and W.A. Harwood. 2019. Creating targeted gene knockouts in barley using CRISPR/Cas9. pp. 217–232. *In*: Barley: Methods and Protocols.

Lee, D., J. Lee, S. Moon, S.Y. Park and G. An. 2007. The rice heterochronic gene SUPERNUMERARY BRACT regulates the transition from spikelet meristem to floral meristem. Plant J. 49: 64–78.

Lee, K., Y. Zhang, B.P. Kleinstiver, J.A. Guo, M.J. Aryee, J. Miller et al. 2019. Activities and specificities of CRISPR/Cas9 and Cas12a nucleases for targeted mutagenesis in maize. Plant Biotechnol. J. 17(2): 362–372.

Li, A., S. Jia, A. Yobi, Z. Ge, S.J. Sato, C. Zhang et al. 2018. Editing of an alpha-kafirin gene family increases digestibility and protein quality in sorghum. Plant Physiol. 177(4): 1425–1438.

Li, C., T. Unver and B. Zhang. 2017. A high-efficiency CRISPR/Cas9 system for targeted mutagenesis in cotton (*Gossypium hirsutum* L.). Scientific Reports 7(1): 43902.

Li, S., L. Shen, P. Hu, Q. Liu, X. Zhu, Q. Qian et al. 2019. Developing disease-resistant thermosensitive male sterile rice by multiplex gene editing. J. Integr. Plant Biol. 61: 1201–1205.

Li, X.T., Y. Jun, M.J. Erickstad, S.D. Brown, A. Parks, D.L. Court et al. 2016. tCRISPRi: Tunable and reversible, one-step control of gene expression. Sci. Rep. 6: 39076.

Li, Z., Q. Cheng, Z. Gan, Z. Hou, Y. Zhang, Y. Li et al. 2021. Multiplex CRISPR/Cas9-mediated knockout of soybean *LNK2* advances flowering time. The Crop J. 9(4): 767–776.

Liang, Z., K. Chen, T. Li, Y. Zhang, Y. Wang, Q. Zhao et al. 2017. Efficient DNA-free genome editing of bread wheat using CRISPR/Cas9 ribonucleoprotein complexes. Nat. Commun. 8(1): 14261.

Liu, M., S. Rehman, X. Tang, K. Gu, Q. Fan, D. Chen et al. 2019. Methodologies for improving HDR efficiency. Frontiers in Genetics 9: 691.

Liu, Z., H. Dong, Y. Cui, L. Cong and D. Zhang. 2020. Application of different types of CRISPR/Cas-based systems in bacteria. Microb. Cell Fact. 19(1): 1–14.

Lou, D., S. Lu, Z. Chen, Y. Lin, D. Yu and X. Yang. 2022. *OsSAPK3* improves drought resistance and grain yield in rice.

Lou, D., Wang, H., Liang, G. and Yu, D. 2017. OSSAPK2 confers abscisic acid sensitivity and tolerance to drought stress in rice. Frontiers in Plant Science 8. https://doi.org/10.3389/fpls.2017.00993.

Lu, K., B. Wu, J. Wang, W. Zhu, H. Nie, J. Qian et al. 2018. Blocking amino acid transporter OsAAP3 improves grain yield by promoting outgrowth buds and increasing tiller number in rice. Plant Biotechnol. J. 16: 1710–1722.

Ma, J., J. Chen, M. Wang, Y. Ren, S. Wang, C. Lei et al. 2018. Disruption of *OsSEC3A* increases the content of salicylic acid and induces plant defense responses in rice. J. Exp. Bot. 69: 1051–1064.

Ma, X., F. Feng, Y. Zhang, I.E. Elesawi, K. Xu, T. Li et al. 2019. A novel rice grain size gene *OsSNB* was identified by genome-wide association study in natural population. PLoS Genet. 15: e1008191.

Macovei, A., N.R. Sevilla, C. Cantos, G.B. Jonson, I. Slamet-Loedin and T. Čermák. 2018. Novel alleles of rice *eIF4G* generated by CRISPR/Cas9-targeted mutagenesis confer resistance to Rice tungro spherical virus. Plant Biotechnol. J. 16: 1918–1927.

Malnoy, M., R. Viola, M.H. Jung, O.J. Koo, S. Kim, J.S. Kim et al. 2016. DNA-free genetically edited grapevine and apple protoplast using CRISPR/Cas9 ribonucleoproteins. Front. Plant Sci. 7: 1904.

Manghwar, H., B. Li, X. Ding, A. Hussain, K. Lindsey, X. Zhang et al. 2020. CRISPR/Cas systems in genome editing: methodologies and tools for sgRNA design, off-target evaluation, and strategies to mitigate off-target effects. Advanced Sci. 7(6): 1902312.

Mei, Y., Y. Wang, H. Chen, Z.S. Sun and J.X. Da. 2016. Recent progress in CRISPR/Cas9 technology. J. Genet. Genomics 43(2): 63–75.

Mercx, S., N. Smargiasso, F. Chaumont, E. De Pauw, M. Boutry and C. Navarre. 2017. Inactivation of the β(1,2)-xylosyltransferase and the α(1,3)-fucosyltransferase genes in Nicotiana tabacum BY-2 cells by a multiplex CRISPR/Cas9 strategy results in glycoproteins without plant-specific glycans. Front. Plant Sci. 8: 403.

Miao, C., L. Xiao, K. Hua, C. Zou, Y. Zhao, R.A. Bressan et al. 2018. Mutations in a subfamily of abscisic acid receptor genes promote rice growth and productivity. Proc. Natl. Acad. Sci. USA 4: 774.

Nekrasov, V., C. Wang, J. Win, C. Lanz, D. Weigel and S. Kamoun. 2017. Rapid generation of a transgene-free powdery mildew resistant tomato by genome deletion. Sci. Rep. 7: 482.

Okada, A., T. Arndell, N. Borisjuk, N. Sharma, N.S. Watson-Haigh, E.J. Tucker et al. 2019. CRISPR/Cas9-mediated knockout of Ms1 enables the rapid generation of male-sterile hexaploid wheat lines for use in hybrid seed production. Plant Biotechnol. J. 17(10): 1905–1913.

Oliva, R., C. Ji, G. Atienza-Grande, J.C. Huguet-Tapia, A. Perez-Quintero, T. Li et al. 2019. Broad-spectrum resistance to bacterial blight in rice using genome editing. Nat. Biotechnol. 37: 1344–1350.

Ordon, J., J. Gantner, J. Kemna, L. Schwalgun, M. Reschke, J. Streubel et al. 2016. Generation of chromosomal deletions in dicotyledonous plants employing a user-friendly genome editing toolkit. The Plant Journal 89(1): 155–168. https://doi.org/10.1111/tpj.13319.

Ortigosa, A., S. Gimenez-Ibanez, N. Leonhardt and R. Solano. 2019. Design of a bacterial speck resistant tomato by CRISPR/Cas9-mediated editing of *Sl JAZ 2*. Plant Biotechnol. J. 17: 665–673.

Pan, C., L. Ye, L. Qin, X. Liu, Y. He, J. Wang et al. 2016. CRISPR/Cas9-mediated efficient and heritable targeted mutagenesis in tomato plants in the first and later generations. Sci. Rep. 6(1): 24765.

Pan, Z., M. Liu, H. Zhao, Z. Tan, K. Liang, Q. Sun et al. 2020. *ZmSRL5* is involved in drought tolerance by maintaining cuticular wax structure in maize. J. Integr. Plant Biol. 62: 1895–1909.

Pawełkowicz, M., B. Zieniuk, P. Staszek and A. Przybysz. 2024. From sequencing to genome editing in Cucurbitaceae: Application of modern genomic techniques to enhance plant traits. Agriculture 14(1): 90.

Peng, A., S. Chen, T. Lei, L. Xu, Y. He, L. Wu et al. 2017. Engineering canker-resistant plants through CRISPR/Cas9-targeted editing of the susceptibility gene *CsLOB1* promoter in citrus. Plant Biotechnol. J. 15: 1509–1519.

Pessina, S., L. Lenzi, M. Perazzolli, M. Campa, L. Dalla Costa, S. Urso et al. 2016. Knockdown of MLO genes reduces susceptibility to powdery mildew in grapevine. Hortic. Res. 3.

Pop-Bica, C., A. Nutu, R. Cojocneanu, S. Chira and I. Berindan-Neagoe. 2022. The CRISPR/Cas9 genome-editing system: principles and applications. pp. 1–22. *In*: CRISPR-/Cas9 Based Genome Editing for Treating Genetic Disorders and Diseases. CRC Press.

Prihatna, C., M.J. Barbetti and S.J. Barker. 2018. A novel tomato Fusarium wilt tolerance gene. Front. Microbiol. 9: 1226.

Pu, Y., C. Liu, J. Li, G.T. Aerzu, Y. Hu and X. Liu. 2018. Different *SlU6* promoters cloning and establishment of CRISPR/Cas9-mediated gene editing system in tomato. Sci. Agric. Sin. 51: 315–326.

Qin, L., J. Li, Q. Wang, Z. Xu, L. Sun, M. Alariqi et al. 2020. High-efficient and precise base editing of C• G to T• A in the allotetraploid cotton (*Gossypium hirsutum*) genome using a modified CRISPR/Cas9 system. Plant Biotechnol. J. 18(1): 45–56.

Raza, H., M.A. Abdullah, S. Khaliq, M.H. Sajjad, M. Elahi, M.R. Khan et al. 2024. Role of CRISPR in crop improvement: A review. Asian J. Res. Agr. For. 10(2): 12–19.

Santosh Kumar, V.V., R.K. Verma, S.K. Yadav, P. Yadav, A. Watts, M.V. Rao et al. 2020. CRISPR-Cas9 mediated genome editing of drought and salt tolerance (*OsDST*) gene in indica mega rice cultivar MTU1010. Physiol. Mol. Biol. Plants 26: 1099–1110.

Santosh Kumar, V.V., S.K. Yadav, R.K. Verma, S. Shrivastava, O. Ghimire, S. Pushkar et al. 2021. The abscisic acid receptor *OsPYL6* confers drought tolerance to indica rice through dehydration avoidance and tolerance mechanisms. J. Exp. Bot. 72(4): 1411–1431.

Scheben, A., F. Wolter, J. Batley, H. Puchta and D. Edwards. 2017. Towards CRISPR/Cas crops–bringing together genomics and genome editing. New Phytologist 216(3): 682–698.

Shen, C., Z. Que, Y. Xia, N. Tang, D. Li, R. He et al. 2017. Knockout of the annexin gene *OsAnn3* via CRISPR/Cas9-mediated genome editing decreased cold tolerance in rice. J. Plant Biol. 60: 539–547.

Shi, J., H. Gao, H. Wang, H.R. Lafitte, R.L. Archibald, M. Yang et al. 2017. *ARGOS8* variants generated by CRISPR-Cas9 improve maize grain yield under field drought stress conditions. Plant Biotechnol. J. 15: 207–216.

Shimatani, Z., S. Kashojiya, M. Takayama, R. Terada, T. Arazoe, H. Ishii et al. 2017. Targeted base editing in rice and tomato using a CRISPR-Cas9 cytidine deaminase fusion. Nat. Biotechnol. 35: 441–443.

Son, N., C.R. Chen and C.H. Syu. 2024. Towards artificial intelligence applications in precision and sustainable agriculture. Agronomy 14(2): 239.

Sun, Y., X. Zhang, C. Wu, Y. He, Y. Ma, H. Hou et al. 2016. Engineering herbicide-resistant rice plants through CRISPR/Cas9-mediated homologous recombination of acetolactate synthase. Mol. Plant 9: 628–631.

Tang, X., L.G. Lowder, T. Zhang, A.A. Malzahn, X. Zheng, D.F. Voytas et al. 2017. A CRISPR-Cpf1 system for efficient genome editing and transcriptional repression in plants. Nat. Plants 3: 17018.

Tashkandi, M., Z. Ali, F. Aljedaani, A. Shami and M.M. Mahfouz. 2018. Engineering resistance against Tomato Yellow Leaf Curl Virus via the CRISPR/Cas9 system in tomato. Plant Signal. Behav. 13: 1525996.

Tian, S., L. Jiang, Q. Gao, J. Zhang, M. Zong, H. Zhang et al. 2017. Efficient CRISPR/Cas9-based gene knockout in watermelon. Plant Cell Rep. 36: 399–406.

Tripathi, J.N., V.O. Ntui, M. Ron, S.K. Muiruri, A. Britt and L. Tripathi. 2019. CRISPR/Cas9 editing of endogenous Banana streak virus in the B genome of Musa spp. overcomes a major challenge in banana breeding. Commun. Biol. 2: 46.

Ueta, R., C. Abe, T. Watanabe, S.S. Sugano, R. Ishihara, H. Ezura et al. 2017. Rapid breeding of parthenocarpic tomato plants using CRISPR/Cas9. Sci. Rep. 7(1): 507.

Voss-Fels, K.P., A. Stahl and L.T. Hickey. 2019. Q&A: Modern crop breeding for future food security. BMC Biol. 17: 18.

Waltz, E. 2016. Gene-edited CRISPR mushroom escapes US regulation. Nature 532: 293.

Wang, C., Q. Liu, Y. Shen, Y. Hua, J. Wang, J. Lin et al. 2019. Clonal seeds from hybrid rice by simultaneous genome engineering of meiosis and fertilization genes. Nat. Biotechnol. 37(3): 283–286.

Wang, C., G. Wang, Y. Gao, G. Lu and T.W. Greene. 2020. A cytokinin-activation enzyme-like gene improves grain yield under various field conditions in rice. Plant Mol. Biol. 102: 373–388.

Wang, F., C. Wang, P. Liu, C. Lei, W. Hao, Y. Gao et al. 2016. Enhanced rice blast resistance by CRISPR/Cas9-targeted mutagenesis of the ERF transcription factor gene *OsERF922*. PLoS ONE 11: e0154027.

Wang, P., J. Zhang, L. Sun, Y. Ma, J. Xu, S. Liang et al. 2018a. High efficient multisites genome editing in allotetraploid cotton (*Gossypium hirsutum*) using CRISPR/Cas9 system. Plant Biotechnol. J. 16(1): 137–150.

Wang, T., Y. Li, Y. Fu, H. Xie, S. Song, M. Qiu et al. 2019. Mutation at different sites of metal transporter gene OsNramp5 affects Cd accumulation and related agronomic traits in rice (*Oryza sativa* L.). Frontiers in Plant Science 10. https://doi.org/10.3389/fpls.2019.01081

Wang, W., S. James, Q. Pan, D. Dwight, F. He, B. Abdulhamit et al. 2018b. Gene editing and mutagenesis reveal inter-cultivar differences and additivity in the contribution of *TaGW2* homoeologues to grain size and weight in wheat. Theor. Appl. Genet. 131: 2463–2475.

Wang, W., W. Wang, Y. Pan, C. Tan, H. Li, Y. Chen et al. 2022a. A new gain-of-function *OsGS2/GRF4* allele generated by CRISPR/Cas9 genome editing increases rice grain size and yield. Crop J. 10: 1207–1212.

Wang, Y., N. Zafar, Q. Ali, H. Manghwar, G. Wang, L. Yu et al. 2022b. CRISPR/Cas genome editing technologies for plant improvement against biotic and abiotic stresses: Advances, limitations, and future perspectives. Cells 11(23): 3928.

Wang, Y., X. Cheng, Q. Shan, Y. Zhang, J. Liu, C. Gao et al. 2014. Simultaneous editing of three homoeoalleles in hexaploid bread wheat confers heritable resistance to powdery mildew. Nat. Biotechnol. 32: 947–951.

Wang, Y.J., L.L. Ma and Z. Liang. 2021. Research progress on CRISPR/Cas9 genome editing technology and its application in crop genetic improvement. J. Shanxi Agric. Sci. 49: 1383–1392.

Wang, Z., T.J. Hardcastle, A.C. Pastor, W.H. Yip, S. Tang and D.C. Baulcombe. 2018c. A novel DCL2-dependent miRNA pathway in tomato affects susceptibility to RNA viruses. Genes Dev. 32: 1155–1160.

Wu, J., C. Chen, G. Xian, D. Liu, L. Lin, S. Yin et al. 2020. Engineering herbicide-resistant oilseed rape by CRISPR/Cas9-mediated cytosine base-editing. Plant Biotechnol. J. 18: 1857–1859.

Xie, K. and Y. Yang. 2013. RNA-guided genome editing in plants using a CRISPR–Cas system. Mol. Plant 6: 1975–1983.

Xu, R., Y. Yang, R. Qin, L. Hao, C. Qiu, L. Li et al. 2016. Rapid improvement of grain weight via highly efficient CRISPR/Cas9-mediated multiplex genome editing in rice. J. Genet. Genom. 43: 529–532.

Xu, W., W. Fu, P. Zhu, Z. Li, C. Wang, C. Wang et al. 2019a. Comprehensive analysis of CRISPR/Cas9-mediated mutagenesis in *Arabidopsis thaliana* by genome-wide sequencing. Int. J. Mol. Sci. 20(17): 4125.

Xu, Z., X. Xu, Q. Gong, Z. Li, Y. Li, S. Wang et al. 2019b. Engineering broad-spectrum bacterial blight resistance by simultaneously disrupting variable TALE-binding elements of multiple susceptibility genes in rice. Mol. Plant 12: 1434–1446.

Yan, F., Y. Kuang, B. Ren, J. Wang, D. Zhang, H. Lin et al. 2018. Highly efficient A·T to G·C base editing by CAS9N-Guided TRNA adenosine deaminase in rice. Molecular Plant 11(4): 631–634. https://doi.org/10.1016/j.molp.2018.02.008.

Yang, H., S. Ren, S. Yu, H. Pan, T. Li, S. Ge et al. 2020. Methods favoring homology-directed repair choice in response to CRISPR/Cas9-induced double-strand breaks. Int. J. Mol. Sci. 21(18): 6461.

Yin, K., T. Han, K. Xie, J. Zhao, J. Song and Y. Liu. 2019. Engineer complete resistance to cotton leaf curl Multan virus by the CRISPR/Cas9 system in *Nicotiana benthamiana*. Phytopathol. Res. 1: 9.

Zhang, S., L. Wang, R. Zhao, W. Yu, R. Li, Y. Li et al. 2018a. Knockout of *SlMAPK3* reduced disease resistance to *Botrytis cinerea* in tomato plants. J. Agric. Food Chem. 66: 8949–8956.

Zhang, Y., Z. Liang, Y. Zong, Y. Wang, J. Liu, K. Chen et al. 2016. Efficient and transgene-free genome editing in wheat through transient expression of CRISPR/Cas9 DNA or RNA. Nature Communications 7(1). https://doi.org/10.1038/ncomms12617.

Zhang, Y., B. Bai, G. Wu, S. Zou, Y. Chen, C. Gao et al. 2017. Simultaneous modification of three homoeologs of *TaEDR1* by genome editing enhances powdery mildew resistance in wheat. Plant J. 91: 714–724.

Zhang, Y., D. Li, D. Zhang, X. Zhao and D. Wang. 2018b. Analysis of the functions of *TaGW2* homoeologs in wheat grain weight and protein content traits. Plant J. Cell Mol. Biol. 94: 857–866.

Zhao, D., J. Li, S. Li, X. Xin, M. Hu, M.A. Price et al. 2021. Publisher correction: Glycosylase base editors enable C-to-A and C-to-G base changes. Nat. Biotechnol. 39: 115.

Zhou, J., Z. Peng, J. Long, D. Sosso, B.O. Liu, J.S. Eom et al. 2015. Gene targeting by the TAL effector *PthXo2* reveals cryptic resistance gene for bacterial blight of rice. Plant J. 82: 632–643.

Zhou, H., M. He, J. Li, L. Chen, Z. Huang, S. Zheng et al. 2016. Development of commercial thermo-sensitive genic male sterile rice accelerates hybrid rice breeding using the CRISPR/Cas9-mediated TMS5 editing system. Sci. Rep. 6(1): 37395.

Chapter 3

Genome Mapping and Gene Sequencing

Applications of CRISPR/Cas9 in Cereal Crops

Samar G. Thabet,[1,*] *Fatmah Ahmed Safhi,*[2] *Amr Elkelish,*[3,4] *Andreas Börner,*[5] *Murad Muhammad,*[6,7] *Mona FA Dawood*[8] and *Ahmad M. Alqudah*[9]

1. Introduction

Genome mapping and gene sequencing are foundational techniques in modern genomics, pivotal to advancing our understanding of the genetic architecture and complexity of organisms (Campbell et al. 2018). These approaches have significantly advanced plant science by enabling researchers to unravel the genetic foundations of important agronomic traits. They have also facilitated the development of crop varieties with improved yield, quality, and resistance to various environmental stresses (Pareek et al. 2011). Among crops, cereals such as rice (*Oryza sativa*), wheat (*Triticum aestivum*), maize (*Zea mays*), and barley (*Hordeum vulgare*) are of paramount importance, providing staple food for a significant proportion of the global population (Luikart et al. 2019). Deciphering the genetic composition of these cereal crops through accurate genome mapping and sequencing is crucial for enhancing food security and tackling the challenges associated with climate change, pests, and diseases (Lesk 2017).

The development of Next-Generation Sequencing (NGS) and Third-Generation Sequencing (TGS) technologies has revolutionized the ability to generate high-quality genomic data at unprecedented speed and scale (Ghosh et al. 2018). However, while sequencing technologies allow for the identification of genomic variants, the functional characterization of these variants,

[1] Department of Botany, Faculty of Science, Fayoum University, 63514 Fayoum, Egypt.
[2] Department of Biology, College of Science, Princess Nourah bint Abdulrahman University, P.O. Box 84428, Riyadh 11671, Saudi Arabia.
[3] Department of Biology, College of Science, Imam Mohammad Ibn Saud Islamic University (IMSIU), P. O. Box: 90950, Riyadh 11623, Kingdom of Saudi Arabia.
[4] Botany and Microbiology Department, Faculty of Science, Suez Canal University, Ismailia 41522, Egypt.
[5] Leibniz Institute of Plant Genetics and Crop Plant Research (IPK), OT Gatersleben, Corrensstr 3, D-06466 Seeland, Germany.
[6] State Key Laboratory of Desert and Oasis Ecology, Key Laboratory of Ecological Safety and Sustainable Development in Arid Lands, Xinjiang Institute of Ecology and Geography, Chinese Academy of Sciences, Urumqi, 830011, PR China.
[7] University of Chinese Academy of Sciences, Beijing 100049, PR China.
[8] Botany and Microbiology Department, Faculty of Science, Assiut University, Assiut 71516, Egypt.
[9] College of Arts and Sciences, Qatar University, Qatar.
* Corresponding author: sgs03@fayoum.edu.eg

DOI: 10.1201/9781003518204-3

particularly those involved in complex traits, remains a challenge (Wang 2021). This is where CRISPR/Cas9 technology comes into play. Initially discovered as a bacterial adaptive immune system, CRISPR/Cas9 has been harnessed as a powerful genome-editing tool capable of precise, targeted modifications within the genome (Singh et al. 2017). Its ability to generate targeted gene knockouts, knock-ins, and other genetic alterations has made it indispensable for functional genomics studies and crop improvement programs (Liu et al. 2022). The importance of CRISPR/Cas9 in cereal crop improvement stems from its ability to expedite the breeding process by precisely editing genes linked to beneficial traits (Rao and Wang 2021). Conventional breeding methods, which depend on crossing and selecting traits, are often time-consuming and constrained by the limited genetic diversity within breeding populations (Liu et al. 2021). In contrast, CRISPR/Cas9 enables targeted editing of specific genes linked to traits such as yield enhancement, disease resistance, and tolerance to abiotic stresses (Ansari et al. 2020). For instance, targeted editing of the *OsSWEET14* gene in rice has provided resistance to bacterial blight, whereas modification of the *ZmDREB2A* gene in maize has improved drought tolerance (Ashaq 2024). These cases highlight the transformative power of CRISPR/Cas9 in creating “next-generation” crops capable of withstanding the challenges of a changing climate while helping to meet the rising global demand for food (Ambika et al. 2024). Despite its vast potential, the application of CRISPR/Cas9 in cereal crops faces several challenges. These include the risk of off-target mutations, complex regulatory frameworks, ethical considerations, and societal concerns, which can hinder the broader acceptance and implementation of CRISPR-edited crops (Razzaq et al. 2021). Therefore, it is crucial to thoroughly assess potential ecological risks, establish comprehensive and transparent regulatory systems, and foster open public dialogue to promote the responsible and equitable integration of CRISPR technology into agricultural practices (Rasheed et al. 2021).

This study provides a comprehensive overview of genome mapping and gene sequencing technologies, with a focus on the applications of CRISPR/Cas9 in cereal crops. It discusses the principles and advancements of genome mapping and sequencing, the mechanisms, and improvements of CRISPR/Cas9 technology, and its transformative applications in enhancing crop resilience and productivity. Additionally, the paper explores the ethical, ecological, and regulatory challenges associated with CRISPR/Cas9, highlighting the need for a balanced approach to harness this technology’s full potential.

2. Fundamentals of Genome Mapping and Gene Sequencing

2.1 Genome Mapping: Definition and Importance

Genome mapping is the process of determining the relative positions of genes and genetic markers within the chromosomes of an organism (Majhi et al. 2024). This process is crucial for understanding the organization of a genome and for locating genes associated with specific traits (Schiessl et al. 2019). In cereal crops, genome mapping is particularly important because these plants often possess large and complex genomes with high levels of polyploidy, repetitive sequences, and structural variations (Gabur et al. 2019). For instance, the wheat genome is hexaploid and contains over 16 billion base pairs, making it significantly more complex than the human genome (Kyriakidou et al. 2018). Mapping such a complex genome requires high-resolution genetic and physical maps that can pinpoint the exact locations of genes and markers associated with traits like grain size, yield, disease resistance, and stress tolerance (Bolger et al. 2014). Two primary types of genome maps are widely used in plant genomics: genetic linkage maps and physical maps. Genetic linkage maps are based on the recombination frequency between markers, providing insights into the relative distances between genes on chromosomes (Garg et al. 2024). These maps are invaluable for identifying Quantitative Trait Loci (QTLs) associated with agronomic traits (Schreiber et al. 2024). Physical maps, on the other hand, provide a more precise and quantitative measure of the distance between genetic elements, often in base pairs (Paterson and Queitsch 2024). Physical maps are constructed using sequencing and assembly techniques, such as Bacterial Artificial Chromosome (BAC) libraries, optical mapping, and more recently, whole-genome sequencing (Song et al. 2023a).

In cereal crop research, comparative mapping is another valuable tool that involves comparing the genetic maps of related species to identify conserved regions and candidate genes (Jeon and Kim 2024). For example, rice, with its relatively small genome and extensive genomic resources, is often used as a reference for mapping genes in other cereal crops like maize and barley (Mascher et al. 2024). Such comparative mapping studies have led to the identification of key genes involved in drought tolerance, disease resistance, and nutrient use efficiency, which are crucial for crop improvement under changing environmental conditions.

2.2 Key Techniques in Gene Sequencing

Gene sequencing refers to the process of identifying the precise sequence of nucleotides adenine, thymine, cytosine, and guanine in a DNA molecule (Mandlik et al. 2024). Sequencing technologies have advanced significantly since the introduction of Sanger sequencing in the 1970s, the first technique capable of generating complete sequences of small genomes (Brlek et al. 2024). However, the low throughput and high cost of Sanger sequencing restrict its use in large-scale genomic studies, particularly when dealing with the complex and sizable genomes of cereal crops (Henry 2022). The advent of Next-Generation Sequencing (NGS) in the early 2000s revolutionized genomics by enabling massively parallel sequencing of millions of DNA fragments, thus reducing the cost and time required to sequence entire genomes (Feuillet et al. 2011; Llaca 2012; Sun et al. 2022). Next-Generation Sequencing (NGS) technologies, including Illumina sequencing, Ion Torrent sequencing, and Pacific Biosciences (PacBio) Single-Molecule Real-Time (SMRT) sequencing, have been extensively employed for decoding the genomes of cereal crops (Li et al. 2017). For instance, the rice genome was sequenced using a combination of next-generation sequencing (NGS) platforms, resulting in a high-quality reference genome that now serves as a foundational model in plant genomics research (Ahmed et al. 2024). Similarly, the International Wheat Genome Sequencing Consortium (IWGSC) employed NGS and optical mapping to produce a reference genome for hexaploid wheat, an achievement that has accelerated gene discovery and functional genomics studies in wheat (Wang et al. 2024). While NGS technologies are highly efficient, they often produce short reads, which can be challenging for assembling repetitive regions in complex cereal genomes (Baum and Kumar 2024). To overcome these limitations, Third-Generation Sequencing (TGS) methods, such as Oxford Nanopore and PacBio's long-read sequencing, have been developed (Garg et al. 2024). These technologies provide much longer reads, enabling more accurate genome assemblies, particularly in regions with high repeat content (Singh et al. 2024). For instance, the maize genome, which contains a significant proportion of repetitive DNA, has been sequenced using TGS methods to provide a more complete picture of its genomic architecture (Aslam et al. 2024).

2.3 Role of Bioinformatics in Genome Mapping and Sequencing

Bioinformatics is crucial in analyzing large-scale genomic data generated by genome mapping and sequencing efforts (Chavhan et al. 2024). In cereal crop research, bioinformatics tools and software are essential for sequence assembly, variant calling, gene annotation, and functional prediction (Selvakumar et al. 2024). Given the complexity of cereal genomes, which often contain high levels of heterozygosity, polyploidy, and repetitive sequences, specialized bioinformatics approaches are required to accurately analyze and interpret the data (Essemine et al. 2024). Several bioinformatics tools are commonly used in cereal crop genomics. For example, TASSEL is widely used for conducting Genome-Wide Association Studies (GWAS) to identify SNPs associated with traits like grain quality and disease resistance in crops like maize and barley (Thabet et al. 2023; Thabet et al. 2020; Thabet et al. 2018). R/qtl is another popular tool for QTL mapping, allowing researchers to link phenotypic traits with specific genomic regions (Thabet et al. 2022). Additionally, platforms like

Galaxy and CerealsDB provide user-friendly interfaces and curated databases for analyzing cereal crop genomic data (Naik et al. 2024). The integration of CRISPR/Cas9 data with bioinformatics pipelines has further enhanced cereal crop research (Vyas et al. 2024). For instance, tools like CRISPResso and CHOPCHOP allow researchers to design guide RNAs (gRNAs) for specific target sites in cereal genomes and to predict off-target effects, thereby ensuring high specificity and efficiency in CRISPR-mediated genome editing (Ali et al. 2024). These integrated approaches are crucial for accelerating functional genomics studies in cereals and for developing improved crop varieties with desirable traits.

3. Overview of CRISPR/Cas9 Technology

3.1 History and Discovery of CRISPR/Cas9

The discovery of the CRISPR/Cas9 system is a groundbreaking milestone in the field of genetic engineering and molecular biology (Gostimskaya 2022). CRISPR (Clustered Regularly Interspaced Short Palindromic Repeats) sequences were initially discovered in *Escherichia coli* in 1987, characterized by short, repetitive DNA segments separated by unique spacer sequences (Han and She 2017). However, it was not until the early 2000s that researchers recognized these sequences as components of an adaptive immune system in bacteria and archaea, enabling them to defend against viral infections by targeting and degrading foreign genetic material (Kumar et al. 2024). The associated Cas (CRISPR-associated) proteins, especially Cas9, were found to act as nucleases that cleave foreign DNA. In 2012, Jennifer Doudna and Emmanuelle Charpentier demonstrated that the CRISPR/Cas9 system could be repurposed for programmable genome editing in eukaryotic cells, enabling precise, targeted modifications of specific DNA sequences (Sarma et al. 2024). This discovery revolutionized genetics, ushering in an era of genome engineering with widespread applications in medicine, agriculture, and biotechnology (Severi and Akbari 2024). The adaptation of CRISPR/Cas9 for plant genome editing came soon after, with the first successful demonstrations in rice and tobacco in 2013 (Quétier 2016). Since then, CRISPR/Cas9 has been rapidly adopted for genome editing in various cereal crops, including wheat, maize, and barley, making it an essential tool for crop improvement.

3.2 Mechanism of CRISPR/Cas9

The CRISPR/Cas9 system is composed of two key components: a guide RNA (gRNA) and the Cas9 nuclease (Raper et al. 2018) as described in Fig. 1. The gRNA is engineered to match the target DNA sequence and directs the Cas9 protein to the specific genomic location where editing is desired (Gupta et al. 2019). The Cas9 protein introduces a Double-Strand Break (DSB) at the designated target site. This break subsequently triggers the cell's intrinsic DNA repair pathways, which are harnessed to achieve the desired genetic modifications (Liu et al. 2017). Double-strand breaks (DSBs) introduced by Cas9 are primarily repaired through two cellular pathways: Non-Homologous End Joining (NHEJ) and Homology-Directed Repair (HDR). NHEJ is an error-prone mechanism that frequently introduces small insertions or deletions (indels) at the break site, potentially disrupting gene function and resulting in gene knockouts (Singh et al. 2017). This method is widely used in cereal crops to disrupt genes that negatively affect yield or quality (Alqahtani et al. 2024). Homology-directed repair (HDR) is a more accurate DNA repair pathway that utilizes a homologous DNA template to guide the repair process, enabling the precise insertion or replacement of specific genetic sequences at the targeted site (Leal et al. 2024). HDR is the preferred pathway for precise gene editing applications, as it allows for the accurate introduction of beneficial traits such as enhanced disease resistance or improved nutritional content into cereal crops (Yuan 2024). CRISPR/Cas9 has been effectively applied to knock out genes responsible for disease susceptibility in wheat, alter genes associated with yield-related traits in maize, and develop rice varieties with enhanced tolerance to environmental stresses (Sami et al. 2021). The simplicity

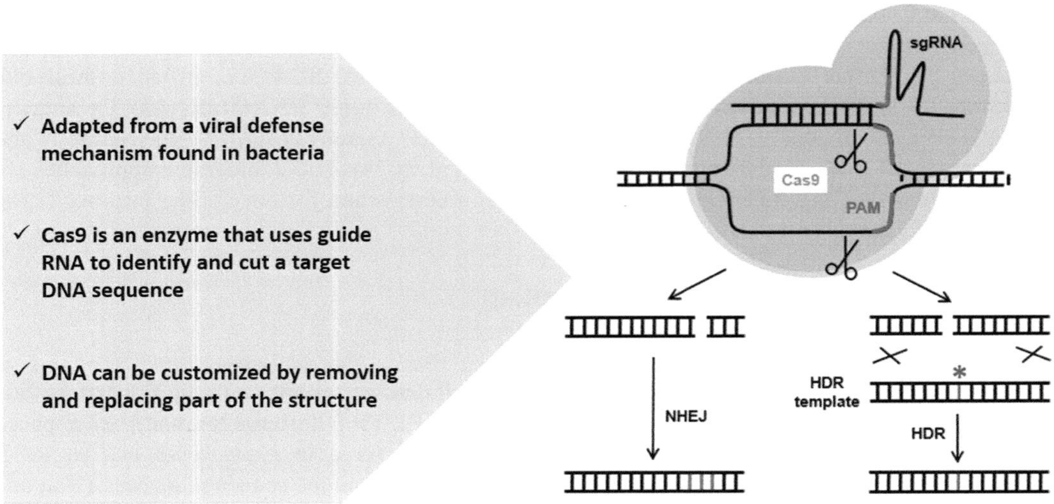

Figure 1. Mechanism of genome editing "How does CRISPR/Cas9 work".

and versatility of the CRISPR/Cas9 system have made it the go-to tool for functional genomics studies and targeted breeding in cereal crops.

3.3 Variants and Improvements of CRISPR Technology

Since its initial development, several variants and improvements of the CRISPR/Cas9 system have been introduced to enhance its efficiency, specificity, and versatility (Yan and Li 2019). One of the most significant improvements is the development of CRISPR-Cas12a (Cpf1). This alternative nuclease recognizes a different Protospacer Adjacent Motif (PAM) sequence and generates staggered cuts in the DNA (Mingarro and del Olmo 2023). This variant offers increased specificity and a reduced risk of off-target effects, making it particularly useful for editing large and repetitive genomes like those of cereal crops (Basu et al. 2023). Other advancements include base editors and prime editors, which enable precise base modifications without inducing double-strand breaks (Safari et al. 2017). Base editors use a deaminase enzyme fused to a catalytically inactive Cas9 to convert one base pair into another, such as cytosine to thymine or adenine to guanine (Mingarro and lí del Olmo 2023). Prime editing, in contrast, utilizes a reverse transcriptase enzyme in conjunction with a prime editing guide RNA (pegRNA) to precisely insert, delete, or replace specific DNA sequences at targeted genomic locations (Barrangou and Doudna 2016). These genome editing tools are particularly useful for correcting point mutations or generating specific allelic variants in cereal crops to improve traits of interest. Beyond these editing techniques, CRISPR interference (CRISPRi) and CRISPR activation (CRISPRa) have also been developed to modulate gene expression levels without introducing DNA breaks (Zegeye et al. 2022). CRISPRi uses a catalytically inactive Cas9 (dCas9) to block transcription, while CRISPRa uses dCas9 fused with activators to enhance gene expression (Murovec et al. 2017). These advanced techniques enable researchers to explore gene functions and regulatory networks in cereal crops with greater depth and precision. However, despite the revolutionary impact of CRISPR/Cas9 on crop improvement, its application also brings forth important ethical and safety concerns that must be carefully addressed (Li et al. 2024). A major concern associated with CRISPR/Cas9 is the risk of off-target effects, where unintended genomic regions are inadvertently modified. Such unintended edits can lead to unexpected mutations that may affect crop safety, reduce productivity, or disrupt ecological balance (Villiger et al. 2024). In cereal crops, reducing off-target effects is particularly important due to their large and complex genomes, as well as the potential ecological risks such as the unintentional transfer of edited genes to wild relatives, which could impact biodiversity and ecosystem stability (Mishra and Pandey 2024). Moreover, the use of CRISPR/Cas9 in food crops raises questions about food safety and public acceptance (Lau et al. 2024). Even though CRISPR-edited crops do not contain foreign DNA, public perception and regulatory policies vary widely across countries (Lu et al. 2024). For instance, while the United States and China have more lenient regulations regarding CRISPR-edited crops, the European Union classifies them as Genetically Modified Organisms (GMOs), subjecting them to strict regulatory scrutiny (Irfan et al. 2024). Addressing these ethical and safety concerns requires transparent communication with the public, robust risk assessment protocols, and the development of international guidelines for the responsible use of CRISPR technology in agriculture.

3.4 Applications of CRISPR/Cas9 in Genome Mapping

CRISPR/Cas9 technology has revolutionized the field of genome mapping by providing tools for the precise manipulation of genomic regions (Ikram et al. 2024). Genome mapping is essential for understanding the genetic architecture of complex traits, identifying key regulatory elements, and locating genes associated with desired phenotypes in cereal crops (Subedi et al. 2024). The ability to edit specific genes or regulatory elements allows researchers to dissect gene function systematically, develop fine-scale genetic maps, and create targeted mutations that aid in the identification of gene-trait associations (Ambika et al. 2024). Below are some of the key applications of CRISPR/Cas9 in genome mapping for cereal crops, as described in Fig. 2.

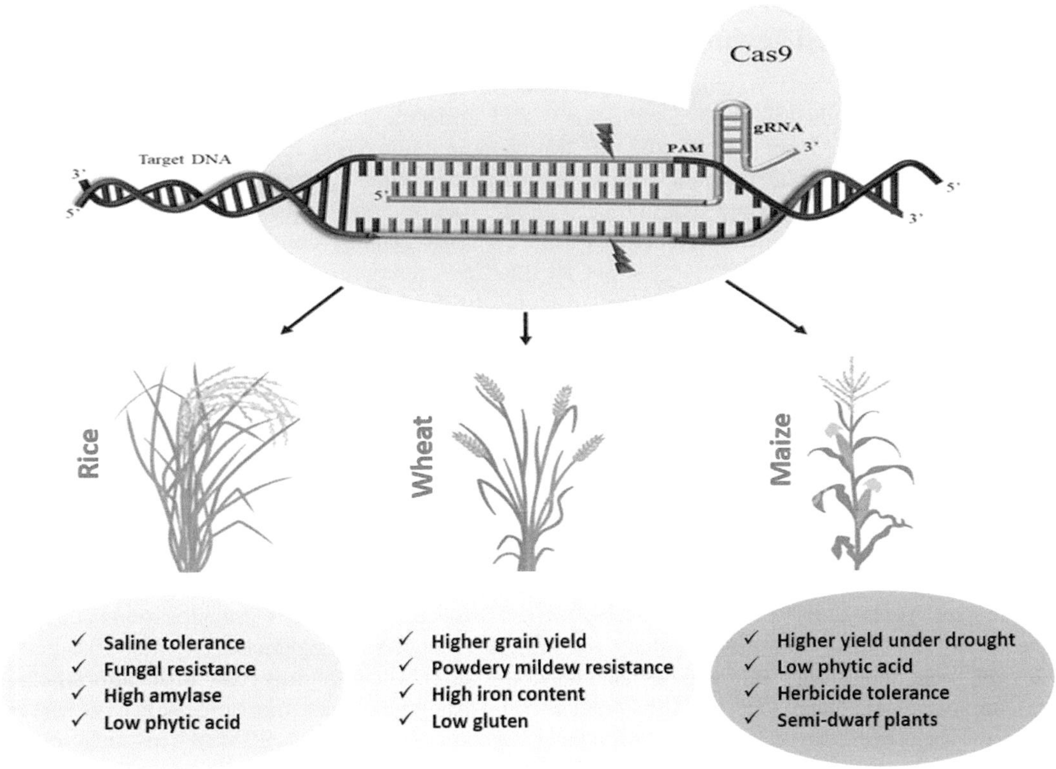

Figure 2. Applications of genome editing in different cereal crops.

4. Role of CRISPR in Identifying Gene Functions and Regulatory Elements

CRISPR/Cas9 identifies gene functions by creating loss-of-function or gain-of-function mutations (Ye and Han 2024). In cereal crops, this is particularly useful for studying genes associated with important agronomic traits such as yield, stress tolerance, and disease resistance (Park et al. 2024). Through the creation of CRISPR knockout libraries, researchers can systematically disrupt thousands of genes across the genome, enabling comprehensive investigations into gene functions and their interactions within complex biological pathways (Kim 2024). In rice, the *GW2* gene functions as a negative regulator of grain size. Using CRISPR/Cas9, researchers successfully knocked out this gene, resulting in increased grain width and weight, which ultimately contributed to improved crop yield (Achary and Reddy 2021). This research not only confirmed the role of *GW2* in regulating grain size but also identified it as a promising target for rice breeding programs focused on enhancing grain yield and quality. Additionally, the *MLO* gene in wheat has been linked to susceptibility to powdery mildew, a widespread fungal disease that significantly impacts wheat production globally (Kang et al. 2020). By knocking out the *MLO* gene using CRISPR/Cas9, researchers were able to develop powdery mildew-resistant wheat lines, providing a sustainable solution for disease management (Kang et al. 2020). In maize, the *BADH2* gene is associated with plant architecture and flowering time, which are critical traits for yield improvement (Zhang et al. 2024b). CRISPR/Cas9-mediated knockout of the *BADH2* gene led to noticeable changes in plant architecture, including reduced height and earlier flowering. These traits can be beneficial in maize breeding programs focused on producing high-yielding, stress-resilient varieties (Haque et al. 2021).

5. CRISPR-Based Techniques for Chromosome Engineering and Mapping

CRISPR technology also provides powerful tools for chromosome engineering, which is essential for constructing high-resolution genetic maps and understanding the structural organization of complex genomes (Mushtaq et al. 2021). Chromosome engineering involves creating large-scale chromosomal rearrangements, such as deletions, duplications, inversions, and translocations, to study their effects on gene expression, genome stability, and trait manifestation (Chaudhary et al. 2021). Wheat is a hexaploid crop with a complex genome comprising three subgenomes (A, B, and D) (Alqudah et al. 2024). Using CRISPR/Cas9, researchers have successfully engineered specific chromosomal deletions in wheat to study the function of multi-copy gene families and their impact on agronomic traits (Wang et al. 2018). For instance, creating large deletions in specific regions of the D subgenome allowed researchers to identify key genes controlling grain protein content, a critical quality trait in wheat (Ye et al. 2022a). In barley, CRISPR/Cas9 was used to induce targeted inversions and duplications in regions associated with disease resistance (Garcia-Gimenez and Jobling 2022). By engineering these chromosomal rearrangements, researchers were able to study how structural variations affect the expression of resistance genes and their ability to confer immunity against pathogens like *Blumeria graminis* (powdery mildew) (Romero 2018; Ternovska et al. 2022). This approach provides insights into the genomic architecture of resistance traits and helps in developing barley varieties with durable resistance. In maize, researchers employed CRISPR/Cas9 to create targeted translocations between chromosomes 1 and 9, aiming to study the effects of such rearrangements on traits like kernel development and drought tolerance (Kushanov et al. 2022). These engineered translocations helped researchers map QTLs associated with kernel size and drought response more precisely by observing how the traits were segregated in subsequent generations (Zhang et al. 2024a). This technique of creating and mapping translocations provides a powerful tool for understanding complex trait loci and their interactions in the maize genome.

6. Use of CRISPR for Generating High-Resolution Genetic Maps

CRISPR/Cas9 technology significantly enhances the generation of high-resolution genetic maps by enabling precise and targeted mutagenesis across the genome (Sadhu et al. 2016). High-resolution genetic maps are essential for the precise localization of quantitative trait loci (QTLs) linked to complex traits such as drought tolerance, disease resistance, grain quality, and nutrient use efficiency in cereal crops (Karmacharya 2022). By creating overlapping deletions or insertions within specific genomic regions, researchers can narrow down QTLs to smaller intervals, facilitating the identification of candidate genes underlying traits of interest. In maize, drought tolerance is a multifaceted trait regulated by numerous genes, making such fine-mapping approaches particularly valuable (He et al. 2021a). Using CRISPR/Cas9, researchers targeted specific QTL regions known to be associated with drought tolerance and created a series of overlapping deletions within these regions (Sahu et al. 2020). By analyzing the phenotypic effects of these deletions, they were able to narrow down the QTL to a small genomic region containing a few candidate genes, including *ZmNAC111*, which was confirmed to play a critical role in drought tolerance (Liu et al. 2020). In rice, researchers have employed CRISPR/Cas9 to fine-map QTLs linked to grain quality traits, including amylose and protein content—critical factors that influence both the nutritional value and market appeal of rice varieties (Wu et al. 2022). By inducing targeted mutations in the QTL regions, they identified candidate genes such as *Wx* (for amylose synthesis) and *ALMT1* (for aluminum tolerance), which significantly contribute to the desired grain quality traits. These fine-mapping studies yield important genetic markers that can be utilized in marker-assisted selection, thereby enhancing the efficiency of breeding programs. Wheat is highly susceptible to several rust diseases namely stem rust, leaf rust, and stripe rust which pose serious threats to yield and global wheat production (Hussain et al. 2022b). Researchers have used CRISPR/Cas9 to fine-map QTLs associated with rust resistance in wheat by targeting specific genomic regions known to harbor resistance genes, such as *Sr22* and *Yr15* (Colasuonno et al. 2021). By generating targeted deletions and studying the phenotypic effects, they were able to identify the functional gene variants responsible for resistance, enabling the development of wheat varieties with enhanced rust resistance through marker-assisted selection.

7. Case Studies Showcasing CRISPR Applications in Genome Mapping

Numerous case studies demonstrate the successful application of CRISPR/Cas9 in genome mapping for cereal crops, providing valuable insights into gene function, trait development, and crop improvement (Xin et al. 2021) as described in Table 1. Nitrogen Use Efficiency (NUE) is a critical trait for sustainable agriculture, as it reduces the need for nitrogen fertilizers and minimizes environmental pollution (Neeraja et al. 2021). In rice, CRISPR/Cas9 was used to map and validate the function of the *NRT1.1B* gene, which is associated with enhanced nitrogen uptake and assimilation (Neeraja et al. 2021). By knocking out the *NRT1.1B* gene and analyzing the resulting phenotypes, researchers confirmed its role in NUE and provided a valuable target for improving nitrogen use efficiency in rice breeding programs (Fiaz et al. 2021). Flowering time is a crucial trait for adaptation to different environments in cereal crops like barley (Mayer 2021). By using CRISPR/Cas9 to target and disrupt key flowering time genes such as *Ppd-H1* and *VRN1*, researchers were able to create mutants with altered flowering times (Hill and Li 2022). This enabled them to map the genetic architecture of flowering time regulation and identify QTLs associated with photoperiod sensitivity and vernalization response, which are critical for optimizing barley growth in diverse agro-climatic zones (Kuijer 2019). Researchers used CRISPR/Cas9 to target genes involved in starch biosynthesis and kernel size regulation, such as *Sh2* and *Bt2* in maize (He et al. 2021b). By creating knockouts and analyzing the resulting phenotypes, they were able to map the genetic basis of these traits and identify new candidate genes for kernel development (Li et al. 2021). This work not only provided insights into the complex genetic networks governing kernel

Table 1. Summarizes the diverse applications of CRISPR/Cas9 in genome mapping for cereal crops. It highlights the significant outcomes achieved by using CRISPR/Cas9 technology to identify gene functions, manipulate chromosomal structures, generate high-resolution genetic maps, and validate candidate genes responsible for key agronomic traits.

Application	**Crop**	**Gene/Target**	**Objective**	**Outcome/Findings**	**Reference/Example**
Identifying Gene Functions and Regulatory Elements	Rice	*GW2*	Increase grain size and yield	Knockout of *GW2* increased grain width and weight, enhancing yield.	Example of rice grain size improvement using CRISPR/Cas9 (Achary and Reddy 2021).
	Wheat	*MLO*	Develop powdery mildew resistance	Knockout of *MLO* gene conferred resistance to powdery mildew, a major fungal disease.	Example of disease resistance in wheat using CRISPR/Cas9 (Kusch and Panstruga 2017).
	Maize	*BADH2*	Alter plant architecture and flowering time for improved yield	CRISPR/Cas9 knockout resulted in reduced plant height and early flowering, beneficial for high-yield breeding programs.	Example of modifying maize plant architecture using CRISPR/Cas9 (Han et al. 2019).
Chromosome Engineering and Mapping	Wheat	Chromosome deletions in D sub genome	Study multi-copy gene functions and their impact on agronomic traits	Large deletions helped identify genes controlling grain protein content, crucial for wheat quality.	Example of chromosome engineering in wheat using CRISPR/Cas9 (Chen et al. 2022)
	Barley	Induced inversions and duplications	Analyze structural variants and their impact on disease resistance	Chromosomal rearrangements provided insights into resistance gene expression and immunity against *Blumeria graminis*.	Example of studying structural variations in barley using CRISPR/Cas9 (Guttman et al. 2014).
	Wheat	Chromosome translocations between 1 and 9	Study effects on kernel development and drought tolerance	Targeted translocations helped map QTLs for kernel size and drought response by observing trait segregation.	Example of mapping trait-associated regions in maize using CRISPR/Cas9 (Kumar et al. 2024).
Generating High-Resolution Genetic Maps	Maize	Overlapping deletions in drought tolerance QTLs	Fine-map QTLs for drought tolerance	Overlapping deletions refined QTL regions and identified *ZmNAC111* as a key gene for drought tolerance.	Example of fine mapping drought tolerance QTLs in maize using CRISPR/Cas9 (Engelhorn et al. 2023).
	Rice	QTLs for amylose and protein content	Identify genes contributing to grain quality traits	Targeted mutations in QTL regions identified *Wx* and *ALMT1* as key genes for desirable grain quality traits.	Example of mapping rice grain quality genes using CRISPR/Cas9 (Chen et al. 2020).
	Wheat	Rust resistance QTLs, e.g., *Sr22*, *Yr15*	Map genes responsible for resistance to rust diseases	Targeted deletions and phenotypic analysis identified functional gene variants, aiding in marker-assisted selection for rust resistance.	Example of fine mapping rust resistance in wheat using CRISPR/Cas9 (Song et al. 2023).

Table 1 contd. ...

...Table 1 contd.

Application	Crop	Gene/Target	Objective	Outcome/Findings	Reference/Example
Case Studies Showcasing CRISPR Applications	Rice	*NRT1.1B*	Enhance nitrogen use efficiency (NUE)	Knockout of *NRT1.1B* confirmed its role in improving NUE, providing targets for NUE breeding programs.	Example of mapping nitrogen use efficiency genes in rice using CRISPR/Cas9 (Liu et al. 2023).
	Barley	*Ppd-H1*, *VRN1*	Understand flowering time regulation for environmental adaptation	Knockouts of flowering time genes helped map QTLs for photoperiod sensitivity and vernalization response, critical for optimizing growth in different climates.	Example of dissecting flowering time regulation in barley using CRISPR/Cas9 (Afsharyan 2020).
	Maize	*Sh2*, *Bt2*	Map genetic basis of kernel development and starch biosynthesis	Mutations provided insights into genetic networks governing kernel traits and facilitated breeding of improved maize varieties.	Example of functional mapping of maize kernel development genes using CRISPR/Cas9 (Li et al. 2025).
	Wheat	*Gpc-B1*	Improve grain protein content for better nutritional value	Targeted mutations narrowed down QTL regions and identified candidate genes to improve protein content in wheat varieties.	Example of mapping grain protein content genes in wheat using CRISPR/Cas9 (Zhai et al. 2018).

traits but also laid the groundwork for breeding maize varieties with improved yield and quality. Grain protein content is a critical quality trait for wheat, affecting its nutritional value and processing quality. CRISPR/Cas9 was used to create targeted mutations in QTL regions associated with high-grain protein content, such as the *Gpc-B1* locus (Subedi et al. 2023). By generating mutants and analyzing their effects on protein accumulation, researchers were able to narrow down the QTL region and identify candidate genes that could be manipulated to improve wheat grain protein content (Roy et al. 2022). These findings provide a genetic basis for breeding wheat varieties with higher protein content and better nutritional value (Gaikwad et al. 2020). These studies underscore the versatility and efficiency of CRISPR/Cas9 in genome mapping for cereal crops, showcasing its potential to rapidly identify critical genes and traits that can drive advancements in crop improvement and sustainable agriculture.

8. Applications of CRISPR/Cas9 in Gene Sequencing

8.1 CRISPR-Based Methods for Targeted Sequencing

CRISPR/Cas9 technology has revolutionized not only genome editing but also introduced innovative approaches for targeted sequencing, enabling researchers to selectively analyze specific genomic regions of interest with high precision and efficiency (Zegeye et al. 2022). In cereal crops, CRISPR-based targeted sequencing has become an invaluable tool for enriching regions that are difficult to sequence using conventional methods due to their repetitive nature or high GC content (Verma et al. 2023). For instance, CRISPR/Cas9-mediated enrichment strategies, such as CRISPR-capture sequencing, utilize Cas9 to cleave the genome at sites flanking regions of interest

(Jangra et al. 2022). The cleaved fragments can then be selectively sequenced, increasing the depth and accuracy of sequencing data in these target regions (Singh et al. 2020). This method is especially valuable for resequencing key agronomic genes and regulatory regions in cereal crops (Nidhi et al. 2021). For instance, in wheat, CRISPR/Cas9-based targeted sequencing has been utilized to analyze the *Rht-B1* gene, known for its role in semi-dwarfism and yield enhancement (Song et al. 2023b). By concentrating on specific genes of interest, researchers can detect allelic variations linked to beneficial traits such as improved drought tolerance or disease resistance, which can subsequently be integrated into breeding strategies.

8.2 Enhancing Sequencing Accuracy and Efficiency with CRISPR Tools

One of the major challenges in sequencing the complex genomes of cereal crops is the abundance of repetitive elements and highly homologous gene families. These features can complicate the accurate assembly and annotation of sequencing data, making it difficult to distinguish between closely related sequences and to identify functional genes with precision (Malekshoar et al. 2023). CRISPR/Cas9 can improve sequencing accuracy and efficiency by reducing background noise from non-target DNA (Hardigan et al. 2019). Techniques such as CRISPR-mediated selective sequencing employ Cas9 to remove unwanted DNA sequences from a sample before sequencing, allowing for deeper coverage of regions of interest and reducing the sequencing cost (Schultzhaus et al. 2021). This method has been applied in cereal crops to achieve high coverage of low-abundance regulatory elements and genes involved in critical pathways (Lee et al. 2019). For example, in barley, CRISPR-based enrichment was used to focus sequencing efforts on low-copy-number genes involved in flowering time regulation, a key trait for adaptation to different climates (Kiran and Abdin 2020).

8.3 Applications in Detecting Genetic Variations and Mutations

Identifying rare genetic variations and mutations is essential for uncovering the genetic foundations of key agronomic traits in cereal crops. These rare variants often play significant roles in trait expression and can provide valuable insights for crop improvement and precision breeding (Sahu et al. 2020). CRISPR/Cas9-based methods, such as CRISPR-DS (Discriminant Sequencing), have been developed to selectively sequence alleles of interest, facilitating the identification of rare variants and mutations that may not be detectable with conventional sequencing approaches (Kaur et al. 2021). This capability is particularly useful in crops like maize, where complex genomic architecture and extensive natural variation can mask the presence of rare but significant mutations (Govindaraj et al. 2015). For instance, CRISPR-mediated sequencing has been applied to identify rare alleles of the *Pi-ta* resistance gene in rice, which is associated with resistance to rice blast, a devastating fungal disease (Georges and Ray 2017).

9. Case Studies on CRISPR/Cas9 in Sequencing Plant Genomes

Several case studies highlight the utility of CRISPR/Cas9 in enhancing gene sequencing efforts in cereal crops (Ansari et al. 2020). One such study involved using CRISPR-Cas9 to sequence the maize genome, focusing on genomic regions associated with kernel development and yield (Bao et al. 2019). By using targeted sequencing approaches, researchers were able to identify novel genetic variants and structural changes that contribute to kernel size and starch content, providing new insights for crop improvement. Another case study in wheat demonstrated the application of CRISPR-mediated enrichment for resequencing genomic regions linked to rust resistance, a major disease affecting wheat yields worldwide (Hussain et al. 2022). By focusing on specific regions harboring resistance genes, researchers could better understand the genetic basis of rust resistance and develop molecular markers for use in breeding programs (Li et al. 2019). These examples illustrate the potential of CRISPR-based sequencing techniques to accelerate the identification of valuable genetic variants and enhance cereal crop breeding efforts.

10. CRISPR/Cas9 in Agricultural Genomics

10.1 Role in Crop Improvement and Genetic Enhancement

CRISPR/Cas9 has emerged as a game-changing tool in agricultural genomics, allowing for precise genome modifications in cereal crops to improve critical traits such as yield, stress resilience, and nutritional value (Biswas et al. 2021). Traditional breeding approaches, though valuable, are typically slow and constrained by the limited genetic variation present within breeding pools (Ahmar et al. 2024). In contrast, CRISPR/Cas9 facilitates direct and targeted modification of genes associated with key agronomic traits, streamlining the improvement process by eliminating the need for prolonged cycles of crossing and selection (Manzoor et al. 2024). For instance, CRISPR/Cas9 has been employed to edit the *Wx* gene in rice, which regulates amylose content in the grains (Li et al. 2020). By knocking out this gene, researchers successfully developed rice varieties with higher amylose levels, leading to a lower Glycemic Index (GI)—a beneficial trait for individuals managing diabetes (Fiaz et al. 2019). Similarly, CRISPR/Cas9 has been applied to edit the *ARGOS8* gene in maize, leading to improved drought tolerance without affecting grain yield (Shi et al. 2017). These examples highlight the potential of CRISPR/Cas9 to accelerate crop improvement programs by generating desirable genetic variants that enhance both productivity and resilience.

10.2 Application in Developing Disease-Resistant and Stress-Tolerant Crops

A particularly compelling use of CRISPR/Cas9 in agricultural genomics is engineering cereal crops with heightened resistance to diseases and various environmental stressors (Erdoğan et al. 2023). Major staples such as wheat, rice, and maize routinely encounter diverse biotic threats like fungal pathogens and insect pests as well as abiotic stresses such as drought and salinity (Maharajan et al. 2022). Thanks to CRISPR/Cas9, scientists can now pinpoint and edit stressrelated genes with great precision, paving the way for more robust and sustainable cultivars (Rao and Wang 2021). In rice, for example, disruption of the susceptibility gene *OsSWEET14* has yielded lines that withstand bacterial blight, a disease notorious for heavy yield losses (Zeng et al. 2020). Similarly, editing the drought responsive regulator *ZmDREB2A* in maize has produced plants that maintain higher yields under water scarcity (Oliva et al. 2019). In wheat, CRISPR/Cas9 generated mutants of *TaERF3* exhibit improved resistance to Fusarium head blight, a fungal disease that compromises grain quality and quantity (Ye et al. 2022). Collectively, these breakthroughs illustrate how CRISPR/Cas9 can accelerate the creation of high yielding cereal varieties capable of withstanding a spectrum of environmental challenges, thereby bolstering food security in the face of climate change.

10.3 CRISPR in Livestock Genomics for Desirable Traits

While the primary focus of this section is on cereal crops, it is essential to recognize the interconnection of plant and livestock improvements in agricultural systems (Rao and Wang 2021). CRISPR/Cas9 is also making significant strides in livestock genomics, where it is used to enhance desirable traits such as disease resistance, growth rate, and feed efficiency (Huitema 2017). These developments in livestock can complement cereal crop improvements, particularly in the context of feed crops (Barrière et al. 2003). For instance, by developing maize and barley varieties with enhanced nutritional content and digestibility, researchers can improve feed efficiency and reduce the environmental footprint of livestock production (Wyngaarden et al. 2020). This synergistic use of CRISPR/Cas9 across plant and animal systems represents a holistic approach to advancing agricultural sustainability.

10.4 Ethical, Ecological, and Regulatory Concerns in Agricultural Use

Despite the promising potential of CRISPR/Cas9 in crop and livestock improvement, several ethical, ecological, and regulatory concerns need to be addressed to ensure the responsible use of this technology (Ahmad et al. 2021). One of the primary concerns is the potential for off-target effects that could result in unintended consequences, such as the creation of new allergens or toxins in edited cereal crops (Zhang et al. 2020). For example, in wheat, off-target mutations could affect gluten proteins, potentially leading to allergenicity issues (Sánchez León 2020). Ecological risks are also a significant concern, particularly the potential for gene flow from CRISPR-edited crops to wild relatives (Camerlengo et al. 2020). This could impact biodiversity and lead to unintended ecological consequences, such as the emergence of new weed species or the disruption of natural ecosystems (Yu et al. 2024). Moreover, the regulatory landscape for CRISPR-edited crops remains fragmented, with different countries adopting varying levels of restriction and oversight (Sánchez-León et al. 2024). While some countries, like the United States, do not classify CRISPR-edited crops as GMOs if no foreign DNA is introduced, the European Union requires extensive safety testing and risk assessment before approval (Kuiken et al. 2021). Addressing these concerns requires a balanced approach that promotes innovation while ensuring safety and sustainability (Távora et al. 2022). This includes developing comprehensive risk assessment protocols, establishing transparent regulatory frameworks, and engaging in public discourse to build trust and acceptance of CRISPR technology in agriculture.

11. Challenges and Ethical Considerations of CRISPR/Cas9

11.1 Potential Off-Target Effects and Unintended Consequences

Although CRISPR/Cas9 has revolutionized genetic engineering by enabling highly precise genome edits (see Fig. 3), the technology is not without drawbacks (Irfan et al. 2024). A major concern is offtarget activity, in which the Cas9 nuclease inadvertently produces double-strand breaks at unintended genomic sites (Doggalli et al. 2024). These unintended alterations can give rise to

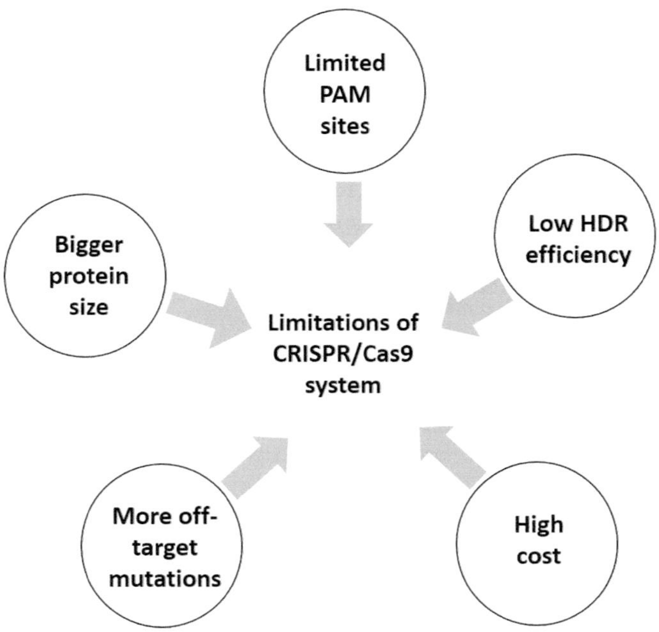

Figure 3. Limitations of genome editing using CRISPR/Cas9 system.

undesirable phenotypes, disrupt essential gene functions, or even introduce novel allergens or toxic compounds in cereal crops (Modrzejewski et al. 2020). This is especially in polyploid cereal crops like wheat, where multiple copies of the same gene may exist, increasing the risk of off-target effects (Awan et al. 2022). To reduce unintended edits, researchers have engineered more selective CRISPR systems—including CRISPRCas12a (Cpf1) and highfidelity Cas9 variants such as eSpCas9, SpCas9HF1, and HypaCas9 (Yin et al. 2018). These nucleases markedly diminish off-target activity while retaining strong on-target performance (Kondrateva et al. 2021). Complementary bioinformatic methods, such as GUIDEseq and CIRCLEseq, help forecast and confirm potential off-target sites, enabling the creation of highly specific guide RNAs for editing cereal crops (Li et al. 2023). Even so, absolute precision is still elusive, highlighting the need for continual enhancements to CRISPR platforms and rigorous screening protocols to ensure the safety of edited plants.

11.2 Bioethical Issues: Gene Editing in Plants and Ecological Concerns

The use of CRISPR/Cas9 in editing the genomes of cereal crops raises significant bioethical issues related to biodiversity, ecological balance, and food safety. One of the main ethical debates centers around "naturalness" and whether CRISPR-edited crops should be considered genetically modified organisms (GMOs) (Shah et al. 2021). Unlike traditional GMOs, CRISPR-edited crops do not necessarily involve the insertion of foreign DNA; instead, they involve precise modifications of native genes (Esha et al. 2021). However, the ethical debate persists, particularly regarding long-term ecological impacts and the potential for "genetic contamination" through gene flow to wild relatives or non-target species (Almeida et al. 2022). For instance, CRISPR-edited wheat or maize plants with herbicide tolerance could potentially crossbreed with wild grasses, leading to the emergence of "superweeds" that are more difficult to control (Pandita and Pandita 2023). Similarly, unintended ecological effects may arise from the introduction of CRISPR-edited cereal crops with novel traits, such as altered flowering times or pest resistance (Martin 2022). These changes could disrupt existing agricultural ecosystems and food webs, affecting both target and non-target organisms (Nawaz et al. 2022). To address these concerns, it is essential to conduct rigorous ecological risk assessments and long-term environmental monitoring of CRISPR-edited crops (Maurya et al. 2024). Additionally, transparent communication with the public and stakeholders is crucial for fostering informed discussions on the ethical implications of deploying genome-edited crops in agriculture.

11.3 Regulatory Frameworks and Governance of Gene Editing Technologies

The regulation of CRISPR/Cas9 and other gene-editing technologies is currently a global patchwork, with different countries adopting varying levels of restriction and oversight (Yotova 2020). In the United States, CRISPR-edited crops are not classified as GMOs if no foreign DNA is introduced, thereby providing a more streamlined path for commercialization (O'Brien-Udry and Pratt 2021). However, in the European Union, all gene-edited organisms are considered GMOs, requiring extensive safety testing, risk assessment, and labeling before approval (Shukla-Jones et al. 2018). This disparity in regulatory approaches creates challenges for international trade and the harmonization of policies governing gene-edited crops (Clark and Hobbs 2024). The regulatory landscape for CRISPR/Cas9 also extends beyond the technical aspects of gene editing to include socio-economic considerations, such as intellectual property rights, seed sovereignty, and access to gene-editing technologies (Boggio et al. 2020). The patenting of CRISPR/Cas9 and edited cereal crops by a few large corporations raises concerns about the concentration of power in the hands of a few entities, potentially limiting the benefits of this technology to smallholder farmers and developing countries (Clark and Hobbs 2024). To ensure equitable access and fair use of CRISPR technology in agriculture, there is a need for international dialogue to establish common standards and guidelines that promote both innovation and safety (Clark and Hobbs 2024). This includes developing frameworks for the transparent evaluation of CRISPR-edited crops, promoting open-access research and collaboration, and addressing socioeconomic disparities in the adoption of new agricultural technologies (Slokenberga et al. 2023).

12. The Future of Genome Editing: Balancing Innovation and Ethics

The future of genome editing with CRISPR/Cas9 in cereal crops holds great promise for addressing global food security challenges and promoting sustainable agriculture (Pacesa et al. 2024). However, balancing the rapid pace of innovation with ethical considerations is essential to ensure public trust and acceptance (Dixit et al. 2024). As new CRISPR tools and techniques continue to emerge, such as prime editing and CRISPR-Cas13, it is important to develop robust governance frameworks that incorporate ethical principles, stakeholder engagement, and public participation (Gupta et al. 2024). Collaborative efforts among scientists, policymakers, farmers, and the public are needed to navigate the complexities of genome editing in agriculture (Puchta 2024). This includes fostering an inclusive dialogue on the benefits and risks of CRISPR technology, developing guidelines for responsible use, and ensuring that technology is accessible to all, particularly in developing regions where food security challenges are most acute.

13. Conclusion

CRISPR/Cas9 has revolutionized genome mapping, gene sequencing, and agricultural genomics by enabling highly accurate and efficient DNA edits. Its use in creating cereal crops that resist diseases and environmental stresses, yield more grain, and possess superior nutritional profiles marks substantial progress toward sustainable farming and food security. This precise gene targeting capability also fuels cutting-edge genomics research, helping scientists decipher complex cereal genomes and accelerate modern breeding efforts. However, the rapid adoption of CRISPR/Cas9 in agriculture is accompanied by significant challenges and ethical concerns. Potential off-target effects, ecological concerns, regulatory disparities, and socio-economic implications must be carefully considered to ensure the responsible use of genome-editing technologies. Progress in agricultural gene editing hinges on developing next-generation CRISPR variants, implementing thorough risk assessment procedures, and establishing transparent, robust regulatory frameworks. With these safeguards, CRISPR/Cas systems can reshape cereal crop improvement and advance sustainability. An open, inclusive dialogue one that weighs innovation alongside ethical and ecological responsibilities will be key to unlocking CRISPR's full potential for bolstering global food security and environmental resilience. Equally important is ensuring that the benefits of this transformative technology are shared equitably, fostering sustainable agricultural development for communities worldwide.

References

Achary, V.M.M. and M.K. Reddy. 2021. CRISPR-Cas9 mediated mutation in GRAIN WIDTH and WEIGHT2 (GW2) locus improves aleurone layer and grain nutritional quality in rice. Sci. Rep. 11: 21941.

Afsharyan, N.P., W. Sannemann, J. Léon and A. Ballvora. 2020. Effect of epistasis and environment on flowering time in barley reveals a novel flowering-delaying QTL allele. J. Exp. Bot. 71: 893–906.

Ahmad, A., M.Z. Ghouri, N. Munawar, M. Ismail, S. Ashraf and S.O. Aftab. 2021. Regulatory, ethical, and social aspects of CRISPR crops. pp. 261–287. *In*: Ahmad, A., S.H. Khan and Z. Khan (eds.). CRISPR Crops: The Future of Food Security. Springer, Singapore.

Ahmar, S., B. Usman, G. Hensel, K.H. Jung and D. Gruszka. 2024. CRISPR enables sustainable cereal production for a greener future. Trends Plant Sci. 29(2): 179–195. https://doi.org/10.1016/j.tplants.2023.10.016.

Ahmed, H.G.M-D., Y. Zeng, X. Yang, N. Fatima and A. Faisal. 2024. Crop genome sequencing and their application for crop improvement. pp. 1–26. *In*: OMICs-based Techniques for Global Food Security.

Akram, F., S. Sahreen, F. Aamir, I.U. Haq, K. Malik, M. Imtiaz et al. 2023. An insight into modern targeted genome-editing technologies with a special focus on CRISPR/Cas9 and its applications. Mol. Biotechnol. 65: 227–242.

Ali, M.A., A. Zahoor, Z. Niaz, M. Jabran, M. Anas, I. Shafique et al. 2024. Bioinformatics and computational biology. pp. 281–334. *In*: Ijaz, S., I. Ul Haq and H. Mohamed Ali (eds.). Trends in Plant Biotechnology. Springer, Singapore. https://doi.org/10.1007/978-981-97-0814-710.

Almeida, M. and R.J. Ranisch. 2022. Beyond safety: mapping the ethical debate on heritable genome editing interventions. SS Commun. 9: 1–14.

Alqahtani, A., E.H.M. Mahmoud, R. Al Deabel, F. Kanwal, Q. Ahmad, M. Naeem et al. 2024. Chapter 3 – CRISPR-Cas: Effectors, mechanism, and classification. pp. 37–50. *In*: Abd-Elsalam, K.A., A. Ahmad and B. Zhang (eds.). CRISPRized Horticulture Crops. Academic Press. https://doi.org/10.1016/B978-0-443-13229-2.00025-9.

Alqudah, A.M., A. Elkelish, E.M. Saied, A.M. Alamri, D.Z. Alomari and S.G. Thabet. 2025. Candidate gene discovery for the biofortification of wheat grains with essential micronutrients for global food security. Cereal Res. Commun. 53: 181–191. https://doi.org/10.1007/s42976-024-00538-y.

Ambika, Bhati, S. and R. Kumar. 2024. Plant breeding using the CRISPR-Cas9 system for food security and facing climate change. pp. 149–181. *In*: Plant Genome Editing Technologies. Springer.

Ansari, W.A., S.U. Chandanshive, V. Bhatt, A.B. Nadaf, S. Vats, J.L. Katara et al. 2020. Genome editing in cereals: approaches, applications and challenges. Int. J. Mol. Sci. 21(11): 4040. https://doi.org/10.3390/ijms21114040.

Ashaq, M. 2024. Global Agri Vision. Global Agri Vision.

Aslam, M., U. Ahmad, R.M. Ahmad and A. Aslam. 2024. Functional characterization of cereal genomics. pp. 27–44. *In*: OMICs-based Techniques for Global Food Security.

Awan, M.J.A., K. Pervaiz, A. Rasheed, I. Amin, N.A. Saeed, K.S. Dhugga et al. 2022. Genome-edited wheat: Current advances for the second green revolution. Biotechnol. Adv. 60: 108006. https://doi.org/10.1016/j.biotechadv.2022.108006.

Bao, A., D.J. Burritt, H. Chen, X. Zhou, D. Cao and L.-S.P. Tran. 2019. The CRISPR/Cas9 system and its applications in crop genome editing. Crit. Rev. Biotechnol. 39(3): 321–336. https://doi.org/10.1080/07388551.2018.1554621.

Barrangou, R. and J.A. Doudna. 2016. Applications of CRISPR technologies in research and beyond. Nat. Biotechnol. 34: 933–941.

Barrière, Y., C. Guillet, D. Goffner and M. Pichon. 2003. Genetic variation and breeding strategies for improved cell wall digestibility in annual forage crops: A review. Anim. Res. 52: 193–228.

Basu, U., S. Riaz Ahmed, B.A. Bhat, Z. Anwar, A. Ali, A. Ijaz et al. 2023. A CRISPR way for accelerating cereal crop improvement: Progress and challenges. Front. Genet. 13: 866976. https://doi.org/10.3389/fgene.2022.866976.

Baum, M. and J. Kumar. 2024. Lentil genome sequencing: Establishing a comprehensive platform for molecular breeding. pp. 217–236. *In*: Kumar, J., D. Sen Gupta and S. Kumar (eds.). The Lentil Genome. Academic Press.

Biswas, S., D. Zhang and J. Shi. 2021. CRISPR/Cas systems: Opportunities and challenges for crop breeding. Plant Cell Rep. 40: 979–998.

Boggio, A., C.P. Romano and J. Almqvist. 2020. Human Germline Modification and the Right to Science: A Comparative Study of National Laws and Policies. Cambridge University Press.

Bolger, M.E., B. Weisshaar, U. Scholz, N. Stein, B. Usadel and K.F.X. Mayer. 2014. Plant genome sequencing—applications for crop improvement. Curr. Opin. Biotechnol. 26: 31–37. https://doi.org/10.1016/j.copbio.2013.08.019.

Brlek, P., L. Bulić, M. Bračić, P. Projić, V. Škaro, N. Shah et al. 2024. Implementing whole genome sequencing (WGS) in clinical practice: Advantages, challenges, and future perspectives. Cells 13: 504. https://doi.org/10.3390/cells13060504.

Camerlengo, F., A. Frittelli, C. Sparks, A. Doherty, D. Martignago, C. Larré et al. 2020. CRISPR-Cas9 multiplex editing of the α-amylase/trypsin inhibitor genes to reduce allergen proteins in durum wheat. Front. Sustain. Food Syst. 4: 104. https://doi.org/10.3389/fsufs.2020.00104.

Campbell, C.R., J. Poelstra and A.D. Yoder. 2018. What is speciation genomics? The roles of ecology, gene flow, and genomic architecture in the formation of species. Biol. J. Linn. Soc. 124: 561–583.

Chaudhary, N., J.-K. Im, S.-H. Nho and H.J. Kim. 2021. Visualizing live chromatin dynamics through CRISPR-based imaging techniques. Mol. Cells 44: 627–636.

Chavhan, R.L., V.R. Hinge, D.J. Wankhade, A.S. Deshmukh, N. Mahajan and U.S. Kadam. 2024. Bioinformatics for molecular breeding and enhanced crop performance: Applications and perspectives. pp. 21–74. *In*: Chen, J.-T. (ed.), Bioinformatics for Plant Research and Crop Breeding. Wiley. https://doi.org/10.1002/9781394209965.ch2.

Chen, Q., Z. Guo, X. Shi, M. Wei, Y. Fan, J. Zhu et al. 2022. Increasing the grain yield and grain protein content of common wheat (*Triticum aestivum*) by introducing missense mutations in the Q gene. Int. J. Mol. Sci. 23(18): 10772. https://doi.org/10.3390/ijms231810772.

Chen, Z., Y. Yang, L. Feng, Y. Sun, C. Zhang, X. Fan et al. 2020. Effects of different combinations of Wx and ALK main alleles on rice grain quality. Chin. J. Rice Sci. 34(3): 228–236. https://doi.org/10.16819/j.1001-7216.2020.9140.

Clark, L.F. and J.E. Hobbs. 2024. Governing the unknown: Regulating future technologies. pp. 79–94. *In*: International Regulation of Gene Editing Technologies in Crops. Springer.

Colasuonno, P., I. Marcotuli, A. Gadaleta and J.M. Soriano. 2021. From genetic maps to QTL cloning: An overview for durum wheat. Plants 10: 315.

Dixit, S., A. Kumar, K. Srinivasan, P.M.D.R. Vincent and N.R. Krishnan. 2024. Advancing genome editing with artificial intelligence: Opportunities, challenges, and future directions. Front. Bioeng. Biotechnol. 11: 1335901. https://doi.org/10.3389/fbioe.2023.1335901.

Doggalli, G., Kavya, O.B. Singh, H.G. Manojkumar, M. Srivastava, G.B. Chaya et al. 2024. Transformative gene editing methods: Precision in genetically modified crops through trait modification. J. Sci. Res. Rep. 30(8): 615–628. https://doi.org/10.9734/jsrr/2024/v30i82283.

Engelhorn, J., S.J. Snodgrass, A. Kok, A.S. Seetharam, M. Schneider, T. Kiwit et al. 2023. Phenotypic variation in maize can be largely explained by genetic variation at transcription factor binding sites. bioRxiv 2023.08.08.551183. https://doi.org/10.1101/2023.08.08.551183.

Erdoğan, İ., B. Cevher-Keskin, Ö. Bilir, Y. Hong and M. Tör. 2023. Recent developments in CRISPR/Cas9 genome-editing technology related to plant disease resistance and abiotic stress tolerance. Biology (Basel) 12(7): 1037. https://doi.org/10.3390/biology12071037.

Esha, S., D. Ludwig and M. Phil. 2021. The complexity of the gene and the precision of CRISPR. Societal Aspects of Advancements in Biotechnology 9: 3327.

Essemine, J., Z. Xu, J.-T. Chen and M. Qu. 2024. Bioinformatics as a powerful tool to foster plant science research and crop breeding through its involvement in a multidisciplinary research activity. pp. 1–19. *In*: Bioinformatics for Plant Research and Crop Breeding. Springer.

Feuillet, C., J.E. Leach, J. Rogers, P.S. Schnable and K. Eversole. 2011. Crop genome sequencing: Lessons and rationales. Trends Plant Sci. 16(2): 77–88. https://doi.org/10.1016/j.tplants.2010.10.005.

Fiaz, S., S. Ahmad, M.A. Noor, X. Wang, A. Younas, A. Riaz et al. 2019. Applications of the CRISPR/Cas9 system for rice grain quality improvement: Perspectives and opportunities. Int. J. Mol. Sci. 20(4): 888. https://doi.org/10.3390/ijms20040888.

Fiaz, S., X. Wang, S.A. Khan, S. Ahmar, M.A. Noor, A. Riaz et al. 2021. Novel plant breeding techniques to advance nitrogen use efficiency in rice: A review. GM Crops Food 12: 627–646. https://doi.org/10.1080/21645698.2021.1921545.

Gabur, I., H.S. Chawla, R.J. Snowdon and I.A.J. Parkin. 2019. Connecting genome structural variation with complex traits in crop plants. Trends Genet. 132: 733–750.

Gaikwad, K.B., S. Rani, M. Kumar, V. Gupta, P.H. Babu, N.K. Bainsla et al. 2020. Enhancing the nutritional quality of major food crops through conventional and genomics-assisted breeding. Front. Nutr. 7: 533453. https://doi.org/10.3389/fnut.2020.533453.

Garcia-Gimenez, G. and S.A. Jobling. 2022. Gene editing for barley grain quality improvement. J. Cereal Sci. 103: 103394.

Garg, V., A. Bohra, M. Mascher, M. Spannagl, X. Xu, M.W. Bevan et al. 2024. Unlocking plant genetics with telomere-to-telomere genome assemblies. Nat. Genet. 56: 1788–1799. https://doi.org/10.1038/s41588-024-01830-7.

Georges, F. and H. Ray. 2017. Genome editing of crops: A renewed opportunity for food security. GM Crops Food 8: 1–12.

Ghosh, M., N. Sharma, A.K. Singh, M. Gera, K.K. Pulicherla and D.K. Jeong. 2018. Transformation of animal genomics by next-generation sequencing technologies: A decade of challenges and their impact on genetic architecture. Crit. Rev. Biotechnol. 38(8): 1157–1175. https://doi.org/10.1080/07388551.2018.1451819.

Gostimskaya, I. 2022. CRISPR–Cas9: A history of its discovery and ethical considerations of its use in genome editing. Biochemistry (Moscow) 87: 777–788.

Govindaraj, M., M. Vetriventhan and M. Srinivasan. 2015. Importance of genetic diversity assessment in crop plants and its recent advances: An overview of its analytical perspectives. Sci. World J. 2015: 431487.

Gupta, D., O. Bhattacharjee, D. Mandal, M.K. Sen, D. Dey, A. Dasgupta et al. 2019. CRISPR-Cas9 system: A new-fangled dawn in gene editing. Life Sci. 232: 116636. https://doi.org/10.1016/j.lfs.2019.116636.

Gupta, M., K. Kumar, A. Singh, B. Singh, A.K. Jha and B. Kumar. 2024. Transgenic and genome-edited maize: Status and prospect. pp. 373–403. *In*: Tiwari, S. and B. Koul (eds.). Genetic Engineering of Crop Plants for Food and Health Security: Volume 2. Springer, Singapore. https://doi.org/10.1007/978-981-97-3119-0_15.

Guttman, D.S., A.C. McHardy and P. SchulzeLefert. 2014. Microbial genomeenabled insights into plant–microorganism interactions. Nat. Rev. Genet. 15(12): 797–813. https://doi.org/10.1038/nrg3884.

Han, W. and Q. She. 2017. CRISPR history: Discovery, characterization, and prosperity. pp. 1–21. *In*: Torres-Ruiz, R. and S. Rodriguez-Perales (eds.). Prog. Mol. Biol. Transl. Sci. Academic Press.

Han, J., B. Guo, Y. Guo, B. Zhang, X. Wang and L.J. Qiu. 2019. Creation of early flowering germplasm of soybean by CRISPR/Cas9 technology. Front. Plant Sci. 10: 1446. https://doi.org/10.3389/fpls.2019.01446.

Haque, M.A., M.Y. Rafii, M.M. Yusoff, N.S. Ali, O. Yusuff, D.R. Datta et al. 2021. Advanced breeding strategies and future perspectives of salinity tolerance in rice. Agronomy 11(8): 1631. https://doi.org/10.3390/agronomy11081631.

Hardigan, A.A., B.S. Roberts, D.E. Moore, R.C. Ramaker, A.L. Jones and R.M. Myers. 2019. CRISPR/Cas9-targeted removal of unwanted sequences from small-RNA sequencing libraries. Nucleic Acids Res. 47(14): e84. https://doi.org/10.1093/nar/gkz425.

He, C., H. Liu, D. Chen, W.-Z. Xie, M. Wang, Y. Li et al. 2021. CRISPR-Cereal: A guide RNA design tool integrating regulome and genomic variation for wheat, maize, and rice. Plant Biotechnol. J. 19(11): 2141–2143. https://doi.org/10.1111/pbi.13675.

He, Y., Q. Yang, J. Yang, Y.-F. Wang, X. Sun, S. Wang et al. 2021. shrunken4 is a mutant allele of ZmYSL2 that affects aleurone development and starch synthesis in maize. Genetics 218(2): iyab070. https://doi.org/10.1093/genetics/iyab070.

Henry, R.J. 2022. Progress in plant genome sequencing. Ann. Bot. 1: 113–128.

Hill, C.B. and C. Li. 2022. Genetic improvement of heat stress tolerance in cereal crops. Agronomy 12: 1205.

Huitema, Y. 2017. CRISPR-Cas9 genome editing in the livestock industry: A promising addition to current breeding programs. Fac. Sci. Eng.

Hussain, B., B.A. Akpınar, M. Alaux, A.M. Algharib, D. Sehgal, Z. Ali et al. 2022. Capturing wheat phenotypes at the genome level. Front. Plant Sci. 13: 851079. https://doi.org/10.3389/fpls.2022.851079.

Ikram, M., A. Rauf, M.J. Rao, M.F.K. Maqsood, M.Z.M. Bakhsh, M. Ullah et al. 2024. CRISPR-Cas9-based molecular breeding in crop plants: A review. Mol. Biol. Rep. 51(1): 227. https://doi.org/10.1007/s11033-023-09086-w.

Irfan, M., H. Majeed, T. Iftikhar and P.K. Ravi. 2024. A review on molecular scissoring with CRISPR/Cas9 genome editing technology. Trends Rev. 13.

Jangra, S., Priti, V. Chaudhary, A. Mishra, R.C. Yadav and N.R. Yadav. 2022. Genomic region analysis and genome editing for grain quality improvement in cereals. pp. 315–345. *In*: Prakash, C.S., S. Fiaz and S. Fahad (eds.). Principles and Practices of OMICS and Genome Editing for Crop Improvement. Springer, Cham. https://doi.org/10.1007/978-3-030-96925-7_14.

Jeon, D. and C.J. Kim. 2024. Polyploids of Brassicaceae: Genomic insights and assembly strategies. Clin. Genet. 13.

Kang, Y., M. Zhou, A. Merry and K. Barry. 2020. Mechanisms of powdery mildew resistance of wheat—a review of molecular breeding. Plant Pathol. 69: 601–617.

Karmacharya, A. 2022. Molecular mapping of QTL for genetic transformation-related trait and CRISPR/Cas9-mediated gene editing in wheat. North Dakota State Univ.

Kaur, B., K.S. Sandhu, R. Kamal, K. Kaur, J. Singh, M.S. Röder et al. 2021. Omics for the improvement of abiotic, biotic, and agronomic traits in major cereal crops: Applications, challenges, and prospects. Plants 10: 1989. https://doi.org/10.3390/plants10101989.

Kim, K.B. 2024. Use of CRISPR-Cas9 in agriculture. J. Hortic. Sci. 8: 212–228.

Kiran, U. and M.Z. Abdin. 2020. Transgenic Technology-Based Value Addition in Plant Biotechnology. Academic Press.

Kondrateva, E., A. Demchenko, A. Lavrov and S. Smirnikhina. 2021. An overview of currently available molecular Cas-tools for precise genome modification. Genes 769: 145225.

Kuijer, H.N. 2019. Roles of LOFSEP genes in barley inflorescence development. Plant Sci.

Kuiken, T., R. Barrangou and K. Grieger. 2021. Promises of sustainable food and agriculture through new biotechnologies: The CRISPR case. Trends Crops 4: 25–31.

Kumar, A., Y.-Y. Yau and V.R. Kumar. 2024. CRISPR-Cas: A history of discovery and innovation. pp. 1–16. *In*: Gene Editing in Plants: CRISPR-Cas and Its Applications. Springer Nature.

Kushanov, F.N., O.S. Turaev and O.A. Muhammadiyev. 2022. Maize (*Zea mays* L.) as a model system for plant genetic, genomic, and applied research. *In*: Abdurakhmonov, I.Y. (ed.). Model Organisms in Plant Genetics. IntechOpen. https://doi.org/10.5772/intechopen.104658.

Kyriakidou, M., H.H. Tai, N.L. Anglin, D. Ellis and M.V. Strömvik. 2018. Current strategies of polyploid plant genome sequence assembly. Front. Plant Sci. 9: 1660. https://doi.org/10.3389/fpls.2018.01660.

Lau, C.-H., Q.-L. Liang and H. Zhu. 2024. Next-generation CRISPR technology for genome, epigenome, and mitochondrial editing. Trends Biotechnol. 1–35.

Leal, A.F., A.M. Herreno-Pachón, E. Benincore-Flórez, A. Karunathilaka and S. Tomatsu. 2024. Current strategies for increasing knock-in efficiency in CRISPR/Cas9-based approaches. Int. J. Mol. Sci. 25: 2456. https://doi.org/10.3390/ijms25052456.

Lee, J., H. Lim, H. Jang, B. Hwang, J.H. Lee, J. Cho et al. 2019. CRISPR-Cap: Multiplexed double-stranded DNA enrichment based on the CRISPR system. Nucleic Acids Res. 47: e1. https://doi.org/10.1093/nar/gky820.

Lesk, A.M. 2017. Introduction to Genomics. Oxford Univ. Press.

Li, C., W. Chu, R.A. Gill, S. Sang, Y. Shi, X. Hu et al. 2023. Computational tools and resources for CRISPR/Cas genome editing. Genomics Proteomics Bioinformatics 21: 108–126. https://doi.org/10.1016/j.gpb.2022.02.006.

Li, C., F. Lin, D. An, W. Wang and R. Huang. 2018. Genome sequencing and assembly by long reads in plants. Genes 9: 6. https://doi.org/10.3390/genes9010006.

Li, H., X. Li, Y. Xu, H. Liu, M. He, X. Tian et al. 2020. High-efficiency reduction of rice amylose content via CRISPR/Cas9-mediated base editing. Rice Sci. 27: 445–448. https://doi.org/10.1016/j.rsci.2020.09.001.

Li, J., Y. Li and L. Ma. 2019. CRISPR/Cas9-based genome editing and its applications for functional genomic analyses in plants. Biotechnol. Genomics 3: 1800473.

Li, R., Y. Tan and H. Zhang. 2021. Regulators of starch biosynthesis in cereal crops. Mol. Biotechnol. 26: 7092.

Li, T., S. Li, Y. Kang, D. Qiu, W. Cao, W. Wu et al. 2024. Harnessing the evolving CRISPR/Cas9 for precision oncology. J. Transl. Med. 22: 749. https://doi.org/10.1186/s12967-024-05570-4.

Li, X., Y. Li, Y. Sun, S. Li, Q. Cai, S. Li et al. 2025. Integrating genetic diversity and agronomic innovations for climate-resilient maize systems. Plants 14: 1552. https://doi.org/10.3390/plants14101552.

Liu, H.-J., L. Jian, J. Xu, Q. Zhang, M. Zhang, M. Jin et al.. 2020. High-throughput CRISPR/Cas9 mutagenesis streamlines trait gene identification in maize. Plant Cell 32: 1397–1413. https://doi.org/10.1105/tpc.19.00906.

Liu, H., W. Chen, Y. Li, L. Sun, Y. Chai, H. Chen et al. 2022. CRISPR/Cas9 technology and its utility for crop improvement. Int. J. Mol. Sci. 23: 10442. https://doi.org/10.3390/ijms231810442.

Liu, Q., F. Yang, J. Zhang, N. Ni, R. Zhan, G. Hu et al. 2021. Application of CRISPR/Cas9 in crop quality improvement. Int. J. Mol. Sci. 22: 4206. https://doi.org/10.3390/ijms22084206q.

Liu, X., S. Wu, J. Xu, C. Sui and J. Wei. 2017. Application of CRISPR/Cas9 in plant biology. Acta Pharm. Sin. B 7: 292–302. https://doi.org/10.1016/j.apsb.2017.01.002.

Liu, Y., B. Hu and C. Chu. 2023. Toward improving nitrogen use efficiency in rice: utilization, coordination, and availability. Curr. Opin. Plant Biol. 71: 102327.

Llaca, V. 2012. Sequencing technologies and their use in plant biotechnology and breeding. Drug Dev. Syst. Med. pp. 35–60.

Lu, Z., L. Zhang, Q. Mu, J. Liu, Y. Chen, H. Wang et al. 2024. Progress in research and prospects for application of precision gene-editing technology based on CRISPR–Cas9 in the genetic improvement of sheep and goats. Agriculture 14(3): 487. https://doi.org/10.3390/agriculture14030487.

Luikart, G., M. Kardos, B.K. Hand, O.P. Rajora, S.N. Aitken and P.A. Hohenlohe. 2018. Population genomics: Advancing understanding of nature. pp. 3–79. *In*: Rajora, O.P. (ed.). Population Genomics: Concepts, Approaches and Applications. Springer, Cham. https://doi.org/10.1007/13836_2018_60.

Maharajan, T., T.P.A. Krishna, K. Rakkammal, S.A. Ceasar and M. Ramesh. 2022. Application of CRISPR/Cas system in cereal improvement for biotic and abiotic stress tolerance. Planta 256(6): 106. https://doi.org/10.1007/s00425-022-04023-w.

Majhi, P.K., A. Guru and S.C. Mogali. 2024. Genetic mapping of valued genes with significant traits in crop plants: Basic principles, current practices, and future perspectives. pp. 99–134. *In*: Bioinformatics for Plant Research and Crop Breeding. Wiley. https://doi.org/10.1002/9781394209965.ch4.

Malekshoar, M., S.A. Azimi, A. Kaki, L. Mousazadeh, J. Motaei and M. Vatankhah. 2023. CRISPR-Cas9 targeted enrichment and next-generation sequencing for mutation detection. J. Mol. Diagn. 25(5): 249–262. https://doi.org/10.1016/j.jmoldx.2023.01.010.

Mandlik, J.S., A.S. Patil and S.J. Singh. 2024. Next-generation sequencing (NGS): Platforms and applications. J. Plant Biotechnol. Sci. 16: S41–S45.

Manzoor, S., S.U. Nabi, T.R. Rather, G. Gani, Z.A. Mir, A.W. Wani et al. 2024. Advancing crop disease resistance through genome editing: A promising approach for enhancing agricultural production. Front. Genome Ed. 6: 1399051. https://doi.org/10.3389/fgeed.2024.1399051.

Martin, A.J. 2022. Regulation of genetically modified and gene-edited foods: An overview. pp. 483–500. *In*: Regulatory Insights Food Biotechnol.

Mascher, M., M.P. Marone, M. Schreiber and N. Stein. 2024. Are cereal grasses a single genetic system? Nat. Plants 10: 719–731. https://doi.org/10.1038/s41477-024-01674-3.

Maurya, B., V. Mishra, K. Divyanshu and J.H. Nishad. 2024. Pros and cons of genetic engineering technologies in reference to biofortified crops. pp. 349–370. *In*: Tiwari, S. and B. Singh (eds.). Harnessing Crop Biofortification for Sustainable Agriculture. Springer Nature, Singapore. https://doi.org/10.1007/978-981-97-3438-2_18.

Mayer, K.S. 2021. Investigating Molecular Mechanisms of Flowering Time Across Plant Lineages. Univ. Wisconsin-Madison.

Mingarro, G. and M.l. del Olmo. 2023. Improvements in genetic editing technologies: CRISPR-Cas and beyond. Gene 852: 147064.

Mishra, A. and V.P. Pandey. 2024. CRISPR/Cas system: A revolutionary tool for crop improvement. Biotechnol. J. 19: 2300298.

Modrzejewski, D., F. Hartung, H. Lehnert, T. Sprink, C. Kohl, J. Keilwagen et al. 2020. Which factors affect the occurrence of off-target effects caused by the use of CRISPR/Cas: A systematic review in plants. Front. Plant Sci. 11: 574959. https://doi.org/10.3389/fpls.2020.574959.

Murovec, J., Ž. Pirc and B.J. Yang. 2017. New variants of CRISPR RNA-guided genome editing enzymes. Plant Biotechnol. J. 15: 917–926.

Mushtaq, M., A.A. Dar, M. Skalicky, A. Tyagi, N. Bhagat, U. Basu et al. 2021. CRISPR-based genome editing tools: Insights into technological breakthroughs and future challenges. Genes 12: 797. https://doi.org/10.3390/genes12060797.

Naik, Y.D., C. Zhao, S. Channale, S.N. Nayak, K.L. Bhutia, A. Gautam et al. 2024. Bioinformatics for plant genetics and breeding research. pp. 35–64. *In*: Pandey, M.K., A. Bentley, H. Desmae and M. Roorkiwal (eds.). Frontier Technologies for Crop Improvement. Springer Nature, Singapore. https://doi.org/10.1007/978-981-99-4673-0_3.

Nawaz, M.A., G. Chung, K.S. Golokhvast and A.M. Tsatsakis (eds.). 2022. GMOs and Political Stance: Global GMO Regulation, Certification, Labeling, and Consumer Preferences. Academic Press (Elsevier), San Diego, CA. ISBN 978-0-12-823903-2.

Neeraja, C.N., K.M. Barbadikar, S.K. Mangrauthia, P.R. Rao, D. Subrahmanyam and R.M. Sundaram. 2021. Genes for NUE in rice: A way forward for molecular breeding and genome editing. Plant Physiol. Rep. 26: 587–599. https://doi.org/10.1007/s40502-021-00610-3.

Nidhi, S., U. Anand, P. Oleksak, P. Tripathi, J.A. Lal, G. Thomas et al. 2021. Novel CRISPR–Cas systems: An updated review of the current achievements, applications, and future research perspectives. Int. J. Mol. Sci. 22: 3327. https://doi.org/10.3390/ijms22073327.

O'Brien-Udry, C. and T. Pratt. 2021. Innovation and interdependence: The case of gene-editing technology. Trends Innov. Biotechnol.

Oliva, R., C. Ji, G. Atienza-Grande, J.C. Huguet-Tapia, A. Perez-Quintero, T. Li et al. 2019. Broad-spectrum resistance to bacterial blight in rice using genome editing. Nat. Biotechnol. 37: 1344–1350. https://doi.org/10.1038/s41587-019-0267-z.

Pacesa, M., O. Pelea and M. Jinek. 2024. Past, present, and future of CRISPR genome editing technologies. Cell 187: 1076–1100.

Pandita, D. and A. Pandita. 2023. CRISPR/Cas-mediated Genome Editing in Plants. CRC Press.

Pareek, C.S., R. Smoczynski and A.J. Tretyn. 2011. Sequencing technologies and genome sequencing. J. Appl. Genet. 52: 413–435.

Park, J.R., E.G. Kim, Y.H. Jang, R. Jan, M. Farooq, S. Asif et al. 2024. CRISPR/Cas9-mediated genome editing of OsCS511 enhances cold tolerance in *Oryza sativa* L. Environ. Exp. Bot. 226: 105932. https://doi.org/10.1016/j.envexpbot.2024.105932.

Paterson, A.H. and C. Queitsch. 2024. Genome organization and botanical diversity. Trends Plant Cell 36: 1186–1204.

Puchta, H. 2024. Regulation of gene-edited plants in Europe: From the valley of tears into the shining sun? Nat. Plants 5: 231–238.

Quétier, F. 2016. The CRISPR-Cas9 technology: Closer to the ultimate toolkit for targeted genome editing. Plant Sci. 242: 65–76.

Rao, M.J. and L. Wang. 2021. CRISPR/Cas9 technology for improving agronomic traits and future prospective in agriculture. Plant Physiol. Biochem. 254: 1–16.

Raper, A.T., A.A. Stephenson and Z. Suo. 2018. Functional insights revealed by the kinetic mechanism of CRISPR/Cas9. J. Am. Chem. Soc. 140: 2971–2984.

Rasheed, A., R.A. Gill, M.U. Hassan, A. Mahmood, S. Qari, Q.U. Zaman et al. 2021. A critical review: Recent advancements in the use of CRISPR/Cas9 technology to enhance crops and alleviate global food crises. Curr. Issues Mol. Biol. 43: 1950–1976.

Razzaq, A., P. Kaur, N. Akhter, S.H. Wani and F. Saleem. 2021. Next-generation breeding strategies for climate-ready crops. Front. Plant Sci. 12: 620420.

Ren, Z., P. Zhang, H. Su, X. Xie, J. Shao, L. Ku et al. 2024. Regulatory mechanisms used by ZmMYB39 to enhance drought tolerance in maize (*Zea mays*) seedlings. Plant Physiol. Biochem. 211: 108696.

Romero, C.C. 2018. Inheritance of nonhost resistance of barley to the powdery mildew fungi of cereals and grasses. Wageningen Univ. Res.

Roy, C., S. Kumar, R.D. Ranjan, S.R. Kumhar and V. Govindan. 2022. Genomic approaches for improving grain zinc and iron content in wheat. Front. Genet. 13: 1045955.

Sadhu, M.J., J.S. Bloom, L. Day and L. Kruglyak. 2016. CRISPR-directed mitotic recombination enables genetic mapping without crosses. Science 352: 1113–1116.

Safari, F., S. Farajnia, Y. Ghasemi and N. Zarghami. 2017. New developments in CRISPR technology: Improvements in specificity and efficiency. Curr. Pharm. Biotechnol. 18: 1038–1054.

Sahu, P.K., R. Sao, S. Mondal, G. Vishwakarma, S.K. Gupta, V. Kumar et al. 2020. Next-generation sequencing-based forward genetic approaches for identification and mapping of causal mutations in crop plants: A comprehensive review. Plants 9: 1355.

Sami, A., Z. Xue, S. Tazein, A. Arshad, Z.H. Zhu, Y.P. Chen et al. 2021. CRISPR–Cas9-based genetic engineering for crop improvement under drought stress. Bioengineered 12: 5814–5829.

Sánchez-León, S., M. Marín-Sanz, M.H. Guzmán-López, M. Gavilán-Camacho, E. Simón and F. Barro. 2024. CRISPR/Cas9-mediated multiplex gene editing of gamma and omega gliadins: Paving the way for gliadin-free wheat. J. Exp. Bot. 75: 7079–7095.

Sánchez León, S. 2020. Identification and Development of Cereals Suitable for People Suffering from Gluten Allergies and Intolerances: Directed Mutagenesis by Specific Nucleases (CRISPR/Cas9) of Immunodominant Genes in Relation to Celiac Disease. Ph.D. diss., Univ. Córdoba, Spain.

Sarma, S., S.C. Thomas and R. Kamat. 2024. It takes two to tango with CRISPR: A history and overview of augmenting the technology for genetic engineering. Proc. Indian Natl. Sci. Acad.

Schiessl, S.-V., E. Katche, E. Ihien, H.S. Chawla and A.S. Mason. 2019. The role of genomic structural variation in the genetic improvement of polyploid crops. Crop J. 7: 127–140.

Schreiber, M., M. Jayakodi, N. Stein and M. Mascher. 2024. Plant pangenomes for crop improvement, biodiversity and evolution. Nat. Rev. Genet. 25: 563–577.

Schultzhaus, Z., Z. Wang and D.J. Stenger. 2021. CRISPR-based enrichment strategies for targeted sequencing. Biotechnol. Adv. 46: 107672.

Selvakumar, D., S.B. Selvamani and J. Mannu. 2024. Overview of the bioinformatics databases and tools for genome research and crop improvement. pp. 229–246. *In*: Genomics Data Analysis for Crop Improvement. Springer Nature.

Severi, A.A. and B. Akbari. 2024. CRISPR-Cas9 delivery strategies and applications: Review and update. Curr. Genet. 62: e23598.

Shah, E., D. Ludwig and P.J. Macnaghten. 2021. The complexity of the gene and the precision of CRISPR: What is the gene that is being edited? Ethics Soc. Adv. Biotech. 9.

Shi, J., H. Gao, H. Wang, H.R. Lafitte, R.L. Archibald, M. Yang et al. 2017. ARGOS8 variants generated by CRISPR-Cas9 improve maize grain yield under field drought stress conditions. Plant Biotechnol. J. 15: 207–216. https://doi.org/10.1111/pbi.12603.

Shukla-Jones, A., S. Friedrichs and D.E. Winickoff. 2018. Gene editing in an international context. OECD Sci. Technol.

Singh, D., P. Chaudhary, J. Taunk, C.K. Singh, V. Chinnusamy, A.M. Sevanthi et al. 2024. Targeting Induced Local Lesions in Genomes (TILLING): Advances and opportunities for fast-tracking crop breeding. Crit. Rev. Biotechnol. 44: 817–836. https://doi.org/10.1080/07388551.2023.2231630.

Singh, R.K., A. Prasad, M. Muthamilarasan, S.K. Parida and M. Prasad. 2020. Breeding and biotechnological interventions for trait improvement: Status and prospects. Planta 252: 54. https://doi.org/10.1007/s00425-020-03465-4.

Singh, V., D. Braddick and P.K. Dhar. 2017. Exploring the potential of genome editing CRISPR-Cas9 technology. Gene 599: 1–18.

Slokenberga, S., T. Minssen and A. Nordberg. 2023. Introduction: The significance of ELSPI perspectives in governing, protecting, and regulating the future of genome editing. pp. 1–14. *In*: Governing, Protecting, and Regulating the Future of Genome Editing. Brill Nijhoff.

Song, B., W. Ning, D. Wei, M. Jiang, K. Zhu, X. Wang et al. 2023. Plant genome resequencing and population genomics: Current status and future prospects. Mol. Plant 16: 1252–1268. https://doi.org/10.1016/j.molp.2023.07.009.

Song, L., J. Liu, B. Cao, B. Liu, X. Zhang, Z. Chen et al. 2023. Reducing brassinosteroid signalling enhances grain yield in semi-dwarf wheat. Nature 617: 118–124. https://doi.org/10.1038/s41586-023-06023-6.

Subedi, M., B. Ghimire, J.W. Bagwell, J.W. Buck and M. Mergoum. 2023. Wheat end-use quality: State of the art, genetics, genomics-assisted improvement, future challenges, and opportunities. Front. Genet. 13: 1032601. https://doi.org/10.3389/fgene.2022.1032601.

Subedi, M., P. Paudel, S. Gopera, A. Sharma and S.S. Sandal. 2024. Site-directed mutagenesis's prospective uses in enhancing cereal crops. J. Adv. Biol. Biotechnol. 27: 868–877. https://doi.org/10.9734/jabb/2024/v27i6950.

Sun, Y., L. Shang, Q.-H. Zhu, J. Fan, S. Ye, K.-C. Lo et al. 2022. Twenty years of plant genome sequencing: Achievements and challenges. Trends Plant Sci. 27: 391–401. https://doi.org/10.1016/j.tplants.2021.10.005.

Ternovska, T.K., T.S. Iefimenko and M.Z. Antonyuk. 2022. Improvement of wheat genetic resistance to powdery mildew: Retrospects and prospects. Open Agric. J. 17.

Thabet, S.G. and A.M. Alqudah. 2023. New genetic insights into improving barley's ability to cope with salt stress via regulating mineral accumulation, cellular ion homeostasis, and membrane trafficking. Environ. Bot. 208: 105252.

Thabet, S.G., Y.S. Moursi, M.A. Karam, O. Graner and A.M. Alqudah. 2018. Genetic basis of drought tolerance during seed germination in barley. PLoS One 13: e0206682. https://doi.org/10.1371/journal.pone.0206682.

Thabet, S.G., Y.S. Moursi, M.A. Karam, O. Graner and A.M. Alqudah. 2020. Natural variation uncovers candidate genes for barley spikelet number and grain yield under drought stress. Genes 11(5): 533. https://doi.org/10.3390/genes11050533.

Thabet, S.G., D.Z. Alomari, A. Börner, H. Brinch-Pedersen and A.M. Alqudah. 2022. Elucidating the genetic architecture controlling antioxidant status and ionic balance in barley under salt stress. Plant Mol. Biol. 110: 287–300. https://doi.org/10.1007/s11103-022-01302-8.

Távora, F.T.P.K., F. de Assis dos Santos Diniz, C. de Moraes Rêgo-Machado, N. Chagas Freitas, F. Barbosa Monteiro Arraes et al. 2022. CRISPR/Cas- and topical RNAi-based technologies for crop management and improvement: Reviewing the risk assessment and challenges toward more sustainable agriculture. Front. Bioeng. Biotechnol. 10: 913728. https://doi.org/10.3389/fbioe.2022.913728.

Verma, V., A. Kumar, M. Partap, M. Thakur and B. Bhargava. 2023. CRISPR-Cas: A robust technology for enhancing consumer-preferred commercial traits in crops. Front. Plant Sci. 14: 1122940. https://doi.org/10.3389/fpls.2023.1122940.

Villiger, L., J. Joung, L. Koblan, A. Lettenberger-Klein, Y. Jin and J.K. Joung. 2024. CRISPR technologies for genome, epigenome, and transcriptome editing. Nat. Rev. Mol. Cell Biol. 25: 464–487. https://doi.org/10.1038/s41580-024-00776-3.

Vyas, B., S. Kumar and M. Akhter. 2024. A bioinformatics insight on agriculture. pp. 191–205. *In*: Technological Approaches for Climate Smart Agriculture. Springer.

Wang, M.J. 2021. Next-generation sequencing (NGS). Curr. Med. Dev. pp. 305–327.

Wang, P., M. Abbas, J. He, Z. Geng, C. Wang, Y. Luo et al. 2024. Advances in genome sequencing and artificially induced mutation provide new avenues for cotton breeding. Front. Plant Sci. 15: 1400201. https://doi.org/10.3389/fpls.2024.1400201.

Wang, W., Q. Pan, F. He, J. Akhunova, J. Chao, P. Trick et al. 2018. Transgenerational CRISPR-Cas9 activity facilitates multiplex gene editing in allopolyploid wheat. CRISPR J. 1: 65–74. https://doi.org/10.1089/crispr.2017.0023.

Wu, B., P. Yun, H. Zhou, Y. He, Y. Peng, Z. Shi et al. 2022. Natural variation in WHITE-CORE RATE 1 regulates redox homeostasis in rice endosperm to affect grain quality. Plant Cell 34: 1912–1932. https://doi.org/10.1093/plcell/koac049.

Wyngaarden, S.L., K.K. Lightburn and R.C. Martin. 2020. Optimizing livestock feed provision to improve the efficiency of the agri-food system. Sustainable Food Syst. 44: 188–214.

Xin, W., J. Wang, J. Li, H. Zhao, H. Liu, H. Zheng et al. 2021. Candidate gene analysis for nitrogen absorption and utilization in japonica rice at the seedling stage based on a genome-wide association study. Front. Plant Sci. 12: 670861. https://doi.org/10.3389/fpls.2021.670861.

Yan, M. and J. Li. 2019. The evolving CRISPR technology. Protein Cell 10: 783–786.

Yan, Z., F. Zhang, C. Mu, Y. Guo, Z. Ding, J. Zhang et al. 2024. The ZmbHLH47-ZmSnRK2.9 module promotes drought tolerance in maize. Int. J. Mol. Sci. 25: 4957. https://doi.org/10.3390/ijms25094957.

Ye, R., X. Yang and Y. Rao. 2022. Genetic engineering technologies for improving crop yield and quality. Agriculture 12: 759.

Ye, X. and F. Han. 2024. Applications of fast breeding technologies in crop improvement and functional genomics study. Front. Plant Sci. 14: 1460642.

Ye, X., K. Wang, H. Liu, L. Zhao, H. Liu and Y. Lin. 2022. Genome editing toward wheat improvement. pp. 241–269. *In*: Genome Editing Technologies for Crop Improvement. Springer. https://doi.org/10.1007/978-981-19-0600-8_12.

Yin, Y., Q. Wang, L. Xiao and H. Zhang. 2018. Advances in the engineering of the gene editing enzymes and the genomes: Understanding and handling the off-target effects of CRISPR/Cas9. J. Biomed. Nanotechnol. 14: 456–476. https://doi.org/10.1166/jbn.2018.2546.

Yotova, R. 2020. Regulating genome editing under international human rights law. Cambridge Law Q. 69: 653–684.

Yu, Z., U. Yunusbaev, A. Fritz, W. Zhao, M. Kemen, P. Rausch et al. 2024. CRISPR-based editing of the ω- and γ-gliadin gene clusters reduces wheat immunoreactivity without affecting grain protein quality. Plant Biotechnol. J. 22: 892–903. https://doi.org/10.1111/pbi.14231.

Yuan, Z. 2024. From origin to present: Establishment, mechanism, evolutions, and biomedical applications of CRISPR/Cas-based macromolecular systems. Preprints.

Zegeye, W.A., M. Tsegaw, Y. Zhang, S. Khan, E. Tesfamichael and B. Kebede. 2022. CRISPR-based genome editing: Advancements and opportunities for rice improvement. Int. J. Mol. Sci. 23: 4454. https://doi.org/10.3390/ijms23094454.

Zeng, X., Y. Luo, N.T.Q. Vu, X. Shen, H. Liu, J. Li et al. 2020. CRISPR/Cas9-mediated mutation of OsSWEET14 in rice cv. Zhonghua11 confers resistance to *Xanthomonas oryzae* pv. oryzae without yield penalty. BMC Plant Biol. 20: 313. https://doi.org/10.1186/s12870-020-02470-3.

Zhai, H., Z. Feng, X. Du, Y. Song, S. Liu, G. Li et al. 2018. A novel allele of TaGW2-A1 is located in a finely mapped QTL that increases grain weight but decreases grain number in wheat (*Triticum aestivum* L.). Theor Appl. Genet. 131: 539–553. https://doi.org/10.1007/s00122-017-3017-y.

Zhang, D., A. Hussain, H. Manghwar, K. Xie, S. Xie, S. Zhao et al. 2020. Genome editing with the CRISPR-Cas system: An art, ethics, and global regulatory perspective. Plant Biotechnol. J. 18: 1651–1669. https://doi.org/10.1111/pbi.13383.

Zhang, D., Y. Gan, L. Le, T. Li, C. Sun, G. Liu et al. 2025 (e-pub 2024). Epigenetic variation in maize agronomical traits for breeding and trait improvement. J. Genet. Genomics 52: 307–318. https://doi.org/10.1016/j.jgg.2024.01.006.

Zhang, S., B. Ma, L. Yi and X. An. 2024. Genome-wide characterization of the aldehyde dehydrogenase gene superfamily in maize and its potential role in anther development. ACS Agric. Sci. Technol. 4(1). https://doi.org/10.1021/acsagscitech.3c00430.

Chapter 4

CRISPR/Cas Design, DNA Repair Pathways, and Mutation Analysis

Abdullah Aydın,[1] *Seher Yerlikaya,*[1] *Nisa Nur Yılmaz,*[1] *Izhar Ullah,*[2] *Bayram Ali Yerlikaya,*[1,*] *Musa Kavas*[1,*] and *Abdul Basit*[3]

1. Introduction

The advent of CRISPR technology has marked a transformative era in genome editing, providing unparalleled precision and efficiency in genetic modification (Bhatia and Yadav 2023; Bhokisham et al. 2023; Meng et al. 2023). This groundbreaking tool capitalizes on the adaptive immune system of bacteria to identify and cleave specific DNA sequences. At the core of this system lies a single guide RNA (sgRNA), which directs the Cas9 nuclease to the desired genetic locus, initiating a double-stranded break in the DNA. Subsequent cellular repair processes, such as Non-Homologous End Joining (NHEJ) or Homology-Directed Repair (HDR), enable the incorporation of targeted genetic alterations. Renowned for its simplicity, affordability, and remarkable accuracy, CRISPR-Cas9 has rapidly emerged as a cornerstone of modern genetic engineering, reshaping the boundaries of scientific innovation (Ilahibaks et al. 2022; Verma et al. 2023).

CRISPR-Cas9 technology has demonstrated remarkable versatility across a broad spectrum of scientific disciplines. In biomedical research, it has become an indispensable tool for unraveling gene functions and elucidating disease mechanisms. By enabling precise gene knockouts and the development of tailored disease models, CRISPR-Cas9 has accelerated the exploration of innovative therapeutic strategies. For example, Hsu et al. successfully leveraged this technology to edit multiple genes simultaneously, highlighting its transformative potential in functional genomics (Hsu et al. 2013). Beyond medicine, CRISPR-Cas9 has revolutionized fields like agriculture, bioinformatics, and biotechnology. In agriculture, it offers a precise and efficient alternative to conventional breeding methods, allowing for targeted modifications in genes linked to disease resistance, yield enhancement, and nutritional quality. A notable application is the work by Li et al., where CRISPR-Cas9 was employed to boost rice grain yield by editing a gene responsible for regulating grain size (Li et al. 2013). This technology's adaptability underscores its promise in driving advances across diverse scientific and industrial domains.

[1] Department of Agricultural Biotechnology, Faculty of Agriculture, Ondokuz Mayıs University, Samsun, Turkey.
[2] Department of Horticulture, Faculty of Agriculture, Ondokuz Mayıs University, Samsun, Turkey.
[3] Department of Horticulture, Graduate School, Kyungpook National University, 41566, Daegu, South Korea.
* Corresponding authors: musa.kavas@omu.edu.tr; bayram.yerlikaya@omu.edu.tr

DOI: 10.1201/9781003518204-4

CRISPR-Cas9 technology has fundamentally transformed the field of genome editing, offering researchers a powerful and flexible tool for genetic modifications. Its broad applications span across biomedical research, agriculture, bioinformatics, and biotechnology, presenting significant potential to tackle diverse challenges and enhance human well-being. This review aims to provide an in-depth examination of the CRISPR-Cas9 mechanism, delving into its various applications, future possibilities, and the ethical and safety concerns that accompany its use.

1.1 Importance of Understanding DNA Repair Pathways in CRISPR-Mediated Mutations

The cleavage of target DNA by CRISPR/Cas9 induces a Double-Strand Break (DSB), triggering the cell's intrinsic repair mechanisms to restore genomic integrity. These repair pathways are vital for resolving DSBs, which occur naturally under both stress and non-stress conditions. Effective repair must balance two competing demands: rapid rejoining of DNA ends to prevent genome instability and the need for high-fidelity restoration. Accurate, error-free repair relies on the presence of a homologous template, which is typically accessible only during specific phases of the cell cycle. DSB repair pathways are classified into four main types based on their molecular mechanisms and associated machinery: Classical Non-Homologous End Joining (c-NHEJ): This pathway directly ligates DNA ends without relying on sequence homology. While c-NHEJ can occasionally produce error-free repairs, it is often error-prone due to the processing (commonly trimming) of DNA ends prior to ligation.

Homologous Recombination (HR): HR utilizes a homologous template for repair, making it inherently precise and largely error-free. Several subtypes exist within this pathway, all requiring homologous sequences for accurate restoration. Single-Strand Annealing (SSA): Sharing key molecular components with HR, SSA repairs breaks through complementarity between homologous sequences (direct repeats) at the DNA ends. However, this process results in the deletion of one repeat and the intervening sequence, making it inherently mutagenic. Polymerase Theta-Mediated End Joining (TMEJ): Also known as Microhomology-Mediated End Joining (MMEJ) or alternative End Joining (alt-EJ), this pathway repairs breaks using short homologous sequences at the DNA ends. TMEJ often introduces deletions and insertions, driven by polymerase theta (Polθ) activity, resulting in less accurate outcomes. These pathways have evolved to ensure robust DSB repair under diverse conditions, often compensating for each other when needed. The factors governing pathway selection remain incompletely understood, but proteins involved in initial DSB recognition may play a pivotal role in this decision-making process. DSBs are most possibly recognized by the KU70/KU80 heterodimer or POLY(ADP-RIBOSE) POLYMERASE 1 (PARP1), both of which localize to damage sites within seconds (Yang et al. 2018). In *Arabidopsis thaliana*, KU70 and KU80 are ubiquitously expressed across tissues and developmental stages, regardless of cell division status (Klepikova et al. 2016; Tamura et al. 2002; Yadav et al. 2009). Similarly, in rice, these proteins are detected in roots, leaves, meristems, and callus cells (Hong et al. 2010). PARP1, however, demonstrates a preference for actively dividing tissues (Klepikova et al. 2016; Tamura et al. 2002; Yadav et al. 2009). Importantly, both KU and PARP1 are upregulated in response to DSBs (Doucet-Chabeaud et al. 2001).

In human cell lines, PARP1 activity is tightly linked to the cell cycle, with expression suppressed during G1/G0 phases but reactivated during the S phase (Pietrzak et al. 2018). PARP1 also serves as an inhibitor of DNA-PKcs, a critical kinase in c-NHEJ. Interestingly, while DNA-PKcs has been identified in green algae, bryophytes, and gymnosperms, it is absent in angiosperms, underscoring species-specific regulation of c-NHEJ (Kumar 2023; Lees-Miller et al. 2021). This divergence necessitates caution when extrapolating findings across taxa. Further, regulatory dynamics within the cell cycle influence c-NHEJ. For example, phosphorylation of KU70 by cyclin-dependent kinases during S, G2, and M phases disrupts KU70/KU80 complex formation, impairing DNA repair (Mukherjee et al. 2016). Contrastingly, PARP1 activity has been reported to facilitate KU70/KU80 binding to DNA in certain contexts (Caron et al. 2019). Such findings highlight the complexity of repair mechanisms, compounded by the fact that many studies utilize cancer-derived cell lines with aberrant cell cycle control, potentially skewing observations. These nuances underscore the need for context-specific research to unravel DSB repair mechanisms in different organisms.

Significant gaps remain in our understanding of the intricate processes involved in genome editing, particularly the repair of DNA breaks induced by CRISPR/Cas9. For instance, the pronounced differences observed in prime editing efficiency between monocots and dicots suggest that the underlying repair pathways for nicked DNA may vary significantly between these two clades. Gaining deeper insights into these pathways and their specific regulatory factors could pave the way for the development of more precise and efficient CRISPR/Cas9 systems tailored for plant breeding and fundamental research. However, advancements in this area require plant-focused studies, as the knowledge derived from animal and yeast models offers limited applicability to plant systems due to inherent biological differences.

The objective of this chapter is to provide a comprehensive understanding of the CRISPR/Cas genome editing technology, with a focus on the underlying DNA repair mechanisms that mediate its efficiency and precision. By examining the components and design strategies of the CRISPR/Cas system, including considerations for off-target effects, we aim to highlight critical factors that influence the success of genome editing applications. This chapter will explore various DNA repair pathways, particularly NHEJ and HDR, and their roles in determining the outcome of CRISPR-induced mutations. Additionally, this chapter will address the current methodologies used for mutation analysis, assessing both on-target and off-target effects, and discuss the tools and technologies that have revolutionized mutation detection. By investigating the applications of CRISPR/Cas in diverse fields such as research, agriculture, biotechnology, nanotechnology, bioinformatics, and medicine, the chapter will also consider the ethical and safety concerns associated with this powerful tool. Looking forward, the review will discuss future innovations in CRISPR/Cas technology, including base editors, prime editors, and the integration of artificial intelligence in the design of guide RNAs. This chapter ultimately seeks to offer a balanced perspective on the state-of-the-art in CRISPR technology, while highlighting the challenges and opportunities that lie ahead in advancing genome editing for various scientific and therapeutic applications.

2. Overview of CRISPR/Cas Systems and Their Applications in Genome Editing

CRISPR-Cas9 technology has emerged as a groundbreaking gene-editing tool, revolutionizing the field of genome modification and capturing widespread scientific attention (Li et al. 2023). Originating from the adaptive immune system of bacteria, this system operates through the interplay of two key components: the Clustered Regularly Interspaced Short Palindromic Repeats (CRISPR) and the CRISPR-associated protein 9 (Cas9) nuclease. The CRISPR component comprises short, repetitive DNA sequences interspersed with unique spacer sequences that function as a molecular archive of previous viral infections. Complementing this, the Cas9 nuclease acts as a molecular scissor, capable of cleaving DNA strands with high precision. Together, these two components operate in harmony to accurately identify and edit specific DNA sequences, enabling targeted genetic modifications (Richardson et al. 2023). The remarkable specificity of CRISPR-Cas9 technology is attributed to the single-guide RNA (sgRNA), which is engineered to be complementary to the desired target DNA sequence. Acting as a molecular guide, the sgRNA directs the Cas9 nuclease to the exact genomic location requiring modification. Upon reaching the target site, the Cas9 nuclease induces a double-stranded break in the DNA, initiating the gene-editing process (Mingarro and lí del Olmo 2023).

CRISPR-Cas9 technology has revolutionized genome editing, garnering widespread acclaim for its simplicity, efficiency, and remarkable adaptability (Naeem and Alkhnbashi 2023). Originally adapted for genetic engineering by Jennifer Doudna and Emmanuelle Charpentier, this system's transformative potential was first demonstrated in a groundbreaking 2014 Science study (Doudna and Charpentier 2014). Their work revealed that the Cas9 nuclease, guided by a synthetic single-guide RNA (sgRNA), could accurately target and cleave specific DNA sequences *in vitro*. This foundational discovery catalyzed the rapid development and broad application of CRISPR-Cas9 technology (Doudna and Charpentier 2014). Since its inception, the system has been widely adopted and meticulously refined for use in diverse organisms and cell types. Ongoing research has significantly expanded its utility and scope. For instance, Mali et al. successfully employed CRISPR-Cas9 to edit human cell genomes, illustrating its profound implications for biomedical research (Mali et al. 2013). Enhancing the specificity of CRISPR-Cas9 has been a critical area of focus, yielding notable improvements in reducing off-target effects. Fu et al. developed SpCas9-HF1, a high-fidelity Cas9 variant with significantly improved target accuracy while retaining high editing efficiency. This innovation underscored the potential of engineering Cas9 to refine its precision (Fu et al. 2014). Beyond its initial use in gene knockout, CRISPR-Cas9 has proven highly versatile, facilitating diverse applications such as gene activation and repression. Gilbert et al. pioneered an approach in which the catalytically inactive Cas9 (dCas9) was fused with transcriptional activators or repressors, enabling precise control of gene expression. This strategy has greatly advanced functional genomics and the exploration of regulatory networks (Gilbert et al. 2014). The profound contributions of Doudna and Charpentier have firmly established CRISPR-Cas9 as a cornerstone of modern genome editing. With ongoing advancements in precision, versatility, and applications, this technology continues to redefine the boundaries of scientific research, therapeutic innovation, and agricultural development (Doudna and Charpentier 2014).

2.1 *Components of the CRISPR/Cas System and Strategies for sgRNA Design*

The CRISPR and associated Cas (CRISPR-associated) systems have transformed genome editing, finding applications across virtually all domains of life. Among the diverse CRISPR/Cas systems, the *Streptococcus pyogenes* Cas9 (SpCas9) system stands out as the most extensively studied and widely adopted, due to its exceptional efficiency and versatility. This system utilizes either the natural crRNA:tracrRNA duplex or a synthetically engineered single guide RNA (sgRNA) to direct DNA DSBs at specific genomic locations, enabling precise editing. The sgRNA itself comprises six functional modules—spacer, lower stem, bulge, upper stem, nexus, and hairpins (Fig. 1)—which collaboratively facilitate RNA-guided DNA cleavage and have been thoroughly characterized to refine editing accuracy and efficiency (Briner et al. 2014).

Extensive engineering efforts have been applied to both key components of the CRISPR/Cas system—Cas9 and sgRNA—to enhance genome editing efficiency, specificity, and functionality. Various approaches have been developed to optimize Cas9, including modifications to its nuclease domains to produce nCas9 (Cas9 nickase) (Jinek et al. 2012) and dCas9 (nuclease-deficient Cas9) (Qi et al. 2013), protein engineering for higher fidelity (Casini et al. 2018; Qi et al. 2013), improved activity (Bao et al. 2015), and the ability to recognize alternative Potospacer-Adjacent-Motif (PAM) sequences (Ma et al. 2019). Additional innovations include fusing Cas9 with effector domains for transcriptional regulation (Maeder et al. 2013), as well as the development of base editing (Komor et al. 2016) and prime editing techniques (Anzalone et al. 2019). Complementary to Cas9 engineering, the refinement of sgRNA design has emerged as a pivotal factor in building more robust and versatile CRISPR/Cas systems. While Cas protein engineering strategies have been well-documented, this chapter focuses on a detailed and structural exploration of sgRNA design and engineering strategies, categorized by spacer, scaffold, and end extensions (Fig. 1).

The CRISPR/Cas system begins with designing sgRNA spacer sequences (e.g., N20 for SpCas9) to achieve high efficiency in targeting specific DNA regions. Due to the frequent occurrence of the NGG PAM sequence for SpCas9 (approximately every 8 bp in the genome), numerous potential sgRNAs can be identified for most target genes, exhibiting varying performance.

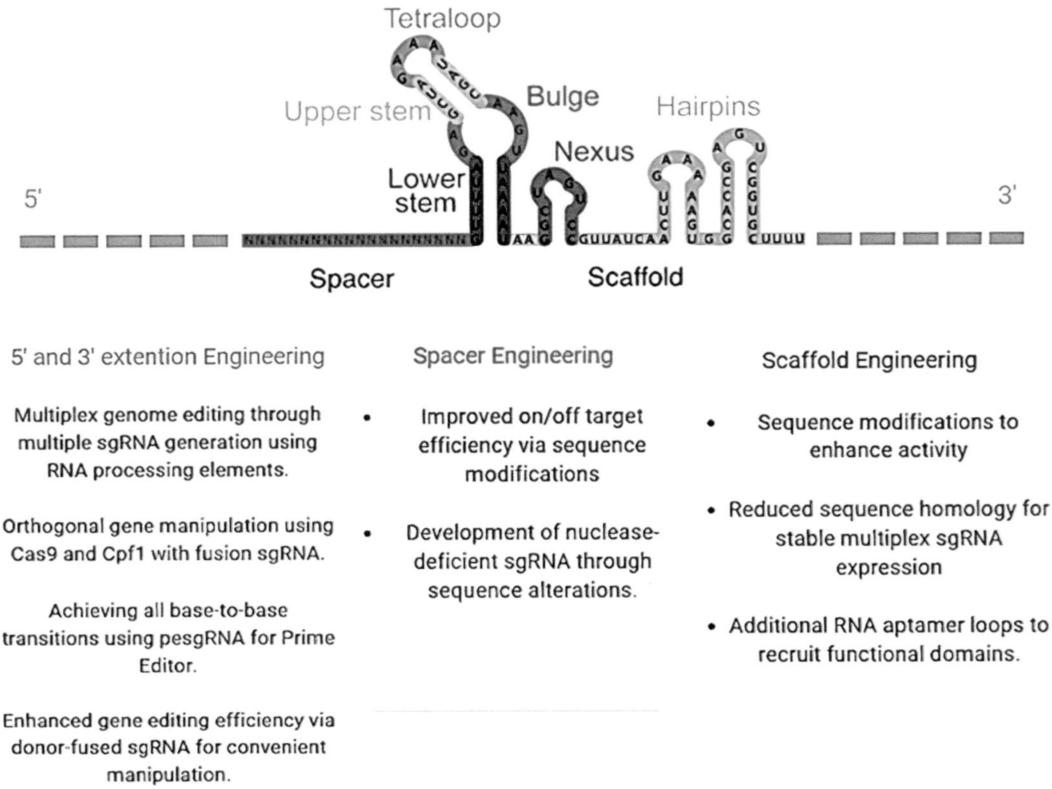

Figure 1. A schematic representation of sgRNA engineering highlights its structural components—spacer, scaffold, and extensions—that have been meticulously modified to improve genome editing efficiency and broaden functional applications.

The efficiency and specificity of these guides (spacer sequences, Fig. 1) are largely influenced by factors such as binding affinity, secondary structures, and chromatin accessibility at the target site. Genome-wide knockout screenings have been instrumental in identifying optimal sgRNA spacer sequences (Doench et al. 2016; Doench et al. 2014; Wu et al. 2014), and numerous public platforms have been developed based on these studies (Doench et al. 2016). For instance, Doench et al. created a library of sgRNAs and assessed their indel-inducing potential using flow cytometry (Doench et al. 2014), which led to predictive models for sgRNA efficiency. This was later refined by testing over 2000 sgRNAs, further enhancing the predictive accuracy (Doench et al. 2016). Advanced methods such as machine learning and deep learning have also been employed to improve sgRNA design (O'Brien et al. 2021). Although the 20-nt spacer length (N20) is standard for SpCas9, engineering sgRNAs with modified spacer lengths has shown promising results. Bao et al. demonstrated similar editing efficiencies using crRNAs with spacers ranging from 20 to 120 nt in yeast (Bao et al. 2015). Conversely, Matson et al. (2019) reported improved specificity with increased spacer lengths for certain sequences. Truncated sgRNAs (e.g., N17–18), studied by Fu et al., significantly reduced off-target effects—up to 5000-fold—while maintaining on-target efficiency (Fu et al. 2014). However, Zhang et al. noted a slight reduction in editing efficiency with N17 sgRNAs, underscoring the need to balance on-target and off-target activities (Zhang et al. 2016). Truncation below N16 impaired Cas9's nuclease activity but allowed targeting and binding, enabling applications in transcriptional regulation when fused with effector domains (Dahlman et al. 2015; Fu et al. 2014; Kiani et al. 2015). Dong et al. developed a tri-functional CRISPR-ARE system using full-length sgRNAs for editing and truncated ones for transcriptional modulation (Dong et al. 2020).

The sgRNA scaffold, comprising functional modules (lower stem, bulge, upper stem, nexus, and hairpins), has also been optimized for enhanced performance and broader applications. Scaffold modifications often target the lower stem, upper stem, and hairpins, as the bulge and nexus regions are crucial for sgRNA functionality (Briner et al. 2014). For instance, Chen et al. improved transcription and stability by modifying the lower stem (A–U flip) and extending the upper stem, enhancing imaging and CRISPR interference efficiencies (Chen et al. 2013). Similarly, Scott et al. showed that specific uridine substitutions (e.g., U13A, U32G, U34A) increased genome editing efficiency (Scott et al. 2019). Riesenberg et al. stabilized sgRNAs by adding high-melting-point RNA hairpins, improving cleavage efficiency by 69%, particularly for challenging targets (Riesenberg et al. 2022). To address challenges in multiplex genome engineering, Reis et al. designed non-repetitive sgRNA scaffolds using biophysical modeling and machine learning, enabling stable expression of up to 22 sgRNAs in a single plasmid to repress 13 genes simultaneously without compromising editing efficiency (Reis et al. 2019).

Furthermore, RNA aptamers integrated into the sgRNA scaffold have expanded its utility. These aptamers, fused at specific scaffold regions (e.g., upper stem, hairpins), recruit effector domains such as fluorescent proteins or transcriptional Regulators via aptamer-Binding Proteins (RBPs). This orthogonal system enables precise activation, repression, or labeling without cross-talk (Ma et al. 2016; Zalatan et al. 2015). For example, Jesse et al. used MS2-MCP, PP7-PCP, and com-Com pairs to independently activate and repress transcription (Zalatan et al. 2015). Ma et al. developed CRISPRainbow for multiplex genome labeling, achieving six distinct color combinations by using various aptamer-protein pairs (Ma et al. 2016). While divalent modifications enhanced performance, excessive hairpins reduced sgRNA activity, emphasizing the importance of balanced scaffold engineering.

2.2 Considerations for Efficient gRNA Design – Tools, Software, and Strategies

The CRISPR/Cas system operates by scanning the PAM sequence, with sgRNA recognizing target loci and triggering endonuclease activity for precise DNA cleavage. Despite this streamlined mechanism, cleavage efficiency varies significantly across different target sites and cell lines, influenced by factors such as gRNA sequence composition, nucleotide positioning, GC content, chromatin accessibility, gene expression levels, and RNA energetics, including secondary structure and melting temperature. To address these complexities, researchers have developed various computational tools designed to optimize gRNA efficacy by integrating these features (Table 1).

The nucleotide composition of a target sequence plays a pivotal role in determining gRNA efficiency (Doench et al. 2014; Wang et al. 2014; Xu et al. 2015). Mammalian CRISPR-based screens have revealed specific preferences, such as guanines at positions 1 and 2 preceding the PAM sequence (Wu et al. 2014), while thymines are disfavored within ±4 nucleotides around the PAM. Downstream sequences of PAMs significantly influence gRNA efficiency, whereas upstream sequences show minimal impact. Cytosine is particularly favored at the CRISPR/Cas9 cleavage site (–3 position relative to PAM), and the GC content in the region 4–13 bases downstream of the PAM is a critical factor (Cong et al. 2013). Leveraging these insights, several predictive models have been developed to enhance gRNA design efficiency. Rule Set 1, a predictive model based on data from 1,841 sgRNAs in humans and mice, utilizes a Support Vector Machine (SVM) and supervised learning methods for general linear classification of sequence features influencing CRISPR efficiency (Doench et al. 2014). It demonstrated a strong correlation with experimental outcomes. To enhance accuracy, Rule Set 2 was introduced in 2016 (Doench et al. 2016), incorporating position-independent nucleotide counts and sgRNA target site location within genes. These refinements supported gRNA design for CRISPR KO and CRISPRa/i experiments and were implemented in the Broad Institute's GPP sgRNA Designer (Fusi et al. 2015). Screenings of CRISPR KO and CRISPRa/i libraries in mammals revealed significant nucleotide preference differences between these editing strategies (Xu et al. 2015). Elastic Net regression, an advanced

Table 1. Tools to select and evaluate the efficiency of gRNA.

Tool	Description	Method	Reference
FlashFry	A swift tool for locating off-target sites, providing essential details on GC content, as well as on- and off-target scores for the target regions.	Alignment	(McKenna and Shendure 2018)
DeepCRISPR	A cutting-edge deep learning-based tool that predicts both on-target and off-target cleavage sites simultaneously, incorporating epigenetic factors.	Scoring	(Chuai et al. 2018)
MIT	A tool that identifies potential off-target sites early in gene editing with 20 bp gRNA, excluding PAM.	Scoring	(Hsu et al. 2013)
Cas-OFFinder	A rapid and highly adaptable computational tool that identifies off-target sites even with unlimited mismatch numbers.	Alignment	(Bae et al. 2014)
CasOT	An online tool designed for detecting off-targets across the genome, capable of handling an unlimited number of mismatches in paired gRNA.	Alignment	(Xiao et al. 2014)
CrisFlash	A tool based on algorithms to identify off-target effects, incorporating user-provided variant data and capable of handling unlimited mismatches.	Alignment	(Jacquin et al. 2019)
Elevation	A machine learning-based tool developed using a two-layer regression model, factoring in epigenetic influences.	Scoring	(Listgarten et al. 2018)
CFD	A widely utilized tool for off-target detection and evaluation, requiring 20 bp gRNA and PAM.	Scoring	(Doench et al. 2016)
CRISTA	A machine learning tool with various features, including GC content, RNA secondary structure, and epigenetic factors.	Scoring	(Abadi et al. 2017)

method outperforming SVM in specific scenarios, was used to construct models for CRISPR KO and CRISPRa/i. These models were integrated into tools like Spacer Scoring for CRISPR (SSC), E-CRISP, CHOPCHOP, and CRISPR-FOCUS (Cao et al. 2017; Heigwer et al. 2014; Labun et al. 2019).

Further advancements were achieved by Moreno-Mateos et al., who linked sgRNA activity to guanine enrichment and adenine depletion (Moreno-Mateos et al. 2015). Analyzing over 1,000 sgRNAs targeting 128 zebrafish genes, they developed a logistic regression-based model incorporated into CRISPRscan. WU-CRISPR enhanced this model with novel features, achieving superior precision (Doench et al. 2014; Wong et al. 2015). Labun et al. (2016) identified GC content-dependent activity at PAM-distal regions, leading to the CRISPRater model, integrated into the CCTop platform (Stemmer et al. 2015). The Church laboratory introduced the sgRNA Scorer software, calculating on-target sgRNA scores via SVM models (Chari et al. 2015). Its updated version supported predictions for additional Cas systems, such as SaCas9 and AsCpf1 (Chari et al. 2017). Housden et al. employed a drug-targeting method to screen CRISPR KO efficiency in Drosophila, creating an efficiency prediction tool (Housden et al. 2015). CASPER combined insights from CRISPRscan with new features to optimize on-target correlations, detect off-target effects, and facilitate multipopulational analyses (Mendoza and Trinh 2018). These tools collectively represent significant progress in refining predictive accuracy for CRISPR applications. Genetic and epigenetic factors, such as chromatin accessibility, gene position, and expression, significantly influence sgRNA binding and Cas nuclease activity. Research has shown that nucleosomes hinder Cas9 target cleavage, while DNase I hypersensitivity and epigenomic markers can modulate gRNA efficacy (Doench et al. 2016; Horlbeck et al. 2016; Uusi-Mäkelä et al. 2018; Yarrington et al. 2018). Leveraging these insights, several computational tools have been developed to predict sgRNA efficiency. For instance, predictSGRNA, an R package inspired by oligonucleotide design principles and nucleosome occupancy models, has demonstrated superior performance compared to other models, including Azimuth and sgRNA Scorer (Chari et al. 2015; Doench et al. 2016; Fusi et al. 2015).

The energetics involved in the interaction between DNA, gRNA, and the Cas protein can provide consistent insights into CRISPR editing efficiency, helping to minimize biases across different predictive models. Certain energetics-based approaches have proven effective in illustrating Cas9 editing efficacy (Szczelkun et al. 2014; Wu et al. 2014; Xu et al. 2017). For example, CRISPRpred incorporates nucleotide positions and sgRNA secondary structures to predict cleavage efficiency, outperforming Rule Set 1 (Doench et al. 2014; Rahman and Rahman 2017). Recently, Zhang et al. introduced a free energy-based model called uCRISPR, designed to evaluate both Cas9 editing efficacy and off-target effects. This versatile model is considered applicable to any dataset involving cleavage-active Cas9 variants (Zhang et al. 2019). The efficiency of guide RNAs (gRNAs) is influenced by various factors, and predictive models for single-guide RNA (sgRNA) design rely on experimental data and established guidelines. However, the on-target scores produced by these models can vary, necessitating careful evaluation during gRNA design. Despite this variability, specific features have consistently been identified as reliable indicators of gRNA efficacy. These include guanine preference, GC content, the seed region's integrity, and the secondary structure of the gRNA. Such characteristics are critical for optimizing the performance of gRNAs in genome editing applications (Liu et al. 2016; Shibata et al. 2017; Xu et al. 2015).

Selecting the appropriate tool is essential for accurate gRNA design. A thorough evaluation of various prediction tools highlighted significant differences in model performance across datasets, with no single model excelling universally (Haeussler et al. 2016). This variability underscores the importance of careful selection and the potential value of cross-evaluating gRNAs with multiple models. Among available tools, Rule Set 2 and DeepCpf1 are widely recognized for their accuracy in assessing Cas9 and Cpf1 cutting efficiency. Meanwhile, uCRISPR shows promise with potentially greater precision, although further experimental validation is required.

2.3 Tools for CRISPR gRNA Design

The CRISPR/Cas complex comprises a Cas protein and a sgRNA, both critical for determining the cutting activity. While protein engineering can enhance the Cas protein's performance, it is often complex and less accessible for many researchers. In contrast, optimizing gRNAs offers a more

Table 2. An overview of web-based CRISPR gRNA design tools is provided, highlighting those marked with an asterisk (*) as they specialize in predicting CRISPR editing outcomes.

Tool	Website	References
CCTop	https://crispr.cos.uni-heidelberg.de	(Stemmer et al. 2017)
CRISPR RGEN Tools	http://www.rgenome.net/	(Hwang et al. 2018; Park et al. 2015)
CHOPCHOP	https://chopchop.cbu.uib.no	(Labun et al. 2016; Labun et al. 2019; Montague et al. 2014)
E-CRISP	http://www.e-crisp.org/E-CRISP	(Heigwer et al. 2014)
CRISPOR	http://crispor.tefor.net	(Concordet and Haeussler 2018; Haeussler et al. 2016)
Lindel*	https://shendurelab.github.io/Lindel/	(Chen et al. 2019)
CRISPETa	http://crispeta.crg.eu	(Pulido-Quetglas et al. 2017)
CRISPRscan	https://www.crisprscan.org	(Moreno-Mateos et al. 2015)
FORECasT*	https://partslab.sanger.ac.uk/FORECasT	(Allen et al. 2019)
CRISPR-ERA	http://CRISPR-ERA.stanford.edu	(Liu et al. 2015a)
CRISPR-PLANT	https://www.genome.arizona.edu/crispr2	(Minkenberg et al. 2019; Xie et al. 2014)
CRISPR-GE	http://skl.scau.edu.cn	(Xie et al. 2017)
inDelphi*	https://www.crisprindelphi.design/	(Shen et al. 2018)
EuPaGDT	http://grna.ctegd.uga.edu	(Peng and Tarleton 2015)
CRISPR-P	http://crispr.hzau.edu.cn/CRISPR2/	(Lei et al. 2014; Liu et al. 2017)

practical and efficient approach. For a gRNA to be effective, it must exhibit high on-target efficiency while minimizing off-target activity. To meet these criteria, numerous computational tools have been developed for designing high-efficiency gRNAs.

Each tool has unique features, and user-friendliness is a crucial factor when selecting the most appropriate one. To assist researchers in navigating these options, we provide an overview of the most widely used gRNA design tools, highlighting their strengths and guiding users in making informed choices. Additionally, we recommend several functional and user-friendly websites for gRNA design, which are detailed in Table 2.

3. DNA Repair Pathways

3.1 Detection of DNA Breaks

Following damage to cellular DNA, a complex and multifaceted response process is involved from the detection of damage to repair. This process is called DNA Damage Response (DDR) (Herbst et al. 2024). Herbst et al. (2024) state that there are three important steps in this process. The first step is the detection of damage and the establishment of the signal cascade, the second is the initiation of signal transmission, and the third is the signal delivery to the effectors. The mechanism involved in damage sensing differs according to the type of DNA Break (DB). Once a DSB occurs in DNA, it is recognized by the complex of MRE11-RAD50-NBS1 proteins (MRN complex). This complex binds the broken ends together by dimerization and recruits the inactive ATAXIA TELANGIECTASIA MUTATED (ATM) (Herbst et al. 2024). However, if the DB is a Single Strand Break (SSB), a different signaling network is involved. If the SSB is detected by the RAD9-RAD1-HUS1 complex (9-1-1 complex), ATAXIA TELANGIECTASIA AND RAD3-RELATED (ATR) is called into the scene and the response continues in this way (Heitzeberg et al. 2004; Zou and Elledge 2003; Zou et al. 2003). In the event of a single strand break due to replication stress, the ATR is recruited to the site by a different complex. Replication Protein A (RPA) binds the SSB and recruits ATR-Interacting Protein (ATRIP) and more stable ATR-ATRIP complex forms (Cortez et al. 2001; Namiki and Zou 2006; Sweeney et al. 2009; Zou and Elledge 2003). Tumor suppressor protein (p53) is phosphorylated in both cases when the signal goes through ATM or ATR. Phosphorylated p53 initiates DNA repair, cell cycle regulation and apoptosis in mammalians (Banin et al. 1998; Canman et al. 1998; Tibbetts et al. 1999). Interestingly, although ATM or ATR homologs exist in plants, there is no protein with a similar sequence to the p53 (Preuss and Britt 2003; Yoshiyama et al. 2009). However, there is SUPPRESSOR OF GAMMA RADIATION 1 (SOG1) in plants, which fulfills the function of p53 (Sjogren et al. 2015; Yoshiyama et al. 2013). SOG1, a transcription factor, activates the transcription of many genes responsible for DNA repair and cell cycle checkpoint activation after phosphorylation by ATM/ATR (Bourbousse et al. 2018; Ogita et al. 2018). Figure 2 shows the signaling network involved in the DDR.

3.1.1 Non-homologous End Joining

DSBs are among the DNA damages that most affect the stability and integrity of genomic DNA (Deriano and Roth 2013; Ferguson and Alt 2001; Jackson and Bartek 2009; Van Gent et al. 2001). The cause of DNA damage in eukaryotes can be endogenous (Replication errors, failure of topoisomerases, accidental nicking by endonucleases) or exogenous (reactive oxygen species, ionizing radiation, radiomimetic chemicals) (Jackson 2002; Mehta and Haber 2014; Sebastian and Raghavan 2016). DSBs can also occur as a result of meiosis or some DNA recombination (Chapman et al. 2012). NHEJ is one of the two major pathways that repairs DSBs (Jackson 2002; Lieber 2010; Lieber et al. 2006; Moore and Haber 1996; Sung and Klein 2006). The NHEJ pathway is the more predominant DSB repair pathway (especially in G1 and G2/M phase of cell cycle) because it does not require homology, despite the fact that NHEJ pathway repairs incorrectly (Chang et al. 2016; Chiruvella et al. 2012; Ghosh and Raghavan 2021; Lieber 2010).

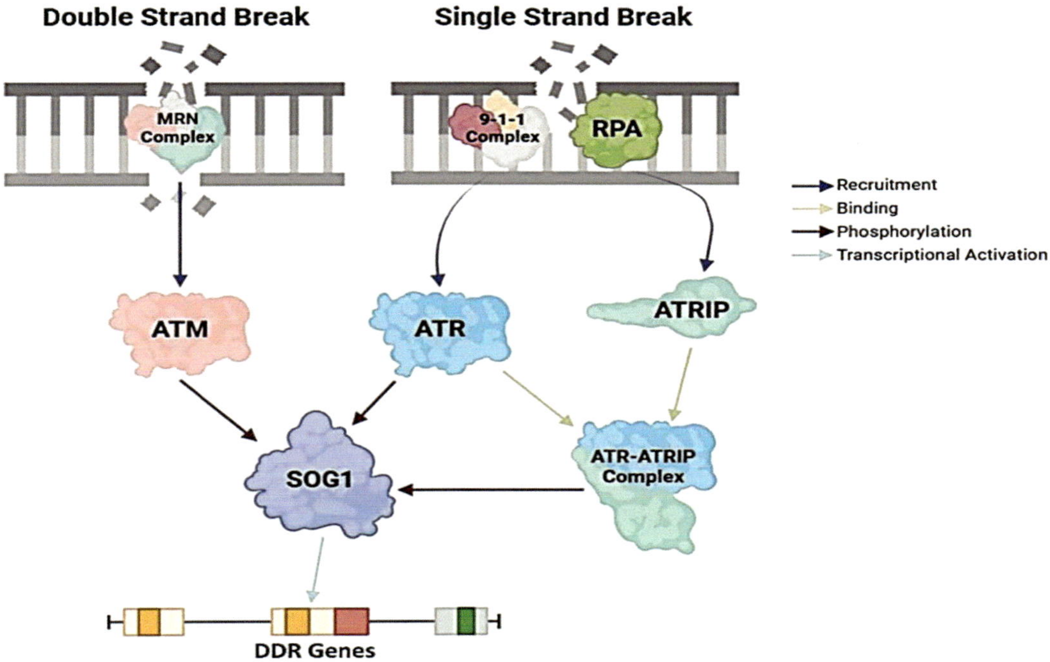

Figure 2. The initial response to DSBs and SSBs. In the case of DSB, ATM is recruited by the MRN complex and the SOG1 protein phosphorylated by ATM transcribes DDR genes. In the case of SSB, if RPA binds to the site of the DNA break, ATRIP is recruited and forms a complex with ATR. With the formation of the ATR-ATRIP complex, SOG1 is phosphorylated and activated, and DDR genes are transcribed. However, when the 9-1-1 complex binds to the site, ATR is recruited and SOG1 is again phosphorylated and activated. As a result, DDR genes are transcribed as well.

Repair of DSB in the NHEJ pathway is initiated by Ku70-Ku80 (Ku heterodimer) binding to the broken ends to prevent further damage (Downs and Jackson 2004; Lee et al. 2016; Mimitou and Symington 2010; Shao et al. 2012). In addition, the Ku heterodimer recruits other proteins required for NHEJ to the relevant DNA region (Ma et al. 2004). The Ku heterodimer then interacts with the kinase protein DNA-PKcs, increasing DNA binding affinity of this protein (West et al. 1998; Yaneva et al. 1997). Another protein recruited by the Ku heterodimer is the inactive Artemis nuclease. After the formation of this complex structure, DNA-PK is activated by autophosphorylation at different sites and transfers a phosphate group to Artemis nuclease, allowing it to switch to the active form (Ding et al. 2003; Goodarzi et al. 2006; Hartley et al. 1995; Niewolik et al. 2006; Niewolik et al. 2017). The 5' and 3' endonuclease activity of activated Artemis is important for the clearance of erroneous structures caused by DNA damage (Goodarzi et al. 2006; Ma et al. 2005; Pawelczak and Turchi 2010). After Artemis cleans up the faulty matches at the damaged site, the gaps are filled by polymerases of Pol X family (Bebenek et al. 2014; Mahajan et al. 2002; Moon et al. 2014). Once the gaps are filled, DNA Ligase IV is recruited to the area to bind to the Ku-DNA complex and join the ends (Ghosh and Raghavan 2021; Wilson et al. 1997). Figure 3 shows the NHEJ pathway assisted DSB repair.

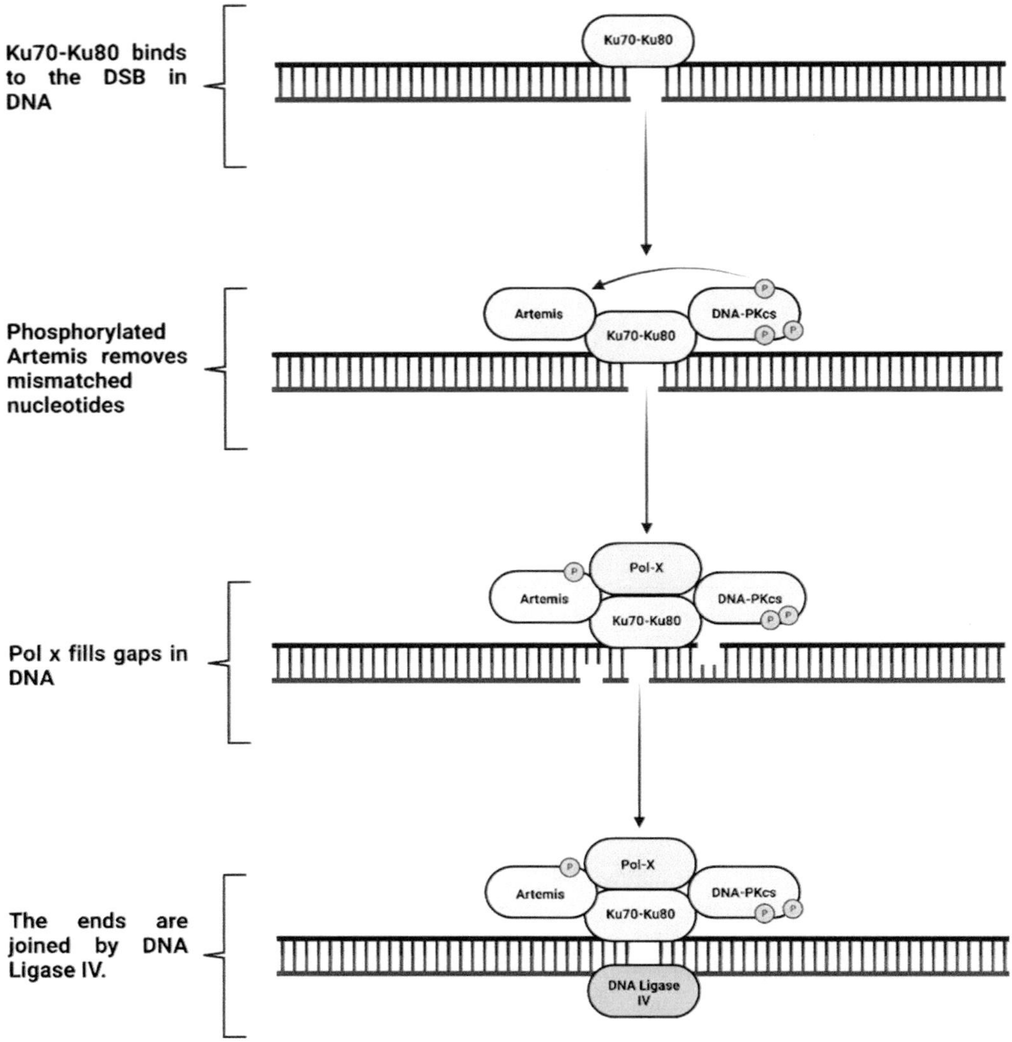

Figure 3. Demonstration of the working process of the NHEJ DNA repair pathway. Following binding of the Ku70-Ku80 heterodimer to the DSB site, DNA-PKcs is recruited to the site and Artemis nuclease removes the mismatches and Pol X fills the gaps, followed by joining of the ends by DNA Ligase IV.

3.1.2 Homology-Directed Repair (HDR)

HDR, another DSB repair mechanism, is less frequent compared to NHEJ (Tang et al. 2019). HDR of a DSB occurs when the damaged DNA and a DNA with approximately 700 bp of homology to the broken DNA are presented (Přibylová and Fischer 2024). In genome editing techniques such as CRISPR, template DNA is exogenously introduced into the cell to increase the probability of HDR. For this pathway to be activated, the MRN complex must bind to the damaged site. After the damaged site is stabilized by MRN complex, C-terminal-binding protein-interacting protein (CtIP) or exonuclease 1-bloom helicase (Exo-1BLM) with 5’ exonuclease activity forms single strand 3’ ends at the damaged site and RPA binds this single stranded DNA (Symington 2014). RAD51 interacts with the molecular chaperones Breast cancer 1 and 2 (BRCA1 and BRCA2) to replace RPA and form filaments on DNA (Salsman et al. 2017). The reconstitution process is initiated by searching for templates or sister chromatids through the 3′ overhang of Rad51 (Buisson et al. 2014). When DNA with homology to the damaged region is found, Proliferating Cell Nuclear Antigen (PCNA) helps to complete the missing fragments (Tang et al. 2019).

3.1.3 Enhancing HDR Efficiency and Precision

The CRISPR system has been improved and used for different purposes through several engineering approaches (Yerlikaya et al. 2024). HDR efficiency can be improved by the use of overlapping sequences (Carroll and Beumer 2014). Appropriate design of homology arms increases HDR efficiency (Renaud et al. 2016). One study in zebrafish found that the use of a linear homologous DNA increased HDR efficiency 10-fold (Irion et al. 2014). There are different approaches to improve HDR effectiveness. One of them is to inhibit key enzymes in the NHEJ pathway. For example, in a study by Srivastava et al. (2012), SCR7, which inhibits DNA Ligase IV, one of the enzymes that play a key role in the NHEJ pathway, was examined and shown to inhibit DNA Ligase IV binding to DNA in human cell lines. In another example, using siRNAs targeting the gene encoding Ku protein, one of the key proteins of the NHEJ pathway, the expression of this protein was reduced and the HDR rate increased up to 1.6–3 fold (Li et al. 2018). In another approach, it may increase HDR activity by inducing the key enzymes of the HDR pathway to perform more efficiently. Overexpression of BRCA1 also increased HDR efficiency (Pinder et al. 2015). There are also studies to increase HDR effectiveness using the CRISPR/Cas system. For example, Charpentier et al. (2018) fused the minimal N-terminal fragment of CtIP with Cas9 and ensured that it was located in the targeted area to be cleaved. As a result, HDR efficiency increased by 2-fold. The CRISPR Primer editing system developed by Anzalone et al. (2019) is one of the newest techniques of homology-based mutation induction. In a study on rice, insertions of up to 11 kb were achieved by Prime-editing using the Lox66-Lox71 or FLP/FRT1 recombinase systems (Sun et al. 2024).

4. Mutation Analysis

There are different approaches to detect mutations generated through CRISPR. Some of these approaches are aimed only at determining whether there is a mutation in the region of interest, while others are aimed at detecting new DNA sequences formed because of the mutation.

4.1 PCR-based Methods

Certain PCR or endonuclease-based methods can be used to detect mutations in DNA. Although these methods do not provide information about the newly formed DNA sequence, they are cost-effective methods in terms of controlling the mutation and proceeding to the next step.

4.1.1 PCR/Restriction Enzyme Assay

Planning for the use of this method should start at the design stage of the CRISPR experiment. The designed sgRNA should be planned to contain the region recognized by a restriction endonuclease enzyme (specifically nucleotides 3 and 4 from the PAM sequence). After CRISPR application, PCR reaction is performed with primers specific to the targeted region and the PCR product obtained can be digested with restriction enzyme targeting the sequence to which sgRNA binds and mutation detection can be performed. Figure 4a shows the PCR/Restriction enzyme assay's workflow. Evans et al. (2018) successfully showed that restriction enzymes can easily be used to screen mutations induced by CRISPR system.

4.1.2 T7E1 Assay

This method was first invented by Mashal et al. (1995). In this method (Fig. 4), after the target regions of wild-type and presumed mutant lines are amplified by PCR, the PCR product of each mutant line is put into the same tube with the WT PCR product, denatured at high temperature and

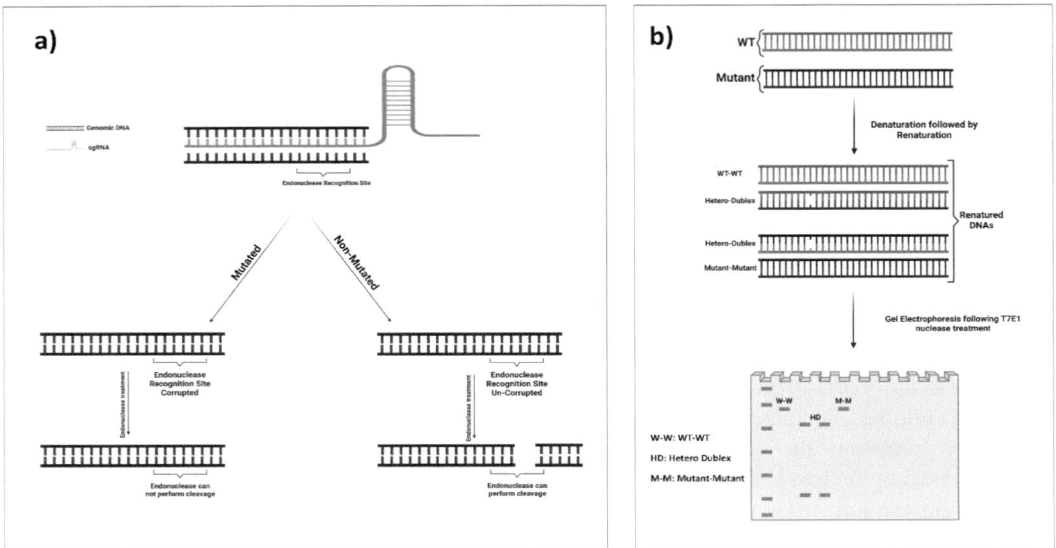

Figure 4. Representation of PCR/Restriction Enzyme Assay (a) and T7E1 Assay (b). In the PCR/restriction assay, the target DNA, which is mutant, is amplified by PCR and then digested with restriction enzyme. However, in the T7E1 assay, after the target regions are amplified in WT and mutant plants, WT and mutant DNA are combined in a tube to form heterodublex structures. Within these structures, there are mismatched DNA regions, and these mismatched regions are cleaved by T7E1 endonuclease.

slowly reduced to room temperature to ensure WT-mutant DNA matches. In case of a mutation in the target region, the mutant DNA that matches the WT DNA will show a mismatch. This mismatch sequence is the target of the T7 endonuclease 1 enzyme and cutting occurs when such sequences are present on the DNA. Figure 4b shows the T7E1 assay's workflow. For example, in a study, Picarella et al. (2024) mutated the *SlHB15* gene via CRISPR/Cas9 and used the T7E1 assay protocol of Mashal et al. (1995) for the detection of mutant lines.

4.2 Sequencing Based Methods

4.2.1 Sanger Sequencing

Sanger sequencing is also a PCR-based method, although it has evolved from its initial application. Sanger and Coulson (1975) introduced the "plus and minus method" for DNA sequencing in 1975. Sanger sequencing is a method for determining the DNA sequence by separating fragments of a specific DNA region of different lengths on a polyacrylamide gel. Since ^{35}S and ^{33}P have lower-energy β emissions, which produce sharper autoradiographic bands and enable the reading of longer sequences, the original radioactive Sanger dideoxy sequencing protocols, which used [α-^{32}P]dATP to label the growing DNA chains, were changed to use [α-^{35}S]dATP or [α-^{33}P]dATP (Shendure et al. 2008). Since then, Sanger sequencing has become safer and a more popular method for sequencing small DNA (< 1 kb) fragments.

4.2.2 High-throughput Methods

In the early 2000s, with the emergence of Next-Generation DNA Sequencing (NGS) methods, the output of genomes has increased considerably and the use of DNA sequencing technologies in various fields has rapidly expanded (Mardis 2013). Illumina technology is one of the most widely used next-generation sequencing technologies today. In this technology, the target DNA is fragmented and fixed to a surface by attaching adapter sequences to the ends. On the surface where it is fixed, the DNA is amplified by a method called bridge amplification. The amplified DNA is sequenced while being amplified by PCR at the fixed site. This method is called sequencing by synthesis.

Over the years, different DNA sequencing technologies have emerged. One of these new sequencing technologies is the Oxford Nanopore technology. In this technology, DNA or RNA, which is single stranded via a motor protein, passes through a reader protein that sits on a conductive membrane (Clarke et al. 2009). During this passage, the ionic current on the membrane is disrupted. By detecting the disturbed ionic current, the sequence of the passing DNA or RNA molecule is determined.

Figure 5 briefly illustrates the working principles of the Illumina and Oxford Nanopore systems. While the Illumina sequencing system sequences by synthesis, the Oxford nanopore system directly sequences the added molecule. These differences have some advantages and disadvantages against each other. For example, the Illumina system provides more accurate results by reading a given DNA fragment more repetitively. However, the library preparation phase is more difficult. Also, the polymerase used to amplify the sequences in the bridge amplification step is prone to errors. Moreover, *de novo* assembly of the short read can be difficult (Hu et al. 2021). The reason for this is the abundance of the repetitive sequences in the eukaryotic genomes. On the other hand, the Oxford nanopore can directly read DNA or RNA molecules. However, since there is no amplification

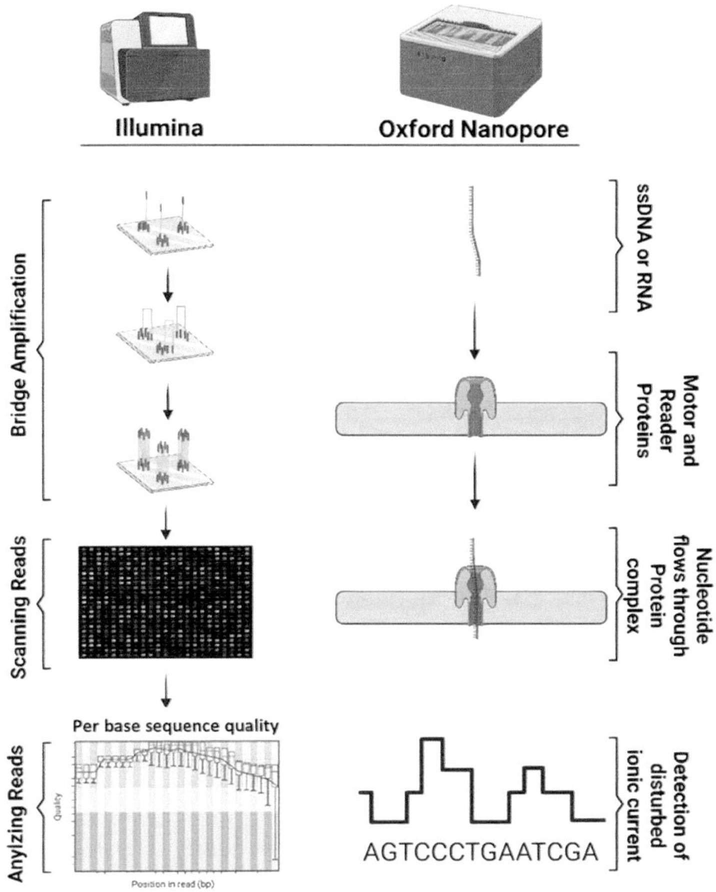

Figure 5. Sequencing workflow of Illumina and Oxford nanopore technologies. Since Illumina is sequencing by synthesis, the DNA or RNA molecule (cDNA) to be sequenced is first amplified on the flow cell to increase signal strength and then the read is performed. However, with Oxford nanopore technology, the molecule is added directly to the flow cell and reading is started.

process, errors can occur during the detection of the ionic current dispersed, resulting in a decrease in reading accuracy (Hu et al. 2021).

4.2.3 Analyzing Sequencing Data

Various tools can be used to detect mutations in DNA fragments sequenced by Sanger or NGS. Which of these bioinformatics tools to use depends on the purpose. In many cases, it may be more efficient to use multiple bioinformatics tools rather than a single tool to analyze the results. For example, DSDecode, developed by Liu et al. (2015b), is a tool that can be used to screen for mutations. This tool has the advantage of showing whether the mutations are homozygous, heterozygous or biallelic. However, it appears to be more primitive than other tools in terms of user interface and presentation of results. There are studies using DSDecode for mutation analysis which increases its reliability (Santosh Kumar et al. 2020; Wang et al. 2016). The Integrative Genomics Viewer (IGV) developed by Robinson et al. (2011). is a tool used for the analysis of NGS data. This tool is particularly useful for analyzing large DNA sequences. IGV has been the subject of numerous studies that are proof of its reliability (Duttke et al. 2024; Janssens et al. 2024; Lucas et al. 2024). For the analysis of sequence files obtained from Sanger sequencing, ICE-Synthego Mutation Analysis Tool is a prominent bioinformatics tool with a robust interface (https://ice.editco.bio/#/). This tool provides the correlation value of the results. It also provides convenience by presenting the results as figures. ICE-Synthego has been used in various studies such as IGV and DSDecode and has proven its reliability (Aydin et al. 2025; Gökdemir et al. 2022; Secgin et al. 2022).

5. Application and Challenges

5.1 Application of CRISPR/Cas Technology in Various Field

ZFN and TALEN techniques have historically been among the most utilized tools in genome editing (Deng et al. 2012). However, since 2013, the CRISPR-Cas system has revolutionized the field, becoming the preferred method due to its superior speed, affordability, and versatility compared to earlier techniques. The integration of genetic engineering principles has propelled the evolution of CRISPR-Cas9 technology, expanding its applications. This versatile tool enables a wide array of genetic modifications, including gene silencing or suppression via knock-out or knockdown, creation of genome-wide loss-of-function libraries, transgenic research through knock-in approaches, gene therapy, chromosomal insertions and deletions, transcriptional regulation, epigenetic modification, and the development of reporter cell lines. Moreover, CRISPR-Cas9 facilitates advanced studies, such as isolating specific DNA regions using engineered DNA-binding molecule-mediated chromatin immunoprecipitation, thus broadening its research potential (Sivanandhan et al. 2016).

5.1.1 Application of CRISPR Technology in Agriculture

CRISPR-Cas9 technology has emerged as a transformative tool for modern agriculture, offering unprecedented precision and efficiency in genetic modifications for crops and livestock (Doudna and Charpentier 2014). This section explores the extensive applications and ongoing advancements of CRISPR-Cas9 within agricultural contexts. One of the primary goals of CRISPR-Cas9 in agriculture is the genetic enhancement of crops to improve key traits such as disease resistance, yield, nutritional value, and resilience to environmental stressors. For instance, Shan et al. successfully employed CRISPR-Cas9 to edit susceptibility genes in rice, resulting in increased resistance to bacterial blight (Shan et al. 2013).

Another significant advantage of CRISPR-Cas9 is its ability to expedite traditional breeding programs. While conventional breeding often requires multiple generations to achieve desired traits, CRISPR-Cas9 enables the precise introduction of targeted genetic changes directly into plant genomes, greatly accelerating the process. A study by Li et al. exemplifies this potential, demonstrating how CRISPR-Cas9 was used to enhance rice grain yield through precise edits to genes regulating grain size (Li et al. 2012; Shan et al. 2013). Beyond individual crop improvement,

CRISPR-Cas9 technology addresses broader global challenges in food security. By engineering crops to withstand pests, diseases, and environmental stresses, CRISPR-Cas9 can significantly boost agricultural productivity while minimizing yield losses. Furthermore, the development of nutritionally fortified crops, such as bioengineered rice with increased vitamin content, could help combat malnutrition and improve global health outcomes (Li et al. 2012; Shan et al. 2013).

The application of CRISPR-Cas9 is not limited to plants; it also holds great promise in livestock breeding. Researchers are exploring its potential to enhance traits such as disease resistance, meat quality, and overall animal welfare. While this field is still in its nascent stages, the technology offers exciting prospects for addressing long-standing challenges in livestock production, including improving sustainability and efficiency (Dawood and Jasim 2021; González et al. 2023). Despite these promising applications, the integration of CRISPR-Cas9 into agriculture is accompanied by challenges. Regulatory hurdles and societal acceptance remain significant barriers, particularly due to concerns surrounding Genetically Modified Organisms (GMOs). Efforts are underway to develop robust regulatory frameworks and engage the public to foster a better understanding of the technology's benefits and safety (Tiwari et al. 2023; Yin et al. 2023). As research and development continue, CRISPR-Cas9 technology stands poised to revolutionize agricultural practices. Its precise genome-editing capabilities pave the way for enhanced crop and livestock traits, improved productivity, and strengthened food security. By addressing regulatory and ethical considerations, CRISPR-Cas9 has the potential to create a more sustainable, resilient, and equitable global food system (Tiwari et al. 2023; Yin et al. 2023).

5.1.2 Application of CRISPR Technology in Medicine

CRISPR-Cas9 technology has become a groundbreaking tool with vast potential to transform medical science (Liang et al. 2015; Mark et al. 2022). This section highlights its diverse applications and notable advancements in medical research and therapeutic strategies. One significant contribution of CRISPR-Cas9 is its utility in exploring gene functions and unraveling disease mechanisms. By enabling precise gene modifications in human cells and model organisms, scientists can identify critical genetic contributors to various disorders. For instance, Shalem et al. used CRISPR-Cas9 to systematically deactivate genes across the human genome, unveiling targets essential for cancer cell survival. The technology also shows great promise in the realm of gene therapy, providing the ability to directly correct disease-causing mutations within patient genomes. For example, CRISPR-Cas9 has been successfully applied to rectify mutations linked to β-thalassemia in stem cells derived from patients (Doudna and Charpentier 2014). These advancements pave the way for targeted, personalized treatments.

In cancer research and therapy, CRISPR-Cas9 has facilitated the disruption of genes driving tumor progression and resistance mechanisms. By silencing specific genes involved in chemotherapy resistance, researchers have demonstrated improved responsiveness of cancer cells to existing treatments (Shalem et al. 2015). This precision-based approach represents a new horizon in combating cancer more effectively. Beyond gene editing, CRISPR-Cas9 has found applications in diagnostics. It has been employed to develop tools like SHERLOCK (Specific High-sensitivity Enzymatic Reporter UnLOCKing) and DETECTR (DNA Endonuclease-Targeted CRISPR Trans Reporter), which provide rapid and accurate detection of viral infections and genetic abnormalities (Shalem et al. 2015). These tools exemplify how CRISPR technology extends beyond therapeutic interventions to enhance disease detection and management. The combination of CRISPR-Cas9 and stem cell technology has also created new opportunities for modeling diseases and investigating therapeutic approaches. For instance, deletion of the PKD2 gene using CRISPR has facilitated the study of Autosomal Dominant Polycystic Kidney Disease (ADPKD) (Trionfini et al. 2020). Similarly, introducing MEN1 gene mutations into induced Pluripotent Stem Cells (iPSCs) has enabled researchers to explore the molecular basis of hypoglycemia (Guo et al. 2017). Additionally, CRISPR-induced knockout of the DGUOK gene in stem cells differentiated into hepatocytes has been utilized to model mitochondrial dysfunction, offering a platform for drug screening and therapeutic discovery (Jing et al. 2018). Despite these advancements, challenges such as off-target effects, efficient delivery mechanisms, and ethical concerns—particularly regarding germline editing—persist. Efforts to enhance the precision and safety of CRISPR-Cas9 are ongoing, while discussions around ethical and regulatory policies continue to shape its application landscape (Doudna and Charpentier 2014). Overall, CRISPR-Cas9 has revolutionized the field of medicine by advancing our understanding of genetic pathways, enabling precise therapeutic interventions, and fostering innovative diagnostic tools. As research progresses and obstacles are addressed, this technology holds immense potential to redefine the treatment of genetic diseases and drive forward personalized medicine (Tsao et al. 2017).

5.1.3 Application of CRISPR Technology in Biotechnology

CRISPR-Cas9 technology has transformed the field of biotechnology by offering versatile tools for genome editing, gene regulation, and synthetic biology (Lander and Chiurillo 2019; Nielsen 2014). This section highlights its wide-ranging applications and ongoing advancements in the field (Nielsen 2014).

One of the most prominent uses of CRISPR-Cas9 in biotechnology lies in precise genome editing. This technology facilitates targeted modifications in the DNA of diverse organisms, including bacteria, yeast, plants, and animals (Mingarro and lí del Olmo 2023; Richardson et al. 2023). Such precision has accelerated the creation of Genetically Modified Organisms (GMOs) with enhanced traits and novel functionalities. For instance, CRISPR-Cas9 was instrumental in engineering a yeast strain capable of producing valuable bio-compounds, exemplifying its utility in industrial biotechnology (Nielsen 2014). In addition to genome editing, CRISPR-Cas9 has revolutionized gene regulation and functional genomics. By enabling the precise modulation of gene expression, it has allowed researchers to delve deeper into gene function and the intricacies of biological systems (Wang and Doudna 2023). Targeting regulatory elements such as promoters and enhancers with CRISPR-Cas9 has provided critical insights into the roles and interactions of specific genes. A notable study demonstrated the technology's utility in activating or silencing genes, thereby advancing the understanding of complex genomic networks (Gilbert et al. 2013). The field of synthetic biology has also greatly benefited from CRISPR-Cas9 innovations. Tools such as CRISPR interference (CRISPRi) and CRISPR activation (CRISPRa) offer unprecedented control over gene expression and cellular pathways. These advancements have paved the way for designing biosensors, optimizing metabolic pathways, and developing groundbreaking biotechnological applications (Gilbert et al. 2013; Nielsen 2014).

CRISPR-Cas9's contributions extend to agricultural biotechnology, where it enhances crop quality and resilience. For example, targeting the OsProDH gene in rice increased proline accumulation, a key factor in protecting plants against abiotic and biotic stresses through the mitigation of Reactive Oxygen Species (ROS). This genetic modification has been shown to improve the thermotolerance of rice, underscoring the potential of CRISPR-Cas9 to address climate-related agricultural challenges (Guo et al. 2020). Furthermore, CRISPR-Cas9 has facilitated the rapid and cost-effective creation of animal models for human disease research. By introducing mutations linked to genetic disorders, researchers can better understand disease mechanisms and develop targeted therapies. A recent study employed CRISPR-Cas9 to generate a pig model of cardiovascular disease, highlighting its role in modeling complex human conditions (Guo et al. 2020). While the potential of CRISPR-Cas9 is vast, its application in biotechnology necessitates careful ethical considerations and adherence to safety and regulatory guidelines. Ongoing discussions aim to address concerns related to unintended effects and the broader societal implications of this transformative technology (Guo et al. 2020; Nielsen 2014). In summary, CRISPR-Cas9 has become an indispensable tool in biotechnology, enabling precise genome modifications, advanced gene regulation, and innovative approaches in synthetic biology. From creating improved GMOs to developing disease models, the technology continues to drive innovation across diverse sectors. As research progresses,

CRISPR-Cas9 is set to redefine the future of biotechnology and its impact on society (Guo et al. 2020).

5.1.4 Application of CRISPR Technology in Nanotechnology

The integration of CRISPR-Cas9 technology with nanotechnology presents significant opportunities for advancing targeted delivery, imaging, and gene regulation applications. By combining the nanoscale precision of nanotechnology with the genome-editing accuracy of CRISPR-Cas9, researchers can explore novel strategies to enhance the efficacy and scope of genetic engineering. Below are key areas where this synergy holds promise: (a) Targeted delivery systems: Nanoparticles can serve as efficient carriers for CRISPR-Cas9 components, including guide RNAs and Cas9 proteins, enabling targeted delivery to specific cells or tissues. This approach not only improves the precision of genome editing but also reduces off-target effects and associated risks. By engineering nanoparticles to respond to specific biological cues, researchers can achieve context-dependent delivery of CRISPR-Cas9, enhancing therapeutic potential (Dilnawaz and Acharya 2023). (b) Imaging and tracking: Nanoparticles embedded with imaging agents offer the ability to monitor the localization and activity of CRISPR-Cas9 components in real-time. This capability facilitates a better understanding of the genome-editing process, providing insights into delivery efficiency, target engagement, and the kinetics of gene modifications. Such advancements could aid in optimizing both therapeutic and research applications (Ma et al. 2022). (c) Gene regulation and epigenetic modifications: Nanotechnology-based platforms provide a versatile toolkit for modulating gene expression and epigenetic states. When integrated with CRISPR-Cas9, these platforms can be tailored to target specific DNA sequences or regulatory factors, enabling precise control over gene activity. This approach has the potential to unlock new therapeutic avenues for treating diseases linked to epigenetic dysregulation and provides a deeper understanding of gene regulatory networks (Xu et al. 2021). Despite the growing interest in this interdisciplinary approach, literature on the combined applications of CRISPR-Cas9 and nanotechnology remains limited. However, emerging studies on CRISPR-based delivery systems, nanoparticle carriers for gene editing, and nanotechnology-enabled gene regulation provide valuable insights into the potential synergies between these fields. Continued research in this domain is essential to realize the full potential of this integration and to address challenges related to delivery efficiency, safety, and scalability (Demirer et al. 2021).

5.1.5 CRISPR/Cas Technology in Bioinformatics and Machine Learning

The convergence of CRISPR-Cas9 technology with bioinformatics and machine learning has catalyzed significant progress in genomic research and data interpretation. By integrating the gene-editing precision of CRISPR-Cas9 with computational methodologies, researchers can explore the intricate structure, function, and regulation of genomes in unprecedented detail. Machine learning, in particular, has transformed the analysis of large-scale genomic datasets, uncovering novel patterns and relationships that were previously inaccessible (Konstantakos et al. 2022; Muhammad Rafid et al. 2020). Predicting Off-Target Effects: Machine learning algorithms have proven invaluable in predicting the off-target activity of CRISPR-Cas9. Using genomic sequence data, these models can identify potential unintended cleavage sites, enhancing the accuracy and specificity of gene-editing experiments. This predictive capability not only refines experimental design but also mitigates risks associated with off-target effects (Alkhnbashi et al. 2020; Makarova and Koonin 2015). CRISPR Screening Data Analysis: High-throughput CRISPR screening generates extensive datasets by perturbing gene functions and observing phenotypic outcomes. Machine learning techniques, such as neural networks and support vector machines, have been applied to these datasets to identify gene functions, map genetic interactions, and provide insights into complex biological pathways (Alipanahi et al. 2023; Kaul et al. 2019; Palermo et al. 2017). Functional Variant Predictions: Machine learning has also advanced the prediction of functional outcomes for genetic variants. By training models on CRISPR-Cas9 experimental data, researchers can assess the impact of specific variants on gene function or disease susceptibility, facilitating the prioritization of disease-associated mutations for further study (Alkhnbashi et al. 2020; Muhammad Rafid et al. 2020).

Optimizing Experimental Design: The application of machine learning to CRISPR-Cas9 experimental planning has streamlined the selection of target sites by considering factors such as on-target efficiency and off-target risk. This approach enhances experimental precision, reduces trial-and-error cycles, and improves success rates, particularly in complex genomic contexts (Alkhnbashi et al. 2020; Makarova and Koonin 2015). The synergy between CRISPR-Cas9, bioinformatics, and machine learning is driving innovation in genomics, paving the way for advancements in precision medicine and targeted therapies. By harnessing computational algorithms to complement experimental data, researchers are expanding the horizons of genomic science and enabling a new era of personalized medical solutions (Konstantakos et al. 2022; Muhammad Rafid et al. 2020).

5.2 Challenges of CRISPR Technology and Mitigation of its Risk

The revolutionary CRISPR-Cas9 system has transformed genetic engineering, offering unprecedented precision and efficiency in editing DNA sequences. However, despite its immense potential, the technology faces several overarching challenges that span across its diverse applications in plants, medicine, clinical research, and biotechnology (Shah 2024; Yerlikaya et al. 2024). One of the most significant obstacles is off-target effects, where unintended genetic modifications may occur, potentially leading to undesired phenotypic outcomes or safety concerns (D'Agostino and D'Aniello 2017). These off-target edits are particularly considered in clinical and medical applications, where the unintended alteration of non-target genes could result in severe side effects or harmful mutations (Gan and Ling 2022). Similarly, in plants, off-target edits might affect traits unrelated to the intended improvement, complicating breeding programs. To mitigate these risks, researchers have developed more precise CRISPR variants such as base editors and prime editors, which allow single-nucleotide modifications with reduced off-target activity (Gan and Ling 2022; Mahfouz et al. 2014).

Another universal challenge is the delivery of CRISPR components to target cells or tissues. Efficient delivery remains a bottleneck across applications, with methods like viral vectors, nanoparticles, and electroporation facing limitations in specificity, efficiency, and scalability (Yang et al. 2022). Recent advances, such as the use of lipid nanoparticles and synthetic RNA delivery systems, have shown promise in improving targeting efficiency and minimizing toxicity. In plants, innovative methods like protoplast transformation, nanotechnology-based delivery, and biolistic approaches are being refined to achieve higher specificity and broader applicability (Li et al. 2015; Yip 2020). Ethical and regulatory considerations also pose significant barriers to CRISPR's broader adoption (Baumann 2016). The rapid pace of CRISPR innovations has outstripped the development of comprehensive regulatory frameworks, leading to uncertainty regarding the approval and deployment of CRISPR-based products. Public perception and ethical debates further complicate the landscape, particularly in clinical and agricultural settings, where the modification of human embryos or food crops raises concerns about safety, equity, and ecological impacts. To address these issues, international guidelines and bioethics committees are actively working to establish transparent policies and engage the public in informed discussions (Shinwari et al. 2018; Tyagi et al. 2020). Finally, scalability and cost-effectiveness remain critical hurdles. While CRISPR shows promise for a wide range of applications, the high costs of research, development, and regulatory approval limit accessibility, particularly in resource-limited settings. Efforts to develop low-cost delivery systems, open-access CRISPR tools, and community-driven research collaborations are helping to make the technology more accessible. Addressing these challenges through ongoing innovation and multidisciplinary collaboration will be key to unlocking the full potential of CRISPR technology in all its applications.

6. Future Directions in DNA Repair Pathway Manipulation for Plant Genome Editing

In plants, DNA repair pathways such as NHEJ, HDR, and alternative end joining mechanisms such as Microhomology-Mediated End Joining (MMEJ) interact in complex and dynamic ways (Manova and Gruszka 2015). These interactions are influenced by various factors, including the developmental stage of the cell, environmental conditions, and the genetic background of the plant. Understanding how these pathways function and interact is critical to improving the precision and efficiency of genome editing in plants. In recent years, advances in molecular biology, genomics, and bioinformatics have significantly deepened our understanding of the complexity of DNA repair mechanisms in plants (Milanowska et al. 2011). These studies have revealed the diversity and functioning of different DNA repair pathways, such as Base Excision Repair (BER), Nucleotide Excision Repair (NER), Homologous Recombination (HR), and NHEJ. However, advances still include some limiting factors (Yeh et al. 2019). Off-target effects, in particular are still ongoing in plant genome editing, as unintended changes can also be found in undesirable genes, leading to unpredictable phenotypes, reduced plant fitness, or even loss of desired traits. These effects are of particular concern in plants due to their large and complex genomes, which often contain repetitive sequences and polyploidy (Graham et al. 2020). Despite the development of advanced genome editing tools such as CRISPR-Cas9, CRISPR-Cas12, and base editors, the specificity of these systems is not absolute. Off-target activity can occur due to: imperfect binding of gRNAs to sequences with partial complementarity, the inherent complexity of Cas nucleases in recognizing similar sequences, and chromatin accessibility that can make certain off-target sites more prone to editing (Asmamaw Mengstie et al. 2024).

NHEJ, on the other hand, is the predominant repair pathway in plants and is active throughout the cell cycle. It is a fast and efficient process, but it often produces small insertions or deletions (indels) that can lead to unintended mutations at the repair site. HDR is a more precise repair mechanism that uses a homologous DNA template to guide the repair process (Que et al. 2019). However, HDR is less efficient in plants and is active mainly in the S and G2 phases of the cell cycle when homologous templates are available. An alternative splicing pathway, MMEJ, relies on short homologous sequences (microhomologies) near the break site and is usually associated with larger deletions or rearrangements (García-Medel et al. 2019; Shen and Li 2022). It is essential to investigate how these pathways compete or cooperate in different plant tissues and under various conditions. For example, HDR is more active in dividing cells, such as in meristematic tissues, while NHEJ is dominant in differentiated cells. Environmental factors such as light, temperature, and stress conditions also affect the activity of these pathways. For example, studies have shown that low temperatures can increase HDR efficiency in some plant species, such as rice (Tang et al. 2020). By understanding the dynamics of these pathways, researchers can develop strategies to promote specific repair mechanisms at the right time or in the right tissue, increasing the precision of genome editing. Identifying and manipulating plant-specific factors that regulate DNA repair pathway selection is another important area of research. Key proteins involved in these pathways, such as KU70/80 and LIG4 for NHEJ or RAD51 and RAD54 for HDR, play critical roles in determining which repair mechanism is used. Modulating the expression or activity of these proteins may help shift the balance toward HDR or other desired pathways (Kelm et al. 2022). For example, inhibiting NHEJ components such as KU70/80 has been shown to increase HDR efficiency in some studies (Li et al. 2020). Additionally, plant-specific factors, such as those affected by hormonal signaling or stress responses, may provide unique opportunities to control repair pathway selection in plants.

6.1 Integration with Other Technologies

Machine learning algorithms have significantly advanced the field of genome editing, particularly in predicting the efficacy and off-target effects of single-guide RNAs (sgRNAs) used in CRISPR/Cas9 systems (Lan et al. 2020). These algorithms analyze large datasets comprising

Table 3. Artificial intelligence tools in genome editing.

Program Name	Purpose	Input	Methodology	Output	Key Features	Source
DeepCRISPR	Predict off-target effects of sgRNAs	Genomic sequences, sgRNA sequences	Deep learning neural networks	Probability of off-target effects, sgRNA design	Sequence feature extraction Thermodynamic analysis	(Chuai et al. 2018)
CRISTA	Predict on-target efficiency of sgRNAs	sgRNA sequences, target site sequences	Gradient boosting machine learning	Efficiency score for sgRNAs	Sequence motif analysis Uses Random Forest or XGBoost	(Abadi et al. 2017)
DeepHF/ DeepCRISTL	Design high-fidelity Cas9 variants	Cas9 protein sequences, sgRNA sequences, target DNA sequences	Deep learning	Recommendations for Cas9 modifications	Mutation impact analysis Trained on CRISPR screen data	(Elkayam and Orenstein 2022)

genomic sequences and the outcomes of editing events. The algorithms and features are described in Table 3. Tools such as DeepCRISPR, CRISTA, and DeepHF leverage sequence-based features to enhance the design of sgRNAs (also see in Table 1), focusing on achieving high on-target activity while reducing unintended off-target cleavage sites (Liu et al. 2020). For instance, DeepCRISPR uses deep learning to predict sgRNA off-target propensity by training on sequence features of sgRNA-DNA pairs, reporting enhancements in prediction accuracy over traditional methods (Chuai et al. 2018). CRISTA, another machine learning model, focuses on improving on-target efficiency through sequence analysis (Abadi et al. 2017), and DeepHF aims at identifying high-fidelity Cas9 variants by learning from large-scale CRISPR screen data (Elkayam and Orenstein 2022).

Machine learning algorithms have significantly advanced the field of genome editing, particularly in predicting the efficacy and off-target effects of sgRNAs used in CRISPR/Cas9 systems (Lan et al. 2020). These algorithms analyze large datasets comprising genomic sequences and the outcomes of editing events. The algorithms and features are described in Table 2. Tools such as DeepCRISPR, CRISTA, and DeepHF leverage sequence-based features to enhance the design of sgRNAs, focusing on achieving high on-target activity while reducing unintended off-target cleavage sites (Liu et al. 2020). For instance, DeepCRISPR uses deep learning to predict sgRNA off-target propensity by training on sequence features of sgRNA-DNA pairs, reporting enhancements in prediction accuracy over traditional methods (Chuai et al. 2018). CRISTA, another machine learning model, focuses on improving on-target efficiency through sequence analysis (Abadi et al. 2017), and DeepHF aims at identifying high-fidelity Cas9 variants by learning from large-scale CRISPR screen data (Elkayam and Orenstein 2022).

AI models have begun to predict how DNA repair pathways might respond to specific CRISPR-induced breaks. This includes determining whether the repair will proceed through non-NHEJ or HDR, which have different implications for the final genetic outcome (Halper et al. 2020). Such predictive capabilities allow for the design of editing strategies that favor a specific repair pathway, enhancing the precision of genome editing. For example, models can suggest sgRNA sequences that might preferentially engage HDR, which is generally considered more precise as it can incorporate a template for repair, thus potentially reducing unwanted mutations (o'Brien et al. 2019). In conclusion, the synergy between machine learning and CRISPR technology has transformed the landscape of genome editing by providing predictive tools that enhance sgRNA design for both efficacy and specificity. By understanding and manipulating DNA repair pathways, AI can facilitate the creation of highly customized editing strategies that account for biological variability, potentially leading to safer and more effective therapeutic interventions.

7. Conclusion

This chapter provided a detailed examination of CRISPR technology as a revolutionary tool for genome editing. Organized into distinct subsections, it explored various facets of the technology, including its principles, applications, and potential across diverse fields such as agriculture, medicine, machine learning, nanotechnology, bioinformatics, and biotechnology. Additionally, the alternative recent sgRNA design strategies and AI tools are also discussed to overcome challenges in CRISPR applications. The discussion underscored not only the transformative potential of CRISPR-Cas9 but also the challenges, ethical dilemmas, and ongoing debates surrounding its use. By synthesizing current knowledge and advancements, this work enhances our understanding of CRISPR-Cas9 and its far-reaching implications, offering valuable insights for future innovations and the responsible integration of this groundbreaking technology into scientific and practical applications.

References

Abadi, S., W.X. Yan, D. Amar and I. Mayrose. 2017. A machine learning approach for predicting crispr-cas9 cleavage efficiencies and patterns underlying its mechanism of action. PLoS Computational Biology 13(10): e1005807.

Alipanahi, R., L. Safari and A. Khanteymoori. 2023. Crispr genome editing using computational approaches: A survey. Frontiers in Bioinformatics 2: 1001131.

Alkhnbashi, O.S., T. Meier, A. Mitrofanov, R. Backofen and B. Voss. 2020. Crispr-cas bioinformatics. Methods 172(3-11).

Allen, F., L. Crepaldi, C. Alsinet, A.J. Strong, V. Kleshchevnikov, P. De Angeli et al. 2019. Predicting the mutations generated by repair of cas9-induced double-strand breaks. Nature Biotechnology 37(1): 64–72.

Anzalone, A.V., P.B. Randolph, J.R. Davis, A.A. Sousa, L.W. Koblan, J.M. Levy et al. 2019. Search-and-replace genome editing without double-strand breaks or donor DNA. Nature 576(7785): 149–157.

Asmamaw Mengstie, M., M. Teshome Azezew, T. Asmamaw Dejenie, A.A. Teshome, F. Tadele Admasu, A. Behaile Teklemariam et al. 2024. Recent advancements in reducing the off-target effect of crispr-cas9 genome editing. Biologics: Targets and Therapy 21–28.

Aydin, A., B.A. Yerlikaya, S. Yerlikaya, N.N. Yilmaz and M. Kavas. 2025. Crispr-mediated mutation of cytokinin signaling genes (slhp2 and slhp3) in tomato: Morphological, physiological, and molecular characterization. The Plant Genome 18(1): e20542.

Bae, S., J. Park and J.-S. Kim. 2014. Cas-offinder: A fast and versatile algorithm that searches for potential off-target sites of cas9 RNA-guided endonucleases. Bioinformatics 30(10): 1473–1475.

Banin, S., L. Moyal, S.-Y. Shieh, Y. Taya, C. Anderson, L. Chessa et al. 1998. Enhanced phosphorylation of p53 by atm in response to DNA damage. Science 281(5383): 1674–1677.

Bao, Z., H. Xiao, J. Liang, L. Zhang, X. Xiong, N. Sun et al. 2015. Homology-integrated crispr–cas (hi-crispr) system for one-step multigene disruption in Saccharomyces cerevisiae. ACS Synthetic Biology 4(5): 585–594.

Baumann, M. 2016. Crispr/cas9 genome editing–new and old ethical issues arising from a revolutionary technology. NanoEthics 10(139-159).

Bebenek, K., L.C. Pedersen and T.A. Kunkel. 2014. Structure–function studies of DNA polymerase λ. Biochemistry 53(17): 2781–2792.

Bhatia, S. and S.K. Yadav. 2023. Crispr-cas for genome editing: Classification, mechanism, designing and applications. International Journal of Biological Macromolecules 238: 124054.

Bhokisham, N., E. Laudermilch, L.L. Traeger, T.D. Bonilla, M. Ruiz-Estevez and J.R. Becker. 2023. Crispr-cas system: The current and emerging translational landscape. Cells 12(8): 1103.

Bourbousse, C., N. Vegesna and J.A. Law. 2018. Sog1 activator and myb3r repressors regulate a complex DNA damage network in arabidopsis. Proceedings of the National Academy of Sciences 115(52): E12453–E12462.

Briner, A.E., P.D. Donohoue, A.A. Gomaa, K. Selle, E.M. Slorach, C.H. Nye et al. 2014. Guide RNA functional modules direct cas9 activity and orthogonality. Molecular Cell 56(2): 333–339.

Buisson, R., J. Niraj, J. Pauty, R. Maity, W. Zhao, Y. Coulombe et al. 2014. Breast cancer proteins palb2 and brca2 stimulate polymerase η in recombination-associated DNA synthesis at blocked replication forks. Cell Reports 6(3): 553–564.

Canman, C.E., D.-S. Lim, K.A. Cimprich, Y. Taya, K. Tamai, K. Sakaguchi et al. 1998. Activation of the atm kinase by ionizing radiation and phosphorylation of p53. Science 281(5383): 1677–1679.

Cao, Q., J. Ma, C.-H. Chen, H. Xu, Z. Chen, W. Li et al. 2017. Crispr-focus: A web server for designing focused crispr screening experiments. PLoS One 12(9): e0184281.

Caron, M.-C., A.K. Sharma, J. O'Sullivan, L.R. Myler, M.T. Ferreira, A. Rodrigue et al. 2019. Poly (adp-ribose) polymerase-1 antagonizes DNA resection at double-strand breaks. Nature Communications 10(1): 2954.

Carroll, D. and K.J. Beumer. 2014. Genome engineering with talens and zfns: Repair pathways and donor design. Methods 69(2): 137–141.

Casini, A., M. Olivieri, G. Petris, C. Montagna, G. Reginato, G. Maule et al. 2018. A highly specific spcas9 variant is identified by *in vivo* screening in yeast. Nature Biotechnology 36(3): 265–271.

Chang, H.H., G. Watanabe, C.A. Gerodimos, T. Ochi, T.L. Blundell, S.P. Jackson et al. 2016. Different DNA end configurations dictate which nhej components are most important for joining efficiency. Journal of Biological Chemistry 291(47): 24377–24389.

Chari, R., P. Mali, M. Moosburner and G.M. Church. 2015. Unraveling crispr-cas9 genome engineering parameters via a library-on-library approach. Nature Methods 12(9): 823–826.

Chari, R., N.C. Yeo, A. Chavez and G.M. Church. 2017. Sgrna scorer 2.0: A species-independent model to predict crispr/cas9 activity. ACS Synthetic Biology 6(5): 902–904.

Charpentier, M., A. Khedher, S. Menoret, A. Brion, K. Lamribet, E. Dardillac et al. 2018. Ctip fusion to cas9 enhances transgene integration by homology-dependent repair. Nature Communications 9(1): 1133.

Chen, B., L.A. Gilbert, B.A. Cimini, J. Schnitzbauer, W. Zhang, G.-W. Li et al. 2013. Dynamic imaging of genomic loci in living human cells by an optimized crispr/cas system. Cell 155(7): 1479–1491.

Chen, W., A. McKenna, J. Schreiber, M. Haeussler, Y. Yin, V. Agarwal et al. 2019. Massively parallel profiling and predictive modeling of the outcomes of crispr/cas9-mediated double-strand break repair. Nucleic Acids Research 47(15): 7989–8003.

Chiruvella, K.K., R. Sebastian, S. Sharma, A.A. Karande, B. Choudhary and S.C. Raghavan. 2012. Time-dependent predominance of nonhomologous DNA end-joining pathways during embryonic development in mice. Journal of Molecular Biology 417(3): 197–211.

Chuai, G., H. Ma, J. Yan, M. Chen, N. Hong, D. Xue et al. 2018. Deepcrispr: Optimized crispr guide RNA design by deep learning. Genome Biology 19(1-18).

Clarke, J., H.-C. Wu, L. Jayasinghe, A. Patel, S. Reid and H. Bayley. 2009. Continuous base identification for single-molecule nanopore DNA sequencing. Nature Nanotechnology 4(4): 265–270.

Concordet, J.-P. and M. Haeussler. 2018. Crispor: Intuitive guide selection for crispr/cas9 genome editing experiments and screens. Nucleic Acids Research 46(W1): W242–W245.

Cong, L., F.A. Ran, D. Cox, S. Lin, R. Barretto, N. Habib et al. 2013. Multiplex genome engineering using crispr/cas systems. Science 339(6121): 819–823.

Cortez, D., S. Guntuku, J. Qin and S.J. Elledge. 2001. Atr and atrip: Partners in checkpoint signaling. Science 294(5547): 1713–1716.

D'Agostino, Y. and S. D'Aniello. 2017. Molecular basis, applications and challenges of crispr/cas9: A continuously evolving tool for genome editing. Briefings in Functional Genomics 16(4): 211–216.

Dahlman, J.E., O.O. Abudayyeh, J. Joung, J.S. Gootenberg, F. Zhang and S. Konermann. 2015. Orthogonal gene knockout and activation with a catalytically active cas9 nuclease. Nature Biotechnology 33(11): 1159–1161.

Dawood, A.A. and B.I. Jasim. 2021. The crispr genome editing process is an effective advancement of short-term cancer treatment. Research Journal of Pharmaceutical Dosage Forms and Technology 13(1): 54–56.

Demirer, G.S., T.N. Silva, C.T. Jackson, J.B. Thomas, D.W. Ehrhardt, S.Y. Rhee et al. 2021. Nanotechnology to advance crispr–cas genetic engineering of plants. Nature Nanotechnology 16(3): 243–250.

Deng, L., C.S. Kenchappa, X. Peng, Q. She and R.A. Garrett. 2012. Modulation of crispr locus transcription by the repeat-binding protein cbp1 in sulfolobus. Nucleic Acids Research 40(6): 2470–2480.

Deriano, L. and D.B. Roth. 2013. Modernizing the nonhomologous end-joining repertoire: Alternative and classical nhej share the stage. Annual Review of Genetics 47(1): 433–455.

Dilnawaz, F. and S. Acharya. 2023. Nanoparticle-based crispr/cas delivery: An emerging tactic for cancer therapy. Current Medicinal Chemistry 30(31): 3562–3581.

Ding, Q., Y.V. Reddy, W. Wang, T. Woods, P. Douglas, D.A. Ramsden et al. 2003. Autophosphorylation of the catalytic subunit of the DNA-dependent protein kinase is required for efficient end processing during DNA double-strand break repair. Molecular and Cellular Biology 23(16): 5836–5848.

Doench, J.G., E. Hartenian, D.B. Graham, Z. Tothova, M. Hegde, I. Smith et al. 2014. Rational design of highly active sgrnas for crispr-cas9–mediated gene inactivation. Nature Biotechnology 32(12): 1262–1267.

Doench, J.G., N. Fusi, M. Sullender, M. Hegde, E.W. Vaimberg, K.F. Donovan et al. 2016. Optimized sgrna design to maximize activity and minimize off-target effects of crispr-cas9. Nature Biotechnology 34(2): 184–191.

Dong, C., L. Jiang, S. Xu, L. Huang, J. Cai, J. Lian et al. 2020. A single cas9-vpr nuclease for simultaneous gene activation, repression, and editing in Saccharomyces cerevisiae. ACS Synthetic Biology 9(9): 2252–2257.

Doucet-Chabeaud, G., C. Godon, C. Brutesco, G. De Murcia and M. Kazmaier. 2001. Ionising radiation induces the expression of parp-1 and parp-2 genes in arabidopsis. Molecular Genetics and Genomics 265(954-963).

Doudna, J.A. and E. Charpentier. 2014. The new frontier of genome engineering with crispr-cas9. Science 346(6213): 1258096.

Downs, J.A. and S.P. Jackson. 2004. A means to a DNA end: The many roles of ku. Nature Reviews Molecular Cell Biology 5(5): 367–378.

Duttke, S.H., C. Guzman, M. Chang, N.P. Delos Santos, B.R. McDonald, J. Xie et al. 2024. Position-dependent function of human sequence-specific transcription factors. Nature 631(8022): 891–898.

Elkayam, S. and Y. Orenstein. 2022. Deepcristl: Deep transfer learning to predict crispr/cas9 functional and endogenous on-target editing efficiency. Bioinformatics 38.

Evans, B.A., O.L. Smith, E.S. Pickerill, M.K. York, K.J. Buenconsejo, A.E. Chambers et al. 2018. Restriction digest screening facilitates efficient detection of site-directed mutations introduced by crispr in c. Albicans ume6. PeerJ. 6(e4920).

Ferguson, D.O. and F.W. Alt. 2001. DNA double strand break repair and chromosomal translocation: Lessons from animal models. Oncogene 20(40): 5572–5579.

Fu, Y., J.D. Sander, D. Reyon, V.M. Cascio and J.K. Joung. 2014. Improving crispr-cas nuclease specificity using truncated guide RNAs. Nature Biotechnology 32(3): 279–284.

Fusi, N., I. Smith, J. Doench and J. Listgarten. 2015. *In silico* predictive modeling of crispr/cas9 guide efficiency. BioRxiv.021568.

Gan, W.C. and A.P. Ling. 2022. Crispr/cas9 in plant biotechnology: Applications and challenges. BioTechnologia 103(1): 81.

García-Medel, P.L., N. Baruch-Torres, A. Peralta-Castro, C.H. Trasviña-Arenas, A. Torres-Larios and L.G. Brieba. 2019. Plant organellar DNA polymerases repair double-stranded breaks by microhomology-mediated end-joining. Nucleic Acids Research 47(6): 3028–3044.

Ghosh, D. and S.C. Raghavan. 2021. Nonhomologous end joining: New accessory factors fine tune the machinery. Trends in Genetics 37(6): 582–599.

Gilbert, L.A., M.A. Horlbeck, B. Adamson, J.E. Villalta, Y. Chen, E.H. Whitehead et al. 2014. Genome-scale crispr-mediated control of gene repression and activation. Cell 159(3): 647–661.

Gilbert, L.A., M.H. Larson, L. Morsut, Z. Liu, G.A. Brar, S.E. Torres et al. 2013. Crispr-mediated modular RNA-guided regulation of transcription in eukaryotes. Cell 154(2): 442–451.

González, M.N., G.A. Massa, M. Andersson, L. Storani, N. Olsson, C.A. Décima Oneto et al. 2023. Crispr/cas9 technology for potato functional genomics and breeding. pp. 333–361. *In*: Plant Genome Engineering: Methods and Protocols. Springer.

Goodarzi, A.A., Y. Yu, E. Riballo, P. Douglas, S.A. Walker, R. Ye et al. 2006. DNA-pk autophosphorylation facilitates artemis endonuclease activity. The EMBO Journal 25(16): 3880–3889.

Gökdemir, G., Z. Seçgin, S. Uluisik and M. Kavas. 2022. Crispr/cas9 knock-out of slphd_ms1 (solyc04g008420) gene results in complete male sterility in tomato. Plant Growth Regulation 98(2): 329–341.

Graham, N., G.B. Patil, D.M. Bubeck, R.C. Dobert, K.C. Glenn, A.T. Gutsche et al. 2020. Plant genome editing and the relevance of off-target changes. Plant Physiology 183(4): 1453–1471.

Guo, D., H. Liu, G. Gao, Y. Liu, Y. Zhuang, F. Yang et al. 2017. Creating a patient carried men1 gene point mutation on wild type ipscs locus mediated by crispr/cas9 and ssodn. Stem Cell Research 18(67-69).

Guo, M., X. Zhang, J. Liu, L. Hou, H. Liu and X. Zhao. 2020. Osprodh negatively regulates thermotolerance in rice by modulating proline metabolism and reactive oxygen species scavenging. Rice 13(1-5).

Haeussler, M., K. Schönig, H. Eckert, A. Eschstruth, J. Mianné, J.-B. Renaud et al. 2016. Evaluation of off-target and on-target scoring algorithms and integration into the guide RNA selection tool crispor. Genome Biology 17(1-12).

Halper, S.M., A. Hossain and H.M. Salis. 2020. Synthesis success calculator: Predicting the rapid synthesis of DNA fragments with machine learning. ACS Synthetic Biology 9(7): 1563–1571.

Hartley, K.O., D. Gell, G.C. Smith, H. Zhang, N. Divecha, M.A. Connelly et al. 1995. DNA-dependent protein kinase catalytic subunit: A relative of phosphatidylinositol 3-kinase and the ataxia telangiectasia gene product. Cell 82(5): 849–856.

Heigwer, F., G. Kerr and M. Boutros. 2014. E-crisp: Fast crispr target site identification. Nature Methods 11(2): 122–123.

Heitzeberg, F., I.P. Chen, F. Hartung, N. Orel, K.J. Angelis and H. Puchta. 2004. The rad17 homologue of arabidopsis is involved in the regulation of DNA damage repair and homologous recombination. The Plant Journal 38(6): 954–968.

Herbst, J., Q.-Q. Li and L. De Veylder. 2024. Mechanistic insights into DNA damage recognition and checkpoint control in plants. Nature Plants 10(4): 539–550.

Hong, J.-P., M.Y. Byun, K. An, S.-J. Yang, G. An and W.T. Kim. 2010. Osku70 is associated with developmental growth and genome stability in rice. Plant Physiology 152(1): 374–387.

Horlbeck, M.A., L.B. Witkowsky, B. Guglielmi, J.M. Replogle, L.A. Gilbert, J.E. Villalta et al. 2016. Nucleosomes impede cas9 access to DNA *in vivo* and *in vitro*. elife 5(e12677).

Housden, B.E., A.J. Valvezan, C. Kelley, R. Sopko, Y. Hu, C. Roesel et al. 2015. Identification of potential drug targets for tuberous sclerosis complex by synthetic screens combining crispr-based knockouts with rnai. Science Signaling 8(393): rs9–rs9.

Hsu, P.D., D.A. Scott, J.A. Weinstein, F.A. Ran, S. Konermann, V. Agarwala et al. 2013. DNA targeting specificity of RNA-guided cas9 nucleases. Nature Biotechnology 31(9): 827–832.

Hu, T., N. Chitnis, D. Monos and A. Dinh. 2021. Next-generation sequencing technologies: An overview. Human Immunology 82(11): 801–811.

Hwang, G.-H., J. Park, K. Lim, S. Kim, J. Yu, E. Yu et al. 2018. Web-based design and analysis tools for crispr base editing. BMC Bioinformatics 19(1-7).

Ilahibaks, N.F., M.J. Hulsbos, Z. Lei, P. Vader and J.P. Sluijter. 2022. Enabling precision medicine with crispr-cas genome editing technology: A translational perspective. pp. 315–339. *In*: Genome Editing in Cardiovascular and Metabolic Diseases. Springer.

Irion, U., J. Krauss and C. Nüsslein-Volhard. 2014. Precise and efficient genome editing in zebrafish using the crispr/cas9 system. Development 141(24): 4827–4830.

Jackson, S.P. 2002. Sensing and repairing DNA double-strand breaks. Carcinogenesis 23(5): 687–696.

Jackson, S.P. and J. Bartek. 2009. The DNA-damage response in human biology and disease. Nature 461(7267): 1071–1078.

Jacquin, A.L., D.T. Odom and M. Lukk. 2019. Crisflash: Open-source software to generate crispr guide rnas against genomes annotated with individual variation. Bioinformatics 35(17): 3146–3147.

Janssens, D.H., J.E. Greene, S.J. Wu, C.A. Codomo, S.S. Minot, S.N. Furlan et al. 2024. Scalable single-cell profiling of chromatin modifications with scicut&tag. Nature Protocols 19(1): 83–112.

Jinek, M., K. Chylinski, I. Fonfara, M. Hauer, J.A. Doudna and E. Charpentier. 2012. A programmable dual-rna–guided DNA endonuclease in adaptive bacterial immunity. Science 337(6096): 816–821.

Jing, R., J.L. Corbett, J. Cai, G.C. Beeson, C.C. Beeson, S.S. Chan et al. 2018. A screen using ipsc-derived hepatocytes reveals nad+ as a potential treatment for mtdna depletion syndrome. Cell Reports 25(6): 1469–1484. e1465.

Kaul, T., N.M. Raman, M. Eswaran, A. Thangaraj, R. Verma, S.K. Sony et al. 2019. Data mining by pluralistic approach on crispr gene editing in plants. Frontiers in Plant Science 10(801).

Kelm, J.M., A. Samarbakhsh, A. Pillai, P.S. VanderVere-Carozza, H. Aruri, D.S. Pandey et al. 2022. Recent advances in the development of non-pikks targeting small molecule inhibitors of DNA double-strand break repair. Frontiers in Oncology 12(850883).

Kiani, S., A. Chavez, M. Tuttle, R.N. Hall, R. Chari, D. Ter-Ovanesyan et al. 2015. Cas9 grna engineering for genome editing, activation and repression. Nature Methods 12(11): 1051–1054.

Klepikova, A.V., A.S. Kasianov, E.S. Gerasimov, M.D. Logacheva and A.A. Penin. 2016. A high resolution map of the arabidopsis thaliana developmental transcriptome based on RNA-seq profiling. The Plant Journal 88(6): 1058–1070.

Komor, A.C., Y.B. Kim, M.S. Packer, J.A. Zuris and D.R. Liu. 2016. Programmable editing of a target base in genomic DNA without double-stranded DNA cleavage. Nature 533(7603): 420–424.

Konstantakos, V., A. Nentidis, A. Krithara and G. Paliouras. 2022. Crispr–cas9 grna efficiency prediction: An overview of predictive tools and the role of deep learning. Nucleic Acids Research 50(7): 3616–3637.

Kumar, K.R.R. 2023. Lost in the bloom: DNA-pkcs in green plants. Frontiers in Plant Science 14(1231678).

Labun, K., T.G. Montague, J.A. Gagnon, S.B. Thyme and E. Valen. 2016. Chopchop v2: A web tool for the next generation of crispr genome engineering. Nucleic Acids Research 44(W1): W272–W276.

Labun, K., T.G. Montague, M. Krause, Y.N. Torres Cleuren, H. Tjeldnes and E. Valen. 2019. Chopchop v3: Expanding the crispr web toolbox beyond genome editing. Nucleic Acids Research 47(W1): W171–W174.

Lan, J., Y. Cui, X. Wang, G. Song and J. Lou. 2020. Optimized sgrna design by deep learning to balance the off-target effects and on-target activity of crispr/cas9. bioRxiv.2020.2003. 2004.976340.

Lander, N. and M.A. Chiurillo. 2019. State-of-the-art crispr/cas9 technology for genome editing in trypanosomatids. Journal of Eukaryotic Microbiology 66(6): 981–991.

Lee, K.-J., J. Saha, J. Sun, K.R. Fattah, S.-C. Wang, B. Jakob et al. 2016. Phosphorylation of ku dictates DNA double-strand break (dsb) repair pathway choice in s phase. Nucleic Acids Research 44(4): 1732–1745.

Lees-Miller, J.P., A. Cobban, P. Katsonis, A. Bacolla, S.E. Tsutakawa, M. Hammel et al. 2021. Uncovering DNA-pkcs ancient phylogeny, unique sequence motifs and insights for human disease. Progress in Biophysics and Molecular Biology 163(87-108).

Lei, Y., L. Lu, H.-Y. Liu, S. Li, F. Xing and L.-L. Chen. 2014. Crispr-p: A web tool for synthetic single-guide RNA design of crispr-system in plants. Molecular Plant 7(9): 1494–1496.

Li, G., D. Liu, X. Zhang, R. Quan, C. Zhong, J. Mo et al. 2018. Suppressing ku70/ku80 expression elevates homology-directed repair efficiency in primary fibroblasts. The International Journal of Biochemistry & Cell Biology 99(154-160).

Li, G., X. Zhang, H. Wang, D. Liu, Z. Li, Z. Wu et al. 2020. Increasing crispr/cas9-mediated homology-directed DNA repair by histone deacetylase inhibitors. The International Journal of Biochemistry & Cell Biology 125(105790).

Li, J.-F., J.E. Norville, J. Aach, M. McCormack, D. Zhang, J. Bush et al. 2013. Multiplex and homologous recombination–mediated genome editing in arabidopsis and nicotiana benthamiana using guide RNA and cas9. Nature Biotechnology 31(8): 688–691.

Li, L., Z.-Y. He, X.-W. Wei, G.-P. Gao and Y.-Q. Wei. 2015. Challenges in crispr/cas9 delivery: Potential roles of nonviral vectors. Human Gene Therapy 26(7): 452–462.

Li, T., B. Liu, M.H. Spalding, D.P. Weeks and B. Yang. 2012. High-efficiency talen-based gene editing produces disease-resistant rice. Nature Biotechnology 30(5): 390–392.

Li, T., Y. Yang, H. Qi, W. Cui, L. Zhang, X. Fu et al. 2023. Crispr/cas9 therapeutics: Progress and prospects. Signal Transduction and Targeted Therapy 8(1): 36.

Liang, P., Y. Xu, X. Zhang, C. Ding, R. Huang, Z. Zhang et al. 2015. Crispr/cas9-mediated gene editing in human tripronuclear zygotes. Protein & Cell 6(5): 363–372.

Lieber, M.R., K. Yu and S.C. Raghavan. 2006. Roles of nonhomologous DNA end joining, v (d) j recombination, and class switch recombination in chromosomal translocations. DNA Repair 5(9-10): 1234–1245.

Lieber, M.R. 2010. The mechanism of double-strand DNA break repair by the nonhomologous DNA end-joining pathway. Annual Review of Biochemistry 79(1): 181–211.

Listgarten, J., M. Weinstein, B.P. Kleinstiver, A.A. Sousa, J.K. Joung, J. Crawford et al. 2018. Prediction of off-target activities for the end-to-end design of crispr guide rnas. Nature Biomedical Engineering 2(1): 38–47.

Liu, H., Y. Ding, Y. Zhou, W. Jin, K. Xie and L.-L. Chen. 2017. Crispr-p 2.0: An improved crispr-cas9 tool for genome editing in plants. Molecular Plant 10(3): 530–532.

Liu, H., Z. Wei, A. Dominguez, Y. Li, X. Wang and L.S. Qi. 2015a. Crispr-era: A comprehensive design tool for crispr-mediated gene editing, repression and activation. Bioinformatics 31(22): 3676–3678.

Liu, M., S. Rehman, X. Tang, K. Gu, Q. Fan, D. Chen et al. 2019. Methodologies for improving hdr efficiency. Front genet 9: 691.

Liu, Q., X. Cheng, G. Liu, B. Li and X. Liu. 2020. Deep learning improves the ability of sgrna off-target propensity prediction. BMC Bioinformatics 21(1-15).

Liu, W., X. Xie, X. Ma, J. Li, J. Chen and Y.-G. Liu. 2015b. Dsdecode: A web-based tool for decoding of sequencing chromatograms for genotyping of targeted mutations. Molecular Plant 8(9): 1431–1433.

Liu, X., A. Homma, J. Sayadi, S. Yang, J. Ohashi and T. Takumi. 2016. Sequence features associated with the cleavage efficiency of crispr/cas9 system. Scientific Reports 6(1): 19675.

Lucas, M.C., L.P. Pryszcz, R. Medina, I. Milenkovic, N. Camacho, V. Marchand et al. 2024. Quantitative analysis of trna abundance and modifications by nanopore RNA sequencing. Nature Biotechnology 42(1): 72–86.

Ma, D., Z. Xu, Z. Zhang, X. Chen, X. Zeng, Y. Zhang et al. 2019. Engineer chimeric cas9 to expand pam recognition based on evolutionary information. Nature Communications 10(1): 560.

Ma, H., L.-C. Tu, A. Naseri, M. Huisman, S. Zhang, D. Grunwald et al. 2016. Multiplexed labeling of genomic loci with dcas9 and engineered sgrnas using crisprainbow. Nature Biotechnology 34(5): 528–530.

Ma, L., C. Dong, M. Yu, X. Song, Y. Yu and Y. Chen. 2022. Nanomedicine enables efficient crispr-cas9 genome editing for disease treatment. Science Bulletin 67(6): 572–576.

Ma, Y., H. Lu, B. Tippin, M.F. Goodman, N. Shimazaki, O. Koiwai et al. 2004. A biochemically defined system for mammalian nonhomologous DNA end joining. Molecular Cell 16(5): 701–713.

Ma, Y., K. Schwarz and M.R. Lieber. 2005. The artemis: DNA-pkcs endonuclease cleaves DNA loops, flaps, and gaps. DNA Repair 4(7): 845–851.

Maeder, M.L., S.J. Linder, V.M. Cascio, Y. Fu, Q.H. Ho and J.K. Joung. 2013. Crispr RNA–guided activation of endogenous human genes. Nature Methods 10(10): 977–979.

Mahajan, K.N., S.A. Nick McElhinny, B.S. Mitchell and D.A. Ramsden. 2002. Association of DNA polymerase μ (pol μ) with ku and ligase iv: Role for pol μ in end-joining double-strand break repair. Molecular and Cellular Biology 22(14): 5194–5202.

Mahfouz, M.M., A. Piatek and C.N. Stewart Jr. 2014. Genome engineering via talens and crispr/cas9 systems: Challenges and perspectives. Plant Biotechnology Journal 12(8): 1006–1014.

Makarova, K.S. and E.V. Koonin. 2015. Annotation and classification of crispr-cas systems. CRISPR: Methods and Protocols 47–75.

Mali, P., L. Yang, K.M. Esvelt, J. Aach, M. Guell, J.E. DiCarlo et al. 2013. RNA-guided human genome engineering via cas9. Science 339(6121): 823–826.

Manova, V. and D. Gruszka. 2015. DNA damage and repair in plants–from models to crops. Frontiers in Plant Science 6(885).

Mardis, E.R. 2013. Next-generation sequencing platforms. Annual Review of Analytical Chemistry 6(1): 287–303.

Mark, J.K.K., C.S.Y. Lim, F. Nordin and G.J. Tye. 2022. Expression of mammalian proteins for diagnostics and therapeutics: A review. Molecular Biology Reports 49(11): 10593–10608.

Mashal, R.D., J. Koontz and J. Sklar. 1995. Detection of mutations by cleavage of DNA heteroduplexes with bacteriophage resolvases. Nature Genetics 9(2): 177–183.

Matson, A.W., N. Hosny, Z.A. Swanson, B.J. Hering and C. Burlak. 2019. Optimizing sgrna length to improve target specificity and efficiency for the ggta1 gene using the crispr/cas9 gene editing system. PloS One 14(12): e0226107.

McKenna, A. and J. Shendure. 2018. Flashfry: A fast and flexible tool for large-scale crispr target design. BMC Biology 16(1-6).

Mehta, A. and J.E. Haber. 2014. Sources of DNA double-strand breaks and models of recombinational DNA repair. Cold Spring Harbor Perspectives in Biology 6(9): a016428.

Mendoza, B.J. and C.T. Trinh. 2018. Enhanced guide-RNA design and targeting analysis for precise crispr genome editing of single and consortia of industrially relevant and non-model organisms. Bioinformatics 34(1): 16–23.

Meng, H., M. Nan, Y. Li, Y. Ding, Y. Yin and M. Zhang. 2023. Application of crispr-cas9 gene editing technology in basic research, diagnosis and treatment of colon cancer. Frontiers in Endocrinology 14(1148412).

Milanowska, K., K. Rother and J.M. Bujnicki. 2011. Databases and bioinformatics tools for the study of DNA repair. Molecular Biology International 2011(1): 475718.

Mimitou, E.P. and L.S. Symington. 2010. Ku prevents exo1 and sgs1-dependent resection of DNA ends in the absence of a functional mrx complex or sae2. The EMBO Journal 29(19): 3358–3369.

Mingarro, G. and M. lí del Olmo. 2023. Improvements in the genetic editing technologies: Crispr-cas and beyond. Gene 852(147064).

Minkenberg, B., J. Zhang, K. Xie and Y. Yang. 2019. Crispr-plant v2: An online resource for highly specific guide RNA spacers based on improved off-target analysis. Plant Biotechnology Journal 17(1): 5.

Montague, T.G., J.M. Cruz, J.A. Gagnon, G.M. Church and E. Valen. 2014. Chopchop: A crispr/cas9 and talen web tool for genome editing. Nucleic Acids Research 42(W1): W401–W407.

Moon, A.F., J.M. Pryor, D.A. Ramsden, T.A. Kunkel, K. Bebenek and L.C. Pedersen. 2014. Sustained active site rigidity during synthesis by human DNA polymerase μ. Nature Structural & Molecular Biology 21(3): 253–260.

Moore, J.K. and J.E. Haber. 1996. Cell cycle and genetic requirements of two pathways of nonhomologous end-joining repair of double-strand breaks in Saccharomyces cerevisiae. Molecular and Cellular Biology.

Moreno-Mateos, M.A., C.E. Vejnar, J.-D. Beaudoin, J.P. Fernandez, E.K. Mis, M.K. Khokha et al. 2015. Crisprscan: Designing highly efficient sgrnas for crispr-cas9 targeting *in vivo*. Nature Methods 12(10): 982–988.

Muhammad Rafid, A.H., M. Toufikuzzaman, M.S. Rahman and M.S. Rahman. 2020. Crisprpred (seq): A sequence-based method for sgrna on target activity prediction using traditional machine learning. BMC Bioinformatics 21(1-13).

Mukherjee, S., P. Chakraborty and P. Saha. 2016. Phosphorylation of ku70 subunit by cell cycle kinases modulates the replication related function of ku heterodimer. Nucleic Acids Research 44(16): 7755–7765.

Naeem, M. and O.S. Alkhnbashi. 2023. Current bioinformatics tools to optimize crispr/cas9 experiments to reduce off-target effects. International Journal of Molecular Sciences 24(7): 6261.

Namiki, Y. and L. Zou. 2006. Atrip associates with replication protein a-coated ssdna through multiple interactions. Proceedings of the National Academy of Sciences 103(3): 580–585.

Nielsen, J. 2014. Synthetic biology for engineering acetyl coenzyme a metabolism in yeast. MBio. 5(6): 10.1128/mbio.02153-02114.

Niewolik, D., U. Pannicke, H. Lu, Y. Ma, L.-C.V. Wang, P. Kulesza et al. 2006. DNA-pkcs dependence of artemis endonucleolytic activity, differences between hairpins and 5′ or 3′ overhangs. Journal of Biological Chemistry 281(45): 33900–33909.

Niewolik, D., I. Peter, C. Butscher and K. Schwarz. 2017. Autoinhibition of the nuclease artemis is mediated by a physical interaction between its catalytic and c-terminal domains. Journal of Biological Chemistry 292(8): 3351–3365.

O'Brien, A.R., G. Burgio and D.C. Bauer. 2021. Domain-specific introduction to machine learning terminology, pitfalls and opportunities in crispr-based gene editing. Briefings in Bioinformatics 22(1): 308–314.

o'Brien, A.R., L.O. Wilson, G. Burgio and D.C. Bauer. 2019. Unlocking hdr-mediated nucleotide editing by identifying high-efficiency target sites using machine learning. Scientific Reports 9(1): 2788.

Ogita, N., Y. Okushima, M. Tokizawa, Y.Y. Yamamoto, M. Tanaka, M. Seki et al. 2018. Identifying the target genes of suppressor of gamma response 1, a master transcription factor controlling DNA damage response in arabidopsis. The Plant Journal 94(3): 439–453.

Palermo, G., Y. Miao, R.C. Walker, M. Jinek and J.A. McCammon. 2017. Crispr-cas9 conformational activation as elucidated from enhanced molecular simulations. Proceedings of the National Academy of Sciences 114(28): 7260–7265.

Park, J., S. Bae and J.-S. Kim. 2015. Cas-designer: A web-based tool for choice of crispr-cas9 target sites. Bioinformatics 31(24): 4014–4016.

Pawelczak, K.S. and J.J. Turchi. 2010. Purification and characterization of exonuclease-free artemis: Implications for DNA-pk-dependent processing of DNA termini in nhej-catalyzed dsb repair. DNA Repair 9(6): 670–677.

Peng, D. and R. Tarleton. 2015. Eupagdt: A web tool tailored to design crispr guide rnas for eukaryotic pathogens. Microbial Genomics 1(4): e000033.

Picarella, M.E., F. Ruiu, L. Selleri, S. Presa, C. Mizzotti, S. Masiero et al. 2024. Genetic and molecular mechanisms underlying the parthenocarpic fruit mutation in tomato. Frontiers in Plant Science 15(1329949).

Pietrzak, J., C.M. Spickett, T. Płoszaj, L. Virág and A. Robaszkiewicz. 2018. Parp1 promoter links cell cycle progression with adaptation to oxidative environment. Redox Biology 18(1-5).

Pinder, J., J. Salsman and G. Dellaire. 2015. Nuclear domain 'knock-in' screen for the evaluation and identification of small molecule enhancers of crispr-based genome editing. Nucleic Acids Research 43(19): 9379–9392.

Preuss, S. and A. Britt. 2003. A DNA-damage-induced cell cycle checkpoint in arabidopsis. Genetics 164(1): 323–334.

Přibylová, A. and L. Fischer. 2024. How to use crispr/cas9 in plants-from target site selection to DNA repair. Journal of Experimental Botany.erae147.

Pulido-Quetglas, C., E. Aparicio-Prat, C. Arnan, T. Polidori, T. Hermoso, E. Palumbo et al. 2017. Scalable design of paired crispr guide rnas for genomic deletion. PLoS Computational Biology 13(3): e1005341.

Qi, L.S., M.H. Larson, L.A. Gilbert, J.A. Doudna, J.S. Weissman, A.P. Arkin et al. 2013. Repurposing crispr as an RNA-guided platform for sequence-specific control of gene expression. Cell 152(5): 1173–1183.

Que, Q., Z. Chen, T. Kelliher, D. Skibbe, S. Dong and M.-D. Chilton. 2019. Plant DNA repair pathways and their applications in genome engineering. Plant Genome Editing with CRISPR Systems: Methods and Protocols 3–24.

Rahman, M.K. and M.S. Rahman. 2017. Crisprpred: A flexible and efficient tool for sgrnas on-target activity prediction in crispr/cas9 systems. PloS One 12(8): e0181943.

Reis, A.C., S.M. Halper, G.E. Vezeau, D.P. Cetnar, A. Hossain, P.R. Clauer et al. 2019. Simultaneous repression of multiple bacterial genes using nonrepetitive extra-long sgrna arrays. Nature Biotechnology 37(11): 1294–1301.

Renaud, J.-B., C. Boix, M. Charpentier, A. De Cian, J. Cochennec, E. Duvernois-Berthet et al. 2016. Improved genome editing efficiency and flexibility using modified oligonucleotides with talen and crispr-cas9 nucleases. Cell Reports 14(9): 2263–2272.

Richardson, C., R.N. Kelsh and R.J. Richardson. 2023. New advances in crispr/cas-mediated precise gene-editing techniques. Disease Models & Mechanisms 16(2): dmm049874.

Riesenberg, S., N. Helmbrecht, P. Kanis, T. Maricic and S. Pääbo. 2022. Improved grna secondary structures allow editing of target sites resistant to crispr-cas9 cleavage. Nature Communications 13(1): 489.

Robinson, J.T., H. Thorvaldsdóttir, W. Winckler, M. Guttman, E.S. Lander, G. Getz et al. 2011. Integrative genomics viewer. Nature Biotechnology 29(1): 24–26.

Salsman, J., J.-Y. Masson, A. Orthwein and G. Dellaire. 2017. Crispr/cas9 gene editing: From basic mechanisms to improved strategies for enhanced genome engineering *in vivo*. Current Gene Therapy 17(4): 263–274.

Sanger, F. and A.R. Coulson. 1975. A rapid method for determining sequences in DNA by primed synthesis with DNA polymerase. Journal of Molecular Biology 94(3): 441–448.

Santosh Kumar, V., R.K. Verma, S.K. Yadav, P. Yadav, A. Watts, M. Rao et al. 2020. Crispr-cas9 mediated genome editing of drought and salt tolerance (osdst) gene in indica mega rice cultivar mtu1010. Physiology and Molecular Biology of Plants 26(1099-1110).

Scott, T., R. Urak, C. Soemardy and K.V. Morris. 2019. Improved cas9 activity by specific modifications of the tracrrna. Scientific Reports 9(1): 16104.

Sebastian, R. and S.C. Raghavan. 2016. Induction of DNA damage and erroneous repair can explain genomic instability caused by endosulfan. Carcinogenesis 37(10): 929–940.

Secgin, Z., S. Uluisik, K. Yıldırım, M.F. Abdulla, K. Mostafa and M. Kavas. 2022. Genome-wide identification of the aconitase gene family in tomato (solanum lycopersicum) and crispr-based functional characterization of slaco2 on male-sterility. International Journal of Molecular Sciences 23(22): 13963.

Shah, S. 2024. Crop improvement using crispr/cas9 systems: Advances and challenges. Plant Biology 1(100004).

Shalem, O., N.E. Sanjana and F. Zhang. 2015. High-throughput functional genomics using crispr–cas9. Nature Reviews Genetics 16(5): 299–311.

Shan, Q., Y. Wang, J. Li, Y. Zhang, K. Chen, Z. Liang et al. 2013. Targeted genome modification of crop plants using a crispr-cas system. Nature Biotechnology 31(8): 686–688.

Shao, Z., A.J. Davis, K.R. Fattah, S. So, J. Sun, K.-J. Lee et al. 2012. Persistently bound ku at DNA ends attenuates DNA end resection and homologous recombination. DNA Repair 11(3): 310–316.

Shen, H. and Z. Li. 2022. DNA double-strand break repairs and their application in plant DNA integration. Genes 13(2): 322.

Shen, M.W., M. Arbab, J.Y. Hsu, D. Worstell, S.J. Culbertson, O. Krabbe et al. 2018. Predictable and precise template-free crispr editing of pathogenic variants. Nature 563(7733): 646–651.

Shendure, J.A., G.J. Porreca and G.M. Church. 2008. Overview of DNA sequencing strategies. Current Protocols in Molecular Biology 81(1): 7.1. 1–7.1. 11.

Shibata, M., H. Nishimasu, N. Kodera, S. Hirano, T. Ando, T. Uchihashi et al. 2017. Real-space and real-time dynamics of crispr-cas9 visualized by high-speed atomic force microscopy. Nature Communications 8(1): 1430.

Shinwari, Z.K., F. Tanveer and A.T. Khalil. 2018. Ethical issues regarding crispr mediated genome editing. Current Issues in Molecular Biology 26(1): 103–110.

Sivanandhan, G., N. Selvaraj, Y.P. Lim and A. Ganapathi. Targeted genome editing using site-specific nucleases, zfns, talens, and the crispr/cas9 system takashi yamamoto (ed.). *In*: Oxford University Press.

Sjogren, C.A., S.C. Bolaris and P.B. Larsen. 2015. Aluminum-dependent terminal differentiation of the arabidopsis root tip is mediated through an atr-, alt2-, and sog1-regulated transcriptional response. The Plant Cell 27(9): 2501.

Srivastava, M., M. Nambiar, S. Sharma, S.S. Karki, G. Goldsmith, M. Hegde et al. 2012. An inhibitor of nonhomologous end-joining abrogates double-strand break repair and impedes cancer progression. Cell 151(7): 1474–1487.

Stemmer, M., T. Thumberger, M. del Sol Keyer, J. Wittbrodt and J.L. Mateo. 2015. Cctop: An intuitive, flexible and reliable crispr/cas9 target prediction tool. PloS One 10(4): e0124633.

Stemmer, M., T. Thumberger, M. del Sol Keyer, J. Wittbrodt and J.L. Mateo. 2017. Correction: Cctop: An intuitive, flexible and reliable crispr/cas9 target prediction tool. PLoS One 12(4): e0176619.

Sun, C., Y. Lei, B. Li, Q. Gao, Y. Li, W. Cao et al. 2024. Precise integration of large DNA sequences in plant genomes using primeroot editors. Nature Biotechnology 42(2): 316–327.

Sung, P. and H. Klein. 2006. Mechanism of homologous recombination: Mediators and helicases take on regulatory functions. Nature Reviews Molecular Cell Biology 7(10): 739–750.

Sweeney, P.R., A.B. Britt and K.M. Culligan. 2009. The arabidopsis atrip ortholog is required for a programmed response to replication inhibitors. The Plant Journal 60(3): 518–526.

Symington, L.S. 2014. End resection at double-strand breaks: Mechanism and regulation. Cold Spring Harbor Perspectives in Biology 6(8): a016436.

Szczelkun, M.D., M.S. Tikhomirova, T. Sinkunas, G. Gasiunas, T. Karvelis, P. Pschera et al. 2014. Direct observation of r-loop formation by single RNA-guided cas9 and cascade effector complexes. Proceedings of the National Academy of Sciences 111(27): 9798–9803.

Tamura, K., Y. Adachi, K. Chiba, K. Oguchi and H. Takahashi. 2002. Identification of ku70 and ku80 homologues in arabidopsis thaliana: Evidence for a role in the repair of DNA double-strand breaks. The Plant Journal 29(6): 771–781.

Tang, X.-D., F. Gao, M.-J. Liu, Q.-L. Fan, D.-K. Chen and W.-T. Ma. 2019. Methods for enhancing clustered regularly interspaced short palindromic repeats/cas9-mediated homology-directed repair efficiency. Frontiers in Genetics 10(551).

Tang, X., S. Sretenovic, Q. Ren, X. Jia, M. Li, T. Fan et al. 2020. Plant prime editors enable precise gene editing in rice cells. Molecular Plant 13(5): 667–670.

Tibbetts, R.S., K.M. Brumbaugh, J.M. Williams, J.N. Sarkaria, W.A. Cliby, S.-Y. Shieh et al. 1999. A role for atr in the DNA damage-induced phosphorylation of p53. Genes & Development 13(2): 152–157.

Tiwari, J.K., A.K. Singh and T.K. Behera. 2023. Crispr/cas genome editing in tomato improvement: Advances and applications. Frontiers in Plant Science 14(1121209).

Trionfini, P., O. Ciampi, E. Romano, A. Benigni and S. Tomasoni. 2020. Generation of two isogenic knockout pkd2 ips cell lines, irfmni003-a-1 and irfmni003-a-2, using crispr/cas9 technology. Stem Cell Research 42(101667).

Tsao, S.W., C.M. Tsang and K.W. Lo. 2017. Epstein–barr virus infection and nasopharyngeal carcinoma. Philosophical Transactions of the Royal Society B: Biological Sciences 372(1732): 20160270.

Tyagi, S., R. Kumar, A. Das, S.Y. Won and P. Shukla. 2020. Crispr-cas9 system: A genome-editing tool with endless possibilities. Journal of Biotechnology 319(36-53).

Uusi-Mäkelä, M.I., H.R. Barker, C.A. Bäuerlein, T. Häkkinen, M. Nykter and M. Rämet. 2018. Chromatin accessibility is associated with crispr-cas9 efficiency in the zebrafish (Danio rerio). PloS One 13(4): e0196238.

Van Gent, D.C., J.H. Hoeijmakers and R. Kanaar. 2001. Chromosomal stability and the DNA double-stranded break connection. Nature Reviews Genetics 2(3): 196–206.

Verma, V., A. Kumar, M. Partap, M. Thakur and B. Bhargava. 2023. Crispr-cas: A robust technology for enhancing consumer-preferred commercial traits in crops. Frontiers in Plant Science 14(1122940).

Wang, F., C. Wang, P. Liu, C. Lei, W. Hao, Y. Gao et al. 2016. Enhanced rice blast resistance by crispr/cas9-targeted mutagenesis of the erf transcription factor gene oserf922. PloS One 11(4): e0154027.

Wang, J.Y. and J.A. Doudna. 2023. Crispr technology: A decade of genome editing is only the beginning. Science 379(6629): eadd8643.

Wang, T., J.J. Wei, D.M. Sabatini and E.S. Lander. 2014. Genetic screens in human cells using the crispr-cas9 system. Science 343(6166): 80–84.

West, R.B., M. Yaneva and M.R. Lieber. 1998. Productive and nonproductive complexes of ku and DNA-dependent protein kinase at DNA termini. Molecular and Cellular Biology.

Wilson, T.E., U. Grawunder and M.R. Lieber. 1997. Yeast DNA ligase iv mediates non-homologous DNA end joining. Nature 388(6641): 495–498.

Wong, N., W. Liu and X. Wang. 2015. Wu-crispr: Characteristics of functional guide rnas for the crispr/cas9 system. Genome Biology 16(1-8).

Wu, X., D.A. Scott, A.J. Kriz, A.C. Chiu, P.D. Hsu, D.B. Dadon et al. 2014. Genome-wide binding of the crispr endonuclease cas9 in mammalian cells. Nature Biotechnology 32(7): 670–676.

Xiao, A., Z. Cheng, L. Kong, Z. Zhu, S. Lin, G. Gao et al. 2014. Casot: A genome-wide cas9/grna off-target searching tool. Bioinformatics 30(8): 1180–1182.

Xie, K., J. Zhang and Y. Yang. 2014. Genome-wide prediction of highly specific guide RNA spacers for crispr–cas9-mediated genome editing in model plants and major crops. Molecular Plant 7(5): 923–926.

Xie, X., X. Ma, Q. Zhu, D. Zeng, G. Li and Y.-G. Liu. 2017. Crispr-ge: A convenient software toolkit for crispr-based genome editing. Molecular Plant 10(9): 1246–1249.

Xu, H., T. Xiao, C.-H. Chen, W. Li, C.A. Meyer, Q. Wu et al. 2015. Sequence determinants of improved crispr sgrna design. Genome Research 25(8): 1147–1157.

Xu, X., D. Duan and S.-J. Chen. 2017. Crispr-cas9 cleavage efficiency correlates strongly with target-sgrna folding stability: From physical mechanism to off-target assessment. Scientific Reports 7(1): 143.

Xu, X., C. Liu, Y. Wang, O. Koivisto, J. Zhou, Y. Shu et al. 2021. Nanotechnology-based delivery of crispr/cas9 for cancer treatment. Advanced Drug Delivery Reviews 176(113891).

Yadav, R.K., T. Girke, S. Pasala, M. Xie and G.V. Reddy. 2009. Gene expression map of the arabidopsis shoot apical meristem stem cell niche. Proceedings of the National Academy of Sciences 106(12): 4941–4946.

Yaneva, M., T. Kowalewski and M.R. Lieber. 1997. Interaction of DNA-dependent protein kinase with DNA and with ku: Biochemical and atomic-force microscopy studies. The EMBO Journal 16(16): 5098–5112.

Yang, G., C. Liu, S.-H. Chen, M.A. Kassab, J.D. Hoff, N.G. Walter et al. 2018. Super-resolution imaging identifies parp1 and the ku complex acting as DNA double-strand break sensors. Nucleic Acids Research 46(7): 3446–3457.

Yang, W., J. Yan, P. Zhuang, T. Ding, Y. Chen, Y. Zhang et al. 2022. Progress of delivery methods for crispr-cas9. Expert Opinion on Drug Delivery 19(8): 913–926.

Yarrington, R.M., S. Verma, S. Schwartz, J.K. Trautman and D. Carroll. 2018. Nucleosomes inhibit target cleavage by crispr-cas9 *in vivo*. Proceedings of the National Academy of Sciences 115(38): 9351–9358.

Yeh, C.D., C.D. Richardson and J.E. Corn. 2019. Advances in genome editing through control of DNA repair pathways. Nature Cell Biology 21(12): 1468–1478.

Yerlikaya, B.A., K. Mostafa, A. Aydin, S. Yerlikaya, M. Mehmood, N.N. Yilm et al. 2024. Innovations in plant genome editing and biotechnology: Addressing global agricultural challenges. Turkish Journal of Botany 48(7): 351–375.

Yin, W., Z. Chen, J. Huang, H. Ye, T. Lu, M. Lu et al. 2023. Application of crispr-cas9 gene editing technology in crop breeding. Sheng wu Gong Cheng xue bao= Chinese Journal of Biotechnology 39(2): 399–424.

Yip, B.H. 2020. Recent advances in crispr/cas9 delivery strategies. Biomolecules 10(6): 839.

Yoshiyama, K., P.A. Conklin, N.D. Huefner and A.B. Britt. 2009. Suppressor of gamma response 1 (sog1) encodes a putative transcription factor governing multiple responses to DNA damage. Proceedings of the National Academy of Sciences 106(31): 12843–12848.

Yoshiyama, K.O., J. Kobayashi, N. Ogita, M. Ueda, S. Kimura, H. Maki et al. 2013. Atm-mediated phosphorylation of sog1 is essential for the DNA damage response in arabidopsis. EMBO Reports 14(9): 817–822.

Zalatan, J.G., M.E. Lee, R. Almeida, L.A. Gilbert, E.H. Whitehead, M. La Russa et al. 2015. Engineering complex synthetic transcriptional programs with crispr RNA scaffolds. Cell 160(1): 339–350.

Zhang, D., T. Hurst, D. Duan and S.-J. Chen. 2019. Unified energetics analysis unravels spcas9 cleavage activity for optimal grna design. Proceedings of the National Academy of Sciences 116(18): 8693–8698.

Zhang, J.-P., X.-L. Li, A. Neises, W. Chen, L.-P. Hu, G.-Z. Ji et al. 2016. Different effects of sgrna length on crispr-mediated gene knockout efficiency. Scientific Reports 6(1): 28566.

Zou, L. and S.J. Elledge. 2003. Sensing DNA damage through atrip recognition of rpa-ssdna complexes. Science 300(5625): 1542–1548.

Zou, L., D. Liu and S.J. Elledge. 2003. Replication protein a-mediated recruitment and activation of rad17 complexes. Proceedings of the National Academy of Sciences 100(24): 13827–13832.

Chapter 5

CRISPR/Cas9 as an Effective Tool Used for the Improvement of Fruit Crops

Zain ul Abideen,[1] *Laiba Chaudhry,*[2] *Mohsin Ali,*[3,*] *Yaqoob Sultan*[4,5] and *Abdul Basit*[5]

1. Introduction

Fruits and vegetables are vital for human nutrition, supplying essential nutrients and components necessary for a well-balanced diet. Their cultivation has a long history, dating back hundreds of years. However, according to (Dalla Costa et al. 2017; Leisner 2020), various biological and environmental challenges hinder their growth, such as lengthy life cycles, diverse genetic makeups, unfavorable climates, and limited fertile land. Therefore, improvements in fruit crop profitability and sustainability via scientific and technological breakthroughs are urgently needed. Cellulose, vitamins, minerals, trace elements, and other essential components needed for human nutrition are abundant in these crops (Giovannoni et al. 2017). However, the growing availability of fruits and vegetables is seriously threatened by changes in temperature and environmental circumstances (Karkute et al. 2017).

Humans have cultivated wild species for a long time. Through crossbreeding techniques, breeders can enhance specific traits by mixing dominant varieties (Meyer et al. 2013). However, traditional breeding methods reveal drawbacks, especially their dependency on the naturally occurring genetic variations, which become evident during long-term artificial selection. This conventional approach can lead to unwanted traits, reducing breeding efficiency, and there are limited genetic resources available to improve desired characteristics (Begna 2021; Tester and Langridge 2010). Even while traditional methods can produce high-yield, quality cultivars resistant to diseases, it is imperative to

[1] Institute of Nutrition, Mahidol University, 999 Phutthamonthon 4 Rd., Salaya, Phutthamonthon, Nakhon Pathom 73170, Thailand.

[2] Institute of Food Science and Nutrition, University of Sargodha, Sargodha 40100, Punjab, Pakistan.

[3] College of Chemistry and Environmental Engineering, Institute for Advanced Study, Shenzhen University, Shenzhen 518055, P. R. China.

[4] Department of Grass Breeding, Institute of Agriculture, Lithuanian Research Centre for Agriculture and Forestry, 58344 Kedainiu̜ r., Lithuania.

[5] Floricultural Biotechnology Lab., Department of Horticulture, Graduate School, Kyungpook National University, 41566, Daegu, South Korea.

* Corresponding author: realmohsinali313@gmail.com

DOI: 10.1201/9781003518204-5

swiftly develop new vegetable varieties to suit market expectations because of the growing world population and demands regarding food (Bigliardi et al. 2013).

Novel crop biotechnology methods can encourage more accurate and effective varietal breeding (Parmar et al. 2017). In addition to improving fruit and vegetable quality, genetic engineering increases the crops' resistance to biotic and abiotic stressors. A transgenic tomato exhibiting storage resistance was approved by the Food and Drug Administration (FDA) in 1994 (Kramer et al. 1994). Furthermore, according to Millstone et al. (2015), 80% of the market share was accounted for by a high-yield transgenic papaya, which showed greater yield and cyclic spot virus resistance.

Nonetheless, the development and use of Genetically Modified (GM) plants face stringent regulations designed to ensure safety during planting and consumption, which significantly delays the market entry of transgenic cultivars (Bawa et al. 2013). Many novel gene-edited agricultural resources emerged in less than two years. Without the lengthy regulatory process usually associated with genetically modified crops, the USFDA approved the market introduction of a gene-edited anti-browning mushroom (*Agaricus bisporus*) and a waxy corn null segregant line in 2016 (Waltz et al. 2013; Waltz et al. 2018).

It shows that the development of crop types has already benefited from CRISPR gene editing. This study discusses the latest developments in CRISPR-Cas systems, examines current uses in fruit and vegetable crops, and gives a description of basic principles of CRISPR technology. We also look at the CRISPR-related regulatory systems that permit the commercial sale of gene-edited crops in various nations. The possibilities and difficulties of adding advantageous alleles and enhancing certain features in the future are also evaluated. Plant genomes can be precisely altered through genome editing, making it easier to achieve desired plant traits. The CRISPR/Cas system is a well-known genetic modification technique nowadays because of its affordability, versatility, and exceptional specificity (Yin et al. 2017). In light of the escalating global agricultural issues, this chapter explores CRISPR/Cas9's transformational potential for editing genes in enhancing fruit

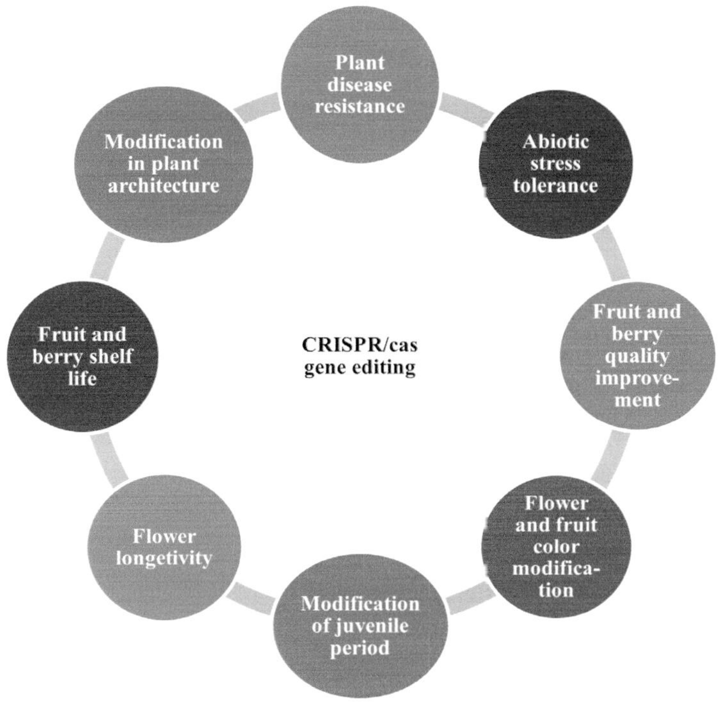

Figure 1. Fruit features that might be improved through editing genes.

crop yield, resilience, and sustainability. It explores the workings of the CRISPR/Cas9 system and highlights its benefits over previous genome editing methods and conventional breeding methods. The chapter highlights recent advancements and practical implications of CRISPR/Cas9 in improving critical traits in commercially significant crops of fruits, focusing on disease resistance, tolerance to stress from abiotic factors, nutritional value, quality of fruits, and shelf-life following harvest. Additionally, it reviews the regulatory landscapes governing gene-edited crops, addresses technological challenges, off-target consequences, and transformation, and considers the future improvements in fruit plants using CRISPR technology, primarily focusing on its likely role in sustainable agriculture, precision breeding, and the creation of customized fruits that fulfill evolving consumer and market demands. Figure 1 explains the features that can be improved through gene editing in fruits including variety of characters. These characters include quality improvement through color modification, juvenile period, flower longevity, shelf-life extension, disease resistance and many more as described in the Fig. 1.

2. Genome Editing in Fruit Plants

The challenge of producing sufficient food, primarily from fruit crops, poses serious socioeconomic issues due to the increase in world population by 2050. Many fruit crops are perennials that require specific methods for handling, transportation, and cultivation (Radha and Mathew 2007; Parajuli et al. 2019).

These crops often face risks from both biological and environmental factors, highlighting the need for careful cultivation and the identification of key traits that require enhancement. To generate varieties with the necessary features, approaches, genetic engineering, interspecific hybridization, mutagenesis, and traditional breeding were used. For example, mutagenesis has been used to improve the color of apples, and genetic engineering has made it easier to produce non-browning apples (Xu 2013; Kumar et al. 2021). Interspecies hybridization has introduced scab resistance in some cultivars (Fedak et al. 2021), while according to Cornille et al. (2014); Kaiser et al. (2020); Pu et al. (2024), natural hybridization has resulted in benefiting sweetness, color, firmness, and size.

Conventional breeding has improved several commercially valuable traits in bananas (Musa spp.), cucumbers (Cucumis sativus), mandarins (Citrus reticulata), peaches (Prunus persica), and strawberries (Fragaria 9 ananassa) (Bouis 2002; Moose and Mumm 2008; Omura and Shimada 2016; Feng et al. 2020; Cockerton et al. 2021).

3. CRISPR-Cas9 Working Mechanism

Genomic editing differs from conventional breeding methods, which mostly rely on recombination and, to a lesser extent, genetic recombination, in that it enables precise targeting of specific DNA sequences. For example, according to Mahfouz et al. (2011), Li et al. (2012), Farinati et al. (2023), Transcription Activator-Like Effector Nucleases (TALENs) (Carroll 2011; Yuan et al. 2024), Zinc-Finger Nucleases (ZFNs) are examples of engineered nuclease enzymes that can make specific genomic changes (Shan et al. 2013). These enzymes are also known as molecular scissors (Punwar et al. 2014; Gibney and Kellett 2024).

The most recent advancement in genome editing is the Clustered Regularly Interspaced Short Palindromic Repeat/Associated Protein-9 Nuclease (CRISPR/Cas9). The base fact about it is that RNA-guided DNA endonucleases allow for precise insertions, substitutions, or changes at specific locations. The efficiency, affordability, and user-friendliness of the CRISPR/Cas9 system make it different from other genome editing techniques (Kaul et al. 2020). According to Jaganathan et al. (2018), these innovative crop development methods can alter any genetic attribute.

Figure 2 describes the working mechanism of gene editing sequence using CRISPR/Cas9. In general, there are three steps to the CRISPR-Cas9 system's response to foreign DNA (Amitai and Sorek 2016; Wang et al. 2016). To create an immunological memory, the invading DNA is

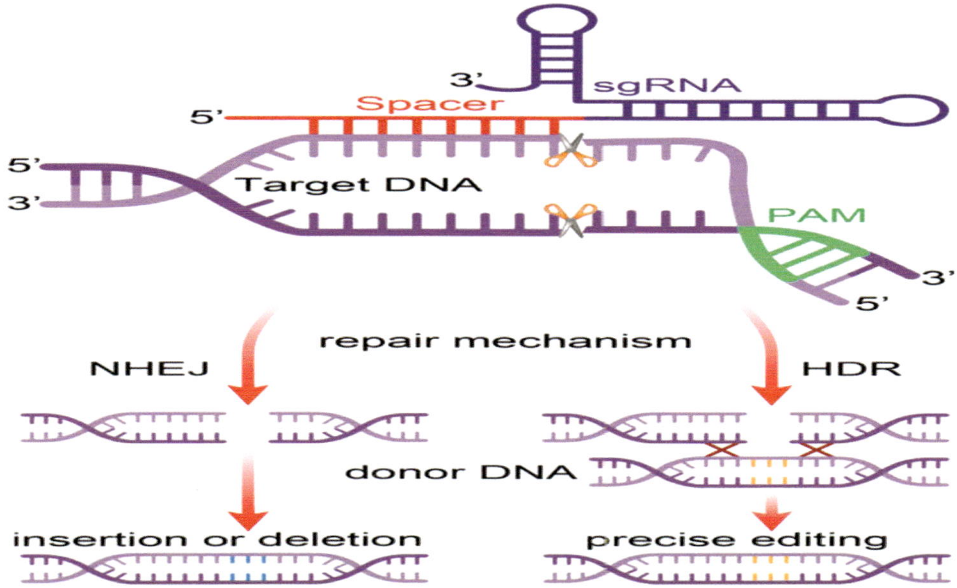

Figure 2. Mechanism of the CRISPR/Cas9 gene editing system.

first identified during the acquisition step, and a spacer sequence generated from it is added to the host's CRISPR array. The CRISPR array is then converted into a precursor RNA transcript (pre-crRNA) during the expression stage, which also involves the synthesis of the Cas9 protein. Mature RNA units known as crRNAs are produced when this pre-crRNA joins forces with a noncoding trans-activating CRISPR RNA (crRNA). Ultimately, the mature crRNA guides the Cas9 protein to find the appropriate DNA target during the interference stage, allowing the foreign DNA to be cleaved and degraded. Cas9 will cause a Double-Strand Break (DSB) when the single guide RNA (sgRNA) guides the Cas9 nuclease to a complementary area of the genome. For Cas9 to work, a 5′-NGG-3′motif—also referred to as a protospacer adjacent motif, or PAM—must come after the target genomic sequence. When a DNA repair template is available, it can be used to incorporate specific genetic alterations or exogenous sequences. When this template is accessible, DSBs are mended by either Non-Homologous End Joining (NHEJ) or Homology-Directed Repair (HDR), as per Wang et al. 2016.

4. Practical Implications of Genome Editing in Fruit Plants

In many nations or areas around the world, fruit plants are the main source of income, which is vital for global economy and human nutrition. According to (Giovannoni 2001), based on varying respiratory characteristics and ethylene production (during fruit ripening) rate, there are two categories for fleshy fruits: climacteric and non-climacteric in climacteric fruits ripening involves use of ethylene, including peaches, apples, bananas, tomatoes, and kiwis (Brumos 2021; Nishitani et al. 2016; Nakajima et al. 2017; Çelik et al. 2024)). In contrast, Cherian et al. (2014) found that abscisic acid (ABA) primarily controls fruit ripening in citrus, watermelon, cucumber, grape, and strawberry, which are non-climacteric fruits. So, ethylene is not necessary for the process of ripening.

4.1 Using Genome Editing on Bananas

In the tropics, bananas rank among the most economically significant fruits. According to Perrier et al. (2011), the fruit crop banana is ranked fourth in cultivation around the world. Its triploid genome and sterility make traditional breeding challenging (Dash and Rai 2016). Perhaps the only way to improve this significant fruit is through genetic engineering. In banana cv. Rasthali embryonic cell suspension cultures, a single guide RNA targeting the preserved area of two RASPDS genes (RAS-PDS1 and RAS-PDS2) was created (Kaur et al. 2018). The regenerated plantlets showed a 59% mutation rate in complete albino and variegated phenotypes. More recently, PDS 1 exon in the Cavendish cultivar was targeted using an integrated strategy for the use of polycistronic tRNA-gRNA system; Among the 19 regenerated plants, a 100% mutation rates, deletions or insertions were discovered and there was a correlation between the genotype and the degree of the albino (Naim et al. 2018). The work shows that bananas can effectively change their genomes using CRISPR/Cas9.

4.2 Genome Editing Application in Citrus

Citrus is one of the most economically significant fruit crops, but because of polyembryonic seeds, pollen incompatibility, parthenocarpic fruit, a lengthy juvenile phase, and resistance to transformation using traditional methods, genetic modification through breeding or gene manipulation is very challenging (Talon and Gmitter 2008). Jia and Wang (2014) effectively inserted Cas9 and a synthetic sgRNA that targets CsPDS into the leaves of Duncan grapefruit and sweet orange (Citrus X sinensis) using Xanthomonas citri subsp.

Citr helped with agroinfiltration (Xcc). Consequently, the mutational rates of invading leaves were low but successful (Jia and Wang 2014a, b). The same team later altered Duncan grapefruit epicotyl explants using Agrobacterium-mediated transformation, producing six transgenic lines with a CRISPR/Cas9 construct that targets the CsLOB1 gene, which controls susceptibility to citrus canker, a disease brought on by the bacterium Xanthomonas axonopodis. Different levels of resistance to citrus canker were displayed by the resultant plants (Jia et al. 2017b). According to Peng et al. (2017), the CsLOB1 promoter, upon successful editing in a subsequent effort, resulted in citrus canker-resistant homozygous plants. Therefore, modifying the promoter region of the genome turns out to be a useful technique for preventing the production of the target gene. These encouraging results show how CRISPR/Cas9-mediated modification will help combat severe diseases like citrus greening.

4.3 High-Efficiency Kiwifruit Editing

Utilizing the PTG/Cas9 technology, is accomplished. Kiwifruit (*Actinidia Lindl.*) retains a higher level of vitamin C along with DF (Huang et al. 2013). CRISPR/Cas9 technology, along with a polycistronic tRNA-sgRNA cassette (PTG)/Cas9 system, was utilized to manipulate the genome of kiwifruit as described by Xie et al. (2015). When editing the PDS gene of kiwifruit the (Wang et al. 2018b) preferred PTG/Cas9 method over conventional CRISPR/Cas9 system.

5. Applications of CRISPR-Cas9 in Fruit Crops

Numerous researchers have attempted to demonstrate the effectiveness of CRISPR/Cas9 technology in the following fruit varieties: apple, banana, Carrizo citrange, Duncan grapefruit, grape, kiwi fruit, and strawberry. To achieve highly efficient genome editing, conditions must be optimized for its application within a specific species. To demonstrate CRISPR/Cas-mediated knockout development, the functional gene Phytoene Desaturase (PDS) was first employed. When the PDS gene is disrupted, albino and dwarf phenotypes follow because the manufacture of gibberellin, carotenoids, and chlorophyll is impaired, which can serve as visual markers.

Citrus sinensis (sweet orange) was the first fruit crop to be modified using CRISPR. Cas9 and a synthetic sgRNA that targeted the CsPDS gene were introduced using agroinfiltration assisted by Xanthomonas citri subsp. citri-Xcc. However, ranging from 3.2 to 3.9%, the efficiency was incredibly low (Jia and Wang 2014). Similarly, it was first demonstrated that CRISPR/Cas editing harms the PDS gene in apples and grapes, as reported by Nakajima et al. (2017).

Similarly, Naim et al. (2018) demonstrated that bananas could undergo multiplex genome editing, focusing on exon 1 of three distinct alleles for the PDS gene, with 100% editing efficiency. AcPDS gene was targeted using PTG driven multiplex genome editing in kiwi fruit, resulting in genome editing increases of ten times compared to traditional Cas9 editing (Wang et al. 2018b). The results of modified profiles and the use of CRISPR in pear revealed that PDS genes primarily experienced chimeric biallelic mutations (Charrier et al. 2019). According to Rieger et al. (2006), Duane Green, a fruit crop is a permanent, edible crop whose main product is either the botanical fruit itself or a derivative of it. Fruit crops also include several plants that are farmed mostly as annuals, like melons, cucumbers, and tomatoes (Miller and Gross 2011; Essiaw-Quayson 2017). Due to its easy access to germplasm supplies, simple diploid inheritance, efficient breeding, short growing season, simplicity of genetic manipulation, and extensive research, the tomato is a model for fruit biology (Vayalil 2012; Varman 2025). Although gene editing is developing quickly, CRISPR/Cas9 is well-reputed as an accurate and efficient editing tool for genes (Javaid et al. 2022). CRISPR/Cas9-edited crops have demonstrated remarkable efficiency. These include different genome equivalencies, such as up to 79% in maize (Wang et al. 2020) and up to 91.5% in rice crop as reported by (Mao et al. 2013; Cong et al. 2013; Abdelrahman et al. 2021).

CRISPR/Cas9 is expected to be a major tool for plant breeding, based on the level of concern developed in the plant breeding community. Its ease of design and ability to multiplex or edit many loci at once by inducing multiple DSBs are part of its appeal (Cong et al. 2013).

Numerous horticulture crops have been modified using CRISPR/Cas9 technology to accomplish a variety of practical breeding objectives and research objectives, including understanding gene function (Table 1). To lay the foundation for a breeding program, some researchers also used CRISPR/Cas9 to identify the genes that lead to a certain characteristic. This makes it possible to employ introgression and controlled crossing techniques. In a similar vein, the breeding procedure can be expedited by directly introducing unique mutations into elite germplasm (Scheben et al. 2016; Ahmar et al. 2020).

Table 1. Horticultural fruit examples where study goals were met using CRISPR/Cas9 technology.

Objective achieved	Fruits	References
Reduction of the fire blight vulnerability	Apple	(Pompili et al. 2019)
Validated the function of the vacuolar sugar transporter gene Obtain gynoecious genotypes Resistance to *Fusarium oxysporum f.* sp. *niveum* Race 1 Characterized the gene involved in the accumulation of sugar in pulp of fruit Resistance development against the herbicide	Water melon	(Ren et al. 2020) (Zhang et al. 2019) (Zhang et al. 2020) (Guo et al. 2019) (Tian et al. 2018)
Streak virus was inactivated Dwarf cultivars and semi-dwarf cultivars basis were generated	Banana	(Tripathi et al. 2019) (Shao et al. 2019)
Transcription factor was characterized for the development of anther Auxin accumulation and biosynthesis genes were identified	Strawberry	(Martin-Pizarro et al. 2019) (Feng et al. 2018)
Canker resistance	Citrus	(Jia et al. 2017; Wang et al. 2019)
Analyzed the oomycetes' evolution in escaping the defensive mechanisms of plants	Papaya	(Gumtow et al. 2018)

5.1 Using CRISPR/Cas9 to Increase Fruit Crop Disease Resistance

Numerous pests and diseases affect fruit crops, impeding their growth and development and resulting in financial losses. Consequently, developing resistant cultivars may be an alternative strategy for addressing these problems. By employing CRISPR/Cas9 genome editing, they can become resistant to biotic stresses that have a major impact on their production; *Citrus xanthomonas* spp. Citrus sinensis Lateral Organ Boundaries (CsLOB1) is the primary gene linked to citrus canker, which is caused by Citri (Xcc) (Hu et al. 2014; Zou et al. 2021). To enhance CsLOB1 induction, the Xcc pathogenicity factor PthA4 binds to a specific location in the promoter region.

The infection on the modified plants was reduced by utilizing the CRISPR/Cas9 system to modify the PthA4 binding element in the promoter of the CsLOB1 gene (Jia et al. 2016; Neto 2022). Numerous changes in the promoter of the plants derived from both CsLOB1 alleles gave them resistance to this illness. Five CRISPR/Cas9 constructs have been used in similar research to modify the PthA4 binding region of the Wanjinchen orange CsLOB1 gene promoter. Although deletion of the PthA4 binding region from both CsLOB1 alleles resulted in a significant tolerance to infection, other mutant lines with increased resistance to citrus canker were developed (Peng et al. 2017; Neto 2022). Mutations in the first exon of the WRKY52 gene resulted in improved resistance to Botrytis cinerea in grapes (Wang et al. 2018a). The biallelic mutants were shown to be more resilient than the monoallelic ones. Strawberries were made resistant to powdery mildew by altering the Mildew-Resistance Locus (MLO), which has been extensively researched in barley. The phylogenetically conservative character of this locus has also produced successful results in strawberries (Jiwan et al. 2013; Rehman et al. 2024).

The Non-Expressor of Pathogenesis-Related Genes3 (NPR3) in cacao was targeted using the CRISPR/Cas9 technology. According to Dorantes-Acosta et al. (2012) and Chandran et al. (2023). NPR3 genes for a repressor protein that is engaged in defensive processes. Therefore, once 27% of the NPR3 copies were eliminated, the modified tissues were resistant to *Phytophthora tropicalis*. Future genome editing events of somatic embryos were conducted in citrus (Dutt et al. 2020) and Theobroma cacao (Fister et al. 2018) to assess the effectiveness of the CRISPR/Cas9 system. Banana Streak Virus (BSV) has a serious effect on banana culture and productivity. Several modifications were made to the BSV sequences incorporated into the DNA of the *Gonja manjaya* cultivar using the CRISPR/Cas9 technique. Seventy-five percent of the altered plants showed no symptoms when subjected to water stress (Tripathi et al. 2019; Zarco-Tejada et al. 2021).

6. Fruit and Vegetable Quality Improvement Using CRISPR/Cas9

Enhancing the quality and prolonged life span after harvest are the main goals of fruit and vegetable breeding. Quality encompasses both internal and exterior elements. Fruit that is visible to the unaided eye in terms of size, color, and texture is said to have external quality. Internal quality, including the concentrations of minerals like sugars, vitamins, and bioactive chemicals like malate, lycopene, and anthocyanins, must be tested with equipment. For example, 50% of the genetic diversity in fruit size in tomatoes is due to the number of ovary locules, Numerous QTLs control the amount of ovary locules, which accounts for 50% of the genetic variability in tomato fruit size (Li et al. 2017). To produce larger and more abundant fruits than wild-type fruits, researchers at Cold Spring Harbor Laboratory created eight sgRNAs and used CRISPR-Cas9 to edit the promoter region of the tomato CLAVATA-WUSCHEL (CLV-WUS) stem cell gene CLV3 (Li et al. 2018). By adjusting fruit-size assessment QTLs, including the QTLs for locule number (lc) and fasciated number (fas), they were able to create germplasm resources with more locules (Rodríguez et al. 2017). For consumers, the color and texture of fruits and vegetables are important factors. For example, European and American consumers like red tomatoes, whereas Asian customers prefer pink tomatoes (Ballester et al. 2010; Lin et al. 2014; Casals et al. 2024).

CRISPR-Cas was used to modify anthocyanin 2 (ANT2), MYB transcription factor 12 (MYB12), and phytoene synthase 1 (PSY1) to generate yellow, pink, and purple tomatoes, respectively (Cermák et al. 2015; Filler et al. 2017; Wan et al. 2021). The transformation of lycopene precursors into lycopene is catalyzed by the Chinese kale carotenoid isomerase gene (BoaCRTISO). When BoaCRTISO was targeted and modified, the color of the mutants changed from green to yellow (Sun et al. 2020). Increasing the nutritional value and bioactive content of fruits and vegetables is the main objective of improving their intrinsic quality. Vitamins and carbohydrates are important for nutrition. The synthesis and metabolism of carotenoids and sucrose include several genes (Kapoor et al. 2022). The human body can absorb provitamin A, one of the carotenoids, and transform it into vitamin A. For example, the transcription and protein products of numerous genes in the sucrose metabolism pathway were disrupted by utilizing CRISPR-Cas to knock out MPK20 (Mitogen-activated Protein Kinase 20) (Chen et al. 2018). When taken regularly, these micronutrients can effectively prevent cancer and cardiovascular disease. According to Kris-Etherton et al. (2020), biofortification, which is the biotechnological improvement of mineral absorption, transport, and metabolism by plants, raises the concentrations of micronutrients that are beneficial to human health.

According to Li et al. (2018), bioactive substances include γ-aminobutyric acid (GABA), lycopene malate, and anthocyanins. These nutrients can be enhanced in fruits by using CRISPRCas9 to modify important genes linked to metabolic pathways. For instance, manipulating the aluminum-activated malate transporter (ALMT9) increased the quantity of malate, whereas changing many genes in the GABA synthesis pathway increased the butylamine content in tomatoes by 19 times (Nonaka et al. 2017).

CRISPR-Cas9 reduces the toxic compounds from vegetables that are bad for human health by concentrating on mutations that make genes in biosynthetic pathways inactive. A low amount is indicative of high quality since, for example, too much steroidal glycoalkaloids (SGAs), in potato tubers change their flavor and make them less safe for human consumption which includes CRISPR-Cas9 reduces the toxic compounds from vegetables that are bad for human health by concentrating on mutations that make genes in biosynthetic pathways inactive. A low amount is indicative of high quality since, for example, too much steroidal glycoalkaloids (SGAs), in potato tubers change their flavor and make them less safe for human consumption, which includes α-solanine and α-chaconine. To remove St16DOX (steroid 16α-hydroxylase) from the potato SGA production pathway, Nakayasu et al. (2018) used CRISPR-Cas9 to create SGA-free potato lines. A long shelf life is an important breeding goal in the production of fruits and vegetables. To remove St16DOX (steroid 16α-hydroxylase) from the potato SGA production pathway, Nakayasu et al. (2018) used CRISPR-Cas9 to create SGA-free potato lines. A long shelf life is an important breeding goal in the production of fruits and vegetables.

CRISPR was used to knock out ripening inhibitors (RIN) or DNA demethylase (DNA demethylase 2, DML2) to delay fruit ripening and extend its shelf life. However, altering the color of the peel and reducing the fruit's flavor and nutritional content significantly reduces the fruit's taste and sensory qualities (Ito et al. 2025; Ling et al. 2017).

6.1 Using CRISPR/Cas9 to Increase the Fruit's Resistance to Stress

The world's food production is seriously threatened by biotic and abiotic pressures, including diseases, pests, drought, and unusually high or low temperatures. These challenges can result in lower agricultural yields and quality, as well as financial losses for farmers. Agriculture now offers tremendous prospects to address these issues thanks to the emergence of gene-editing tools. Researchers can target specific genes associated with desired traits or stress tolerance by meticulously altering an organism's DNA using gene-editing methods like CRISPR/Cas9 (Zaidi and Saud 2020).

According to recent studies, resistance to a range of environmental difficulties can now be conferred using CRISPR/Cas technology (Joshi et al. 2020). It has been discovered that a Transcription Factor (TF) called Dehydration-Responsive Element Binding (DREB) is crucial for regulating several stress-induced genes. A subtype of DREB proteins called DREB2-type protein has been demonstrated to enhance drought, salinity, and heat tolerance in a variety of plants, including fruit-bearing trees (Lian et al. 2021). CRISPR/Cas9 has evolved into a potent tool for inserting potent transcriptional regulatory components into the gene promoter region in order to regulate the expression of stress-responsive genes. Thus, this augmentation boosts plant stress resistance by increasing their expression. The role of the Calcium-Dependent Protein Kinase (CDPK) in response to environmental stressors has been extensively shown by Wang and associates (Wang et al. 2016). Overexpression of the apple CDPK gene, also called MdCIPK6L, has greatly enhanced resistance to cold, drought, osmotic stress, and salt without influencing root growth (Kaur et al. 2020).

6.2 Using Genome Editing as a Strategy to Improve Fruit Crops

New technologies known as GE have emerged as a result of genetic engineering advancements during the last ten years. Site-directed insertion, deletion, or substitution at a particular locus in an organism's genome is referred to as GE (Zhang et al. 2017).

Three primary categories make up the GE toolbox: (1) CRISPR/Cas9 systems, (2) TALENs, and (3) ZFNs. These instruments use designed endonucleases that cause a double-strand break in the DNA at a specified place. DNA Double-Strand Breaks (DSBs) then trigger cellular DNA repair mechanisms such Homology-Directed Repair (HDR) and error-prone non-homologous end joining (NHEJ) (Xiong et al. 2015). While ZFNs and TALEN tools have been successfully used on plants, the introduction of CRISPR/Cas9 has fundamentally altered the way that GE is used on a range of plant species, including fruit crops. The Cas9 gene and sgRNA are the two parts of the designed CRISPR/Cas9 tool (Zhang et al. 2017).

The widely used Cas9 (SpCas9) gene (4107 bp) is a Type II gene that originates in *S. pyogenes* (Adli 2018; Wang et al. 2016a). The Cas9 protein contains two nuclear domains: RuvC and HNH. Each nuclear domain cuts one of the target DNA strands, resulting in a double-strand break in the target DNA sequence. The designed non-coding RNA known as sgRNA is produced by fusing two RNAs: the transactivating crRNA (tracrRNA) and the protospacer-matching crRNA. The CRISPR/Cas9 system is simpler to use and implement since the single gene encoding the sgRNA mimics the original crRNA: tracrRNA duplex formation (Jinek et al. 2012; Doudna and Charpentier 2014; Hsu et al. 2014; Liu et al. 2022). Table 2 explains the previous research from different authors about the changes in different fruit crops like banana, sweet variety of orange and apple which were driven by genome editing from various parts of fruits using CRISPR/Cas9.

Table 2. Changes in fruit crops driven by genome editing.

Fruit type	Genome editing tool	Targeted loci	Genotyped subject	Transformation mode	Feature	References
Banana	CRISPR/ Cas9	maPDS	Embryo genic cell suspension	Agrobacterium mediated	Synthesis of pigmented carotenoids	Kaur et al. (2018)
Sweet variety of orange	CRISPR/ Cas9	CsPDS	Leaf	Agrobacterium mediated	Synthesis of pigmented carotenoids	Jia and Wang (2014)
Apple	CRISPR/ Cas9	DIPM-1, DIPM-2, and DIPM-4 PDS uidA	Protoplasm Leaf leaf	Agrobacterium mediated Agrobacterium mediated Protoplast transformation	Fire blight disease resistance Carotenoid biosynthesis Beta-glucuronidase activity	Malnoy et al. (2016) Nishitani et al. (2016) Peer et al. (2014)

7. Fruit Growth and Development

One of the crucial quantitative characteristics that CRISPR/Cas can potentially improve crop development is the fruit size (Li et al. 2020). But it is crucial to remember that changing specific genes could have several effects. For instance, fruit weight and the *SlKLUH* gene's copy number are positively correlated. However, *SlKLUH* deletion or decrease often results in smaller fruits and can also lead to other growth problems, including smaller inflorescences and sterile flowers (Alonge et al. 2020).

Similarly, it has been demonstrated that the KLUH promoter contains conserved motifs associated with the promotion of plant organ growth in crops such as pepper and sweet cherry. This implies that *KLUH* gene can be utilized in the same way to create variations in fruit and seed size (Li et al. 2022). These results highlight the promise of genetic editing to improve specific features in crops, but they also highlight the significance of comprehending the genetic complexities and related negative effects of this technique.

7.1 Shelf-Life and Fruit Ripening

The fruit's shelf life, which is a crucial aspect of post-harvest fruit quality, depends on ethylene for both ripening and softening the fruit. According to recent studies, ethylene-related chemicals, ethylene biosynthesis, and signal transduction can either increase or reduce ethylene production in plums and apricots, hence prolonging the fruit's shelf life (Salazar et al. 2022).

Modifying methylation patterns or inhibiting important genes related to ripening, ethylene production, or their transmission of signals in fruit crops can result in fruits with a longer shelf life. To illustrate, MA-ACO1 (aminocyclopropane-1-carboxylateoxidase1), the gene encoding the enzyme that transforms ACC into ethylene, was deleted using the CRISPR/Cas9 gene-editing technology. When compared to bananas of the natural type, this genetic change produced fruit with a longer shelf life of up to 40 days (Hu et al. 2022).

7.2 Color, Flavor, and Bioactive Compounds of Fruits

The nutritional content and quality of fruits could be significantly increased by producing fruit cultivars with higher concentrations of beneficial metabolites and pigments by the modification of key genes. Research has focused on red fruits, which are abundant in a variety of nutrients and bioactive components, including vitamins, minerals, antioxidants, and dietary fiber (Skrovankova et al. 2015; Cosme et al. 2022). Accordingly, CRISPR/Cas9 technology can alter specific fruit traits. Using this technology, for example, to alter MdMKK9 led to a favorable rise in anthocyanin expression levels, which is what gives apples their distinctive red color (Sun et al. 2022).

Furthermore, a variety of secondary metabolites regarded as critical for component regeneration, growth regulation, and nutrition enhancement can be produced by plants. For example, carotenoids contribute to the vibrant hues of fruits and other plant organs by having antioxidant properties and taking part in processes like photosynthesis (Dall et al. 2012). The carotenoid production pathway's rate-limiting enzyme has been identified as phytoene desaturase (PDS) (Min et al. 2022).

The basic importance of this gene for the synthesis of carotenoids was highlighted by the mutants with an albino phenotype that were produced when PDS was eliminated by utilizing the CRISPR/Cas9. Altering the lycopene epsilon cyclase (LCYε) gene in bananas enriched in β-carotene using CRISPR/Cas9 (Sattar et al. 2021). The LCYε gene has insertion-deletion mutations, according to sequence analysis, which greatly increased β-carotene accumulation without negatively influencing agro-morphological traits. Pomegranates, a fruit rich in phenolic metabolites, had their levels of phenolic anthocyanins reduced when two UDP-glycosyl transferase genes, PgUGT84A23 and PgUGT84A24, were removed using CRISPR/Cas9 (Chang et al. 2019). Additionally, CRISPR/Cas9 was used to genetically eliminate the L-idonate dehydrogenase (*IdnDH*) gene, which regulates the production of Tartaric Acid (TA) in grapes (Ren et al. 2016). The MdAAT1 gene has been connected to the synthesis of esters in apples (Goulet et al. 2015). Furthermore, the same regulatory function of NAC transcription factors in the production of flavor ester in apples emphasizes the importance of NAC1 homologs in other fruit species (Cao et al. 2021).

8. Challenges in Applying Genome Editing in Fruit Crops

Over the past ten years, genome editing has been used to increase plant immunity and development, but there are still numerous obstacles to overcome before genome editing systems can be improved. Genome editing in fruit crops is hampered by six main issues: (1) Predicting high-efficiency sgRNAs is difficult; (2) Cas protein variants' protein expression as well as enzyme activity are not entirely optimized; (3) The effectiveness of sgRNA and Cas delivery systems remains low; (4) More improvement is required in genome editing; (5) Gene-edited plantlets require a time-consuming and ineffective regeneration method; (6) Another significant obstacle is transgene-free genome editing, particularly for these polyploid fruit crops.

Through the HDR mechanism, CRISPR/Cas9 may accurately add or swap out DNA from a specific location on the genome. Compared to knock-out by NHEJ, knock-in via HDR occurs much less frequently. Thus, the CRISPR/Cas9 knock-out method has been utilized for crop plants to acquire physiologically important characters, resistance against disease, male sterility, drought tolerance, altered ripening, and other agronomic properties. Nevertheless, knock-in has a lot of untapped potential.

Plant transgene and/or native gene activity may be hampered by the random insertions caused by classic Agrobacterium-mediated overexpression, which has a high likelihood of multi-copy integration at various genomic sites. Increasing knock-in-based delivery of the donor template along with the CRISPR/Cas9 element is a significant problem. Few studies show that the knock-in-based method works less effectively in plants like moss, wheat, maize, tobacco, and *Arabidopsis* (Svitashev et al. 2015; Cermák et al. 2015; Humanes et al. 2017; Miki et al. 2018).

To improve the knock-in efficiency, the length of the homology arms is an important consideration. A recent study in Arabidopsis found that knock-in efficiency was higher when the donor and CRISPR constructs were used independently rather than together (Miki et al. 2018). Additionally, it has been shown that knock-in efficiency is increased in *Arabidopsis* when Cas9 expression is germline-driven (Miki et al. 2018). Given the variety of uses for CRISPR/Cas9 knock-in, it is imperative to optimize this approach in fruit plants.

9. Economic Trait Improvement in Fruit Crops

Devastating diseases pose a threat to fruit crops, and traditional methods of disease control, quite successful when used depending on resistant cultivars and agrochemicals. However, recombination or mutation may quickly produce new genotypes that are susceptible to the resistance gene or the phytosanitary agent because many plant diseases have a high capacity for evolution. A recent attempt was made to generate resistance in major fruit crops using CRISPR. Citrus canker, a devastating disease that causes significant financial losses worldwide, affects most commercial citrus cultivars. It is caused by Xanthomonas citri subspecies citri (Xcc) (Barbieri et al. 2023; Theivanai et al. 2020). Citrus canker can be effectively and sustainably controlled by creating citrus types that are resistant to the disease (Shahbaz et al. 2022; Ali et al. 2023).

The genes involved in susceptibility are discovered similarly to those responsible for resistance to a certain disease. The pathogenicity factor PthA4 activated the CsLOB1 (Citrus sinensis Lateral Organ Boundaries) gene during infection, which was found to be a source of vulnerability for citrus cancer (Cardoso et al. 2022; Shahbaz et al. 2022). Canker resistance can be attained by altering the PthA4 Effector Binding Elements (EBEs) in the CsLOB1 Promoter (EBEPthA4-CsLOBP) of the *CsLOB1* gene. The Duncan grapefruit's canker resistance was caused by the disruption of Type I and

Type II alleles of CsLOB1, and the editing efficiency has been reported to range from 15 to 81% (Jia et al. 2016; Jia et al. 2017b; Peng et al. 2017; Su et al. 2024).

The powdery mildew susceptibility gene *MLO-7* of grapes and the fire blight susceptibility genes *DIPM-1, DIPM-2*, and *DIPM-4* of apples are also altered to create resistant cultivars (Malnoy et al. 2016; Siva et al. 2023).

9.1 Improved Nutritional Content

Genome editing can also be used to increase the nutritional value of crops by increasing the levels of essential components such as oil, phenolic acid, vitamins, and minerals (Kumar et al. 2022). This is particularly important in developing countries where vitamin deficiencies are prevalent and access to a diverse diet is limited. To change the amount of oleic acid in soybean seeds, the *GmFAD2-1A* and *GmFAD2* genes of the Fatty Acid Desaturase (FAD2) family were knocked out (Wu et al. 2020). The loss of function of these genes has allowed for the generation of soybean genotypes that have a significant oleic acid content.

The *SmLACs* genes of the laccase family, which consists of 29 highly similar family members, were concurrently silenced in *Salvia miltiorrhiza* by concentrating on their conserved sequences, revealing their role in the manufacture of medicinal phenolic acid compounds (Zhou et al. 2021). *SmLACs* have been found to contribute to lignin formation and phenolic acid biosynthesis in the roots because of decreased gene target expression, retarded hairy growth of the roots, wider and more flexible xylem cells, minimal RA and SAB storage, and extremely low lignin content in edited lines compared to wild types.

By utilizing a single sgRNA to knock out six genes, which as *BBLa, BBLc, BBLd.2, BBLb, BBLd.1,* and *BBLe* that belong to the Berberine Bridge enzyme-Like (BBL) family, a nicotine-free tobacco variety was produced, and the amount of nicotine was reduced (Jouanin et al. 2019). This is one way that using tobacco without nicotine may protect smokers' health.

10. Utilization and Difficulties with the CRISPR/Cas9 Technique for Improving Fruit Crops

Crop genome editing presents several difficulties despite its possible advantages. The regulatory environment is a great challenge. The usage of Genetically Modified Organisms (GMOs) is governed by laws in many nations, and it is currently unclear how gene-edited crops will be governed. Even if the modifications made to the DNA sequence are much more subtle and accurate, gene-edited crops may occasionally be ruled using the same laws as GMOs.

Current plant biotechnology methods have been completely transformed utilizing the CRISPR/Cas9 tool due to excellent efficiency, simplicity of design, and flexibility, which makes this the most widely used toolkit for plant GE today. The plants that are altered using this method are identical to those produced through natural or artificial mutations or traditional breeding. When compared to breeding or mutagenesis techniques, CRISPR/Cas9-based gene editing is extremely accurate. Avoiding the costly and time-consuming regulations required for transgenic evaluation would also be advantageous because GE-modified plants are thereafter considered non-transgenic. Aside from several encouraging comments, several issues must be resolved when using CRISPR/Cas9 to boost fruit crops as shown in Fig. 3. The figure explains the applications and challenges of suing CRISPR/CAS9 to improve fruit crops. It states that the resistance against biotic and abiotic stress can be improved, enrichment of micronutrients, development of seedless fruits, and reduction in antinutritional substances are among the primary benefits that are emphasized at the top. These developments can significantly improve the nutritional value, resilience, and quality of fruit crops. However, the lower part of the figure talks about several technical problems that limit its full potential, including incomplete genomic sequences, off-target effects, difficulties delivering CRISPR, and the need for efficient tissue culture systems. These combined advantages

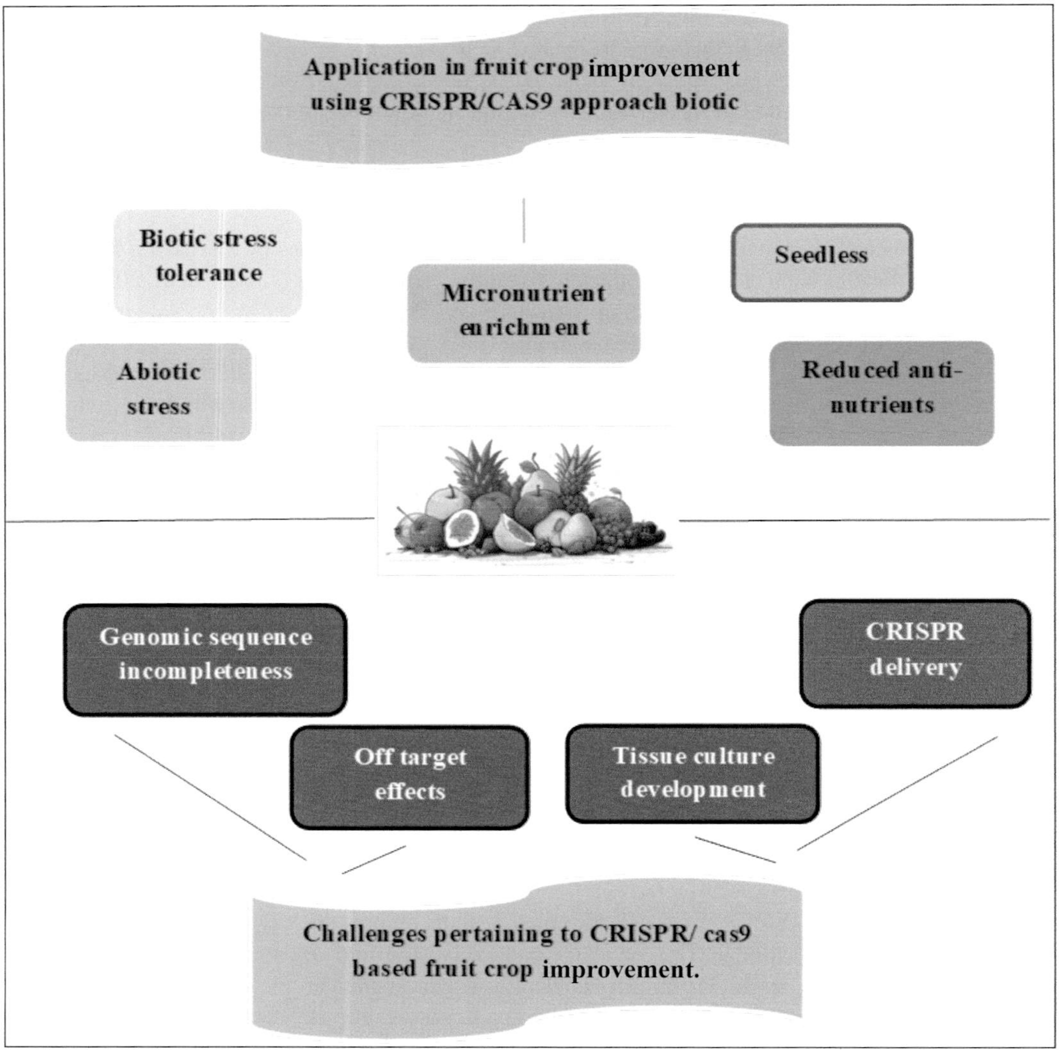

Figure 3. A description of the CRISPR-Cas9 gene-editing technology for fruit plants, including its applications and problems.

and disadvantages have shaped the current state of CRISPR/Cas9 in fruit crop biotechnology. All these features and characteristics are clearly mentioned in the Fig. 3.

11. Future Direction

CRISPR/Cas9 genome editing is a groundbreaking approach for both scientific and practical studies in plants and animals. However, introducing CRISPR/Cas9 to most fruit crops is still a laborious and challenging process because of transformational obstacles, complex genomes, lengthy development cycles, and a lack of genomic information. Some of the difficulties in horticulture crops may be resolved with the rapid advancement of new and enhanced CRISPR/Cas9-based instruments and delivery systems. Techniques that boost editing efficiency or lessen the requirement for multi-generation cycles are likewise very sought after. There is an urgent need for quick and inexpensive mutation screening.

LOB1 gene in citrus fruit and *CLV3* gene in tomato serve as examples of how to cleverly create an allelic series by targeting cis-elements to change gene regulation. Commercialization and consumer acceptability of trans gene-free modified fruit crops will depend largely on production strategies. Furthermore, it is impossible to overstate the significance of investing in consumer-friendly features like longer shelf life and healthier products, as well as promoting public discussion on genome editing.

The CRISPR/Cas system was successful for producing loss-of-function mutants and mutations. However, plant breeders often find it more appealing to overexpress genes to obtain advantageous agronomic features. Many biological and ethical problems arise with conventional gene overexpression systems, which often require the introduction of an exogenous transgene. Therefore, it is necessary to create novel techniques for introducing foreign transgenic components to stimulate the gene expression newly created method that combines chemically modified donor DNA with CRISPR/Cas has been said to be able to insert sequences up to 2049 base pairs long. To stabilize donor DNA and the CRISPR/Cas system, a 5′-phosphorylated double-stranded oligodeoxynucleotide (dsODN) was developed especially for this method. Translational enhancers could be inserted into the target gene's 5′ untranslated region to boost its expression (Lu et al. 2020). More flexible than traditional gene knockout, precise genome editing provides advanced techniques to remove, alter, or modify gene function by targeted nucleotide substitutions, gene insertions/deletions, and gene replacements (Gao 2021).

References

Abdelrahman, M., Z. Wei, J.S. Rohila and K. Zhao. 2021. Multiplex genome-editing technologies for revolutionizing plant biology and crop improvement. Front. Plant Sci. 12: 721203.

Adli, M. 2018. The CRISPR tool kit for genome editing and beyond. Nat. Commun. 9(1): 1911.

Ahmar, S., R.A. Gill, K.H. Jung, A. Faheem, M.U. Qasim, M. Mubeen et al. 2020. Conventional and molecular techniques from simple breeding to speed breeding in crop plants: recent advances and future outlook. Int. J. Mol. Sci. 21(7): 2590.

Ali, S., A. Hameed, G. Muhae-Ud-Din, M. Ikhlaq, M. Ashfaq, M. Atiq et al. 2023. Citrus canker: a persistent threat to the worldwide citrus industry—An analysis. Agro. 13(4): 1112.

Alonge, M., X. Wang, M. Benoit, S. Soyk, L. Pereira, L. Zhang et al. 2020. Major impacts of widespread structural variation on gene expression and crop improvement in tomato. Cell 182: 145–161.

Amitai, G. and R. Sorek. 2016. CRISPR–Cas adaptation: insights into the mechanism of action. Nat. Rev. Microbiol. 14(34): 67–76.

Anand, T., G.K. Ingle, G.S. Meena, J. Kishore and R. Kumar. 2014. Hypertension and its correlates among school adolescents in Delhi. Int. J. Prev. Med. 5: 65e70.

Ballester, A.R., J. Molthoff, R. de Vos, B.T.L. Hekkert, D. Orzaez, J.P. Fernandez-Moreno et al. 2010. Biochemical and molecular analysis of pink tomatoes: deregulated expression of the gene encoding transcription factors SlMYB12 leads to pink tomato fruit color. Plant Physiol. 152: 71–84.

Barbieri, H.B., L.S. Fernandes, J.G.D.M. Pontes, A.K. Pereira and T.P. Fill. 2023. An overview of the most threating diseases that affect worldwide citriculture: Main features, diagnose, and current control strategies. Frontiers in Nat. Pro. 2: 1045364.

Bawa, A.S. and K.R. Anilakumar. 2013. Genetically modified foods: Safety, risks and public concerns—a review. J. Food Sci. Technol. 50: 1035–1046.

Begna, T. 2021. Conventional breeding methods widely used to improve self-pollinated crops. Int. J. of Research 7(1): 1–16.

Bigliardi, B. and F. Galati. 2013. Innovation trends in the food industry: The case of functional foods. Trends Food Sci. Technol. 31: 118–129.

Bouis, H.E. 2002. Plant breeding: a new tool for fighting micronutrient malnutrition. J. Nutr. 132: 491–494. https://doi.org/10.1093/jn/132.3.491s.

Brumos, J. 2021. Gene regulation in climacteric fruit ripening. Curr. Opin. Plant Biol. 63: 102042.

Cao, X., C. Wei, W. Duan, Y. Gao, J. Kuang, M. Liu et al. 2021. Transcriptional and epigenetic analysis reveals that NAC transcription factors regulate fruit flavor ester biosynthesis. Plant J. 106: 785–800.

Cardoso, J.L.S. 2022. *De novo* transcriptome assembly, functional annotation and expression profiling of sweet passion fruit (Passiflora alata) in response to Xanthomonas axonopodis pv. passiflorae infection (Doctoral dissertation, Universidade de São Paulo).

Carroll, D. 2011. TALE–based-targeted genome editing: prospects and perspectives. Nature Biotechnology 29: 135–136. https://doi.org/10.1038/nbt.1835 (often cited as discussion/editorial).

Casals, J., R.R. del Castillo, C. Pons, A. Mazzucato, I. Tringovska, G. Pasev et al. 2024. European fresh-market tomato sensory ideotypes based on consumer preferences. Scientia Horticulturae 335: 113351.

Çelik, S., U. Deniz, O. Gencer and E. Aksoy. 2024. Editing of banana, apple, and grapevine genomes using the CRISPR-Cas9 system. Horticulture Crops 349–364.

Cermák, T., N.J. Baltes, R. Cegan, Y. Zhang and D.F. Voytas. 2015. High-frequency, precise modification of the tomato genome. Genome Biol. 16: 232.

Chandran, S., V. Muthu, T. Umapathy, S. Jayakumar and S. Chokkalingam. 2023. CRISPR/Cas9 assisted genome editing technology for the improvement of Horticultural crops. J. Phytopharmacol. 12(2): 127–134.

Chang, L., S. Wu and L. Tian. 2019. Effective genome editing and identification of a regiospecific gallic acid 4-O-glycosyltransferase in pomegranate (*Punica granatum* L.). Hortic. Res. 6: 123.

Charrier, A., E. Vergne, N.J.P. Dousset, A. Richer, A. Petiteau and E. Chevreau. 2019. Efficient targeted mutagenesis in apple and first time edition of pear using the CRISPR-Cas9 system. Front. Plant Sci. 10(40): 1–12.

Chen, L., D. Yang and Y. Zhang. 2018. Evidence for a specific and critical role of mitogen-activated protein kinase 20 in uni-to-binucleate transition of microgametogenesis in tomato. Plant Biol. 219: 176–194.

Cherian, S., C.R. Figueroa and H. Nair. 2014. 'Movers and shakers' in the regulation of fruit ripening: a cross-dissection of climacteric versus non-climacteric fruit. J. Exp. Bot. 65: 4705–22.

Cockerton, H., A. Karlström, A.W. Johnson, B. Li, E. Stavridou, K.J. Hopson et al. 2021. Genomic informed breeding strategies for strawberry yield and fruit quality traits. Front. Plant Sci. 12: 724847.

Cong, L., F.A. Ran, D. Cox, S. Lin, R. Barretto, N. Habib et al. 2013. Multiplex Genome Engineering Using CRISPR/Cas Systems. Science 339: 819–823.

Cornille, A., T. Giraud and M.J.M. Smulders. 2014. The domestication and evolutionary ecology of apples. Trends Genet. 30: 57–65. https://doi.org/10.1016/j.tig.2013.10.002.

Cosme, F., T. Pinto, A. Aires, M.C. Morais, E. Bacelar, R. Anjos et al. 2022. Red fruits composition and their health benefits—A review. Foods 11(5): 644.

Crosby, J.A., J. Janick and P.C. Pecknold. 1992. Breeding apples for scab resistance: 1945–1990. pp. 43–70. *In*: Acta horticulturae. Int. Soc. Hortic. Sci. (ISHS), Leuven.

Dall'Osto, L., N.E. Holt, S. Kaligotla, M. Fuciman, S. Cazzaniga, D. Carbonera et al. 2012. Zeaxanthin protects plant photosynthesis by modulating chlorophyll triplet yield in specific lightharvesting antenna subunits. Journal of Biological Chemistry 287(50): 41820–41834. https://doi.org/10.1074/jbc.M112.405498.

Dall'Osto, L., N.E. Holt, S. Kaligotla, M. Fuciman, S. Cazzaniga, D. Carbonera et al. 2012. Zeaxanthin protects plant photosynthesis by modulating chlorophyll triplet yield in specific light-harvesting antenna subunits. J. Biol. Chem. 287: 41820–41834.

Dalla C.L., M. Malnoy and I. Gribaudo. 2017. Breeding next generation tree fruits: technical and legal challenges. Hortic Res. 4: 17067.

Dalla Costa, L., M. Malnoy and I. Gribaudo. 2017. Breeding next generation tree fruits: Technical and legal challenges. Horticulture Research 4: 17067. https://doi.org/10.1038/hortres.2017.67.

Dash, P.K. and R. Rai. 2016. Translating the "banana genome" to delineate stress resistance, dwarfing, parthenocarpy and mechanisms of fruit ripening. Front. Plant Sci. 7: 1543.

Deng, L., H. Wang, C. Sun, Q. Li, H. Jiang, M. Du et al. 2018. Efficient generation of pink-fruited tomatoes using CRISPR/Cas9 system. J. Genet. Genom. 45: 51–54.

Dorantes-Acosta, A.E., C.V. Sánchez-Hernández and M.A. Arteaga-Vazquez. 2012. Biotic stress in plants: Life lessons from your parents and grandparents. Front. Genet. 3: 256.

Doudna, J.A. and E. Charpentier. 2014. The new frontier of genome engineering with CRISPR-Cas9. Science 346(6213): 1258096.

Dutt, M., Z. Mou, X. Zhang, S.E. Tanwir and J.W. Grosser. 2020. Efficient CRISPR/Cas9 genome editing with Citrus embryogenic cell cultures. BMC Biotechnol. 20(58): doi:org/10.1186/s12896020-00652-9.

Essiaw-Quayson, G. 2017. Parasitological and microbial quality of some fresh-cut ready-to-eat fruits and vegetables sold in some markets in the Kumasi Metropolis, Ghana. J. Food Qual. 2023: 9495339.

Farinati, S., S. Draga, A. Betto, F. Palumbo, A. Vannozzi, M. Lucchin et al. 2023. Current insights and advances into plant male sterility: new precision breeding technology based on genome editing applications. Front. Plant Sci. 14: 1223861.

Fedak, G., D. Chi, C. Hiebert, T. Fetch, B. McCallum, A. Xue et al. 2021. Capturing multiple disease resistance in wheat through intergeneric hybridization. Biology 10(7): 631.

Feng, J., C. Dai, H. Luo, Y. Han, Z. Liu and C. Kang. 2018. Reporter gene expression reveals precise auxin synthesis sites during fruit and root development in wild strawberry. J. Exp. Bot. 70: 563–574.

Feng, S., J. Zhang and Z. Mu. 2020. Recent progress on the molecular breeding of *Cucumis sativus* L. in China. Theor. Appl. Genet. 133: 1777–1790. https://doi.org/10.1007/s00122-019-03484-0.

Filler Hayut, S., C. MelamedBessudo and A.A. Levy. 2017. Targeted mutagenesis in tomato using CRISPR/Cas9 system. Plant Biotechnology Journal 15(8): 1019–1023. https://doi.org/10.1111/pbi.12604.

Fister, A.S., L. Landherr, S.N. Maximova and M.J. Guiltinan. 2018. Transient expression of CRISPR/Cas9 machinery targeting TcNPR3 enhances defense response in Theobroma cacao. Front. Plant Sci. 9: 268.

Gao, C. 2021. Genome engineering for crop improvement and future agriculture. Cell 184: 1621–1635.

Gibney, A. and A. Kellett. 2024. Gene editing with artificial DNA scissors. Chem. Eur. J. 30(57): e202401621.

Giovannoni, J. 2001. Molecular biology of fruit maturation and ripening. Annu. Rev. Plant Physiol. Plant Mol. Biol. 52: 725–749.

Giovannoni, J., C. Nguyen, B. Ampofo, S. Zhong and Z. Fei. 2017. The epigenome and transcriptional dynamics of fruit ripening. Annu. Rev. Plant Biol. 68: 61–84.

Goulet, C., Y. Kamiyoshihara, N.B. Lam, T. Richard, M.G. Taylor, D.M. Tieman et al. 2015. Divergence in the enzymatic activities of a tomato and Solanum pennellii alcohol acyltransferase impacts fruit volatile ester composition. Mol. Plant 8: 153–162.

Gumtow, R., D. Wu, J.Y. Uchida and M. Tian. 2018. A Phytophthora palmivora extracellular cystatin-like protease inhibitor targets papain to contribute to virulence on papaya. Mol. Plant Microbe Interact. 31: 363–373.

Guo, S., S. Zhao, H. Sun, X. Wang, S. Wu, T. Lin et al. 2019. Resequencing of 414 cultivated and wild watermelon accessions identifies selection for fruit quality traits. Nat. Genet. 51: 1616–1623.

Hayut, S., C. Melamed Bessudo and A. Levy. 2017. Targeted recombination between homologous chromosomes for precise breeding in tomato. Nat. Commun. 8: 15605.

Hsu, P.D., E.S. Lander and F. Zhang. 2014. Development and applications of CRISPR-Cas9 for genome engineering. Cell 157(6): 1262–1278.

Hu, H., A. Scheben, B. Verpaalen, S. Tirnaz, P.E. Bayer, R.G.J. Hodel et al. 2022. Amborella gene presence/absence variation is associated with abiotic stress responses that may contribute to environmental adaptation. New Phytol. 233: 1548–1555.

Hu, Y., J. Zhang, H. Jia, D. Sosso, T. Li, W.B. Frommer et al. 2014. Lateral organ boundaries 1 is a disease susceptibility gene for citrus bacterial canker disease. Proc. Natl. Acad. Sci. U.S.A. 111: E521–E529.

Huang, S., J. Ding, D. Deng, W. Tang, H. Sun, D. Liu et al. 2013. Draft genome of the kiwifruit Actinidia chinensis. Nat. Commun. 4: 2640.

Humanes, J., Y. Wang, Z. Liang, Q. Shan, C.V. Ozuna, S. Sánchez-León et al. 2017. High-efficiency gene targeting in hexaploid wheat using DNA replicons and CRISPR/Cas9. Plant J. 89(6): 1251–1262.

Ito, Y., A. Nishizawa-Yokoi, M. Endo, M. Mikami and S. Toki. 2015. CRISPR/Cas9-mediated mutagenesis of the RIN locus that regulates tomato fruit ripening. Biochem. Biophys. Res. Commun. 467: 76–82.

Jaganathan, D., K. Ramasamy, G. Sellamuthu, S. Jayabalan and G. Venkataraman. 2018. CRISPR for crop improvement: An update review. Front. Plant Sci. 9: 985.

Javaid, D., S.Y. Ganie, Y.A. Hajam and M.S. Reshi. 2022. CRISPR/Cas9 system: a reliable and facile genome editing tool in modern biology. Mol. Biol. Rep. 49(12): 12133–12150.

Jia, H. and N. Wang. 2014. Targeted genome editing of sweet orange using Cas9/sgRNA. PLoS One 9(4): e93806.

Jia, H. and N. Wang. 2014. Xcc-facilitated agroinfiltration of citrus leaves: a tool for rapid functional analysis of transgenes in citrus leaves. Plant Cell Rep. 33(12): 1993–2001.

Jia, H., V. Orbovic, J.B. Jones and N. Wang. 2016. Modification of the PthA4 effector binding elements in Type I CsLOB1 promoter using Cas9/sgRNA to produce transgenic Duncan grapefruit alleviating XccΔpthA4:dCsLOB1.3 infection. Plant Biotechnol. J. 14(5): 1291–1301.

Jia, H., Y. Zhang, V. Orbović, J. Xu, F.F. White, J.B. Jones et al. 2017. Genome editing of the disease susceptibility gene CsLOB1 in citrus confers resistance to citrus canker. Plant Biotechnol. J. 15: 817–823.

Jinek, M., K. Chylinski, I. Fonfara, M. Hauer, J.A. Doudna and E. Charpentier. 2012. A programmable dual-RNA–guided DNA endonuclease in adaptive bacterial immunity. Science 337(6096): 816–821.

Jiwan, D., E.H. Roalson, D. Main and A. Dhingra. 2013. Antisense expression of peach mildew resistance locus O (PpMlo1) gene confers cross-species resistance to powdery mildew in Fragaria × ananassa. Transgenic Res. 22: 1119–1131.

Joshi, R.K., S.S. Bharat and R. Mishra. 2020. Engineering drought tolerance in plants through CRISPR/Cas genome editing. 3 Biotech 10: 400.

Jouanin, A., F. Gilard, S. Renouard, G. Tcherkez, C. Rihouey, B. Gakière et al. 2019. The Berberine Bridge enzyme-like gene family is required for a nicotine-free tobacco plant obtained by CRISPR-Cas9. Scientific Reports 9: 11434. https://doi.org/10.1038/s41598-019-47850-6.

Kaiser, N., D. Douches, A. Dhingra, K.C. Glenn, P.R. Herzig, E.C. Stowe et al. 2020. The role of conventional plant breeding in ensuring safe levels of naturally occurring toxins in food crops. Trends Food Sci. Technol. 100: 51–66

Kapoor, L., A.J. Simkin, C. George Priya Doss and R. Siva. 2022. Fruit ripening: dynamics and integrated analysis of carotenoids and anthocyanins. BMC Plant Biol. 22(1): 27.

Karkute, S.G., A.K. Singh, O.P. Gupta, P.M. Singh and B. Singh. 2017. CRISPR/Cas9 mediated genome engineering for improvement of horticultural crops. Front. Plant Sci. 8: 1635.

Kaul, T., S.K. Sony, R. Verma, K.F.A. Motelb, A.T. Prakash, M. Eswaran et al. 2020. Revisiting CRISPR/Cas-mediated crop improvement: special focus on nutrition. J. Biosci. 45: 137.

Kaur, N., A. Alok, Shivani, N. Kaur, P. Pandey, P. Awasthi et al. 2018. CRISPR/Cas9-mediated efficient editing in phytoene desaturase (PDS) demonstrates precise manipulation in banana cv. Rasthali genome. Funct. Integr. Genomics 18(1): 89–99.

Kaur, N., P. Awasthi and S. Tiwari. 2020. Fruit crops improvement using CRISPR/Cas9 system. *In*: Elsevier Inc., Amsterdam, ISBN 9780128181409.

Kramer, M.G. and K. Redenbaugh. 1994. Commercialization of a tomato with an antisense polygalacturonase gene—The Flavr Savr™ tomato story. Euphytica 79(3): 293–297. https://doi.org/10.1007/BF00022530.

Kramer, M.G. and K. Redenbaugh. 1994. Commercialization of a tomato with an antisense polygalacturonase gene: the FLAVR SAVR™ tomato story. Euphytica 79: 293–297.

Kris-Etherton, P.M., K.D. Hecker, A. Bonanome, S.M. Coval, A.E. Binkoski, K.F. Hilpert et al. 2002. Bioactive compounds in foods: their role in the prevention of cardiovascular disease and cancer. Am. J. Med. 113: 71–88.

Kumar, D., A. Yadav, R. Ahmad, U.N. Dwivedi and K. Yadav. 2022. CRISPR-based genome editing for nutrient enrichment in crops: a promising approach toward global food security. Front. Genet. 13: 932859.

Kumar, K. and R. Kaur. 2021. Genetic engineering in fruit crops. pp. 191–223. *In*: Winemaking. CRC Press.

Lang, Z., Y. Wang, K. Tang, D. Tang, T. Datsenka, J. Cheng et al. 2017. Critical roles of DNA demethylation in the activation of ripening-induced genes and inhibition of ripening-repressed genes in tomato fruit. Proc. Natl. Acad. Sci. U.S.A. 114: E4511–E4519.

Leisner, C.P. 2020. Review: climate change impacts on food security—focus on perennial cropping systems and nutritional value. Plant Sci. 293: 110412.

Li, H., M. Qi, M. Sun, Y. Liu, Y. Liu, T. Xu et al. 2017. Tomato transcription factor SlWUS plays an important role in tomato flower and locule development. Front. Plant Sci. 8: 457.

Li, L., M.J. Piatek, A. Atef, A. Piatek, A. Wibowo, X. Fang et al. 2012. Rapid and highly efficient construction of TALE-based transcriptional regulators and nucleases for genome modification. Plant Mol. Biol. 78: 407–416.

Li, Q., M. Sapkota and E. van der Knaap. 2020. Perspectives of CRISPR/Cas-mediated cis-engineering in horticulture: Unlocking the neglected potential for crop improvement. Hortic. Res. 7: 36.

Li, Q., Q. Feng, A. Snouffer, B. Zhang, G.R. Rodríguez and E. van der Knaap. 2022. Increasing fruit weight by editing a cis-regulatory element in tomato KLUH promoter using CRISPR/Cas9. Front. Plant Sci. 13: 879642.

Li, T., X. Yang, Y. Yu, X. Si, X. Zhai, H. Zhang et al. 2018. Domestication of wild tomato is accelerated by genome editing. Nat. Biotechnol. 36: 1160–1163.

Li, X., Y. Wang, S. Chen, H. Tian, D. Fu, B. Zhu et al. 2018. Lycopene is enriched in tomato fruit by CRISPR/Cas9-mediated multiplex genome editing. Front. Plant Sci. 9: 559.

Lian, X., X. Zhao, Q. Zhao, G. Wang, Y. Li and Y. Hao. 2021. MdDREB2A in apple is involved in the regulation of multiple abiotic stress responses. Hortic. Plant J. 7: 197–208.

Lin, T., G. Zhu, J. Zhang, X. Xu, Q. Yu, Z. Zheng et al. 2014. Genomic analyses provide insights into the history of tomato breeding. Nat. Genet. 46: 1220–1226.

Liu, Y., F. Pinto, X. Wan, Z. Yang, S. Peng, M. Li et al. 2022. Reprogrammed tracrRNAs enable repurposing of RNAs as crRNAs and sequence-specific RNA biosensors. Nat. Commun. 13: 1937.

Lu, Y., Y. Tian, R. Shen, Q. Yao, M. Wang, M. Chen et al. 2020. Targeted, efficient sequence insertion and replacement in rice. Nat. Biotechnol. 38: 1402–1407.

Ma, X., Q. Zhu, Y. Chen and Y.G. Liu. 2016. CRISPR/Cas9 platforms for genome editing in plants: Developments and applications. Mol. Plant 9(7): 961–974.

Mahfouz, M.M., L. Li, M. Shamimuzzaman, A. Wibowo, X. Fang and J.K. Zhu. 2011. *De novo*-engineered transcription activator-like effector (TALE) hybrid nuclease with novel DNA binding specificity creates double-strand breaks. Proc. Natl. Acad. Sci. USA 108: 2623–2628.

Malnoy, M., R. Viola, M.-H. Jung, O.-J. Koo, S. Kim, J.-S. Kim et al. 2016. DNA-free genetically edited grapevine and apple protoplast using CRISPR/Cas9 ribonucleoproteins. Front. Plant Sci. 7: 1904.

Mao, Y., H. Zhang, N. Xu, B. Zhang, F. Gou and J.-K. Zhu. 2013. Application of the CRISPR–Cas system for efficient genome engineering in plants. Mol. Plant 6: 2008–2011.

Martín-Pizarro, C., J.C. Triviño and D. Posé. 2019. Functional analysis of the TM6 MADS-box gene in the octoploid strawberry by CRISPR/Cas9-directed mutagenesis. J. Exp. Bot. 70: 885–895.

Meng, X., D. Yang, X. Li, S. Zhao, N. Sui and Q. Meng. 2015. Physiological changes in fruit ripening caused by overexpression of tomato SlAN2, an R2R3-MYB factor. Plant Physiol. Biochem. 89: 24–30.

Meyer, R.S. and M.D. Purugganan. 2013. Evolution of crop species: Genetics of domestication and diversification. Nat. Rev. Genet. 14: 840–852.

Miao, J., D. Guo, J. Zhang, Q. Huang, G. Qin, X. Zhang et al. 2013. Targeted mutagenesis in rice using CRISPR-Cas system. Cell Res. 23: 1233–1236.

Miki, D., W. Zhang, W. Zeng, Z. Feng and J.-K. Zhu. 2018. CRISPR/Cas9-mediated gene targeting in Arabidopsis using sequential transformation. Nat. Commun. 9: 1967.

Mikulic-Petkovsek, M., A. Ivancic, V. Schmitzer, R. Veberic and F. Stampar. 2016. Comparison of major taste compounds and antioxidative properties of fruits and flowers of different Sambucus species and interspecific hybrids. Food Chem. 200: 134–140.

Miller, A.J. and B.L. Gross. 2011. From forest to field: Perennial fruit crop domestication. Am. J. Bot. 98: 1389–1414.

Millstone, E., A. Stirling and D. Glover. 2015. Regulating genetic engineering: The limits and politics of knowledge. Issues Sci. Technol. 31: 23–26.

Min, T., D. Hwarari, D. Li, A. Movahedi and L. Yang. 2022. CRISPR-based genome editing and its applications in woody plants. Int. J. Mol. Sci. 23: 10175.

Moose, S.P. and R.H. Mumm. 2008. Molecular plant breeding as the foundation for 21st century crop improvement. Plant Physiol. 147: 969–977 [https://doi.org/10.1104/pp.108.118232].

Naim, F., B. Dugdale, J. Kleidon, A. Brinin, K. Shand, P. Waterhouse et al. 2018. Gene editing the phytoene desaturase alleles of Cavendish banana using CRISPR/Cas9. Transgenic Res. 27(5): 451–460.

Naim, F., B. Dugdale, J. Kleidon, A. Brinin, K. Shand, P. Waterhouse et al. 2018. Gene editing the phytoene desaturase alleles of Cavendish banana using CRISPR/Cas9. Transgenic Research 27(5): 451–460. https://doi.org/10.1007/s11248-018-0083-0.

Nakajima, I., Y. Ban, A. Azuma, N. Onoue, T. Moriguchi, T. Yamamoto et al. 2017. CRISPR/Cas9-mediated targeted mutagenesis in grape. PLoS One 12(5): 1–7.

Nakayasu, M., R. Akiyama, H.J. Lee, K. Osakabe, Y. Osakabe, B. Watanabe et al. 2018. Generation of α-solanine-free hairy roots of potato by CRISPR/Cas9-mediated genome editing of the St16DOX gene. Plant Physiol. Biochem. 131: 70–77.

Neto, R.R.D.S. 2022. Characterization of CsLIEXP1 as a new susceptibility gene to Xanthomonas citri subsp. citri and gene editing by CRISPR/Cas9 for citrus canker tolerance.

Nishitani, C., N. Hirai, S. Komori, M. Wada, K. Okada, K. Osakabe et al. 2016. Efficient genome editing in apple using a CRISPR/Cas9 system. Sci. Rep. 6: 31481.

Nonaka, S., C. Arai, M. Takayama, C. Matsukura and H. Ezura. 2017. Efficient increase of γ-aminobutyric acid (GABA) content in tomato fruits by targeted mutagenesis. Sci. Rep. 7: 7057.

Omura, M. and T. Shimada. 2016. Citrus breeding, genetics and genomics in Japan. Breeding Science 66(1): 3–17. https://doi.org/10.1270/jsbbs.66.3.

Parajuli, R., G. Thoma and M.D. Matlock. 2019. Environmental sustainability of fruit and vegetable production supply chains in the face of climate change: A review. Sci. Total Environ. 650: 2863–2879.

Parmar, N., K.H. Singh, D. Sharma, L. Singh, P. Kumar, J. Nanjundan et al. 2017. Genetic engineering strategies for biotic and abiotic stress tolerance and quality enhancement in horticultural crops: A comprehensive review. 3 Biotech. 7: 239.

Peer, R., G. Rivlin, S. Golobovitch, M. Lapidot, A. Gal-On, A. Vainstein et al. 2014. Targeted mutagenesis using zinc-finger nucleases in perennial fruit trees. Planta 241(4): 941–951.

Peng, A., S. Chen, T. Lei, L. Xu, Y. He, L. Wu et al. 2017. Engineering canker-resistant plants through CRISPR/Cas9 targeted editing of the susceptibility gene CsLOB1 promoter in citrus. Plant Biotechnol. J. 15(12): 1509–1519.

Peng, A.H., S.C. Chen, T.G. Lei, L.Z. Xu, Y.R. He, L. Wu et al. 2017. Engineering canker-resistant plants through CRISPR/Cas9-targeted editing of the susceptibility gene CsLOB1 promoter in citrus. Plant Biotechnol. J. 15: 1509–1519.

Perrier, X., E. De Langhe, M. Donohue, C. Lentfer, L. Vrydaghs, F. Bakry et al. 2011. Multidisciplinary perspectives on banana (Musa spp.) domestication. Proc. Natl. Acad. Sci. USA 108: 11311–11318.

Pompili, V., L.D. Costa, S. Piazza, M. Pindo and M. Malnoy. 2019. Reduced fireblight susceptibility in apple cultivars using a high-efficiency CRISPR/Cas9-FLP/FRT-based gene editing system. Plant Biotechnol. J. 18: 845–858.

Pu, X., I. Khan, T. Zhang, G. Huang, J. Chen, Y. Ding et al. 2024. Genetic analyses of flower, fruit, and stem traits of intergeneric hybrids between 'Honghuagqinglong' and 'Heilong' pitayas. Plants 13(24): 3546.

Punwar, B.S., C. Ram, A. Singh and A. Vala. 2014. Genome editing: Conceptual introduction. Tech. Rep. [https://doi.org/10.13140/RG.2.1.1169.5203].

Radha, T. and L. Mathew. 2007. Fruit Crops, vol. 3. New India Publishing.

Rehman, A.U., T. Iso-Touru, J. Junkers, M. Rantanen, S. Karhu, D. Fischer et al. 2024. Multi-model GWAS reveals key loci for horticultural traits in reconstructed garden strawberry. Physiol. Plant. 176(4): e14440.

Ren, C., X. Liu, Z. Zhang, Y. Wang, W. Duan, S. Li et al. 2016. CRISPR/Cas9-mediated efficient targeted mutagenesis in Chardonnay (Vitis vinifera L.). Sci. Rep. 6(1): 32289.

Ren, F., C. Ren, Z. Zhang, W. Duan, D. Lecourieux, S. Li et al. 2019. Efficiency optimization of CRISPR/Cas9-mediated targeted mutagenesis in grape. Front. Plant Sci. 10: 612.

Ren, Y., H. Sun, M. Zong, S. Guo, Z. Ren, J. Zhao et al. 2020. Localization shift of a sugar transporter contributes to phloem unloading in sweet watermelons. New Phytol. 227: 1858–1871.

Rieger, M. 2006. Introduction to Fruit Crops (ed. Rieger, M.), Ch. 1. Food Products Press, Boca Raton.

Rodríguez-Leal, D., Z.H. Lemmon, J. Man, M.E. Bartlett and Z.B. Lippman. 2017. Engineering quantitative trait variation for crop improvement by genome editing. Cell 171: 470–480.

Salazar, J.A., D. Ruiz, P. Zapata, P.J. Martínez-García and P. Martínez-Gómez. 2022. Whole transcriptome analyses of apricots and Japanese plum fruits after 1-MCP (ethylene-inhibitor) and Ethrel (ethylene-precursor) treatments reveal new insights into the physiology of the ripening process. Int. J. Mol. Sci. 23: 11045.

Sattar, M.N., Z. Iqbal, J.M. Al-Khayri and S.M. Jain. 2021. Induced genetic variations in fruit trees using new breeding tools: Food security and climate resilience. Plants 10: 1347.

Schachtsiek, J. and F. Stehle. 2019. Nicotine-free, nontransgenic tobacco (Nicotiana tabacum L.) edited by CRISPR-Cas9. Plant Biotechnol. J. 17: 2228–2230.

Scheben, A., Y. Yuan and D. Edwards. 2016. Advances in genomics for adapting crops to climate change. Curr. Plant Biol. 6: 2–10.

Shahbaz, E., M. Ali, M. Shafiq, M. Atiq, M. Hussain, R.M. Balal et al. 2022. Citrus canker pathogen, its mechanism of infection, eradication, and impacts. Plants 12(1): 123.

Shan, Q., Y. Wang, J. Li, Y. Zhang, K. Chen, Z. Liang et al. 2013. Targeted genome modification of crop plants using a CRISPR-Cas system. Nat. Biotechnol. 31: 8.

Shao, X., S. Wu, T. Dou, H. Zhu, C. Hu, H. Huo et al. 2019. Using CRISPR/Cas9 genome editing system to create MaGA20ox2 gene-modified semi-dwarf banana. Plant Biotechnol. J. 18: 17–19.

Siva, M., D. Shanmugapriya and E. Santhoshini. 2023. Management of bacterial diseases by CRISPR-Cas9 system. Arch. Agric. Allied Sci.

Skrovankova, S., D. Sumczynski, J. Mlcek, T. Jurikova and J. Sochor. 2015. Bioactive compounds and antioxidant activity in different types of berries. Int. J. Mol. Sci. 16: 24673–24706.

Su, H., Y. Wang, J. Xu, A.A. Omar, J.W. Grosser and N. Wang. 2024. Cas12a/3 crRNAs RNP transformation enables transgene-free multiplex genome editing, long deletions, and inversions in citrus chromosome in the T0 generation. bioRxiv 2024-06.

Sun, X., X. Li, Y. Wang, J. Xu, S. Jiang and Y. Zhang. 2022. MdMKK9-mediated the regulation of anthocyanin synthesis in red-fleshed apple in response to different nitrogen signals. Int. J. Mol. Sci. 23: 7755.

Svitashev, S., J.K. Young, C. Schwartz, H. Gao, S.C. Falco and A.M. Cigan. 2015. Targeted mutagenesis, precise gene editing, and site-specific gene insertion in maize using Cas9 and guide RNA. Plant Physiol. 169(2): 931–945.

Talon, M. and F.G. Gmitter Jr. 2008. Citrus genomics. Int. J. Plant Genomics 2008: 528361.

Tester, M. and P. Langridge. 2010. Breeding technologies to increase crop production in a changing world. Science 327: 818–822.

Theivanai, M., O.P. Awasthi, R. Prusty and R.K. Pandey. 2020. CRISPR/Cas: A modern fruit breeding tool. Prog. Hortic. 52(2): 213–219.

Tian, S., L. Jiang, X. Cui, J. Zhang, S. Guo, M. Li et al. 2018. Engineering herbicide-resistant watermelon variety through CRISPR/Cas9-mediated base-editing. Plant Cell Rep. 37: 1353–1356.

Tripathi, J.N., V.O. Ntui, M. Ron, S.K. Muiruri, A. Britt and L. Tripathi. 2019. CRISPR/Cas9 editing of endogenous banana streak virus in the B genome of Musa spp. overcomes a major challenge in banana breeding. Commun. Biol. 2: 46.

Varman, C. 2025. Plant Genetics and Breeding Simplified. Educohack Press.

Vayalil, P.K. 2012. Date fruits (Phoenix dactylifera Linn): an emerging medicinal food. Crit. Rev. Food Sci. 52: 249–271.

Voytas, D.F. and C. Gao. 2014. Precision genome engineering and agriculture: opportunities and regulatory challenges. PLoS Biol. 12: e1001877.

Waltz, E. 2016. Gene-edited CRISPR mushroom escapes US regulation. Nature 7599: 293.

Waltz, E. 2018. With a free pass, CRISPR-edited plants reach market in record time. Nat. Biotechnol. 36: 6–7.

Wan, L., Z. Wang, M. Tang, D. Hong, Y. Sun, J. Ren et al. 2021. CRISPR-Cas9 gene editing for fruit and vegetable crops: strategies and prospects. Horticulturae 7(7): 193.

Wang, H., M. La Russa and L.S. Qi. 2016. CRISPR/Cas9 in genome editing and beyond. Annu. Rev. Biochem. 85: 227–264.

Wang, K. and B. Yang. 2020. A high-efficiency CRISPR platform for maize improvement. Available online: https://vtechworks.lib.vt.edu/handle/10919/78865 (accessed on 2 October 2020).

Wang, L., S. Chen, A. Peng, Z. Xie, Y. He and X. Zou. 2019. CRISPR/Cas9-mediated editing of CsWRKY22 reduces susceptibility to Xanthomonas citri subsp. citri in Wanjincheng orange (Citrus sinensis (L.) Osbeck). Plant Biotechnol. Rep. 13: 501–510.

Wang, X., M. Tu, D. Wang, J. Liu, Y. Li, Z. Li et al. 2018a. CRISPR/Cas9-mediated efficient targeted mutagenesis in grape in the first generation. Plant Biotechnol. J. 16: 844–855.

Wang, Y., T. Sun, T. Li, M. Wang, G. Yang and G. He. 2016. A CBL-interacting protein kinase TaCIPK2 confers drought tolerance in transgenic tobacco plants through regulating the stomatal movement. PLoS ONE 11: e0167962.

Wang, Z., S. Wang, D. Li, Q. Zhang, L. Li, C. Zhong, Y. Liu, H. Huang. 2018b. Optimized paired-sgRNA/Cas9 cloning and expression cassette triggers high-efficiency multiplex genome editing in kiwifruit. Plant Biotechnol. J. 16: 1424–1433.

Wu, N., Q. Lu, P. Wang, Q. Zhang, J. Zhang, J. Qu et al. 2020. Construction and analysis of GmFAD2-1A and GmFAD2-2A soybean fatty acid desaturase mutants based on CRISPR/Cas9 technology. Int. J. Mol. Sci. 21: 1104.

Xie, K.B. and Y.N. Yang. 2013. RNA-guided genome editing in plants using a CRISPR/Cas system. Mol. Plant 6: 1975–1983.

Xiong, J.-S., J. Ding and Y. Li. 2015. Genome-editing technologies and their potential application in horticultural crop breeding. Horticulture Res. 2(1): 15019.

Xu, K. 2013. An overview of arctic apples: basic facts and characteristics. N.Y. State Hortic. Soc. 21: 8–10.

Ye, J., X. Wang, T. Hu, F. Zhang, B. Wang, C. Li et al. 2017. An InDel in the Promoter of Al-ACTIVATED MALATE TRANSPORTER9 selected during tomato domestication determines fruit malate contents and aluminum tolerance. Plant Cell 29: 2249.

Yin, K., C. Gao and J.L. Qiu. 2017. Progress and prospects in plant genome editing. Nat. Plants 3: 17107.

Yuan, Y.G., S.Z. Liu, M. Farhab, M.Y. Lv, T. Zhang and S.X. Cao. 2024. Genome editing: An insight into disease resistance, production efficiency, and biomedical applications in livestock. Functional & Integrative Genomics 24(3): 81.

Zaidi, S.M.S. and A. Saud. 2020. Future of US-China relations: conflict, competition or cooperation? Asian Soc. Sci. 16: 1–14.

Zarco-Tejada, P.J., T. Poblete, C. Camino, V. González-Dugo, R. Calderon, A. Hornero et al. 2021. Divergent abiotic spectral pathways unravel pathogen stress signals across species. Nature Communications 12(1): 6088.

Zhang, H., J. Zhang, Z. Lang, J.R. Botella and J.-K. Zhu. 2017. Genome editing—principles and applications for functional genomics research and crop improvement. Crit. Rev. Plant Sci. 36(4): 291–309.

Zhang, J., S. Guo, G. Ji, H. Zhao, H. Sun, Y. Ren et al. 2019. A unique chromosome translocation disrupting ClWIP1 leads to gynoecy in watermelon. Plant J. 101: 265–277.

Zhang, M., Q. Liu, X. Yang, J. Xu, G. Liu, X. Yao et al. 2020. CRISPR/Cas9-mediated mutagenesis of Clpsk1 in watermelon to confer resistance to Fusarium oxysporum f.sp. niveum. Plant Cell Rep. 39: 589–595.

Zhao, X., Z. Tan, H. Feng, R. Yang, M. Li, J. Jiang et al. 2011. Microsatellites in different Potyvirus genomes: survey and analysis. Gene 488(2): 52–56.

Zhou, Z., Q. Li, L. Xiao, Y. Wang, J. Feng, Q. Bu, et al. 2021. Multiplexed CRISPR/Cas9-mediated knockout of Laccase genes in Salvia miltiorrhiza revealed their roles in growth, development, and metabolism. Front. Plant Sci. 12: 647768.

Zou, X., M. Du, Y. Liu, L. Wu, L. Xu, Q. Long et al. 2021. CsLOB1 regulates susceptibility to citrus canker through promoting cell proliferation in citrus. The Plant Journal 106(4): 1039–1057.

CHAPTER 6

Ornamental Crops Improvement Using CRISPR/Cas9

Nouraiz Ahmed Tanveer,[1,*] *Momna Mehmood,*[1] *Zain Ul Abadin,*[2] *Noor ul Ain*[1] and *Abdul Basit*[3]

1. Background

1.1 Genetic Engineering and Its Relevance

The changing climate patterns, the fast-growing worldwide human population pose extensive strain on agricultural systems that include staple as well as ornamental crops. Due to these challenges, crops experience different types of abiotic stress events combined with biotic stress, culminating in drought along with heat and salt exposure. Crops suffer from pest and pathogen attacks, which make it difficult to fulfill growing demands through conventional breeding techniques (Andualem and Seid 2021). Genetic engineering enables scientists to perform direct genomic modification on crops to develop plants with better attributes. Through genome manipulation, characteristics such as resistance to diseases and tolerance to environmental stresses are achievable (Liu et al. 2022). This valuable approach significantly benefits the fields of agriculture and medicine, along with environmental conservation. Traditional gene modification practices have been executing genetic engineering since the 1970s, but scientists continue practicing gene cloning with the implementation of Transcription Activator Like Effector Nucleases (TALENs) and Zinc-Finger Nucleases (ZFNs) systems. The process of genetic engineering using TALENs stands out as prolonged and expensive, with unreliable precision outcomes (Shamshirgaran et al. 2022). Complex processes that use custom protein generation along with pre-designed proteins serve to implement these methods. Genetic materials inserted randomly may trigger unanticipated genetic changes against the original sequence and disrupt other important genes (Table 1). The CRISPR/Cas9 method has proven to be a breakthrough technology for genetic reprogramming.

The bacterial immune system provided the basis for this technology through guide RNA (gRNA) that enables CRISPR/Cas9 technology can home in on designated DNA sequences to conduct accurate modifications with decreased side effects. The CRISPR technique has become the favored method by research scientists because of its economical nature and the ease of designing guide RNAs

[1] Department of Agricultural Biotechnology, Faculty of Agriculture, Ondokuz Mayis University, Samsun Türkiye.
[2] Department of Plant Physiology, Biochemistry and Genetics, Agricultural University, Plovdiv, Bulgaria.
[3] Floricultural Biotechnology Lab., Department of Horticulture, Graduate School, Kyungpook National University, 41566, Daegu, South Korea.
* Corresponding author: nouraizahmed10@gmail.com

DOI: 10.1201/9781003518204-6

Table 1. CRISPR vs traditional genome editing techniques.

Aspect	**CRISPR/Cas9**	**Traditional Genome Editing Techniques**	**Reference**
Gene Knockout	High precision with targeted DSBs, minimal off-target effects	Random mutations, lengthy, imprecise, high off-target risk	Shamshirgaran et al. (2022)
Gene Insertion	Highly precise with specific integration sites, efficient homologous recombination	Random integration with potential disruption of other genes	
Gene Regulation	Fine-tuned control via dCas9 for upregulation/downregulation, minimal off-target effects	Indirect and often imprecise silencing methods like RNAi	

(Mushtaq et al. 2021). Unlike previous methods, the technology does not need complicated protein design unless combined with hard-to-achieve laboratory work, thus making it more accessible and practical for a wide range of applications.

1.2 Mechanism of the CRISPR/Cas9 System in Plants

The CRISPR/Cas9 system operates using two essential components that consist of Cas9 nuclease and the guide RNA (gRNA). The system makes use of Cas9 nuclease and guide RNA (gRNA), where the RNA directs Cas9 toward a specific target sequence. The guide RNA works through a DNA sequence match with its complementary target position, which typically finds itself near PAM. A Protospacer Adjacent Motif (PAM) serves as the targeting requirement for precise DNA sequence recognition. The Cas9-gRNA complex finds its target site in DNA before Cas9 creates DNA double-strand breaks, which activate the plant's DNA repair system, natural DNA repair mechanisms. The plant cell's DNA repair machinery initiates work following the introduction of DSBs. The plant cells use two DNA repair pathways, named Non-Homologous End Joining (NHEJ) and Homologous Recombination (HR) pathways to repair the damage. Plants mainly use NHEJ as their dominant DNA repair pathway while generating breaks at the site. Cas9 enables DNA

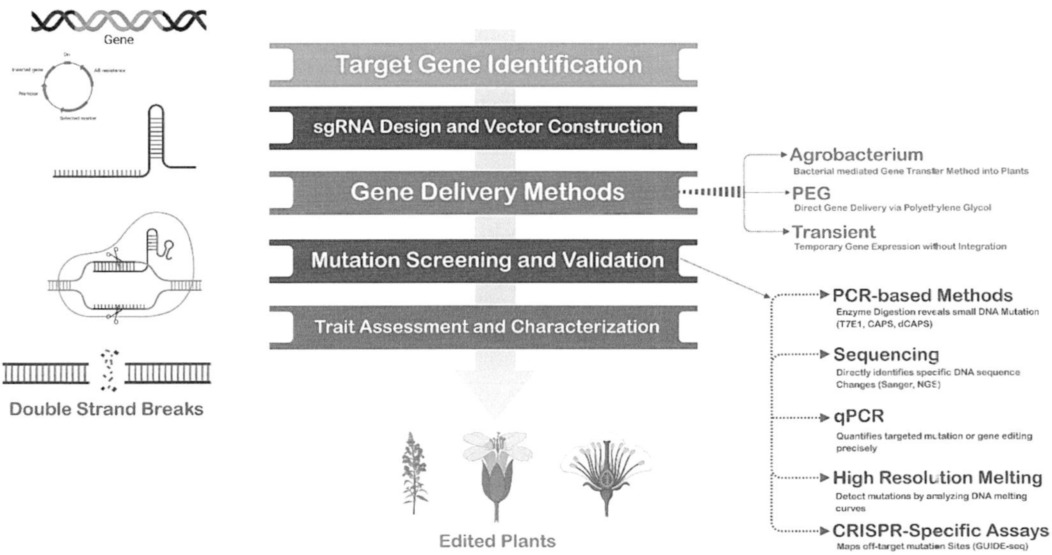

Figure 1. Illustrative workflow of CRISPR/Cas9 from target identification to trait assessment.

insertions or deletions thus turning off gene expression. HR, on the other hand, uses plant donor templates for performing precise gene modifications through the HR process. Researchers use Agrobacterium-mediated transformation as well as biolistics and electroporation techniques to introduce Cas9-gRNA complex into plant cells (Mehta et al. 2022). Edited cells are then regenerated through tissue culture methods that transform plant cells into whole plants after which scientists conduct genetic modification searches using specific laboratory methods. The analysis of cells needs molecular tests including PCR (Polymerase Chain Reaction) and sequencing approaches. The detailed workflow the workflow sketch of CRISPR technology appears in Fig. 1.

2. Introduction to Ornamental Plants and CRISPR/Cas9

2.1 Significance of Ornamental Plants

Ornamental plants maintain their worth through their aesthetic qualities derived from their morphological attributes. The important characteristics of ornamental plants include their patterns together with colors and shapes alongside their dimensions and their aromatic qualities. The worldwide cultivation of these plants occurs mainly for landscaping purposes as well as for gardening and interior decoration applications. Scenic elements including gardening and interior decoration use ornamental plants for enhancing the visual attractiveness of designed spaces. The growing demand for ornamental plants continues to increase based on the society's better standard of living as well as their growing love for greenery in urban environments (Indurthi et al. 2024). There are substantial barriers meeting the rising demands because traditional breeding methods fall short of providing adequate solutions. While conventional traditional breeding techniques successfully enhanced flower qualities but require extended generation times because these methods are slow. Several generational steps through extensive time and manual labor efforts are needed to reach the final desired results with traditional breeding approaches. The floriculture industry maintains rapid growth as an agricultural sector because people seek ornamental plants both for beauty and monetary purposes. The industry encompasses the production, distribution, and sale of ornamental plants. This field features all types of flowers together with decorative plants that include cut flowers and both potted plants and landscaping plants. Colombia, the Netherlands, Kenya along with Ethiopia operate as primary exporters in the global market. Ornamental crops obtain the major market share in international trade (Anumala and Kumar 2021). With growing customer interest for distinct colorful floral products drives the expansion of the floriculture industry, making it a multi-billion-dollar enterprise. Through floriculture operations major contributions occur in both national and international economies. The employment sector gains importance from floriculture by offering opportunities through production and marketing while providing logistics services.

2.2 Challenges in Traditional Breeding of Ornamentals

The ancient breeding techniques use crossbreeding together with mutagenesis and selection processes which provide very poor precision and slow results. The development of flower scents along with colors typically demands extensive backcrossing over decades. Traditional plant breeding through hybridization together with mutagenesis techniques have proven the key for developing new plant varieties. Polymorphic traits are difficult to fix with conventional breeding strategies. Hence, traditional breeding strategies lack the capability to work with this complexity. It fails to yield precise control over complex genetic interactions at this time (Caruso et al. 2020). Additionally, the combination of pests with pathogens together with changes in climate conditions intensifies the challenges of traditional methods. The existing approach of developing new plant varieties using conventional methods imposes substantial barriers. The production of desirable traits through these methods takes numerous years or stretches into multiple decades. Additionally, they contain random genetic variations during this process that contain numerous damaging effects alongside pointless changes that cause ineffective results (Frachon et al. 2021). Changes in flower-related traits typically require the modification of more than one gene. The breeding process becomes more complicated because of environmental stress conditions. Standard genetic methods present limitations when researchers need to address stress tolerance because the stress-related gene basis remains beyond their reach effectively. For instance, biotic stresses such as pest attacks and abiotic stresses like drought and traditional methods cannot make precise genetic changes needed for plants that grow in salinity environments. The development of CRISPRCas9 tools becomes essential because they enable precise adaptation of ornamental plant genetics. In contrast, CRISPR/Cas9 provides researchers with the capability to identify and label genes responsible for stress responses. The method enables researchers to create ornamental plants that become more resistant to environmental challenges.

2.3 The Need for Genome Editing in Floriculture

The emerging trends in consumer preferences require floriculture to transform its operations so it can fulfill present market requirements, sustainability, novelty, and quality. Standard breeding approaches are notable, but they alone do not suffice and are insufficient to address these challenges. Genome editing, particularly through CRISPR/Cas9, accelerates the process of developing ornamental plants with targeted essential characteristics. For instance, CRISPR has a gene modification technique that has recently transformed rose flowers through its successful alteration of pigment genes, which led to blue-colored petals. Scientists achieved the previously inaccessible development of blue-pigmented flowers through CRISPR/Cas9 technology, according to Sirohi et al. (2022). Through CRISPR/Cas9 technology, growers can make their ornamental plants resilient, alongside creating unique, attractive colors. The efficiency and durability of the floriculture industry reinforce the competitiveness established by CRISPR/Cas9. CRISPR/Cas9 obtained prominence as an efficient tool for precise gene editing through its adaptation in 2016.

Researchers now benefit from an effective and accurate gene editing instrument for decorative plants used to drive progress in ornamental science, its fundamental research, and new phenotypic trait creation (Fig. 2). It functions as a method that delivers powerful and very specific services, alternative to traditional breeding techniques for ornamental plants. By enabling targeted genome this technology expedites the evolution of features like enlarged flower colors through targeted genome modification procedures, improved fragrances, and increased resistance to environmental stresses. Scientists employed CRISPR/Cas9 technology to edit ethylene-related genes thus expanding flower growing periods alongside stress defense improvements in ornamental plants (Xu et al. 2021). The process of utilizing CRISPR/Cas9 technology achieves extended flowering duration together with enhanced stress resistance in ornamental crops (Xu et al. 2021). Unlike conventional CRISPR has the capability to address polygenic traits through exact gene modifications to various genetic targets. Modern floriculture depends on this technology due to its ability to operate simultaneously. CRISPR/Cas9 provides numerous advantages over traditional breeding techniques. Unlike conventional methods, random genetic variations get substituted by the CRISPR technology which performs targeted precise gene modifications. The exact nature of CRISPR edits lowers the probability of unintentional genetic changes while protecting against unintended genetic alterations. Additionally, research with CRISPR scientists can expedite trait breeding because they bypass extensive backcrossing to bring favorable characteristics into one generation (Lyzenga et al. 2021). Through its efficient approach, new ornamental plant breeding reduces the duration needed, along with labor costs and necessary resources. CRISPR enables researchers to resolve problems dealing with polygenic traits that cannot be easily controlled through traditional methods. Scientific experts now achieve simultaneous gene targeting to accomplish their research goals. Scientists can use CRISPR technology to develop advanced traits, including flower color, fragrance, as well as disease resistance. Furthermore, the CRISPR method enables scientists to bring advantageous traits, diversity, and sustainability from wild relatives into their creations. Scientists now have the capability of developing non-transgenic plants through accurate genetic modifications.

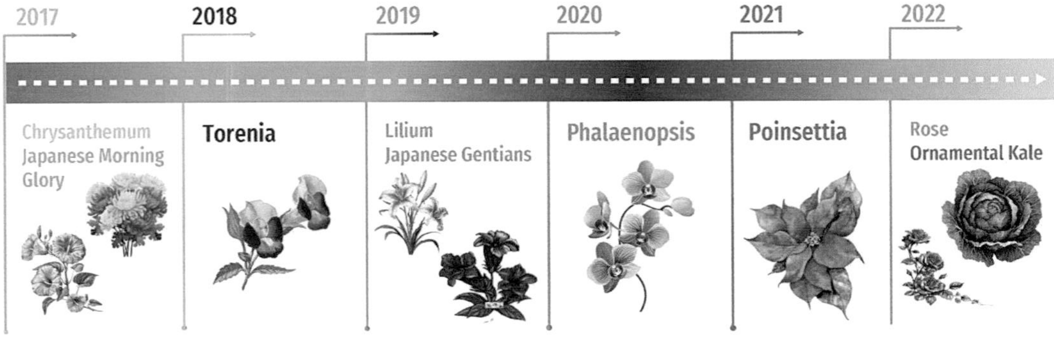

Figure 2. Year-Wise Timeline of genome editing in ornamental plants by CRISPR/Cas9.

CRISPR-edited crops benefit from reduced regulations that speed up the process of market access, according to Rasheed et al. (2021).

3. CRISPR and Its Application in Floriculture: A Focus on Key Traits

Continuous efforts to uncover new floral traits including flower shape, color, fragrance, longevity, and durability are driving the demand for floriculture. Such traits have been introduced since long applying traditional breeding techniques that include mutagenesis, polyploidization, and hybridization. Progress in the development of ornamental plants has accelerated due to the ability of CRISPR/Cas9 technology to create precise genome-targeted modifications. This technology allows for gene editing of various genes that control key traits like early flowering time, flower number, flower duration, flower shape and size and flower color and fragrance (Ashwini et al. 2024; Ge et al. 2019; Kaur et al. 2021). Flower color is still a key focus trait due to its substantial economic importance. CRISPR has facilitated the production of non-existing flower colors such as blue roses, blue *chrysanthemum*, red irises, and yellow cyclamen, achieved by targeting pigment biosynthetic pathways (Tanaka et al. 2009). In another study on petunia, an ornamental model plant, six genes associated with plant architecture were targeted using CRISPR/Cas system. Edited plants with paralogous of *FLOWERING LOCUS T* (PhFT) resulted in early flowering and mutants with alterations in *TERMINAL FLOWER-LIKE1* (PhTFL1) paralogous genes led to shorter internodes compared to wild-type (Abdulla et al. 2024). A detailed overview of ornamental pants edited by CRISPR/Cas9 system has been described in Table 2.

3.1 Floral Color Modification

Flower color is a major commercial trait in ornamental plants, determined by the biosynthesis of pigments such as flavonoids, carotenoids, and betalains. In the past, flower color modification was mainly improved through traditional breeding due to limited pigment pathway availability (Sirohi et al. 2022). The pathway in anthocyanin biosynthesis is composed of numerous genes, such as *dihydroflavonol-4-reductase* (DFR), *anthocyanidin synthase* (ANS), and chalcone synthase (CHS). These are the genes that play an important role in the production of anthocyanins. Early genetic approaches were used in inserting antisense genes like *dihydrofavonol-4-reductase* (DFR) and *Chalcone synthase* (CHS), which resulted in prominent success rates in altering pigmentation. However, RNA interference (RNAi) techniques provided scientists more stable outcomes, where researchers were able to achieve over 50% success in generating white flowers by targeting pigmentation genes, including *chalcone synthase* (CHS) and *anthocyanidin synthase* (ANS) (Nakamura et al. 2006; Ryutaro Aida et al. 2000; Sirohi et al. 2022). Moreover, recent advances in

Table 2. Detailed overview of CRISPR/Cas9 edited Ornamental Plants.

Species	Gene	Function of Gene	Transformation Mode	Explant Type	Mutation Detection	Changes in Phenotype	Reference
Chrysanthemum morifolium	CpYGFP	Yellowish-Green Fluorescence, Visualization of Gene Editing Progress	Agrobacterium-Mediated	Leaf	PCR, DNA Sequencing	Disruption in Fluorescent Protein, Fluorescence Reduction	Kishi-Kaboshi et al. (2017)
Dendrobium officinale	IRX, C4H, C3H, CCR, 4Cl	Biosynthesis Pathway of Lignocelluloses	Agrobacterium-Mediated	Protocorm	PCR, DNA Sequencing	Ligno-Cellulose Biosynthesis Reduction	Kui et al. (2017)
Petunia hybrid	PhN	Nitrogen Fixation	PEG-Mediated Transfection of RNP	Leaf protoplast	T7E1 Assay, PCR, Illumina Sequencing	Site-Directed Mutagenesis in NR Genes	Subburaj et al. (2016)
Petunia hybrid	NR	Controls Nitrate Metabolism	PEG-Mediated Protoplast	Protoplast	T7E1 Assay, PCR, Illumina Sequencing	Nitrate Assimilation Deficiency	Subburaj et al. (2016)
Petunia hybrid	PDS	Codes for the Carotenoid Biosynthesis Enzyme	Agrobacterium-Mediated	Leaf	PCR, DNA Sequencing	Mosaic Shoots and Albino Flowers	Zhang et al. (2016)
Petunia inflate	PiSSK1	Self-Incompatibility	Agrobacterium-Mediated	Leaf	PCR, DNA Sequencing	RNAse-Based Self-Incompatibility	Sun and Kao (2018)
Petunia hybrid	F3HA, F3HB	Flavonol Pigments and Anthocyanidin synthesis	PEG-Mediated	Protoplast	PCR, Illumina Sequencing	Change of Floral Color to Purplish Pink	Yu et al. (2021)
Petunia hybrid	ACO3, ACO1, ACO4	Ethylene Biosynthesis Pathway	PEG-Mediated	Leaf	PCR, Deep Sequencing	Reduction of Ethylene Production and Flower Longevity Enhancement, Delayed Senescence	Xu et al. (2021)
Petunia hybrid	DPL	Floral Color	Agrobacterium-Mediated	Leaves	PCR, Sanger Sequencing	Lack of Anthocyanin in Vain-Associated Region	Zhang et al. (2021)
Phalaenopsis equestris	MADS	Floral Development and Initiation	Agrobacterium-Mediated	Protocorm	PCR, Sequencing	Long Juvenile Period	Tong et al. (2020)
Torenia fournieri	TfRAD1	Corolla Pigmentation and Petal Shape Regulation	Agrobacterium-Mediated	Leaf	PCR, DNA Sequencing	Violet Coloration Pattern on Dorsal Petals	Su et al. (2017)
Torenia fournieri	F3H	Flavonoid Biosynthesis	Agrobacterium-Mediated	Leaf	PCR, Sanger and NGS	Change in Floral Color to Pale Blue	Nishihara et al. (2018)

Table 2 contd. ...

...Table 2 contd.

Species	**Gene**	**Function of Gene**	**Transformation Mode**	**Explant Type**	**Mutation Detection**	**Changes in Phenotype**	**Reference**
Lilium longiflorum	LlPDS	Carotenoid Synthesis Enzymes	Agrobacterium-Mediated	Tissue Culture, Seedling Scales	GUS Assay, PCR and Southern Blot	Pale Yellow, Albino Green, and Albino Flowers	Yan et al. (2019)
Lilium pumilum	LpPDS	Chlorophyll Biosynthesis	Agrobacterium-Mediated	Embryogenic Calli	GUS Assay, PCR and Southern Blot	Pale Yellow, Albino Green, and Albino Flowers	Yan et al. (2019)
Japanese gentian	GST1	Anthocyanin Accumulation in Petals	Agrobacterium-Mediated	Leaf	Sanger, NGS	Pale Blue and White Flower due to Reduced Anthocyanin	Tasaki et al. (2020)
Japanese Morning Glory	CCD4	Carotenoid Degradation	Agrobacterium-Mediated	Secondary Embryo	CAPS analysis, DNA Sequencing	White Petals Change Color to Pale Yellow	Watanabe et al. (2018)
Japanese Morning Glory	EPH1	Key Regulator of Petal Senescence	Agrobacterium-Mediated	Secondary Embryo	CAPS analysis, DNA Sequencing	Delayed Petal Senescence	Shibuya et al. (2018)
Japanese Morning Glory	DFR-B	Anthocyanin Biosynthesis Enzyme	Agrobacterium-Mediated	Secondary Embryo	CAPS analysis, DNA Sequencing	Influenced Synthesis of Anthocyanin, resulting in a Floral Color shift to White	Watanabe et al. (2017b)
Ornamental kale	BoDFR1, BoDFR2	Floral Color	Agrobacterium-Mediated	Cotyledon	PCR, Sequencing, NGS	Affected Anthocyanin Accumulation	Zhang et al. (2022)
Ornamental kale	BoORP3a	Shelf Life	Agrobacterium-Mediated	Cotyledon	PCR, Sequencing, NGS	Reduced Cuticular Wax Deposition	Zhang et al. (2022)
Rosa hybrid	RhEIN1, RhEIN2	Shelf Life	Agrobacterium-Mediated	Suspension Cells	Sanger Sequencing	Affected Ethylene Signaling Pathway	Wang et al. (2023)

CRISPR/Cas9 technology have facilitated more precise gene edits that modify flower colors with higher efficiency and precision.

Watanabe et al. 2018 used CCD gene alteration in Ipomoea nil to produce pale yellow petals at a 55.5% mutation success rate. Scientists obtained white-flowering plants (75% biallelic mutations: white-flowering) through biallelic mutations in the *dihydroflavonol-4-reductase-B* gene (DFR-B), which regulates anthocyanin biosynthesis (Watanabe et al. 2017a). The mutation of the *flavone 3-hydroxylase 3* gene using CRISPR/Cas9 technology in the flowering plant *Torenia fournieri* produced white flowers from blue flowers in 80% of cases (Nishihara et al. 2018). In addition, albino and pale-yellow mutants have been generated through silencing of the phytoene desaturase (PDS) gene, which is a key rate-limiting enzyme in the carotenoid biosynthesis pathway, in petunia and lilies, where mutation frequencies ranged between 55–87% (Zhang et al. 2016). In petunias, knocking out genes involved in the flavonoid biosynthetic pathway, such as f3ha and f3hb, resulted in mutants with pale purple-pink flowers (Yu et al. 2021). Further research demonstrated that the R2R3 MYB transcription factor DPL does not regulate corolla tube venation, as mutations caused by CRISPR/Cas9 eliminated vein-associated anthocyanin patterns without altering venation (Zhang et al. 2021).

3.2 Floral Longevity

Flower longevity, especially the longevity of cut flowers, is an important commercial trait in floriculture. Ethylene is a plant hormone that regulates the flowering process and has been implicated in accelerating senescence in floral tissues (Ashwini et al. 2024). The shelf life of cut flowers increases when plants become less sensitive to ethylene and produce it in less amounts. CRISPR/Cas9 demonstrates its potential to enhance floral durability through mutation of ethylene production genes according to Xu et al. (2021). The research by Xu et al. (2020) proved that editing the aminocyclopropane-1-carboxylate oxidase (ACO) genes in petunia, which included *PhACO1*, *PhACO3*, and *PhACO4*, reduced ethylene production, thus extending flower longevity. Research conducted with "*Mirage Rose*" petunia accession proved the extension of flower duration through genetic manipulation of *ACO* genes, which lowered ethylene production levels (Xu et al. 2021). Liu et al. (2016) used a virus-induced gene silencing approach to silence ethylene signaling components *EIL1* and *EIL2* that extended flower duration while maintaining plant characteristics intact. Research breakthroughs have produced better waste management methods that extend flower visual appeal while boosting profits (Noman et al. 2017).

3.3 Floral Morphology and Size

The market value and economic potential of flowers largely depend on two significant aspects which are their flower size and morphology together with their color. The development of floral organs depends on multiple homeotic genes which include *AP1, PI, AG,* and *SEP* genes from the *MADS* Box gene family (Tong et al. 2020). CRISPR/Cas9 provides researchers with an accurate solution to modify flower morphology attributes including floral shape and symmetry together with flower size development through the control of these essential genetic factors.

Numerous studies have recently demonstrated how CRISPR can edit genomes in different ornamental plants. The orchid flower produces larger blooms when researchers knocked out two essential *MADS* genes from the plants (Tong et al. 2020). Multiplying *chrysanthemum* flower petals after suppressing the *AGAMOUS* gene with CRISPR technology improved floral attractiveness (Klap et al. 2017). Historical research has revealed that CRISPR/Cas industry opportunities exist for floral applications.

4. CRISPR for Disease Resistance and Stress Tolerance in Ornamental Plants

4.1 Disease Resistance

The pathogenic agents encompass viruses, fungi, and bacteria, together with viroids (Mekapogu et al. 2021). In ornamentals, fungal diseases are either soil-borne caused by *Fusarium, Rhizoctonia, Pythium, Phytophthora, Sclerotium, Sclerotinia,* or air-borne such as black spots, canker, powdery mildews, anthracnosis, downy mildews, and leaf blights (Agrios 2005). Different symptoms arising from host-pathogen interactions mainly comprise wilting, followed by crown gall, then root rot and canker, together with leaf blights, while also resulting in necrotic lesions and spots among other visible signs (Traversari et al. 2021). Plant bacterial infections cause leaf spots along with rotting, while producing symptoms of blights and galls in addition to wilting, yellowing, and canker (Agrios 2005). Bacterial agents like *Xanthomonas* spp. (Mirik et al. 2018), and *Ralstonia* spp. (Hammond et al. 2023), *Agrobacterium tumefaciens* and *Rhodococcus* are important in this regard. Numerous viruses like Tobacco Streak Virus (TSV), Impatiens Necrotic Spot Virus (INSV), Tobacco Mosaic Virus (TMV), Turnip Mosaic Virus (TuMV), Tomato Spotted Wilt Virus (TSWV), Alternanthera mosaic virus (AltMV), helenium virus S. (HelVS), sugarcane mosaic virus (SCMV), Phlox virus S (PhlVS), Cucumber Mosaic Virus (CMV), jasmine virus C (JaVC), and Tobacco Ringspot Virus (TRV), etc., have been documented infecting different ornamental plants like *Gomphrena globosa, Marigold, Dahlia, Impatiens balsamina, Gladiolus, Canna indica* L., *Crossandra infundibuliformis, Phlox., Zinnia, Jasminum sambac* L., and *Catharanthus roseus.* Symptoms like chlorosis, vein yellowing, mottling, ringspots, mosaic, stunting, defoliation, browning, and enation are noticed in the host plants depending upon the host and pathogen combination (Shafiq et al. 2024). Various viroids, such as Chrysanthemum stunt, Chrysanthemum chlorotic mottle viroid (CChMVd), Potato spindle tuber viroid (PSTVd), Chrysanthemum stunt viroid (CSVd), Citrus exocortis viroid (CEVd), Tomato apical stunt viroid (TASVd), Iresine viroid (IrVd), and Tomato Chlorotic Dwarf Viroid (TCDVd), are also reported to cause diseases in different ornamental plants (Flores 2008). These ornamental plants include *chrysanthemum*, *Argyranthemum frutescens, Ageratum* spp., *Solanum jasminoides, Dahlia* spp., *Petunia* spp., and *Vinca major*, etc. No visible symptoms like mottling, chlorosis, dwarfing, or delayed blossoms can be noticed on infection (Dimock et al. 1971). Different approaches have been adopted to control these pathogens, particularly chemical control, but the residual effects of such proportional to the powdery mildew disease severity in petunia. The mutants that lacked the expression of the gene showed resistance to the powdery mildew disease caused by *Sphaerotheca fusca* (Xu et al. 2023). Chrysanthemum White Rust (CWR) is influenced by the transcription rate of the gene CmTGA1 in *Chrysanthemum morifolium*. Mutants with this particular gene knockout through the CRISPR/Cas9 system show less resistance to the fungus *Puccinia horiana* (Chen et al. 2024).

4.2 Abiotic Stress Tolerance

Abiotic stresses like drought, heat, and low temperature are important factors that can significantly limit plant growth (Gupta et al. 2013). It is not possible to avoid or manually control such adverse conditions. In ornamentals, abiotic stress management has been attempted through different breeding strategies such as mutation, hybridization, and selection (Sirohi et al. 2022). However, the prolonged timeframe requirements, polyploidy and complex genomics of ornamentals have made the outcome of such approaches less productive (Sharma and Messar 2017). The need to deliver the results in a short time has led scientists and researchers to explore new, modern approaches that can provide more promising results. CRISPR/Cas genome editing technique is one such technological revelation that has the promising potential of improving abiotic stresses (Erdoğan et al. 2023). Genes that are involved in abiotic stress responses are targeted using the CRISPR/Cas technique. Such as, drought tolerance has increased by editing the *TaERF3* (*Ethylene responsive factor 3*) and *TaDREB2* (*Dehydration responsive element binding 2*) genes in wheat (Kim et al. 2018). Similarly,

SAPK2 gene mutation through CRISPR/Cas results in loss of function for the gene in rice, conferring tolerance to drought and salt (Lou et al. 2017). An increased expression of stress-response genes and antioxidant enzyme genes is noticed in mutant lines when exposed to drought conditions. In tomato, *SLNPR1* gene mutation improves drought tolerance (Li et al. 2019). Overexpression of a maize gene *ZmWRKY40* helps enhancing the drought tolerance in transgenic Arabidopsis plants (Wang et al. 2018). *OsRAV2* gene mutation through CRISPR/Cas results in increased salt tolerance in rice (Duan et al. 2016).

Salt and drought tolerance of the soybean plant was enhanced by CRISPR/Cas modified *GmMYB118* gene overexpression (Du et al. 2018). Researchers improved cold resistance by modifying the *OsAnn3* gene using CRISPR/Cas9 technology in rice mutants (Shen et al. 2017). Rice cold tolerance was improved by CRISPR/Cas9 targeting of the *OsMYB30* gene (Zeng et al. 2020). Heat tolerance is achieved by the CRISPR/Cas edited *BRAZZINAZOLE RESISTANT 1* (*BZR1*) gene in tomato, which are useful for modulating heat tolerance. Klap et al. (2017) report that heat-tolerant tomatoes are created wherein the *SlAGL6* (*SlAGAMOUS-LIKE 6*) gene is mutated. *HSPs* gene expressions are utilized by plants reacting to external abiotic stress stimuli (Wei et al. 2020). Transgenic *Chrysanthemum*s show increased thermotolerance due to the insertion of *AtDREB1A* into the genome (Hong et al. 2009). Transcription levels of *HSP70* heat shock protein genes were also upregulated directly following heat exposure. Similarly, Cloning of the *CmDREB6* gene was done in *Chrysanthemum morifolium* (Du et al. 2018). This gene set shows specific transcriptomic responses, especially when the plants are exposed to multiple environmental stresses (Liu et al. 1998). Heat stress tolerance capacity continues in *CmDREB6* gene-induced *Chrysanthemum morifolium* plants. The outputs are gene expression of *heat shock proteins* and *ROS* (*Reactive Oxygen Species*). Expression of the *CmCPL1* gene in plant cells of *Chrysanthemum morifolium* results in the formation of *RNAPII CTD phosphatase-like 1*. When *CmCPL1* is genetically overexpressed, temperature tolerance in *Chrysanthemum morifolium* is mainly regulated. According to scientific analyses, the transgenic plants showed their highest expression of heat-stress-related proteins (Qi et al. 2018). Following the expression of *AtDREB1A* transcriptional factor in the round-cover *Chrysanthemum* (*Dendranthema grandiflorum*), tolerance to cold stress was still unaltered (Hong et al. 2009). Huang et al. (2021) showed that enhancing the production of *temperature-induced lipocalin-1-like gene* (*DgTIL1*) increases resistance to cold temperatures. The transgenic *Chrysanthemum* lines exhibited enhanced POD (promote peroxidase) activity together with higher ROS accumulation, which resulted in superior cold tolerance.

5. Novel CRISPR Strategies for Enhancing Aesthetic and Functional Traits

Floriculture products contribute substantially to global agricultural production, making up a market worth $43.23 billion in 2018. Europe leads the total market space with a 40.3% share, and this business sector is predicted to grow by $11.68 billion by 2027 (Kaur et al. 2021). The expansion of the floriculture sector depends on rising market demand for decorative plants that exhibit superior traits such as prolonged blooming time, distinctive color, and adaptable scented flowers (Giovannini et al. 2021). The failure of traditional breeding approaches using polyploidization and hybridization originates from ornamental species having high levels of heterozygosity, prolonged life cycles, and large chromosomal numbers and restricted genetic diversity (Regalado González et al. 2017).

Ornamental crops such as roses, *chrysanthemums*, orchids, tulips, lilies, and daffodils represent a multibillion-dollar global market due to their widespread use in floriculture and landscaping (Chandler and Brugliera 2011). Consumer demand is driven largely by aesthetic traits, including flower color, fragrance, plant architecture, and overall form. However, traditional breeding methods—such as hybridization and mutation breeding—are often hindered by lengthy breeding cycles, low trait precision, and the genetic complexity of polyploid species (Xiong et al. 2015). To overcome these limitations, researchers have increasingly turned to genetic engineering and genome-editing technologies, particularly CRISPR/Cas9, which allow for more efficient and targeted trait modification.

5.1 CRISPR/Cas9 in Floriculture: A Game-Changer for Plant Breeding

The advancement of ornamental plants through CRISPR/Cas9 technology became faster because scientists gained control over precise genome-targeted modifications. The technology enables gene editing of essential floral traits regarding early flowering time and flower number and duration and shape and size and color, and fragrance (Ashwini et al. 2024; Ge et al. 2019; Kaur et al. 2021). CRISPR/Cas9 continues to concentrate on flower color modifications since the trait holds major economic value. Scientists utilized CRISPR to generate flower colors that previously did not exist by altering pigment biosynthetic pathways in flowers (Tanaka et al. 2009). Therefore, they produced blue roses as well as blue *chrysanthemum* and red irises, and yellow cyclamen. Scientists used the CRISPR/Cas system to target six architectural genes within the petunia model organism that serves as an ornamental plant. Edited plants with paralogues of FLOWERING LOCUS T (PhFT) resulted in early flowering, and mutants with alterations in TERMINAL FLOWER-LIKE1 (PhTFL1) paralogous genes led to shorter internodes compared to wild-type (Abdulla et al. 2024).

5.2 Multiplex Genome Editing for Stacking Multiple Traits

Multiplex genome editing is a way of editing two or more genes at a time, newly opened a new concept of editing in the plant biotechnology field. Using cutting-edge CRISPR systems will enable this strategy to stack multiple desirable traits in one breeding cycle, which will likely change the face of ornamental plant breeding (Abdulla et al. 2024). This method enables an increase in sets of traits together, including changes in flower colors, larger size, or higher disease-hardiness, etc. To efficiently co-edit numerous genomic loci in rice, Wu et al. 2024 generated a CRISPR/Cas9 knockout vector with 49 sgRNAs targeting 40 genomic loci and reported remarkable co-editing efficiency.

CRISPR/Cas9 along with *CRISPR-Cpf1* systems, were discovered first, and researchers use multiple guide RNAs (gRNAs) to target separate gene regions during single transformation events (Rukavtsova et al. 2022; Zhang et al. 2018). The technique enables horticulturists in ornamental plant breeding to combine desirable market attributes into one plant, including bright petals and long blooming times or unique fragrance. Researchers can combine CRISPR technology to edit genes relating to pigment pathway production alongside disease-resistance genes that result in specific flower color patterns (Mekapogu et al. 2023).

Research focusing on multiplex genome editing in plants exists beyond floriculture limitations although exclusive floriculture case studies are currently scarce. Researchers used CRISPR/Cas9 on *Solanum lycopersicum* to modify fruit ripening related genes which improved both storage length and taste attributes according to Li et al. (2020) and Yang et al. (2023). Multiplex CRISPR technology has improved rice (*Oryza sativa*) plants because researchers used it to target genes that affect both nutrient absorption and drought defense mechanisms (Zafar et al. 2020). Such methods applied to floriculture breeding enable developers to create ornamental plants containing both attractive features with functional traits like pest immunity and drought tolerance.

6. Challenges and Limitations in Applying CRISPR to Floriculture

CRISPR/Cas9 technology has transformative potential for breeding ornamental plants, but improvements and resolutions for several technical, genetic, and societal and regulatory challenges are needed to ensure its full application. All these challenges have ramifications for how viable, stable, and accepted CRISPR-edited ornamental plants might be in the commercial floriculture market. Figure 3 describes the summary of challenges and limitations of CRISPR technology in floriculture.

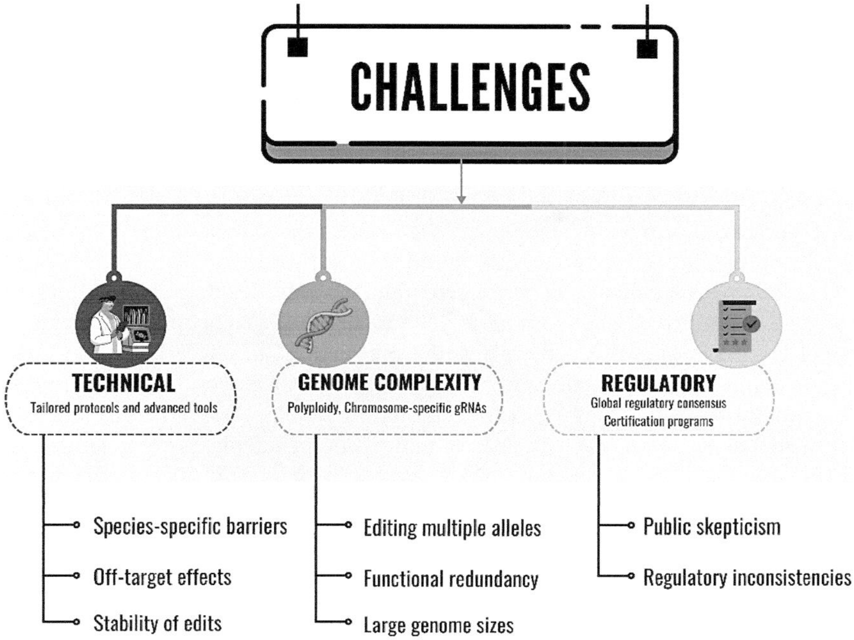

Figure 3. Summary of challenges and limitations in applying CRISPR to floriculture:

6.1 Technical Challenges in Genome Editing

The use of the CRISPR/Cas9 tool in ornamentals is usually limited due to the specific technical barriers of the plants. One of the major challenges is the variable efficiency depending on species-specific features, including cell architecture and regeneration ability of the organism (Zafar et al. 2020). Conventional transformation methods, such as *Agrobacterium*-mediated gene transfer or particle bombardment, often fail to successfully introduce and integrate CRISPR/Cas components into recalcitrant ornamental plants (Mekapogu et al. 2023). Despite the advancement of these approaches with the aid of protoplast-primarily based editing and tailored *in vitro* regeneration protocols, emergence of stable and efficient transformation continues to be difficult for many ornamental species, especially woody and polyploid plants (Li et al. 2019). The second major technical challenge is the off-target effects, in which the modification at genomic sites other than the target locus (Sirohi et al. 2022). These off-target changes can cause phenotypic alterations that are often undesirable and question the predictability and stability of edited phenotypes. Though high-fidelity Cas variants, including *Cas9-HF1* and *eSpCas9*, and improved design methodologies for guide RNA (gRNA) have made considerable progress in reducing off-target risks, absolute specificity has not yet been provided (Ge et al. 2019). In addition, due to the complicated genome constructions that might simply be composed of equivalent series of SNPs (Single Nucleotide Polymorphism), ornamental plants are more vulnerable to these kinds of off target impacts and the chance amongst equivalent series of SNPs for the potential gRNA binding would also more easily occur (Hua et al. 2019). Another technical hurdle is monitoring the stable transmission of desirable traits across generations: inheritable gene edits must not only be accurate, but they must also go through successive generations without reversion (Li et al. 2019; Sretenovic et al. 2021).

6.2 Overcoming the Challenges of Polyploidy and Genetic Complexity

Many ornamental species are polyploid (i.e., the presence of more than two homologous sets of chromosomes within a single organism), creating challenges for their genome editing. In plants, polyploidy raises the genomic complexity and makes it challenging to attain uniform and stable editing of all alleles of a target gene (Azadi et al. 2016). For instance, Rosa spp. is mainly tetraploid with four sets of chromosomes, and thus, the genome editing tools that function is complex and would not perform well within a tetraploid genome (Chang Ho Ahn et al. 2020).

New strategies are required to overcome challenges posed by polyploidy, including simultaneous targeting of multiple alleles of the same gene or generating chromosome-specific guide RNAs that would selectively edit alleles within individual chromosome copies (Li et al. 2019). This method is essential to generate uniform and phenotypically stable results on every gene copy. The situation is even worse for polyploids, where knockout of one gene copy may be compensated for by another copy due to functional redundancy, as entire gene families are often targeted (Abdulla et al. 2024). Thus, exhaustive studies would be required to determine which gene copies are the most regulated for successful editing, especially considering the presence of multiple copies of one gene with overlapping or compensatory functions in polyploid plants. Moreover, the high genome sizes and high heterozygosity values of many ornamentals, besides polyploidy, add challenges to genome editing (Traversari et al. 2021). The genomes of many ornamental plants are not only large but also enriched with repetitive DNA sequences, which creates difficulties in gRNA design and enhances the risk of off-target effects (Kaur et al. 2021). Certain tools like CROPSR, a genome-scale gRNA sequence design program, have substantial promise as they specifically streamline the process of troubleshooting CRISPR experiments, particularly when targeting polyploid genomes or target regions with a high AT content in plants (Rukavtsova et al. 2022).

6.3 Societal and Regulatory Challenges in the Adoption of CRISPR-Edited Ornamentals

There are also the major societal and regulatory hurdles involved with commercializing CRISPR-edited ornamental plants. Public attitudes towards GMOs often include issues of possible environmental harm, food safety (where relevant), and morality (Giovannini et al. 2021; Sirohi et al. 2022). Despite ornamental plants being grown for visual and not for consumption functions, consumer fear of genetic transformation is an important obstacle to acceptance, especially when transformation does not result in visible phenotypes (Lee et al. 2020). Conveying the benefits of genetic modifications can be more difficult with ornamentals than with edible crops, for example, because the purposes of GM crops often involve pesticide reduction or nutritional improvements (Zhang et al. 2021).

The public requires solutions such as maintaining transparency and implementing certification plans and educational initiatives about CRISPR-edited ornamental sustainability (Souza et al. 2021). Through certification programs, one can verify that CRISPR-edited plants fulfill environmental sustainability requirements and pose no health risks while using fewer chemical pesticides through disease resistance programs (Kaur et al. 2021). Public campaign initiatives would help customers understand how genome-edited ornamentals fulfill environmental goals such as drought resistance, which appeals to environmentally conscious purchasers (Lee et al. 2020).

The regulatory procedures for CRISPR-edited plants persist in multiple complex types because national regulations across different countries maintain significant differences between them. CRISPR-modified plants without foreign DNA operate under distinct U.S. regulatory requirements from genetically modified organisms which results in reduced market access limitations (Ashwini et al. 2024). The strict safety verification procedures of the European Union for altered organisms result in GMO classification, through which it demands prolonged regulatory approvals for all genome-edited lifeforms, even without introducing foreign DNA (Rukavtsova et al. 2022). The diversity in regulations across markets presents a significant barrier for global ornamental plant industries trying to bring CRISPR-edited varieties to their main markets (Lee et al. 2020). A consensus in network regulation would establish simplified market opportunities for genome-edited plants worldwide while maintaining consumer protection as floriculture stakeholders aim to achieve this goal.

7. Sustainable Floriculture Through Genome Editing

CRISPR has emerged as a powerful tool to specifically improve disease resistance, water efficiency, and fertility in flowering plants. The use of the CRISPR/Cas9 technique brings significant improvements in protecting floriculture crops from diseases. The CRISPR method enables scientists to accurately modify target pathogen-related genes using CRISPR technology. Several studies show that CRISPR proves to be effective in developing tomato plants that resist pathogens. Most importantly, immunity against diseases like *powdery mildew* and *the Tomato Yellow Leaf Curl Virus* (Pramanik et al. 2021). The implemented approach leads to better plant survival against chemical pesticides, enabling industry to transition toward sustainable floriculture practices (Wani et al. 2023). Scientists can develop disease-resistant plant cultivars through a single gene locus editing, thus improving the breeding process (Liu et al. 2021). Through CRISPR technology, the floriculture industry obtains various opportunities that allow it to create disease-resistant and water-efficient plant cultivars by targeting DNA sequences. Genetic transfers that affect stomatal response enable plants to sustain higher moisture levels by minimizing water loss during periods of low water availability (Sirohi et al. 2022). The significance of CRISPR technology has increased due to the combined threats of climate change and worsening water scarcity, which endanger agricultural sustainability. These controlled crop systems, combined with CRISPR technology and modern cultivation practices of hydroponics and vertical farming, achieve water conservation by reducing waste (Wani et al. 2023).

CRISPR also may be part of the solution for flowering plants' fertilizer-driven desires. Genetic modifications that alter metabolic pathways can optimize nutrient utilization for the nutrients which are present. CRISPR was also used to modify genes associated with nitrogen uptake in other pathways to enhance the growth and flowering of plants, possibly creating transgenic crops that required less nitrogen fertilization (Giovannini et al. 2021). This reduction in fertilizer use reduces the negative environmental consequences that may come because of over fertilization (i.e., soil degradation and nutrient runoff) while also lowering production costs (Regmi et al. 2022).

8. Future Directions and Emerging Trends in CRISPR Applications for Floriculture Synthetic Biology and Designer Flowers

Today modern CRISPR and synthetic biology are effective tools for engineering flowers with novel desired aesthetic characteristics including remarkable color patterns as well as photopatterning-induced cell-cycle arrest leading to a darker or lighter flower.

Recently, two advancements have been made using the potentiality of CRISPR/Cas9 genome editing to change genes associated with flower configuration and color. Targeting the editing of biosynthetic pathways to alter flower color is a major function that CRISPR has in floral modification. Its ability, for example to modify the color of flowers in species such as *Torenia fournieri* by silencing homologous genes playing a role within flavonoid biosynthesis has effectively demonstrated this outcome. Gene silencing of choice such as *DFR* (*dihydroflavonol reductase*) or *CHS* (*chalcone synthase*), can transform the accommodative feature into flowers even making minor changes in floral color leaving CRISPR a powerful tool to bring about such custom modifications and yield suitable results (Nishihara et al. 2018; Noman et al. 2017). More recently, CRISPR/Cas9 technology was employed to create white anthocyanin-free morning glory flowers by chemically shocking the DFR-B locus (Watanabe et al. 2017a). Many more approaches are possible with CRISPR for a specific color to be altered in any other ornamental plant. In addition to color changes, CRISPR technology allows for genes that cause the petals of flowers to glow, such as a bioluminescent flower. The genetic code of plants could be used to insert genes from organisms such as fireflies or specific strains of fungi that give off bioluminescent light. Whilst direct examples of CRISPR-made bioluminescent flowers are still some way off, indeed, the wider successes behind such advances owe much to more fundamental daily work in synthetic biology (Sirohi et al. 2022). CRISPR can also be a useful tool in facilitating flowers to create variations in their color depending on environmental conditions. In stress response pathways, genes can be modified to induce flowers that change color by environmental conditions (e.g., temperature or light levels). Perhaps other specific illustrations will be helpful as well; altering the expression of genes responsible for anthocyanin biosynthesis may show improvements in producing flowers with environmental adaptability and good decorative color tuning (Chang Ho Ahn et al. 2020; Khan et al. 2022). In addition to being a pivotal step forward for the ornamental sector, this activity is also the best sustainable technique, which becomes more important as horticulture grows around the world. With current advances in synthetic biology and CRISPR technology, ornamental plants with improved aesthetic characteristics can be produced. It is possible to make marketplace novelty flowers by simply changing the morphological features of their genes. For instance, shifting these genes working on flower color and shape so that this type changes its phenotypic quality from small, frequent blossoms.

But chances are that the more industry gets involved, new breeding materials may end up delivering flowers with different properties than we have grown accustomed to.

8.1 Integration with Other Omics Technologies

The integrated use of these technologies provides an unprecedented opportunity to revolutionize floriculture through CRISPR technology in combination with state-of-the-art omics approaches, viz., transcriptomics, metabolomics or phenomics, and genomics. This, combined with a better understanding of how any change in one gene can affect plant metabolism and growth more broadly, makes uncovering new gene targets easier. CRISPR technology, especially the CRISPR/Cas9 and Cpf1 systems, has revolutionized plant breeding in precision genome editing. These systems have been shown to generate transgene-free plants in several crops vital for floriculture, as, e.g., cotton and maize (Gong et al. 2021). Multiplex gene editing directly accelerates the breeding process by changing many genes (Chen et al. 2019; Gong et al. 2021), and can improve one or more traits at once, such as flower color, size, and disease resistance. A study utilized the CRISPR/Cas9 system to modify the flower colors of Torenia fournieri (Nishihara et al. 2018). Scientists can evaluate gene expression modifications through the combination of transcriptomics and CRISPR technology. CROP-seq technology enables researchers to analyze complete transcriptomes from massive numbers of single cells, which helps them identify unique responses between cell types after perturbing genes (Datlinger et al. 2017). This single question provides understanding of how genetic modification affects floral qualities and the metabolic process, as well as developing programs within plants. Genetic modifications require additional metabolomic along transcriptomic studies to understand their metabolic effects so scientists can understand changes in floral attributes together with plant health (Rehman et al. 2022). Phenomics enables the study of phenotypic characteristics across entire genomes, which provides crucial insights into CRISPR-induced modifications of plant growth and structural development. High-throughput phenotyping strategies emerged recently to measure floriculture characteristics by tracking specific gene modifications in real time (Chen et al. 2019; Hassan et al. 2021). This complete method enables researchers to determine both singular traits and their cumulative progress throughout plant development as it functions as an environmental detector. The use of CRISPR technology enables researchers to detect previously unseen gene targets that remained hidden during previous experimental examinations. CRISPR-based screening identifies genes connected to stress responses as well as metabolic pathways responsible for flower quality and health (Rehman et al. 2022). The knowledge of genetic bases allows breeders to choose target genes for genetic modification that result in better cultivars satisfying market consumer needs. The combination of CRISPR with molecular biology tools creates a highly effective method for modern floriculture development, which operates in the current genomics, transcriptomics metabolomics and phenomics era. A cross-field analysis speeds up gene target discovery while providing valuable perception regarding how genetic adjustments influence metabolic processes and plant evolution. Science can constantly improve which will lead to larger potential in accurate genetic ornamental plant design and joint resolution of big-scale floriculture agricultural problems together with aesthetic market needs.

9. Conclusion

CRISPR/Cas9 is a new milestone for ornamental plant breeding, which supports a novel age in floricultural improvement. Genome editing has facilitated the accurate manipulation of specific genes, leading to the fast generation of new traits, e.g., flower color, size, and stress resistance. "Such breakthroughs of generating non-existing floral pigments and prolonging the vase life of cut flowers highlight the power of CRISPR in addressing market demands and the needs of green cities." Further, this technology is successful in addressing issues like biotic and abiotic stress tolerance, like resistance against environmental stresses or plant pathogens. The combination of CRISPR/Cas9 technology with multi-omics technology platforms along with multiplexed genome editing methods has enhanced its capabilities by enabling simultaneous modification of multiple traits while revealing detailed insights about genetic and molecular processes. Such cross-disciplinary parcels will not only speed up breeding but could also produce flowers with new properties, providing novel applications for the scientific and commercial landscapes. Although the use of CRISPR/Cas9 is just beginning to transform the way we conduct plant science, it is not without its challenges, specifically species-specific challenges, polyploidy pain points, and public perception of genome-edited plants. Better methodologies, alongside public awareness and streamlined regulatory frameworks, must be developed to achieve the full potential of genome editing.

References

Abdulla, M.F., K. Mostafa and M. Kavas. 2024. CRISPR/Cas9-mediated mutagenesis of FT/TFL1 in petunia improves plant architecture and early flowering. Plant Mol. Biol. 114(3): 1–13. https://doi.org/10.1007/s11103-024-01454-9.

Agrios, G.N. 2005. Plant Pathology. 5th edn. Elsevier Academic Press: 79–103.

Ali, Z., A. Abulfaraj, A. Idris, S. Ali, M. Tashkandi and M.M. Mahfouz. 2015. CRISPR/Cas9-mediated viral interference in plants. Genome Biol. 16: 1–11. https://doi.org/10.1186/s13059-015-0799-6.

Andualem, B. and A. Seid. 2021. The role of green biotechnology through genetic engineering for climate change mitigation and adaptation, and for food security: current challenges and future perspectives. J. Adv. Biol. Biotechnol. 24(1): 1–11. https://doi.org/10.9734/jabb/2021/v24i130192.

Anumala, N.V. and R. Kumar. 2021. Floriculture sector in India: current status and export potential. J. Hortic. Sci. Biotechnol. 96(5): 673–680. https://doi.org/10.1080/14620316.2021.1902863.

Ashwini, T., V.B. Prathyusha, N. Sahu, D. Divya and G. Mallikarjuna. 2024. CRISPR edited floriculture crops: A revolutionary technique to increase flower production, their color and longevity. pp. 381–400. *In*: CRISPRized Horticulture Crops. Elsevier. https://doi.org/10.1016/B978-0-443-13229-2.00022-3.

Azadi, P., H. Bagheri, A.M. Nalousi, F. Nazari and S.F. Chandler. 2016. Current status and biotechnological advances in genetic engineering of ornamental plants. Biotechnol. Adv. 34(6): 1073–1090. https://doi.org/10.1016/B978-0-443-13229-2.00022-3.

Baltes, N.J., A.W. Hummel, E. Konecna, R. Cegan, A.N. Bruns, D.M. Bisaro et al. 2015. Conferring resistance to geminiviruses with the CRISPR–Cas prokaryotic immune system. Nat. Plants 1(10): 1–4. https://doi.org/10.1038/nplants.2015.145.

Caruso, M., M.W. Smith, Y. Froelicher, G. Russo and F.G. Gmitter Jr. 2020. Traditional breeding. pp. 129–148. *In*: The Genus Citrus. Elsevier. https://doi.org/10.1016/B978-0-12-812163-4.00007-3.

Chandler, S.F. and F. Brugliera. 2011. Genetic modification in floriculture. Biotechnol. Lett. 33: 207–214. https://doi.org/10.1007/s10529-010-0424-4.

Chandrasekaran, J., M. Brumin, D. Wolf, D. Leibman, C. Klap, M. Pearlsman et al. 2016. Development of broad virus resistance in non-transgenic cucumber using CRISPR/Cas9 technology. Mol. Plant Pathol. 17: 1140–1153. https://doi.org/10.1111/mpp.12375.

Chang Ho Ahn, M. Ramya, H.R. An, P.M. Park, Y.-J. Kim, S.Y. Lee et al. 2020. Progress and challenges in the improvement of ornamental plants by genome editing. Plants 9: 687. https://doi.org/10.3390/plants9060687.

Chen, K., Y. Wang, R. Zhang, H. Zhang and C. Gao. 2019. CRISPR/Cas genome editing and precision plant breeding in agriculture. Annu. Rev. Plant Biol. 70: 667–697. https://doi.org/10.1146/annurev-arplant-050718-100049.

Chen, Q., X. Zhang, R. Jin and H. Mao. 2024. Establishment of a CRISPR/Cas9 gene-editing system for Chrysanthemum morifolium. Ornam. Plant Res. 4: 1. https://doi.org/10.48130/opr-0024-0012.

Damalas, C.A. and I.G. Eleftherohorinos. 2011. Pesticide exposure, safety issues, and risk assessment indicators. Int. J. Environ. Res. Public Health 8: 1402–1419. https://doi.org/10.3390/ijerph8051402.

Datlinger, P., A.F. Rendeiro, C. Schmidl, T. Krausgruber, P. Traxler, J. Klughammer et al. 2017. Pooled CRISPR screening with single-cell transcriptome readout. Nat. Methods 14: 297–301. https://doi.org/10.1038/nmeth.4177.

Dimock, A.W., C.M. Geissinger and R.K. Horst. 1971. Chlorotic mottle: A newly recognized disease of chrysanthemum. Phytopathology 61: 415–419.

Du, X., W. Li, L. Sheng, Y. Deng, Y. Wang, W. Zhang et al. 2018. Over-expression of chrysanthemum CmDREB6 enhanced tolerance of chrysanthemum to heat stress. BMC Plant Biol. 18: 140. https://doi.org/10.1186/s12870-018-1400-8.

Du, Y.-T., M.-J. Zhao, C.-T. Wang, Y. Gao, Y.-X. Wang, Y.-W. Liu et al. 2018. Identification and characterization of GmMYB118 responses to drought and salt stress. BMC Plant Biol. 18: 155. https://doi.org/10.1186/s12870-018-1551-7.

Duan, Y.-B., J. Li, R.-Y. Qin, R.-F. Xu, H. Li, Y.-C. Yang et al. 2016. Identification of a regulatory element responsible for salt induction of rice OsRAV2 through *ex situ* and *in situ* promoter analysis. Plant Mol. Biol. 90: 49–62. https://doi.org/10.1007/s11103-015-0393-z.

Erdoğan, İ., B. Cevher-Keskin, Ö. Bilir, Y. Hong and M. Tör. 2023. Recent developments in CRISPR/Cas9 genome-editing technology related to plant disease resistance and abiotic stress tolerance. Biology 12: 1037. https://doi.org/10.3390/biology12071037.

Flores, R. 2008. Viroids in ornamentals. Acta Hortic. 901: 23–34. https://doi.org/10.17660/ActaHortic.2011.901.2.

Frachon, L., S.A. Stirling, F.P. Schiestl and N. Dudareva. 2021. Combining biotechnology and evolution for understanding the mechanisms of pollinator attraction. Curr. Opin. Biotechnol. 70: 213–219. https://doi.org/10.1016/j.copbio.2021.06.004.

Ge, Z., L. Zheng, Y. Zhao, X. Li, Q. Zhang, M. Chen et al. 2019. Engineered xCas9 and SpCas9-NG variants broaden PAM recognition sites to generate mutations in Arabidopsis plants. Plant Biotechnol. J. 17: 1865. https://doi.org/10.1111/pbi.13148.

Giovannini, A., M. Laura, B. Nesi, M. Savona and T. Cardi. 2021. Genes and genome editing tools for breeding desirable phenotypes in ornamentals. Plant Cell Rep. 40: 461–478. https://doi.org/10.1007/s00299-020-02632-x.

Gong, C., S. Huang, R. Song and W. Qi. 2021. Comparative study between the CRISPR/Cpf1 (Cas12a) and CRISPR/Cas9 systems for multiplex gene editing in maize. Agriculture 11: 429. https://doi.org/10.3390/agriculture11050429.

Gupta, K., A. Dey and B. Gupta. 2013. Polyamines and their role in plant osmotic stress tolerance. pp. 1053–1072. *In*: Climate Change Abiotic Stress Tolerance. Wiley-VCH. https://doi.org/10.1002/9783527675265.ch40.

Hammond, J., Q. Huang, R. Jordan, T. Miller, S. Davis, K. Patel et al. 2023. International trade and local effects of viral and bacterial diseases in ornamental plants. Annu. Rev. Phytopathol. 61: 73–95. https://doi.org/10.1146/annurev-phyto-021621-114618.

Hassan, M.M., Y. Zhang, G. Yuan, L. Zhong, H. Feng, B. Wang et al. 2021. Construct design for CRISPR/Cas-based genome editing in plants. Trends Plant Sci. 26(11): 1133–1152. https://doi.org/10.1016/j.tplants.2021.06.015.

Hong, B., C. Ma, Y. Yang, Z. Wang, X. Cheng and J. Li. 2009. Over-expression of AtDREB1A in chrysanthemum enhances tolerance to heat stress. Plant Mol. Biol. 70: 231–240. https://doi.org/10.1007/s11103-009-9468-z.

Hua, K., X. Tao, P. Han, R. Wang and J.-K. Zhu. 2019. Genome engineering in rice using Cas9 variants that recognize NG PAM sequences. Mol. Plant 12(7): 1003–1014.

Huang, Q., X. Liao, X. Yang, C. Zhang, L. Wang and Y. Liu. 2021. Lysine crotonylation of DgTIL1 at K72 modulates cold tolerance by enhancing DgnsLTP stability in chrysanthemum. Plant Biotechnol. J. 19(6): 1125–1140. https://doi.org/10.1111/pbi.13533.

Indurthi, S., I. Sarma and D.V. Vinod. 2024. Horticultural innovations elevating crop yields and agricultural sustainability for a flourishing future. Plant Cell Biotechnol. Mol. Biol. 25(1–2): 22–44. https://doi.org/10.56557/PCBMB/2024/v25i1-28560.

Kaur, H., D.K. Pandey, U. Goutam and V. Kumar. 2021. CRISPR/Cas9-mediated genome editing is revolutionizing the improvement of horticultural crops: Recent advances and future prospects. Sci. Hortic. 289: 110476. https://doi.org/10.1016/j.scienta.2021.110476.

Khan, Z., A. Razzaq, T. Sattar, M.A. Malik, M. Shakeel, M. Raza et al. 2022. Understanding floral biology for CRISPR-based modification of color and fragrance in horticultural plants. F1000Research 11: 854. https://doi.org/10.12688/f1000research.122453.1.

Kim, D., B. Alptekin and H. Budak. 2018. CRISPR/Cas9 genome editing in wheat. Funct. Integr. Genomics 18: 31–41. https://doi.org/10.1007/s10142-017-0572-x.

Kishi-Kaboshi, M., R. Aida and K. Sasaki. 2017. Generation of gene-edited Chrysanthemum morifolium using multicopy transgenes as targets and markers. Plant Physiol. Biochem. 58(2): 216–226.

Klap, C., E. Yeshayahou, A.M. Bolger, E. Arazi, A. Shavrukov and A. Eshed. 2017. Tomato facultative parthenocarpy results from Sl AGAMOUS-LIKE 6 loss of function. Plant Biotechnol. J. 15(5): 634–647. https://doi.org/10.1111/pbi.12662.

Kui, L., H. Chen, W. Zhang, Q. He, Y. Liu and X. Liu. 2017. Building a genetic manipulation tool box for orchid biology: identification of constitutive promoters and application of CRISPR/Cas9 in the orchid, Dendrobium officinale. Front. Plant Sci. 7: 2036.

Lee, H., S. Subburaj, L. Tu, K.-Y. Lee, G. Park and G.-J. Lee. 2020. Overview of CRISPR/Cas9: A chronicle of the CRISPR system and application to ornamental crops. Korean J. Agric. Sci. 47(4): 903–920. https://doi.org/10.7744/kjoas.20200075.

Li, J., J. Luo, M. Xu, S. Li, J. Zhang, H. Li et al. 2019. Plant genome editing using xCas9 with expanded PAM compatibility. J. Genet. Genomics. https://doi.org/10.1016/j.jgg.2019.03.004.

Li, S., B. Zhu, J. Pirrello, C. Xu, B. Zhang, M. Bouzayen et al. 2020. Roles of RIN and ethylene in tomato fruit ripening and ripening-associated traits. New Phytol. 226(2): 460–475. https://doi.org/10.1111/nph.16362.

Liu, C., Y. Zhang, Y. Tan, T. Zhao, X. Xu, H. Yang et al. 2021. CRISPR/Cas9-mediated SlMYBS2 mutagenesis reduces tomato resistance to Phytophthora infestans. Int. J. Mol. Sci. 22(21): 11423. https://doi.org/10.3390/ijms222111423.

Liu, F., L. Hu, Y. Cai, H. Lin, J. Liu and Y. Yu. 2016. Molecular characterization and functional analysis of two petunia PhEILs. Front. Plant Sci. 7: 1606. https://doi.org/10.3389/fpls.2016.01606.

Liu, Q., M. Kasuga, Y. Sakuma, H. Abe, S. Miura, K. Yamaguchi-Shinozaki et al. 1998. Two transcription factors, DREB1 and DREB2, with an EREBP/AP2 DNA binding domain separate two cellular signal transduction pathways in drought- and low-temperature-responsive gene expression, respectively, in Arabidopsis. Plant Cell 10(8): 1391–1406. https://doi.org/10.1105/tpc.10.8.1391.

Liu, Y., J. Feng, H. Pan, X. Zhang and Y. Zhang. 2022. Genetically engineered bacterium: Principles, practices, and prospects. Front. Microbiol. 13: 997587. https://doi.org/10.3389/fmicb.2022.997587.

Lou, D., H. Wang, G. Liang and D. Yu. 2017. OsSAPK2 confers abscisic acid sensitivity and tolerance to drought stress in rice. Front. Plant Sci. 8: 993. https://doi.org/10.3389/fpls.2017.00993.

Lyzenga, W.J., C.J. Pozniak and S. Kagale. 2021. Advanced domestication: Harnessing the precision of gene editing in crop breeding. Plant Biotechnol. J. 19(4): 660–670. https://doi.org/10.1111/pbi.13576.

Malnoy, M., R. Viola, M.-H. Jung, O.-J. Koo, S. Kim, J.-S. Kim et al. 2016. DNA-free genetically edited grapevine and apple protoplast using CRISPR/Cas9 ribonucleoproteins. Front. Plant Sci. 7: 1904. https://doi.org/10.3389/fpls.2016.01904.

Mehta, R., T. Radhakrishnan, A. Kumar et al. 2013. Coat protein-mediated transgenic resistance of peanut (*Arachis hypogaea* L.) to peanut stem necrosis disease through *Agrobacterium*-mediated genetic transformation. Indian J. Virol. 24: 205–213. https://doi.org/10.1007/s13337-013-0157-9.

Mekapogu, M., J.-A. Jung, O.-K. Kwon, M.-S. Ahn, H.-Y. Song and S. Jang. 2021. Recent progress in enhancing fungal disease resistance in ornamental plants. Int. J. Mol. Sci. 22(15): 7956. https://doi.org/10.3390/ijms22157956.

Mekapogu, M., H.-Y. Song, S.-H. Lim and J.-A. Jung. 2023. Genetic engineering and genome editing advances to enhance floral attributes in ornamental plants: An update. Plants 12(23): 3983. https://doi.org/10.3390/plants12233983.

Mirik, M., Y. Aysan and F. Baysal-Gurel. 2018. Bacterial spot and blight diseases of ornamental plants caused by different Xanthomonas species in Turkey. Plant Prot. Sci. 54(4): 240–247. https://doi.org/10.17221/10/2017-PPS.

Mushtaq, M., A. Ahmad Dar, M. Skalicky, A. Tyagi, N. Bhagat, U. Basu et al. 2021. CRISPR-based genome editing tools: Insights into technological breakthroughs and future challenges. Genes 12(6): 797. https://doi.org/10.3390/genes12060797.

Nakamura, N., M. Fukuchi-Mizutani, K. Miyazaki, K. Suzuki and Y. Tanaka. 2006. RNAi suppression of the anthocyanidin synthase gene in Torenia hybrida yields white flowers with higher frequency and better stability than antisense and sense suppression. Plant Biotechnol. 23(1): 13–17. https://doi.org/10.5511/plantbiotechnology.23.13.

Nishihara, M., A. Higuchi, A. Watanabe and K. Tasaki. 2018. Application of the CRISPR/Cas9 system for modification of flower color in Torenia fournieri. BMC Plant Biol. 18: 1–9. https://doi.org/10.1186/s12870-018-1539-3.

Noman, A., M. Aqeel, J. Deng, N. Khalid, T. Sanaullah and H. Shuilin. 2017. Biotechnological advancements for improving floral attributes in ornamental plants. Front. Plant Sci. 8: 530. https://doi.org/10.3389/fpls.2017.00530.

Peng, A., S. Chen, T. Lei, L. Xu, Y. He, L. Wu et al. 2017. Engineering canker-resistant plants through CRISPR/Cas9-targeted editing of the susceptibility gene CsLOB1 promoter in citrus. Plant Biotechnol. J. 15(12): 1509–1519. https://doi.org/10.1111/pbi.12733.

Pompili, V., L. Dalla Costa, S. Piazza, M. Pindo and M. Malnoy. 2020. Reduced fire blight susceptibility in apple cultivars using a high-efficiency CRISPR/Cas9-FLP/FRT-based gene editing system. Plant Biotechnol. J. 18(3): 845–858. https://doi.org/10.1111/pbi.13253.

Pramanik, D., R.M. Shelake, J. Park, M.J. Kim, I. Hwang, Y. Park et al. 2021. CRISPR/Cas9-mediated generation of pathogen-resistant tomato against tomato yellow leaf curl virus and powdery mildew. Int. J. Mol. Sci. 22(4): 1878. https://doi.org/10.3390/ijms22041878.

Qi, Y., Y. Liu, Z. Zhang, J. Gao, Z. Guan, W. Fang et al. 2018. The over-expression of a chrysanthemum gene encoding an RNA polymerase II CTD phosphatase-like 1 enzyme enhances tolerance to heat stress. Hortic. Res. 5: 37. https://doi.org/10.1038/s41438-018-0037-y.

Rasheed, A., R.A. Gill, M.U. Hassan, A. Mahmood, S. Qari, Q.U. Zaman et al. 2021. A critical review: recent advancements in the use of CRISPR/Cas9 technology to enhance crops and alleviate global food crises. Curr. Issues Mol. Biol. 43(3): 1950–1976. https://doi.org/10.3390/cimb43030135.

Regalado González, J.J., E. Carmona Martín, V. Querol, C.G. Vélez, C.L. Encina and S.I. Pitta-Alvarez. 2017. Production of compact petunias through polyploidization. Plant Cell Tissue Organ Cult. 129: 331–342. https://doi.org/10.1007/s11240-016-1156-5.

Regmi, A., S. Singh, N. Moustaid-Moussa, C. Coldren and C. Simpson. 2022. The negative effects of high rates of biochar on violas can be counteracted with fertilizer. Plants 11(4): 491. https://doi.org/10.3390/plants11040491.

Rehman, R.S., M. Ali, S.A. Zafar, M. Ahmad, A.N. Pasha, H. Bashir et al. 2022. Tapping into the unsung potential of CRISPR/CAS technology in agriculture. Asian J. Biochem. Genet. Mol. Biol. 10(4): 1–26. https://doi.org/10.9734/ajbgmb/2022/v10i430249.

Ren, C., Y. Lin and Z. Liang. 2022. CRISPR/Cas genome editing in grapevine: recent advances, challenges and future prospects. Fruit Res. 2(1): 1–9. https://doi.org/10.48130/FruRes-2022-0007.

Aida, R., K. Yoshida, T. Kondo, S. Kishimoto and M. Shibata. 2000. Copigmentation gives bluer flowers on transgenic torenia plants with the antisense dihydroflavonol-4-reductase gene. Plant Sci. 160: 49–56. https://doi.org/10.1016/s0168-9452(00)00364-2.

Rukavtsova, E.B., N.S. Zakharchenko, V.G. Lebedev and K.A. Shestibratov. 2022. CRISPR-Cas genome editing for horticultural crops improvement: advantages and prospects. Horticulturae 9: 38. https://doi.org/10.3390/horticulturae9010038.

Schwartz, A.R., N. Potnis, S. Timilsina, M. Wilson, J. Patané, J. Martins Jr et al. 2015. Phylogenomics of Xanthomonas field strains infecting pepper and tomato reveals diversity in effector repertoires and identifies determinants of host specificity. Front. Microbiol. 6: 535. https://doi.org/10.3389/fmicb.2015.00535.

Shafiq, M., M.T. Abbas, S. Mushtaq, R. Khaliq, H. Arshad, R. Haroon et al. 2024. Viral diseases of ornamental plants-I. pp. 777–787. *In*: Viral Dis. Field Hortic. Crops. Elsevier. https://doi.org/10.1016/B978-0-323-90899-3.00045-8.

Shamshirgaran, Y., J. Liu, H. Sumer, P.J. Verma and A. Taheri-Ghahfarokhi. 2022. Tools for efficient genome editing; ZFN, TALEN, and CRISPR. pp. 29–46. *In*: Appl. Genome Modulation. https://doi.org/10.1007/978-1-0716-2301-5_2.

Sharma, R. and Y. Messar. 2017. Transgenics in ornamental crops: creating novelties in economically important cut flowers. Curr. Sci. 113: 43–52. https://doi.org/10.18520/cs/v113/i01/43-52.

Shen, C., Z. Que, Y. Xia, N. Tang, D. Li, R. He et al. 2017. Knock out of the annexin gene OsAnn3 via CRISPR/Cas9-mediated genome editing decreased cold tolerance in rice. J. Plant Biol. 60: 539–547. https://doi.org/10.1007/s12374-016-0400-1.

Shibuya, K., K. Watanabe and M. Ono. 2018. CRISPR/Cas9-mediated mutagenesis of the EPHEMERAL1 locus that regulates petal senescence in Japanese morning glory. Plant Physiol. Biochem. 131: 53–57. https://doi.org/10.1016/j.plaphy.2018.04.036.

Sirohi, U., M. Kumar, V.R. Sharma, S. Teotia, D. Singh, V. Chaudhary et al. 2022. CRISPR/Cas9 system: A potential tool for genetic improvement in floricultural crops. Mol. Biotechnol. 64: 1303–1318. https://doi.org/10.1007/s12033-022-00523-y.

Souza, J.L., V.V. Nunes, C.C. Calazans, M.F.O. Torres, B.A.L. de Freitas and R. Silva-Mann. 2021. CRISPR-based solutions for agriculture: A systematic. Rev. Cienc. Agrar. 64: 1. https://doi.org/10.51722/S2318-3403202100011356.

Sretenovic, S., D. Yin, A. Levav, J.D. Selengut, S.M. Mount and Y. Qi. 2021. Expanding plant genome-editing scope by an engineered iSpyMacCas9 system that targets A-rich PAM sequences. Plant Commun. 2: 100101. https://doi.org/10.1016/j.xplc.2020.100101.

Subburaj, S., S.J. Chung, C. Lee, S.-M. Ryu, D.H. Kim, J.-S. Kim et al. 2016. Site-directed mutagenesis in Petunia× hybrida protoplast system using direct delivery of purified recombinant Cas9 ribonucleoproteins. Plant Cell Rep. 35: 1535–1544. https://doi.org/10.1007/s00299-016-1937-7.

Su, S., W. Xiao, W. Guo, X. Yao, J. Xiao, Z. Ye et al. 2017. The CYCLOIDEA–RADIALIS module regulates petal shape and pigmentation, leading to bilateral corolla symmetry in Torenia fournieri (Linderniaceae). New Phytol. 215: 1582–1593. https://doi.org/10.1111/nph.14673.

Sun, L. and T. Kao. 2018. CRISPR/Cas9-mediated knockout of PiSSK1 reveals essential role of S-locus F-box protein-containing SCF complexes in recognition of non-self S-RNases during cross-compatible pollination in self-incompatible Petunia inflata. Plant Reprod. 31: 129–143. https://doi.org/10.1007/s00497-017-0314-1.

Tanaka, Y., F. Brugliera and S. Chandler. 2009. Recent progress of flower colour modification by biotechnology. Int. J. Mol. Sci. 10: 5350–5369. https://doi.org/10.3390/ijms10125350.

Tasaki, K., M. Yoshida, M. Nakajima, A. Higuchi, A. Watanabe and M. Nishihara. 2020. Molecular characterization of an anthocyanin-related glutathione S-transferase gene in Japanese gentian with the CRISPR/Cas9 system. BMC Plant Biol. 20: 1–14. https://doi.org/10.1186/s12870-020-02565-3.

Tong, C., F. Wu, Y. Yuan, Y. Chen and C. Lin. 2020. High-efficiency CRISPR/Cas-based editing of Phalaenopsis orchid MADS genes. Plant Biotechnol. J. 18: 889. https://doi.org/10.1111/pbi.13264.

Traversari, S., S. Cacini, A. Galieni, B. Nesi, N. Nicastro and C. Pane. 2021. Precision agriculture digital technologies for sustainable fungal disease management of ornamental plants. Sustainability 13(7): 3707. https://doi.org/10.3390/su13073707.

Wang, C., Y. Li, N. Wang, Q. Yu, Y. Li, J. Gao et al. 2023. An efficient CRISPR/Cas9 platform for targeted genome editing in rose (Rosa hybrida). J. Integr. Plant Biol. 65(4): 895–899. https://doi.org/10.1111/jipb.13421.

Wang, C.-T., J.-N. Ru, Y.-W. Liu, J.-F. Yang, M. Li and Z.-S. Xu. 2018. The maize WRKY transcription factor ZmWRKY40 confers drought resistance in transgenic Arabidopsis. Int. J. Mol. Sci. 19(9): 2580. https://doi.org/10.3390/ijms19092580.

Wang, F., C. Wang, P. Liu, C. Lei, W. Hao, Y. Gao et al. 2016. Enhanced rice blast resistance by CRISPR/Cas9-targeted mutagenesis of the ERF transcription factor gene OsERF922. PLoS One 11(4): e0154027. https://doi.org/10.1371/journal.pone.0154027.

Wang, Y., X. Cheng, Q. Shan, Y. Zhang, J. Liu, C. Gao et al. 2014. Simultaneous editing of three homoeoalleles in hexaploid bread wheat confers heritable resistance to powdery mildew. Nat. Biotechnol. 32(9): 947–951. https://doi.org/10.1038/nbt.2969.

Wani, M.A., A. Din, I.T. Nazki, T.U. Rehman, J.M. Al-Khayri, S.M. Jain et al. 2023. Navigating the future: Exploring technological advancements and emerging trends in the sustainable ornamental industry. Front. Environ. Sci. 11: 1188643. https://doi.org/10.3389/fenvs.2023.1188643.

Watanabe, K., A. Kobayashi, M. Endo, K. Sage-Ono, S. Toki and M. Ono. 2017a. CRISPR/Cas9-mediated mutagenesis of the dihydroflavonol-4-reductase-B (DFR-B) locus in the Japanese morning glory Ipomoea (Pharbitis) nil. Sci. Rep. 7(1): 10028. https://doi.org/10.1038/s41598-017-10715-1.

Watanabe, K., A. Kobayashi, M. Endo, K. Sage-Ono, S. Toki and M. Ono. 2017b. CRISPR/Cas9-mediated mutagenesis of the dihydroflavonol-4-reductase-B (DFR-B) locus in the Japanese morning glory Ipomoea (Pharbitis) nil. Sci. Rep. 7(1): 10028. https://doi.org/10.1038/s41598-017-10715-1.

Watanabe, K., C. Oda-Yamamizo, K. Sage-Ono, A. Ohmiya and M. Ono. 2018. Alteration of flower colour in Ipomoea nil through CRISPR/Cas9-mediated mutagenesis of carotenoid cleavage dioxygenase 4. Transgenic Res. 27: 25–38. https://doi.org/10.1007/s11248-017-0051-0.

Wei, Y., W. Liu, W. Hu, Y. Yan and H. Shi. 2020. The chaperone MeHSP90 recruits MeWRKY20 and MeCatalase1 to regulate drought stress resistance in cassava. New Phytol. 226(2): 476–491. https://doi.org/10.1111/nph.16346.

Wu, S., H. Kyaw, Z. Tong, Y. Yang, Z. Wang, L. Zhang et al. 2024. A simple and efficient CRISPR/Cas9 system permits ultra-multiplex genome editing in plants. Crop J. 12(2): 569–582. https://doi.org/10.1016/j.cj.2024.01.010.

Xiong, J.-S., J. Ding and Y. Li. 2015. Genome-editing technologies and their potential application in horticultural crop breeding. Hortic. Res. 2: 15019. https://doi.org/10.1038/hortres.2015.19.

Xu, J., B. Kang, A.H. Naing, S. Bae, J. Kim, H. Kim et al. 2020. CRISPR/Cas9-mediated editing of 1-aminocyclopropane-1-carboxylate oxidase1 enhances Petunia flower longevity. Plant Biotechnol. J. 18(1): 287–297. https://doi.org/10.1111/pbi.13197.

Xu, J., A.H. Naing, H. Bunch, J. Jeong, H. Kim and C.K. Kim. 2021. Enhancement of the flower longevity of Petunia by CRISPR/Cas9-mediated targeted editing of ethylene biosynthesis genes. Postharvest Biol. Technol. 174: 111460. https://doi.org/10.1016/j.postharvbio.2020.111460.

Xu, D., J. Liang, T. Zhou, Y. Liu, Q. Wang, X. Mu et al. 2024. First report of powdery mildew caused by *Podosphaera fusca* on Coreopsis tinctoria in China. Plant. Disease 108(2): 532.

Yan, R., Z. Wang, Y. Ren, H. Li, N. Liu and H. Sun. 2019. Establishment of efficient genetic transformation systems and application of CRISPR/Cas9 genome editing technology in *Lilium pumilum* DC. Fisch. and *Lilium longiflorum* White Heaven. Int. J. Mol. Sci. 20(12): 2920. https://doi.org/10.3390/ijms20122920.

Yang, T., M. Ali, L. Lin, P. Li, H. He, Q. Zhu et al. 2023. Recoloring tomato fruit by CRISPR/Cas9-mediated multiplex gene editing. Hortic. Res. 10(1): uhac214. https://doi.org/10.1093/hr/uhac214.

Yin, K. and J.-L. Qiu. 2019. Genome editing for plant disease resistance: applications and perspectives. Philos. Trans. R. Soc. Lond. B Biol. Sci. 374(1767): 20180322. https://doi.org/10.1098/rstb.2018.0322.

Yin, Y., K. Qin, X. Song, Q. Zhang, Y. Zhou, X. Xia et al. 2018. BZR1 transcription factor regulates heat stress tolerance through FERONIA receptor-like kinase-mediated reactive oxygen species signaling in tomato. Plant Cell Physiol. 59(11): 2239–2254. https://doi.org/10.1093/pcp/pcy146.

Yu, J., L. Tu, S. Subburaj, S. Bae and G.-J. Lee. 2021. Simultaneous targeting of duplicated genes in Petunia protoplasts for flower color modification via CRISPR/Cas9 ribonucleoproteins. Plant Cell Rep. 40(6): 1037–1045. https://doi.org/10.1007/s00299-020-02593-1.

Zafar, K., M.Z. Khan, I. Amin, Z. Mukhtar, S. Yasmin, M. Arif et al. 2020. Precise CRISPR/Cas9 mediated genome editing in super basmati rice for resistance against bacterial blight by targeting the major susceptibility gene. Front. Plant Sci. 11: 575. https://doi.org/10.3389/fpls.2020.00575.

Zeng, Y., J. Wen, W. Zhao, Q. Wang and W. Huang. 2020. Rational improvement of rice yield and cold tolerance by editing the three genes OsPIN5b, GS3, and OsMYB30 with the CRISPR–Cas9 system. Front. Plant Sci. 10: 1663. https://doi.org/10.3389/fpls.2019.01663.

Zhang, B., X. Yang, C. Yang, M. Li and Y. Guo. 2016. Exploiting the CRISPR/Cas9 system for targeted genome mutagenesis in petunia. Sci. Rep. 6: 20315. https://doi.org/10.1038/srep20315.

Zhang, B., X. Xu, R. Huang, S. Yang, M. Li and Y. Guo. 2021. CRISPR/Cas9-mediated targeted mutation reveals a role for AN4 rather than DPL in regulating venation formation in the corolla tube of Petunia hybrida. Hortic. Res. 8: 55. https://doi.org/10.1038/s41438-021-00555-6.

Zhang, M., Q. Liu, X. Yang, J. Xu, G. Liu, X. Yao et al. 2020. CRISPR/Cas9-mediated mutagenesis of Clpsk1 in watermelon to confer resistance to Fusarium oxysporum f. sp. niveum. Plant Cell Rep. 39: 589–595. https://doi.org/10.1007/s00299-020-02516-0.

Zhang, S., F. Zhou, Z. Liu, X. Feng, Y. Li and P. Zhu. 2022. Inactivation of BoORP3a, an oxysterol-binding protein, causes a low wax phenotype in ornamental kale. Hortic. Res. 9: uhac219. https://doi.org/10.1093/hr/uhac219.

Zhang, T., Q. Zheng, X. Yi, H. An, Y. Zhao, S. Ma et al. 2018. Establishing RNA virus resistance in plants by harnessing CRISPR immune system. Plant Biotechnol. J. 16(8): 1415–1423. https://doi.org/10.1111/pbi.12881.

Zhang, Y., X. Feng, Y. Liu, F. Zhou and P. Zhu. 2022. A single-base insertion in BoDFR1 results in loss of anthocyanins in green-leaved ornamental kale. Theor. Appl. Genet. 135(6): 1855–1865. https://doi.org/10.1007/s00122-022-04079-y.

CHAPTER 7

Manipulating the CRISPR/Cas Genome Editing Technology for Targeted Genome Mutagenesis in Ornamental Plants

Irum Saadia Khan,[1] *Khurram Shahzad,*[2] *Fahim Ullah Khan,*[3] *Abdul Basit,*[4] *Yaqoob Sultan*[5] and *Attiq ur Rehman*[6,*]

1. Introduction

The field of biological research has significantly expanded with recent technological advancements, inventions, and discoveries. Various areas of biological research have become increasingly efficient, cost-effective, and reliable. In this era of a rapidly changing environment, increased food demand, disease prevalence, and survival challenges, CRISPR-Cas has become a leading-edge technology with numerous applications in research and development (Noman et al. 2016). There have been many areas of research where scientists have previously faced limitations, but now, with this novel tool, the potential for gene editing in plants, humans, and animals has been greatly enhanced (Li et al. 2019; Wang 2020). CRISPR-Cas has been extensively utilized in many aspects of genome editing owing to its exceptional precision and accuracy. Disease resistance, gene knockout, trait enhancement, drought resistance, stress resistance, inducing and correcting mutations, addressing genetic disorders, producing desired characteristics, and many other objectives can be achieved through CRISPR-Cas application. It has granted scientists the ability to provide rapid solutions to contemporary challenges (Mao et al. 2013; Ma et al. 2016).

The explanation of CRISPR can vary in complexity, depending on the level of detail provided. As previously noted, CRISPR was initially studied in *Escherichia coli*. The CRISPR system comprises two main components: first, the repeats, which are short sections of DNA between

[1] Department of Plants and Crops, Faculty of Bioscience Engineering, Ghent University, Ghent, Belgium.
[2] Department of Plant Breeding and Genetics, The University of Haripur, Haripur, Pakistan.
[3] Department of Agriculture, Hazara University Mansehra, 21120 Mansehra, Pakistan.
[4] Floricultural Biotechnology Lab., Department of Horticulture, Kyungpook National University, South Korea.
[5] Department of Grass Breeding, Institute of Agriculture, Lithuanian Research Centre for Agriculture and Forestry, Kėdainių r., 58344, Lithuania.
[6] Department of Agricultural Sciences, Faculty of Agriculture and Forestry, University of Helsinki, FI-00790, Helsinki, Finland.
* Corresponding author: attiq.rehman@helsinki.fi

DOI: 10.1201/9781003518204-7

20 and 30 nucleotides in length, referred to as palindromes. A palindrome is a sequence of nucleotides that repeats identically in both strands of DNA when read from left to right. These palindromes are interspersed with DNA segments called "spacer DNA" (Semenova et al. 2011). Spacer DNA has garnered significant interest because of its heterogeneous nature and the fact that each portion of spacer DNA is unique (Garneau et al. 2010). Subsequent research revealed that spacer DNA exhibits perfect similarity with viral DNA, particularly that of bacteriophages (viruses that infect bacteria) (Bolotin et al. 2005). Researchers have pinpointed multiple genes linked to CRISPR, referred to as CRISPR-associated genes (or cas-genes). These cas genes code for cas-proteins, which function as helicases responsible for unwinding and cutting DNA. This system serves as an immune mechanism for bacterial organisms, providing a defensive strategy against bacteriophages (van der Oost et al. 2014). When a bacteriophage introduces its DNA into *E. coli*, it typically hijacks cellular machinery and causes cell death. However, the presence of the CRISPR system enables cells to transcribe CRISPR RNA and encode Cas proteins. Consequently, the cell inactivates viral DNA before infection can occur (Rollie et al. 2015). In the event of a novel viral attack, bacteria encode another type of protein called cas1. This cas1 protein captures viral DNA, fragments it, and importantly, incorporates it into the CRISPR system of the bacterial genome as an additional spacer DNA. Thus, these spacer DNA sequences effectively represent a historical record of prior viral infections (Fagerlund et al. 2017).

Scientists employed CRISPR to induce precise editing of various genomic regions for altered phenotypes that gained attention from researchers across the world to explore the technology for their specific use. The two-component gRNA system is used to precisely edit the targeted DNA. It consists of two elements: the CRISPR RNA (crRNA), which provides the sequence-specific guide for locating the target DNA site, and the Cas protein, which carries out the actual cleavage of the DNA (Jinek et al. 2012). Additionally, there is a sequence called the Protospacer Adjacent Motif (PAM), which assists the protein in locating the exact position on the DNA where it is designed to cut. The PAM is always adjacent to the targeted sequence. It consists of a short sequence of three or more nucleotides and frequently begins with NGG, NRG, NGCG (among various other PAM types), where N can represent any nucleotide (Kleinstiver et al. 2015) and G represents guanine, C represents cytosine, and R represents a purine nucleotide (adenine or guanine). The guide RNA scans the entire DNA strand, and upon identifying the matching PAM sequence, it creates a double-stranded cut in the DNA (Jinek et al. 2012). Consequently, once the cut is made, the DNA separates, and the cell attempts to repair it, resulting in mutations. Through this process, a gene can be silenced or knocked out (Schiml et al. 2014). Alternatively, when a gene is to be added to a DNA strand, a host nucleotide sequence is provided to the cell, allowing for gene insertion (Ding et al. 2016; Song et al. 2016b). Figure 1 illustrates that the gRNA, in conjunction with Cas9, binds to the target DNA at the PAM site and creates a Double-Stranded Break (DSB). This break is subsequently repaired by the cell through two mechanisms. In the case of gene knock-out, the cell repairs it via the Non-Homologous End Joining (NHEJ) method by introducing indels (Kleinstiver et al. 2015). Conversely, in the case of gene knock-in, an exogenous DNA donor template is inserted into the cell, facilitating the repair, through the Homology-Directed Repair (HDR/HR repair) method. Both methods are utilized within the scientific community for various objectives related to gene knock-out or gene insertion (Kleinstiver et al. 2015; Ding et al. 2016).

1.1 History of CRISPR

The CRISPR-Cas system, first identified in *Escherichia coli* in 1987, has been extensively developed as a versatile tool for genome editing in a variety of organisms, including ornamental plant species (Ishino et al. 2018). The application of CRISPR-Cas in ornamental plants began gaining traction in the mid-2010s, when researchers started exploring its potential for targeted mutagenesis to enhance traits such as fragrance, disease resistance, and flower color. Studies on Petunia and Torenia showed the feasibility of CRISPR-Cas for creating novel flower pigmentation patterns by targeting

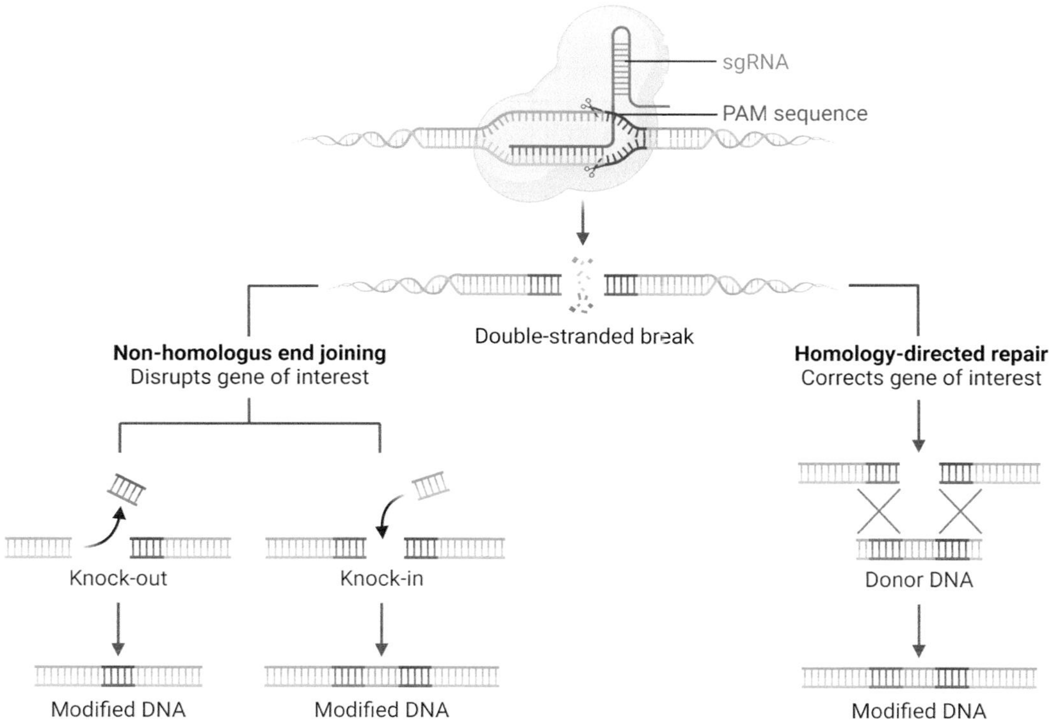

Figure 1. Visual demonstration of the working of Cas9.

the genes encoding the anthocyanin biosynthesis pathway (Nishihara et al. 2018; Yu et al. 2021). These pioneering efforts set the foundation for the use of CRISPR-Cas for ornamental plant genome manipulation and breeding, enabling efficient and precise genome modifications. On one hand, adoption of the technology has been slower compared to other crops due to the non-edible nature of ornamentals, while on the other hand, it offers considerable opportunities to researchers to explore vast research avenues in non-food crops, where the regulatory considerations are strict.

Biological life is characterized by biomolecules based on chains of nucleic acids called DNA. DNA structures define the core biological attributes and features of an organism. DNA is transcribed and translated to produce proteins involved in nearly all cellular processes. Alterations in genomic sequences can be beneficial or detrimental and can occur naturally or be induced artificially (Singh et al. 2016). CRISPR regions were subsequently identified in numerous other organisms, specifically bacteria and archaea, and have been defined as their defense mechanism (Han and She 2017). CRISPR regions have not garnered significant attention from the scientific community until the past 5 to 10 years, when they emerged as a revolutionary technology for precise site-specific genome editing (Han and She 2017; Ishino et al. 2018). Since then, this novel technology has attracted the interest of both manufacturers and scholars (Sherkow 2015). It is a highly efficient, accurate, and continuously evolving technique that is currently used for genomic modifications in both animals and plants (Song et al. 2016a). Genomic modifications include gene knockout, gene insertions, mutation, gene multiplication, and CRISPR has demonstrated its effectiveness as a tool for these objectives (Endo et al. 2015).

1.2 Mechanism of CRISPR/Cas

CRISPR/Cas can assist researchers in eliminating genes responsible for undesirable attributes in plants, particularly in inducing resistance against pathogens (Andolfo et al. 2016). CRISPR/Cas is a potent tool for deploying genes that confer beneficial traits. CRISPR/Cas mechanisms and structures are both intriguing and complex (Liu et al. 2012; Mao et al. 2013; Xing et al. 2014; Belhaj et al. 2015a). CRISPR can be excessively complex or overly simplistic, depending on the details provided. CRISPR was initially studied in *E. coli*, as previously described. Two fundamental components have been identified in the CRISPR system: first, there are repeats, which are short sections of DNA between 20 to 30 nucleotides in length, called palindromes. A palindrome is a sequence of nucleotides that repeats identically from left to right in both strands of DNA. These palindromes are indistinguishable, and there are DNA segments called "spacer DNA" (Amitai and Sorek 2016). Spacer DNA is of particular interest because it is not homogeneous, and each segment is unique. Researchers subsequently discovered that spacer DNA is very similar to viral DNA, primarily bacteriophage DNA (Ran et al. 2013; Maeder and Gersbach 2016). They also identified several genes linked with CRISPR, termed CRISPR-associated or Cas-associated genes. These Cas-genes encode Cas proteins, which are typically classified as helicases. These proteins disentangle and cleave the DNA. Thus, this system is essentially the bacterial immune system, which acts as a protective mechanism against bacteriophages. When a bacteriophage inserts its DNA into *E. coli*, it typically attempts to command and destroy the cellular machinery and destroy it (Schaeffer and Nakata 2015). However, as the cell possesses the CRISPR framework, Cas-proteins are encoded, and CRISPR RNA is transcribed. Prior to viral DNA initiation, cells deactivate it. In the event of a novel viral attack, bacteria encode another type of protein called cas1 (Doudna and Charpentier 2014; Scheben et al. 2017). This cas1 captures the viral DNA, degrades it, and, more importantly, replicates it as a new spacer DNA within the CRISPR array of the bacterial genome. Consequently, all these spacer DNAs essentially represent a historical record of previous viral infections (Jiang and Doudna 2017). The entire CRISPR/Cas editing process in plants is illustrated in Fig. 2.

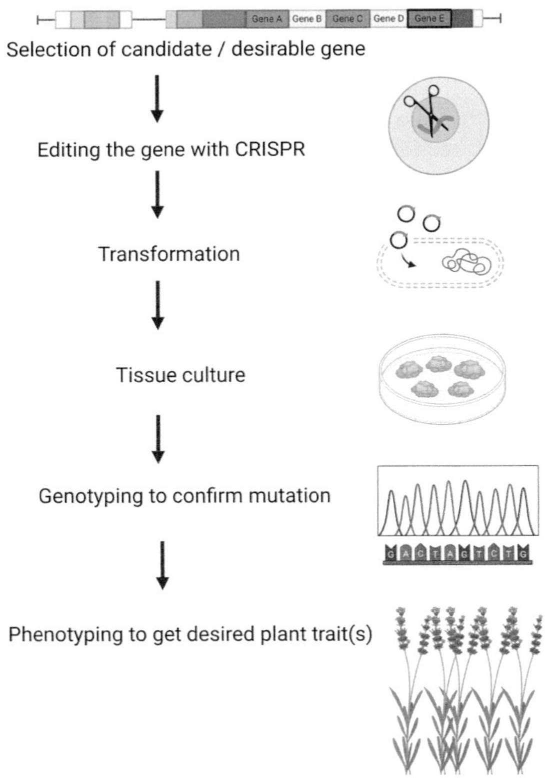

Figure 2. A schematic workflow of gene editing with CRISPR/Cas system in ornamental plants.

1.3 Gene Knock-out and Knock-in

To induce specific mutations and remove or insert a gene, researchers have employed a similar approach utilizing CRISPR. This CRISPR system was initially discovered in the laboratory of Doudna and Charpentier during the investigation of Streptococcus pyogenes. They developed a CRISPR-Cas9 system with their guide RNA (gRNA) specific to the DNA segment of interest (Doudna and Charpentier 2014).

gRNA has two main components: the CRISPR segment, which contains the RNA instructions for making the incision, and the protein that performs DNA cleavage (Bouché and Bouchez 2001). The Protospacer Adjacent Motif (PAM) is an additional sequence that facilitates protein localization to a precise site intended for DNA cleavage. PAM was consistently proximal to the targeted sequence. It comprises a short sequence of at least three nucleotides, and frequently begins with NGG, NRG, and NGCG (among numerous other PAM variants), where N represents any nucleotide, G represents guanine, C represents cytosine, and R represents a purine nucleotide (adenine or guanine). The guide RNA scans the entire DNA strand and creates a double-stranded break upon identification of the matching PAM sequence (Tian et al. 2017). Consequently, when the incision is made, the DNA is disrupted and the cell attempts to repair it, resulting in a mutation. Through this process, genes can be silenced or removed. Conversely, when a gene is added to a DNA strand, a host nucleotide sequence is provided to the cell into which the gene is inserted. The gRNA, in conjunction with Cas9, binds to the targeted DNA at the PAM site and creates a Double-Strand Break (DSB). This disruption is then repaired by cells through two distinct mechanisms. When it comes to gene knockout, the cell repairs via the Non-Homologous End Joining (NHEJ) method by introducing indels. For gene knock-in, a donor DNA template was provided to the cell and repaired using the Homology-Directed Repair (HDR/HR repair) method. Both techniques are utilized within the scientific community for various objectives of gene knockout or knock-in (Tsien et al. 1996; Belhaj et al. 2015b; Ma et al. 2016; Andolfo et al. 2016).

Agrobacterium Transformation is a crucial procedure in CRISPR/Cas9 methodology. Since its discovery, genetic modifications have garnered significant research attention. Agrobacterium tumefaciens is capable of transferring its vector to plant cells (Rashid et al. 1996). Agrobacterium-mediated transformation involves isolating a gene of interest from the source plant and inserting it into a vector containing a promoter to enable gene transcription in the plant (Pickardt et al. 1991). The vector is subsequently inserted into a bacterium, such as Agrobacterium, using various transformation techniques, including electroporation and heat shock. The bacteria containing the transgenic vector were then allowed to proliferate in growth medium (LB medium) (Ding et al. 2009). Subsequently, leaf sections were cultivated with this bacterial medium, enabling vector transfer to leaf cells. These leaf sections were then cultured in the dark on a co-culture medium to facilitate vector transfer to plant cells (Walkerpeach and Velten 1994). The leaves were subsequently transferred to a medium containing antibiotics to prevent bacterial growth and maintain sterility of the transformed leaves (Hensel et al. 2009). Gene transfer is observed as plants develop, and the leaves grow similarly to normal *in vitro* plants (Ding et al. 2009).

Two primary methods exist for introducing the CRISPR/Cas protocol into plants: leaf infiltration and co-culture growth. Leaf infiltration has gained popularity owing to its rapid interpretation of results. The co-culture growth method is essential for efficient Agrobacterium transformation; however, it is a time-consuming process for result visualization and successful evaluation. Transformation via transient mechanisms allows for temporary gene addition or deletion. This method offers greater flexibility, reduced labor requirements, and increased speed compared to traditional stable co-culture growth transformation. It involves the direct injection of transformed Agrobacterium into plant leaves using a needle-free syringe (Fig. 3). The syringe was filled with the Agrobacterium solution, which was then injected into the plant under minimal pressure. The leaf air spaces absorb this suspension. This method facilitates the transfer of multiple Agrobacterium strains and enables comparison of multiple genes (Berestovoy et al. 2018).

Figure 3. Agrobacterium infiltration of CRISPR-Cas for transient gene expression analysis (Photo credits: Irum Saadia Khan).

2. Model Plant in the Ornamental Sector

Despite their potential, relatively few studies have focused on gene editing in ornamental flowers. This chapter reviews the advancements of gene editing technology, particularly CRISPR/Cas9, in the context of ornamental plants. It discusses advances related to stress resistance, growth, and developmental traits, and explores the future prospects of this modern tool in ornamental plant breeding. In doing so, it seeks to provide valuable insights and references for the efficient creation of new flower varieties using CRISPR/Cas9 technology. Plant scientists have focused on model plants because they facilitate research, especially molecular research. For an extended period, Arabidopsis has served as a model plant because of its compact genome, rapid maturation, brief life cycle, and simplicity in transformation. Numerous experiments have been conducted on this plant (Meinke et al. 1998; Koornneef and Meinke 2010). In the past, when sequencing was relatively expensive, options for alternative model plants were limited. Arabidopsis is the predominant choice because of the aforementioned characteristics. Arabidopsis has also been a suitable host for functional genomics, genetic interactions, mutations, and overall genetic studies (Meinke et al. 1998; Cheng et al. 2017). With recent advances in bioinformatics, Arabidopsis has been adapted for modern research. However, as Next-Generation Sequencing (NGS) has become more cost-effective and other technologies have become readily accessible, even to individual research teams, scientists have gained confidence in exploring alternative model plants (Goodwin et al. 2016). Owing to its various characteristics, Petunia has the potential to serve as a model plant (van der Krol and Immink 2016).

3. Utilization of CRISPR/Cas9 in Ornamental Plant Breeding

CRISPR/Cas9 technology is being more widely applied in ornamental plant breeding to incorporate desirable traits, including flower color alterations, fragrance, disease resistance, and size. The primary advantage of CRISPR over traditional breeding methods is its precision and efficiency in editing the specific genes responsible for these traits. There are various examples of CRISPR/Cas9 application in *Torenia fournieri*, where researchers have successfully modified flower color by targeting the *F3H* gene contributing to anthocyanin biosynthesis. This resulted in transgenic plants with varied flower colors, such as violet and pale blue, demonstrating stable inheritance of this trait under greenhouse conditions (Nishihara et al. 2018a). CRISPR has also been employed to enhance traits in other ornamental species, such as Chrysanthemum, where editing the *GA20ox* gene affects plant growth and architecture; it is a crucial gene involved in gibberellin biosynthesis. The modified plants exhibited reduced plant height and altered flowering time, illustrating how this technology can be applied to control flower development and plant architecture (Liu et al. 2024a). Another application involves altering flowering time by targeting genes that control the plant's biological clock (Ahn et al. 2020a). CRISPR/Cas9 technology has become an essential tool in ornamental plant breeding, revolutionizing the ability to enhance aesthetic traits and improve resilience to environmental stress. Unlike traditional breeding methods, CRISPR allows precise and targeted genetic modifications, creating new opportunities for developing ornamental plants with desirable traits such as improved flower color, extended vase life, and increased disease resistance (Mekapogu et al. 2022a). Here are relevant studies showcasing CRISPR's applications in different ornamental species:

3.1 Modification of Flower Color

Color is one of the most fundamental and commercially significant traits in ornamental plants and is primarily determined by pigments, such as betaines, anthocyanins, carotenoids, and flavonoids. However, certain colors are not present naturally in specific species, which has led to numerous studies aimed at generating novel and vibrant flower colors. CRISPR-Cas9 ribonucleoproteins (RNPs) are a novel methodology for modifying flower color by targeting duplicated genes in *Petunia hybrida*. Researchers successfully edited both copies of the F3H (Flavanone 3-Hydroxylase) gene, which is crucial for the anthocyanin biosynthesis pathway. Through the inactivation of both gene copies, they achieved a significant color alteration, yielding pale purplish-pink flowers instead of the typical deep-purple flowers. The CRISPR-Cas9 method employed in this study, which utilizes ribonucleoproteins rather than DNA, circumvents certain regulatory restrictions related to Genetically Modified Organisms (GMOs), offering a potential advantage in the ornamental plant industry. Among the regenerated plants, only those with mutations in both F3H genes exhibited an altered pale pink flower color, demonstrating the precision of the method for flower color manipulation. This DNA-free gene-editing approach is a pioneering advancement in ornamental crop engineering (J et al. 2021).

A comparable investigation to the one in question focused on the utilization of CRISPR/Cas9 to modify the *F3H* gene in *Torenia fournieri*, which resulted in a successful alteration of flower color from violet to pale blue or white. This study showed the application of CRISPR/Cas9 to specifically target genes involved in the anthocyanin biosynthesis pathway. Genetic modifications introduced mutations in both alleles of the F3H gene, which disrupted pigment production and resulted in observable color changes (Nishihara et al. 2018b).

In *Ipomoea nil*, the Carotenoid Cleavage Dioxygenase 4 (*CCD4*) gene is involved in the regulation of carotenoid biosynthesis, which contributes to the yellow pigmentation of flowers. Through the application of CRISPR-Cas9 to knock out this gene, researchers have been able to significantly alter the levels of carotenoids, resulting in a change in flower color from yellow to white. Disruption of the CCD4 gene resulted in a decrease in carotenoid accumulation, resulting in characteristic white flowers instead of the usual yellow flowers. This study elucidates the potential of CRISPR technology for modifying non-anthocyanin pigments, demonstrating its broader applications in flower color manipulation and ornamental plant breeding. This study underscores the capacity of targeted mutagenesis to enhance aesthetic diversity in horticulture (Watanabe et al. 2018). This study has proven the effectiveness of CRISPR-Cas9 for modulating anthocyanin content for flower color improvement in high-value ornamental species such as orchids. These examples illustrate how CRISPR-Cas9 technology enables the precise editing of genes involved in pigment biosynthesis pathways to modify flower color. This technology presents new possibilities for creating custom-colored ornamental plants, thereby adding commercial value, and enhancing aesthetic diversity in horticulture. CRISPR/Cas9 has been successfully used to edit the complex genomes of these orchids, thereby facilitating potential enhancements in traits such as flower aesthetics and resistance to environmental stress. This study underscores that CRISPR is a valuable tool for advancing ornamental species with intricate genomes (Ahn et al. 2020b).

3.2 Fragrance Modifications through CRISPR-Cas Technology

CRISPR/Cas9 technology offers substantial potential for enhancing the floral fragrance in ornamental plants through precise genetic modifications within the biosynthetic pathways responsible for scent production. By focusing on specific genes associated with volatile compound synthesis, researchers can effectively modulate fragrance profiles to align with consumer preferences and market demand. This genetic engineering approach not only expands the range of floral scents available in the horticultural market but also facilitates the development of unique cultivars that may exhibit novel aromatic properties, thereby increasing their commercial value (Noman et al. 2017). CRISPR/Cas9 technology has been utilized effectively to modify the fragrance of ornamental plants, enabling researchers to target and edit genes responsible for floral scent production. This approach can tailor the fragrance characteristics and enhance the commercial appeal of flowers (Mekapogu et al. 2023).

One example involves the manipulation of key enzymes such as monoterpenes, sesquiterpenes, and benzenoids in the biosynthesis of volatile compounds responsible for fragrance. By employing CRISPR to disrupt or enhance specific genes in these pathways, scientists can increase or decrease the production of certain scent compounds. For instance, in roses, genes responsible for emitting desirable scents, such as geraniol, can be adjusted to intensify their fragrance, while in petunias, the scent profile can be modulated to favor specific floral notes. Another avenue explored through CRISPR is to prolong the longevity of fragrance in flowers by reducing the expression of genes related to flower senescence. For example, by targeting ethylene-related pathways, CRISPR can delay floral aging, leading to flowers that not only last longer, but also retain their scent for extended periods (Tang et al. 2023).

3.3 Modification of Leaf Shape

The leaf shape has long been a significant aspect of plant science. In numerous crops, leaf shape has economic value and ornamental significance, and/or serves as a readily identifiable characteristic for plant recognition. However, the mechanism underlying leaf shape determination remains unclear. Current literature and research suggests that shape of a leaf is a very complex trait with diverse controlling factors among different plant groups (Tsukaya 2006). Figure 4 illustrates examples of leaf shapes in various plant groups. Leaf shape can be modified through natural mutations, evolution, or artificial means, including genetic mutations, controlled breeding, or alterations in the plant's growth environment (Mohan Jain 2006). The modification of leaf shape is of interest to both scientists and growers for its economic benefits, enhanced visual appeal, improved climate adaptation, and scientific research to elucidate the underlying mechanisms (Huang et al. 2021).

Approximately 310,000–420,000 plant species have been identified by science and assigned scientific names. Given this extensive diversity of plants, leaf shape is often the primary visual characteristic used to identify a plant species (Du et al. 2007). Leaf shape is a crucial feature of the overall growth success of a plant. Numerous plant species have been classified into different divisions, based on their characteristics. The first division is comprised of bryophytes, which are simple plants with rudimentary leaves. They lack a vascular transport system, do not produce flowers, and do not contain lignin in their cell walls. These plants have a significant role in maintaining the soil structure and preventing soil erosion (Vanderpoorten and Goffinet 2009). The second division is pteridophytes. These plants possess true compound leaves and active vascular systems. They utilize spores for reproduction, but do not produce flowers (Mirek et al. 2002). The subsequent division was referred to as Spermatophyta. These plants have green photosynthetic leaves, and this division is further subdivided into two major groups: angiosperms and gymnosperms (Rothwell and Serbet 1994). Gymnosperms exhibit thick needle-shaped leaves, have a waxy coating, and possess submerged stomata. They produce seeds, but not fruits (Chamberlain 1935). The final and

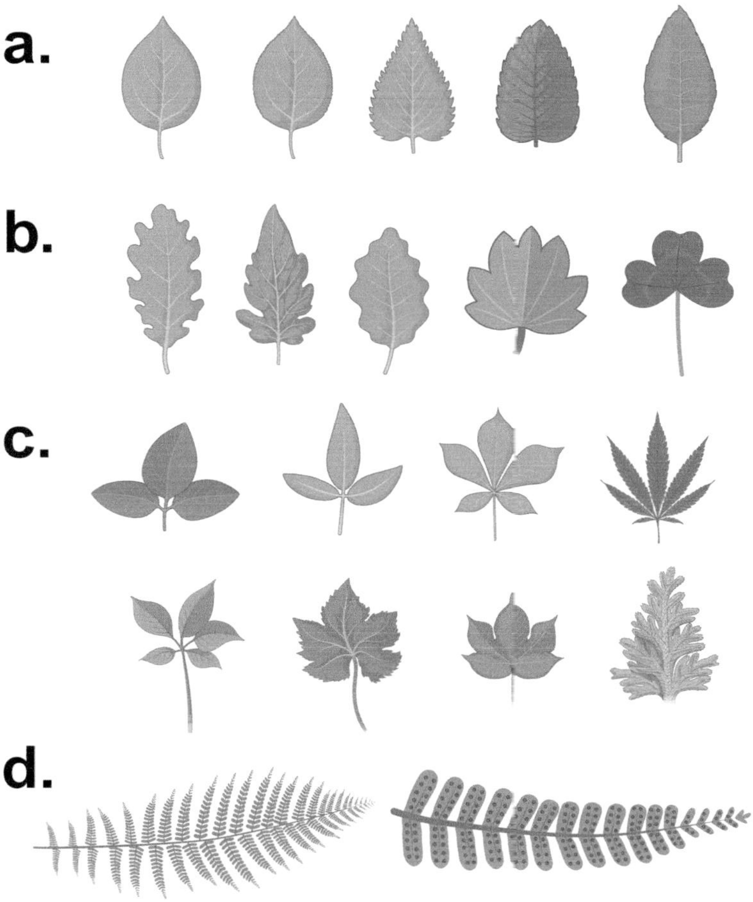

Figure 4. Different leaf shapes existing among various plant species; (a) Simple leaves with entire or slightly serrated margins. (b) Lobed leaves with pinnate or palmate lobation. (c) Compound leaves, including pinnately and palmately compound structures. (d) Highly divided and pinnately compound leaves, such as those found in ferns.

most advanced divisions are angiosperms. Angiosperms primarily display two types of leaves: monocotyledonous and dicotyledonous leaves. Monocots have long, thin leaves with parallel leaf veins, whereas dicots have branched leaf veins. Angiosperms produce seeds within enclosed fruits (Eames 1961). Within this classification system based on leaf shape, petunia falls under the category of dicot angiosperms because of its flower production, branched leaf venation, and fruit and seed formation (Sink 2013). Similar to other physical features of living organisms, leaf shape is primarily determined by the genetic composition of the organism. Various leaf shapes have evolved in plants throughout their evolution. The primary reason for this evolutionary change in leaf shape is to enable better adaptation to the environment, primarily as a survival mechanism adopted by plants (Tsukaya 2018). However, although the genetics of a plant largely determines its leaf shape, it is not the sole parameter influencing this characteristic. Numerous other factors can also affect leaf shape (Tsukaya 2006). For instance, variations in light intensity can cause plants to develop larger or smaller leaves (depending on the species) (Njoku 1956). Similarly, temperature plays a crucial role in the morphogenesis of leaf shape, as the plant may require a greater or lesser number of stomata according to the temperature (Njoku 1957). Leaf shape is also influenced by plant nutrition. If plants are subjected to drought stress, salinity stress, or disease attack, or if certain essential nutrients are less available to the plant, they may respond to the stress, potentially resulting in a change in leaf shape. The change in leaf shape due to such stresses is not substantial, and leaf shape generally remains similar to the original form (Farris 1984; Brodribb and Holbrook 2005). Genetic mutations can also alter leaf shape. This type of change depends on the nature of the mutation and can range from minor modifications to significant differences. Inter and intraspecies crosses can also develop new leaf shapes. Plant breeders have created numerous ornamental plants by implementing such changes and enhancing their visual appeal of the plants (Ashby 1948). The primary contributing factor to the shape of a leaf in a plant species is undoubtedly its genetics, which is preserved by the plant generation after generation and confers its identity. Multiple genes are involved in controlling leaf shape morphogenesis. Through the study of leaf shape genetics, it can be concluded that determining leaf shape is not a simple trait but rather a complex one that, as discussed above, is dependent on numerous factors (Chitwood et al. 2014). Considering the genetics of plants, cup-shaped cotyledons (*CUC*) 1, 2, and 3 play a predominant role in defining the leaf margins in Arabidopsis. The expression of *CUC2* effectively renders serrated leaf margins. Conversely, the expression of a microRNA, miR164, functions antagonistically with CUC2. Overexpression of miR164 results in smooth leaf margins. The equilibrium between the expression of both genes determines the final form factor of Arabidopsis leaves in Arabidopsis (Nikovics et al. 2006). Other genes, such as pin formed-1 (*PIN1*), asymmetric rough phantastica (*ARP*), clavata (*CLC*), and wuschel (*WUS*), were found to be involved in controlling leaf initiation in different plant species. PIN1 regulates auxin efflux, whereas ARP influences the differentiation of stem cells (Okada et al. 1991; Waites et al. 1998). CLC assists the plant in maintaining cell size, while WUS manages the identity of the floral and shoot meristem (Clark et al. 1995; Laux et al. 1996). In Zea mays, proximodistal leaf outgrowth is controlled by liguleless narrow-reference (*Lgn-R*) and Knotted1 (*Kn1*) genes. *LGN-R* plays a role in creating the boundary of the sheath, while Kn1 aids the plant in maintaining the identity of the stem cells (Ramirez et al. 2009; Moon et al. 2013). Dorsoventral leaf outgrowth in Arabidopsis is influenced by genes such as RNA-dependent RNA polymerase6 (*RDR6/DCL4*), Auxin Response Factor (ARF), Kanadi (KAN), and asymmetric leaves2 (*AS2*). *DCL4, ARF*, and *KAN* regulate leaf polarity specifications of the leaf (Eshed et al. 2001; Pekker et al. 2005; Fahlgren et al. 2006), whereas AS2 contributes to maintaining the patterns of venation and the development of the lamina (Xu et al. 2003). The Maewest (*MAW*) gene in Petunia × hybrida is responsible for organ fusion and lateral expansion (Vandenbussche et al. 2009). In Arabidopsis, the mediolateral outgrowth patterning of the leaf is regulated by Yucca (YUC), narrow sheath (ns), Yabby (*YAB*), and pressed flower (*PRS*) genes. *YUC* regulates vascular development and flower patterning (Cheng et al. 2006). The expansion of the leaf is controlled by the ns gene, whereas the expansion of the lamina and specification of leaf polarity are regulated by the YAB gene (Kumaran et al. 2002; Nardmann et al. 2004). *PRS* primarily mediates the proliferation of marginal cells (Matsumoto and Okada 2001). Leaf expansion and maturation are regulated by Cincinnata (*CIN*) (Nath et al. 2003), Angustifolia interacting factor (*GIF1*), and Growth Regulating Factor5 (*GRF5*) genes in Arabidopsis. These genes are involved in cell proliferation and expansion (Kim and Kende 2004; Horiguchi et al. 2005). Leaf margin morphology, whether lobed, serrated, or smooth, is controlled by the Development PcG Apex (*DPA*) and Pin-formed1 (PIN1) genes. *DPA* is involved in organ initiation and leaf development, whereas PIN1 regulates auxin efflux in Arabidopsis (Hay et al. 2006; Julia et al. 2012). These findings are derived from various studies conducted across a variety of plant species. The expression of these genes varies significantly among species; therefore, to elucidate the effects of a specific gene in a particular plant, a targeted study must be conducted.

A substantial amount of genetic data about various plant traits is available. However, it is virtually impossible to study an organism comprehensively. Knotted1 (*Kn1*) is a homeobox gene that regulates several distinct traits and influences leaf morphology. Research has demonstrated that under-expression of this gene in tomatoes alters the normal leaf phenotype, resulting in the production of multiple phenotypes. Furthermore, the same study yielded different results for Arabidopsis (Hareven et al. 1996). In a separate investigation involving tobacco plants, overexpression of the same gene group was found to modify leaf shape (Nishimura et al. 2000). Another study examined the effects of *Kn1* overexpression on leaf senescence in tobacco plants. These findings indicated that overexpression delayed senescence, proving that these homeobox genes play a crucial role in leaf morphology (Ori et al. 1999). An additional study compared the expression *Kn1* in tomato and pea leaves. The results revealed varying outcomes, suggesting that the genes regulate different aspects of leaf growth in both species (Hofer et al. 2001). Research on maize suggests that the Kn1 homeobox may assist researchers in further elucidating the mechanism of cell division control, while a dominant mutation in Kn1 demonstrated its role in regulating cells in leaf veins (Vollbrecht et al. 1991; Smith et al. 1992). In summary, another study discussed the role of *Kn1* in determining the proximal end of the leaf (Bolduc et al. 2012). These studies collectively indicate that disruption of the gene produces diverse phenotypic traits depending on the species and nature of the disruption. Further investigation of this gene is necessary to enhance leaf characterization in ornamental plants.

3.4 Prolonged Vase Life and Delayed Senescence

CRISPR/Cas9 has been extensively utilized to modify ethylene-related genes in ornamental plants with the objective of reducing ethylene sensitivity and synthesis. This genetic modification can significantly extend the flowering period and provide more precise control over flowering timing. In commercial floriculture, post-harvest longevity of cut flowers is a critical trait that affects market value and consumer satisfaction. CRISPR/Cas9 can extend the vase life of flowers by delaying the onset of senescence (aging process) (Ahn et al. 2020c). A notable example is the application of CRISPR in *Chrysanthemum morifolium*, where researchers edited senescence-associated genes. By reducing the expression of these genes, they successfully delayed flower aging, thereby extending the vase life of cut flowers. This advancement is particularly valuable for commercial growers as it enhances the shelf life of flowers, making them more desirable in retail settings (Mekapogu et al. 2022b).

3.5 Broadening the Application Scope and Enhancing Ornamental Plant Traits

CRISPR/Cas9 technology is a potent tool for advancing the genetic improvement of ornamental plants through precise gene editing. This method facilitates the knockout of deleterious genes or the induction of beneficial mutations, resulting in the generation of gene-edited mutants that promote the growth, development, and desirable traits of ornamental plants. This genetic refinement not only enhances plant quality but also augments yield and economic value. Several proof-of-concept examples are listed in Table 1.

4. Outlook

CRISPR/Cas9 technology has demonstrated significant promise for advancements in plant breeding, exhibiting its potential to enhance various plant traits, including biotic and abiotic stress resistance, as well as improve overall crop quality. While the majority of successful applications of CRISPR/Cas9 have been observed in diploid plants, such as rice (Shimatani et al. 2017) and tomato (Ortigosa et al. 2019), where genome editing is less complex, ornamental plants present distinct challenges. These challenges large genome size and high heterozygosity which complex traditional breeding approaches. Consequently, ornamental plants have become a critical focus for gene editing research with the aim of improving traits such as flowering time, morphology, color and stress resistance.

Despite the potential of CRISPR/Cas9 technology, several obstacles remain. The primary challenge is the effective targeted delivery of CRISPR/Cas9 components. Conventional methods, such as tissue culture and Agrobacterium-mediated transformation and at times are inefficient and time-consuming, particularly with regard to callus induction and plant regeneration. This limitation impedes the application of CRISPR/Cas9 in ornamental plants. The development of tissue

Table 1. A summary of CRISPR/Cas9 studies on target genes of different ornamental plant species.

Plant Species	Target Gene/Pathway	Goal	Outcome/Impact	Reference
Alfalfa	U6 promoter, MtPDS gene	Mutant creation	Produced homozygous mutants for both single and double alleles in T0 generation.	(Meng et al. 2017)
Dwarf Morning Glory	SSK1 gene (self-incompatibility mechanism)	Understand and manipulate self-incompatibility	Identified T0 plant with Cas9-containing T-DNA, aiding in self-incompatibility research.	(Sun and Kao 2018)
	PH3 gene (pigmentation regulation)	Knockout/frameshift mutations	Induced base insertions and knockouts, reducing downstream gene expression.	(Liu et al. 2024b)
	PhHMA5II-1 gene (metal transport)	Mutant development	Created homozygous PhHMA5II-1 mutant using leaf disc transformation.	(Wang et al. 2020a)
Chrysanthemum	YGFP gene	First successful CRISPR gene editing in chrysanthemums	Edited YGFP gene in transgenic plants, showcasing the technology's potential in ornamental plant species.	(Kishi-Kaboshi et al. 2017)
	DgGA20ox gene (gibberellin synthesis enzyme)	Growth regulation	Introduced frameshift mutation, impacting growth and dwarfing traits.	(Verma et al. 2023)
	eYGFPuv and target gene knockout	Optimization of gene-editing systems	Improved screening efficiency and reduced chimeric plants in edited chrysanthemum.	(Liu et al. 2023)
Roses and Strawberries	TFL1 gene (flowering regulator)	Continuous flowering enhancement	Mutations in TFL1 promoted continuous flowering in both species.	(Iwata et al. 2012)
Arabidopsis	TCP4 gene (flower morphology)	Enhancing ornamental traits	Altered flower morphology with narrow petals, defected stamens.	(Nag et al. 2009)
Cymbidium sinense	PI gene (labellum formation)	Understanding floral morphology	Successfully validated CRISPR/Cas9 multi-sgRNA vector system, targeting the PI gene.	(Wu et al. 2018)
Roses	RhEIN2 gene (ethylene signaling pathway)	Modify flower longevity and senescence	Edited RhEIN2 gene in cell cultures, affecting the ethylene pathway, important for flower longevity traits.	(Wang et al. 2023)

culture-independent delivery methods, such as *de novo* induction of gene editing meristems, can address these limitations (Maher et al. 2020). Another major concern is the potential for off-target effects associated with CRISPR/Cas9 that can result in unintended mutations, deletions, or rearrangements in the genome. To mitigate these off-target effects, direct delivery of Cas9 protein and sgRNA through methods such as particle bombardment and protoplast transformation has proven to be an effective strategy (Wang et al. 2020b).

Recent advancements in gene editing have introduced novel tools, such as single-base editing technology, which enables precise modifications to individual DNA bases. As previously reported that this technique facilitates critical amino acid changes with high specificity, providing a more precise approach to modifying biological traits and addressing genetic diseases (Komor et al. 2016). This innovation presents strong prospects for advancing gene editing technologies in ornamental plants, offering a refined tool for achieving desired traits. The continuous advancement of gene editing technologies, optimization of genetic transformation methods, and expansion of functional genome research on flowers are anticipated to establish gene editing as a fundamental component of future flower breeding endeavors. For instance, through inhibition of the *AGAMOUS* gene via transgenic methods, researchers have successfully induced double-flower formation in *Arabidopsis thaliana* (Coen and Meyerowitz 1991). Comparable advancements have been achieved in other species, such as the transformation of tobacco with the petunia homeotypic gene fbp2, which results in significant floral alterations (Jing-Juan et al. 1999). Furthermore, fusion of the *WRKY71* gene has demonstrated the potential to accelerate flowering in Arabidopsis (Yu et al. 2016). Additional gene targets, including the scent gene *BADH2* (Sun et al. 2017), *NUDX1* (Conart et al. 2023), color genes such as PDS (Kaur et al. 2018), and stress-resistance genes such as *DREB2* (Izadi-Darbandi et al. 2023), are currently being investigated for their potential to improve ornamental plant traits through CRISPR/Cas9 editing. Moreover, root growth and development genes, such as *AhHDA1* (Su et al. 2018) and *PIN1/PIN2* have been examined for their roles in auxin transport and distribution, which could potentially lead to enhanced stress resistance and improved flower traits.

5. Limitations in Adapting CRISPR/Cas Technology in Ornamental Plants: CRISPR Under Controversies

Despite its high accuracy, low cost, and precise operation, CRISPR has been subject to debate regarding its classification as a GMO and its associated ethical challenges (Polcz and Lewis 2016). Although CRISPR has demonstrated significant prospects in addressing food and medicinal issues, it has also raised concerns within the scientific community, as is common with novel scientific innovations. The primary concern stems from the fact that previous genome-editing practices are traceable and distinguishable, while CRISPR-cas will result in imperceptible changes to the genome, or more precisely, appear as mutants (Butler 2016). However, CRISPR's ability to make changes at the nucleotide level resembles natural mutations, causing apprehension among food regulation authorities worldwide (Ledford 2019). In Europe, GMOs are defined as organisms, either plants or animals, in which DNA is manipulated through synthetic or human-directed approaches (Plan et al. 2010). Given that GMOs are prohibited in Europe, food regulation authorities face a significant challenge in determining whether the genetic changes in an organism are natural or CRISPR-induced. Conversely, in countries such as the United States, where GMOs are permitted, concerns primarily revolve around the potential unethical applications of CRISPR (Shew et al. 2018). A notable example is the use of CRISPR to alter the human genome in China, which has been rejected for publication by Nature and other prominent journals (Jiang and Stevens 2015). Despite its benefits, CRISPR poses substantial challenges to the ethical values of countries, cultures, and belief systems. The case in China exemplifies this issue, as many individuals from diverse backgrounds oppose the use of CRISPR in humans. However, these controversies predominantly pertain to animals and humans, and the application of CRISPR in plants remains less contentious (Hefferon and Herring 2017; Benston 2017).

6. Conclusions

Although gene editing in ornamental plants remains at its nascent stages, the progress achieved thus far provides a robust foundation for future applications. Nevertheless, challenges persist, particularly in the areas of efficient transformation methods, the management of off-target effects, and ethical considerations. As research advances, resolving these challenges is necessary to fully utilize the potential of CRISPR/Cas9 technology in ornamental plant breeding (Gehrke et al. 2022). The future of ornamental plant breeding looks promising, with gene editing playing a crucial role in the development of new and improved varieties (Rönspies et al. 2021).

References

Ahn, C.H., M. Ramya, H.R. An, P.M. Park, Y.-J. Kim, S.Y. Lee et al. 2020a. Progress and challenges in the improvement of ornamental plants by genome editing. Plants 9: 687.

Ahn, C.H., M. Ramya, H.R. An, P.M. Park, Y.-J. Kim, S.Y. Lee et al. 2020b. Progress and challenges in the improvement of ornamental plants by genome editing. Plants 9: 687.

Ahn, C.H., M. Ramya, H.R. An, P.M. Park, Y.-J. Kim, S.Y. Lee et al. 2020c. Progress and challenges in the improvement of ornamental plants by genome editing. Plants 9: 687.

Amitai, G. and R. Sorek. 2016. CRISPR-Cas adaptation: insights into the mechanism of action. Nature Reviews. Microbiology 14: 67–76.

Andolfo, G., P. Iovieno, L. Frusciante and M.R. Ercolano. 2016. Genome-editing technologies for enhancing plant disease resistance. Frontiers in Plant Science 7.

Ashby, E. 1948. Studies in the Morphogenesis of leaves. I. An essay on leaf shape. The New Phytologist 47: 153–176.

Belhaj, K., A. Chaparro-Garcia, S. Kamoun, N.J. Patron and V. Nekrasov. 2015a. Editing plant genomes with CRISPR/Cas9. Current Opinion in Biotechnology 32: 76–84.

Belhaj, K., A. Chaparro-Garcia, S. Kamoun, N.J. Patron and V. Nekrasov. 2015b. Editing plant genomes with CRISPR/Cas9. Current Opinion in Biotechnology 32: 76–84.

Benston, S. 2017. Everything in moderation, even hype: learning from vaccine controversies to strike a balance with CRISPR. Journal of Medical Ethics 43: 819–823.

Berestovoy, M., A. Tyurin, K. Kabardaeva, Y. Sidorchuk, A. Fomenkov, A. Nosov et al. 2018. Transient gene expression for the characteristic signal sequences and the estimation of the localization of target protein in plant cell. Bio-Protocol 8: e2738.

Bolduc, N., D. O'Connor, J. Moon, M. Lewis and S. Hake. 2012. How to pattern a leaf. Cold Spring Harbor Symposia on Quantitative Biology 77: 47–51.

Bolotin, A., B. Quinquis, A. Sorokin and S.D. Ehrlich. 2005. Clustered regularly interspaced short palindrome repeats (CRISPRs) have spacers of extrachromosomal origin. Microbiology 151: 2551–2561.

Bouché, N. and D. Bouchez. 2001. Arabidopsis gene knockout: phenotypes wanted. Current Opinion in Plant Biology 4: 111–117.

Brodribb, T.J. and N.M. Holbrook. 2005. Water stress deforms tracheids peripheral to the leaf vein of a Tropical Conifer. Plant Physiology 137: 1139–1146.

Butler, K. 2016. Reviews of science for science librarians: CRISPR-Cas9 revolutionizes gene editing. Science and Technology Libraries 35: 221–227.

Chamberlain, C.J. 1935. Gymnosperms. Structure and evolution. Gymnosperms. Structure and Evolution.

Cheng, Y., X. Dai and Y. Zhao. 2006. Auxin biosynthesis by the YUCCA flavin monooxygenases controls the formation of floral organs and vascular tissues in Arabidopsis. Genes and Development 20: 1790–1799.

Cheng, C.-Y., V. Krishnakumar, A.P. Chan, F. Thibaud-Nissen, S. Schobel and C.D. Town. 2017. Araport11: a complete reannotation of the Arabidopsis thaliana reference genome. The Plant Journal 89: 789–804.

Chitwood, D.H., A. Ranjan, C.C. Martinez, L.R. Headland, T. Thiem, R. Kumar et al. 2014. A Modern Ampelography: A genetic basis for leaf shape and venation patterning in grape. Plant Physiology 164: 259–272.

Clark, S.E., M.P. Running and E.M. Meyerowitz. 1995. CLAVATA3 is a specific regulator of shoot and floral meristem development affecting the same processes as CLAVATA1. Development 121: 2057–2067.

Coen, E.S. and E.M. Meyerowitz. 1991. The war of the whorls: genetic interactions controlling flower development. Nature 353: 31–37.

Conart, C., D.P. Bomzan, X.-Q. Huang, J.-E. Bassard, S.N. Paramita, D. Saint-Marcoux et al. 2023. A cytosolic bifunctional geranyl/farnesyl diphosphate synthase provides MVA-derived GPP for geraniol biosynthesis in rose flowers. Proceedings of the National Academy of Sciences of the United States of America 120: e2221440120.

Ding, L., S. Li, J. Gao, Y. Wang, G. Yang and G. He. 2009. Optimization of Agrobacterium-mediated transformation conditions in mature embryos of elite wheat. Molecular Biology Reports 36: 29–36.

Ding, Y., H. Li, L.-L. Chen and K. Xie. 2016. Recent advances in genome editing using CRISPR/Cas9. Frontiers in Plant Science 7.

Doudna, J.A. and E. Charpentier. 2014. The new frontier of genome engineering with CRISPR-Cas9. Science 346: 1258096.
Du, J.-X., X.-F. Wang and G.-J. Zhang. 2007. Leaf shape based plant species recognition. Applied Mathematics and Computation 185: 883–893.
Eames, A.J. 1961. Morphology of the angiosperms. Morphology of the Angiosperms.
Endo, M., M. Mikami and S. Toki. 2015. Multigene knockout utilizing off-target mutations of the CRISPR/Cas9 system in rice. Plant and Cell Physiology 56: 41–47.
Eshed, Y., S.F. Baum, J.V. Perea and J.L. Bowman 2001. Establishment of polarity in lateral organs of plants. Current Biology 11: 1251–1260.
Fagerlund, R.D., M.E. Wilkinson, O. Klykov, A. Barendregt, F.G. Pearce, S.N. Kieper et al. 2017. Spacer capture and integration by a type I-F Cas1–Cas2-3 CRISPR adaptation complex. Proceedings of the National Academy of Sciences: 201618421.
Fahlgren, N., T.A. Montgomery, M.D. Howell, E. Allen, S.K. Dvorak, A.L. Alexander et al. 2006. Regulation of AUXIN RESPONSE FACTOR3 by TAS3 ta-siRNA Affects Developmental Timing and Patterning in Arabidopsis. Current Biology 16: 939–944.
Farris, M.A. 1984. Leaf size and shape variation associated with drought stress in Rumex acetosella L. (Polygonaceae). American Midland Naturalist 111: 358.
Garneau, J.E., M.-È. Dupuis, M. Villion, D.A. Romero, R. Barrangou, P. Boyaval et al. 2010. The CRISPR/Cas bacterial immune system cleaves bacteriophage and plasmid DNA. Nature 468: 67–71.
Gehrke, F., A. Schindele and H. Puchta. 2022. Nonhomologous end joining as key to CRISPR/Cas-mediated plant chromosome engineering. Plant Physiology 188: 1769–1779.
Goodwin, S., J.D. McPherson and W.R. McCombie. 2016. Coming of age: ten years of next-generation sequencing technologies. Nature Reviews Genetics 17: 333–351.
Han, W. and Q. She. 2017. CRISPR History: Discovery, characterization, and prosperity. pp. 1–21. *In*: Progress in Molecular Biology and Translational Science. Elsevier.
Hareven, D., T. Gutfinger, A. Parnis, Y. Eshed and E. Lifschitz. 1996. The making of a compound leaf: genetic manipulation of leaf architecture in tomato. Cell 84: 735–744.
Hay, A., M. Barkoulas and M. Tsiantis. 2006. ASYMMETRIC LEAVES1 and auxin activities converge to repress BREVIPEDICELLUS expression and promote leaf development in Arabidopsis. Development 133: 3955–3961.
Hefferon, K.L. and R.J. Herring. 2017. The End of the GMO? Genome Editing, Gene Drives and New Frontiers of Plant Technology 32.
Hensel, G., C. Kastner, S. Oleszczuk, J. Riechen and J. Kumlehn. 2009. Agrobacterium-mediated gene transfer to cereal crop plants: current protocols for barley, wheat, triticale, and maize. International Journal of Plant Genomics 2009: 835608.
Hofer, J., C. Gourlay, A. Michael and T.H.N. Ellis. 2001. Expression of a class 1 knotted1-like homeobox gene is down-regulated in pea compound leaf primordia. Plant Molecular Biology 45: 387–398.
Horiguchi, G., G.-T. Kim and H. Tsukaya. 2005. The transcription factor AtGRF5 and the transcription coactivator AN3 regulate cell proliferation in leaf primordia of Arabidopsis thaliana. The Plant Journal 43: 68–78.
Huang, X., J. Hilscher, E. Stoger, P. Christou and C. Zhu. 2021. Modification of cereal plant architecture by genome editing to improve yields. Plant Cell Reports 40: 953–978. https://doi.org/10.1007/s00299-021-02668-7.
Ishino, Y., M. Krupovic and P. Forterre. 2018. History of CRISPR-Cas from Encounter with a Mysterious Repeated Sequence to Genome Editing Technology (W Margolin, Ed.). Journal of Bacteriology 200.
Iwata, H., A. Gaston, A. Remay, T. Thouroude, J. Jeauffre, K. Kawamura et al. 2012. The TFL1 homologue KSN is a regulator of continuous flowering in rose and strawberry. The Plant Journal: For Cell and Molecular Biology 69: 116–125.
Izadi-Darbandi, A., H. Alameldin, N. Namjoo and K. Ahmad. 2023. Introducing sorghum DREB2 gene in maize (Zea mays L.) to improve drought and salinity tolerance. Biotechnology and Applied Biochemistry 70: 1480–1488.
J, Y., T. L, S. S, B. S and L. Gj. 2021. Simultaneous targeting of duplicated genes in Petunia protoplasts for flower color modification via CRISPR-Cas9 ribonucleoproteins. Plant Cell Reports 40.
Jiang, F. and J.A. Doudna. 2017. CRISPR–Cas9 structures and mechanisms. Annual Review of Biophysics 46: 505–529.
Jiang, L. and H. Stevens. 2015. Chinese biotech versus international ethics? Accounting for the China–America CRISPR ethical divide. BioSocieties 10: 483–488.
Jinek, M., K. Chylinski, I. Fonfara, M. Hauer, J.A. Doudna and E. Charpentier. 2012. A programmable dual-RNA–guided DNA endonuclease in adaptive bacterial immunity. Science 337: 816–821.
Jing-Juan, Y.U., G.U.O. Feng-Li, Z. De-Gang, F.U. Yong-Fu, H.a.N. Yu-Zhen, A.G.-M. Fan-Jing et al. 1999. Cloning of the Homeotic Gene fbp2 from Petunia hybrida and its effects in tobacco flower. Journal of Integrative Plant Biology 41.
Julia, E., J.R. Julia, L. Iris, G. Ulrike, H. Bruno, F. Sara et al. 2012. DEVELOPMENT-RELATED PcG TARGET IN THE APEX 4 controls leaf margin architecture in Arabidopsis thaliana | Development.
Kaur, N., A. Alok, Shivani null, N. Kaur, P. Pandey, P. Awasthi et al. 2018. CRISPR/Cas9-mediated efficient editing in phytoene desaturase (PDS) demonstrates precise manipulation in banana cv. Rasthali genome. Functional and Integrative Genomics 18: 89–99.
Kim, J.H. and H. Kende. 2004. A transcriptional coactivator, AtGIF1, is involved in regulating leaf growth and morphology in Arabidopsis. Proceedings of the National Academy of Sciences 101: 13374–13379.

Kishi-Kaboshi, M., R. Aida and K. Sasaki. 2017. Generation of gene-edited chrysanthemum morifolium using multicopy transgenes as targets and markers. Plant and Cell Physiology 58: 216–226.

Kleinstiver, B.P., M.S. Prew, S.Q. Tsai, V.V. Topkar, N.T. Nguyen, Z. Zheng et al. 2015. Engineered CRISPR-Cas9 nucleases with altered PAM specificities. Nature 523: 481–485.

Komor, A.C., Y.B. Kim, M.S. Packer, J.A. Zuris and D.R. Liu. 2016. Programmable editing of a target base in genomic DNA without double-stranded DNA cleavage. Nature 533: 420–424.

Koornneef, M. and D. Meinke. 2010. The development of Arabidopsis as a model plant. The Plant Journal 61: 909–921.

van der Krol, A.R. and R.G.H. Immink. 2016. Secrets of the world's most popular bedding plant unlocked. Nature Plants 2.

Kumaran, M.K., J.L. Bowman and V. Sundaresan. 2002. YABBY polarity genes mediate the repression of KNOX homeobox genes in Arabidopsis. The Plant Cell 14: 2761–2770.

Laux, T., K.F. Mayer, J. Berger and G. Jurgens. 1996. The WUSCHEL gene is required for shoot and floral meristem integrity in Arabidopsis. Development 122: 87–96.

Ledford, H. 2019. CRISPR conundrum: Strict European court ruling leaves food-testing labs without a plan. Nature 572: 15–15.

Li, Y., S. Li, J. Wang and G. Liu. 2019. CRISPR/Cas Systems towards Next-Generation Biosensing. Trends in Biotechnology 37: 730–743.

Liu, L., C. Liu, X. Hou, W. Xi, L. Shen, Z. Tao et al. 2012. FTIP1 is an essential regulator required for florigen transport. PLOS Biology 10: e1001313.

Liu, H., Z. Wang, Y. Liu, Y. Wei, Z. Hu, X. Wu et al. 2024a. Applications of CRISPR/Cas9 technology in ornamental plants. Plant Molecular Biology Reporter 42: 193–200.

Liu, H., Z. Wang, Y. Liu, Y. Wei, Z. Hu, X. Wu et al. 2024b. Applications of CRISPR/Cas9 technology in ornamental plants. Plant Molecular Biology Reporter 42: 193–200.

Liu, L., Y. Xue, J. Luo, M. Han, X. Liu, T. Jiang et al. 2023. Developing a UV–visible reporter-assisted CRISPR/Cas9 gene editing system to alter flowering time in Chrysanthemum indicum. Plant Biotechnology Journal 21: 1519–1521.

Ma, X., Q. Zhu, Y. Chen and Y.-G. Liu. 2016. CRISPR/Cas9 platforms for genome editing in plants: developments and applications. Molecular Plant 9: 961–974.

Maeder, M.L. and C.A. Gersbach. 2016. Genome-editing technologies for gene and cell therapy. Molecular Therapy: The Journal of the American Society of Gene Therapy 24: 430–446.

Maher, M.F., R.A. Nasti, M. Vollbrecht, C.G. Starker, M.D. Clark and D.F. Voytas. 2020. Plant gene editing through *de novo* induction of meristems. Nature Biotechnology 38: 84–89.

Mao, Y., H. Zhang, N. Xu, B. Zhang, F. Gou and J.-K. Zhu. 2013. Application of the CRISPR–Cas system for efficient genome engineering in plants. Molecular Plant 6: 2008–2011.

Matsumoto, N. and K. Okada. 2001. A homeobox gene, PRESSED FLOWER, regulates lateral axis-dependent development of Arabidopsis flowers. Genes and Development 15: 3355–3364.

Meinke, D.W., J.M. Cherry, C. Dean, S.D. Rounsley and M. Koornneef. 1998. Arabidopsis thaliana: A model plant for genome analysis. Science 282: 662–682.

Mekapogu, M., O.-K. Kwon, H.-Y. Song and J.-A. Jung. 2022a. Towards the improvement of ornamental attributes in chrysanthemum: recent progress in biotechnological advances. International Journal of Molecular Sciences 23: 12284.

Mekapogu, M., O.-K. Kwon, H.-Y. Song and J.-A. Jung. 2022b. Towards the improvement of ornamental attributes in chrysanthemum: recent progress in biotechnological advances. International Journal of Molecular Sciences 23: 12284.

Mekapogu, M., H.-Y. Song, S.-H. Lim and J.-A. Jung. 2023. Genetic engineering and genome editing advances to enhance floral attributes in ornamental plants: an update. Plants 12: 3983.

Meng, Y., Y. Hou, H. Wang, R. Ji, B. Liu, J. Wen et al. 2017. Targeted mutagenesis by CRISPR/Cas9 system in the model legume Medicago truncatula. Plant Cell Reports 36: 371–374.

Mirek, Z., H. Piekos-Mirkowa, A. Zajac and M. Zajac. 2002. Flowering plants and Pteridophytes of Poland. A Checklist. Ró¿norodnoSæ biologiczna Polski.

Mohan Jain, S. 2006. Mutation-assisted breeding for improving ornamental plants. Acta Horticulturae: 85–98.

Moon, J., H. Candela and S. Hake. 2013. The Liguleless narrow mutation affects proximal-distal signaling and leaf growth. Development 140: 405–412.

Nag, A., S. King and T. Jack. 2009. miR319a targeting of TCP4 is critical for petal growth and development in Arabidopsis. Proc. Natl. Acad. Sci. U.S.A. 106: 22534–22539. doi: 10.1073/pnas.0908718106.

Nardmann, J., J. Ji, W. Werr and M.J. Scanlon. 2004. The maize duplicate genes narrow sheath1 and narrow sheath2 encode a conserved homeobox gene function in a lateral domain of shoot apical meristems. Development 131: 2827–2839.

Nath, U., B.C.W. Crawford, R. Carpenter and E. Coen. 2003. Genetic control of surface curvature. Science 299: 1404–1407.

Nikovics, K., T. Blein, A. Peaucelle, T. Ishida, H. Morin, M. Aida et al. 2006. The Balance between the MIR164A and CUC2 genes controls leaf margin serration in Arabidopsis. The Plant Cell 18: 2929–2945.

Nishihara, M., A. Higuchi, A. Watanabe and K. Tasaki. 2018a. Application of the CRISPR/Cas9 system for modification of flower color in Torenia fournieri. BMC Plant Biology 18: 331.

Nishihara, M., A. Higuchi, A. Watanabe and K. Tasaki. 2018b. Application of the CRISPR/Cas9 system for modification of flower color in Torenia fournieri. BMC Plant Biology 18: 331.

Nishimura, A., M. Tamaoki, T. Sakamoto and M. Matsuoka. 2000. Over-expression of tobacco knotted 1-Type Class1 Homeobox genes alters various leaf morphology. Plant and Cell Physiology 41: 583–590.

Njoku, E. 1956. Studies in the Morphogenesis of Leaves XI. The effect of light intensity on leaf shape in Ipomea caerulea. New Phytologist 55: 91–110.

Njoku, E. 1957. The effect of mineral nutrition and temperature on leaf shape in Ipomoea caerulea. New Phytologist 56: 154–171.

Noman, A., M. Aqeel, J. Deng, N. Khalid, T. Sanaullah and H. Shuilin. 2017. Biotechnological advancements for improving floral attributes in ornamental plants. Frontiers in Plant Science 8.

Noman, A., M. Aqeel and S. He. 2016. CRISPR-Cas9: Tool for qualitative and quantitative plant genome editing. Frontiers in Plant Science 7.

Okada, K., J. Ueda, M.K. Komaki, C.J. Bell and Y. Shimura. 1991. Requirement of the auxin polar transport system in early stages of Arabidopsis floral bud formation. The Plant Cell 3: 677–684.

van der Oost, J., E.R. Westra, R.N. Jackson and B. Wiedenheft. 2014. Unravelling the structural and mechanistic basis of CRISPR–Cas systems. Nature Reviews Microbiology 12: 479–492.

Ori, N., M.T. Juarez, D. Jackson, J. Yamaguchi, G.M. Banowetz and S. Hake. 1999. Leaf senescence is delayed in tobacco plants expressing the maize homeobox gene knotted1 under the control of a senescence-activated promoter. The Plant Cell 11: 1073–1080.

Ortigosa, A., S. Gimenez-Ibanez, N. Leonhardt and R. Solano. 2019. Design of a bacterial speck resistant tomato by CRISPR/Cas9-mediated editing of SlJAZ2. Plant Biotechnology Journal 17: 665–673.

Pekker, I., J.P. Alvarez and Y. Eshed. 2005. Auxin response factors mediate Arabidopsis organ asymmetry via modulation of KANADI activity. The Plant Cell 17: 2899–2910.

Pickardt, T., M. Meixner, V. Schade and O. Schieder. 1991. Transformation of Vicia narbonensis via Agrobacterium-mediated gene transfer. Plant Cell Reports 9: 535–538.

Plan, D., G. Van den Eede, Institute for Health and Consumer Protection. 2010. The EU Legislation on GMOs: An Overview. Luxembourg: Publications Office.

Polcz, S. and A. Lewis. 2016. CRISPR-Cas9 and the Non-Germline Non-Controversy. Rochester, NY: Social Science Research Network.

Prof Andy Lowe. 2017. Why is leaf shape important to plants? Biodiversity Revolution.

Ramirez, J., N. Bolduc, D. Lisch and S. Hake. 2009. Distal expression of knotted1 in maize leaves leads to reestablishment of proximal/distal patterning and leaf dissection. Plant Physiology 151: 1878–1888.

Ran, F.A., P.D. Hsu, J. Wright, V. Agarwala, D.A. Scott and F. Zhang. 2013. Genome engineering using the CRISPR-Cas9 system. Nature Protocols 8: 2281–2308.

Rashid, H., S. Yokoi, K. Toriyama and K. Hinata. 1996. Transgenic plant production mediated by Agrobacterium in Indica rice. Plant Cell Reports 15: 727–730.

Rollie, C., S. Schneider, A.S. Brinkmann, E.L. Bolt and M.F. White. 2015. Intrinsic sequence specificity of the Cas1 integrase directs new spacer acquisition (TW Nilsen, Ed.). eLife 4: e08716.

Rönspies, M., A. Dorn, P. Schindele and H. Puchta. 2021. CRISPR–Cas-mediated chromosome engineering for crop improvement and synthetic biology. Nature Plants 7: 566–573.

Rothwell, G.W. and R. Serbet. 1994. Lignophyte phylogeny and the evolution of spermatophytes: a numerical cladistic analysis. Systematic Botany 19: 443.

Schaeffer, S.M. and P.A. Nakata. 2015. CRISPR/Cas9-mediated genome editing and gene replacement in plants: Transitioning from lab to field. Plant Science: An International Journal of Experimental Plant Biology 240: 130–142.

Scheben, A., F. Wolter, J. Batley, H. Puchta and D. Edwards. 2017. Towards CRISPR /Cas crops – bringing together genomics and genome editing. New Phytologist 216: 682–698.

Schiml, S., F. Fauser and H. Puchta. 2014. The CRISPR/Cas system can be used as nuclease for in planta gene targeting and as paired nickases for directed mutagenesis in Arabidopsis resulting in heritable progeny. The Plant Journal 80: 1139–1150.

Semenova, E., M.M. Jore, K.A. Datsenko, A. Semenova, E.R. Westra, B. Wanner et al. 2011. Interference by clustered regularly interspaced short palindromic repeat (CRISPR) RNA is governed by a seed sequence. Proceedings of the National Academy of Sciences 108: 10098–10103.

Sherkow, J.S. 2015. Law, history and lessons in the CRISPR patent conflict. Nature Biotechnology 33: 256–257.

Shew, A.M., L.L. Nalley, H.A. Snell, R.M. Nayga and B.L. Dixon. 2018. CRISPR versus GMOs: Public acceptance and valuation. Global Food Security 19: 71–80.

Shimatani, Z., S. Kashojiya, M. Takayama, R. Terada, T. Arazoe, H. Ishii et al. 2017. Targeted base editing in rice and tomato using a CRISPR-Cas9 cytidine deaminase fusion. Nature Biotechnology 35: 441–443.

Singh, A., D. Chakraborty and S. Maiti. 2016. CRISPR/Cas9: a historical and chemical biology perspective of targeted genome engineering. Chemical Society Reviews 45: 6666–6684.

Sink, K.C. 2013. Petunia. Springer Science and Business Media.

Smith, L.G., B. Greene, B. Veit and S. Hake. 1992. A dominant mutation in the maize homeobox gene, Knotted-1, causes its ectopic expression in leaf cells with altered fates. Development 116: 21–30.

Song, G., M. Jia, K. Chen, X. Kong, B. Khattak, C. Xie et al. 2016a. CRISPR/Cas9: A powerful tool for crop genome editing. The Crop Journal 4: 75–82.

Song, G., M. Jia, K. Chen, X. Kong, B. Khattak, C. Xie et al. 2016b. CRISPR/Cas9: A powerful tool for crop genome editing. The Crop Journal 4: 75–82.

Su, L., S. Liu, X. Liu, B. Zhang, M. Li, L. Zeng et al. 2021. Transcriptome profiling reveals histone deacetylase 1 gene overexpression improves flavonoid, isoflavonoid, and phenylpropanoid metabolism in Arachis hypogaea hairy roots. PeerJ 9: e10976. doi: 10.7717/peerj.10976.

Sun, Y., G. Jiao, Z. Liu, X. Zhang, J. Li, X. Guo et al. 2017. Generation of high-amylose rice through CRISPR/Cas9-mediated targeted mutagenesis of starch branching enzymes. Frontiers in Plant Science 8.

Sun, L. and T.-H. Kao. 2018. CRISPR/Cas9-mediated knockout of PiSSK1 reveals essential role of S-locus F-box protein-containing SCF complexes in recognition of non-self S-RNases during cross-compatible pollination in self-incompatible Petunia inflata. Plant Reproduction 31: 129–143.

Tang, J., J. Ye, P. Liu, S. Wang, F. Chen and A. Song. 2023. Ornamental plant gene editing: Past, present and future. Ornamental Plant Research 3: 1–6.

Tian, S., L. Jiang, Q. Gao, J. Zhang, M. Zong, H. Zhang et al. 2017. Efficient CRISPR/Cas9-based gene knockout in watermelon. Plant Cell Reports 36: 399–406.

Tsien, J.Z., D.F. Chen, D. Gerber, C. Tom, E.H. Mercer, D.J. Anderson et al. 1996. Subregion- and cell type-restricted gene knockout in mouse brain. Cell 87: 1317–1326.

Tsukaya, H. 2006. Mechanism of leaf-shape determination. Annual Review of Plant Biology 57: 477–496.

Tsukaya, H. 2018. A consideration of leaf shape evolution in the context of the primary function of the leaf as a photosynthetic organ. pp. 1–26. *In*: Adams III W.W. and I. Terashima (eds.). Advances in Photosynthesis and Respiration. The Leaf: A Platform for Performing Photosynthesis. Cham: Springer International Publishing.

Vandenbussche, M., A. Horstman, J. Zethof, R. Koes, A.S. Rijpkema and T. Gerats. 2009. Differential recruitment of WOX transcription factors for lateral development and organ fusion in Petunia and Arabidopsis. The Plant Cell 21: 2269–2283.

Vanderpoorten, A. and B. Goffinet. 2009. Introduction to Bryophytes. Cambridge University Press.

Verma, V., A. Kumar, M. Partap, M. Thakur and B. Bhargava. 2023. CRISPR-Cas: A robust technology for enhancing consumer-preferred commercial traits in crops. Frontiers in Plant Science 14.

Vollbrecht, E., B. Veit, N. Sinha and S. Hake. 1991. The developmental gene Knotted-1 is a member of a maize homeobox gene family. Nature 350: 241–243.

Waites, R., H.R.N. Selvadurai, I.R. Oliver and A. Hudson. 1998. The PHANTASTICA Gene Encodes a MYB transcription factor involved in growth and dorsoventrality of lateral organs in antirrhinum. Cell 93: 779–789.

Walkerpeach, C.R. and J. Velten. 1994. Agrobacterium-mediated gene transfer to plant cells: cointegrate and binary vector systems. pp. 33–51. *In*: Gelvin, S.B. and R.A. Schilperoort (eds.). Plant Molecular Biology Manual. Dordrecht: Springer Netherlands.

Wang, X. 2020. A review on the molecular mechanism, superiorities, applications, limitations and experimental workflow of CRISPR/Cas-9 system, and the future of gene engineering. IOP Conference Series: Materials Science and Engineering 729: 012044.

Wang, Y., G. Huang, H. Yang and X. Zhang. 2020a. Advances on nonviral vectors of CRISPR/Cas9 system for genome editing. Acta Pharmaceutica Sinica: 2606–2617.

Wang, C., Y. Li, N. Wang, Q. Yu, Y. Li, J. Gao et al. 2023. An efficient CRISPR/Cas9 platform for targeted genome editing in rose (Rosa hybrida). Journal of Integrative Plant Biology 65: 895–899.

Wang, M., Z. Xu, G. Gosavi, B. Ren, Y. Cao, Y. Kuang et al. 2020b. Targeted base editing in rice with CRISPR/ScCas9 system. Plant Biotechnology Journal 18: 1645–1647.

Watanabe, K., C. Oda-Yamamizo, K. Sage-Ono, A. Ohmiya and M. Ono. 2018. Alteration of flower colour in Ipomoea nil through CRISPR/Cas9-mediated mutagenesis of carotenoid cleavage dioxygenase 4. Transgenic Research 27: 25–38.

Wu, L., X. Lan, Y. Zhu, S. Hu, Y. Tian, L. Feng et al. 2018. Construction of multi-target CRISPR/Cas9 vector for PI gene in Cymbidium sinense. Molecular Plant Breeding 16: 7027–7031

Xing, H.-L., L. Dong, Z.-P. Wang, H.-Y. Zhang, C.-Y. Han, B. Liu et al. 2014. A CRISPR/Cas9 toolkit for multiplex genome editing in plants. BMC Plant Biology 14.

Xu, L., Y. Xu, A. Dong, Y. Sun, L. Pi, Y. Xu et al. 2003. Novel as1 and as2 defects in leaf adaxial-abaxial polarity reveal the requirement for ASYMMETRIC LEAVES1 and 2 and ERECTA functions in specifying leaf adaxial identity. Development 130: 4097–4107.

Yu, J., L. Tu, S. Subburaj, S. Bae and G. Lee. 2021. Simultaneous targeting of duplicated genes in Petunia protoplasts for flower color modification via CRISPR-Cas9 ribonucleoproteins. Plant Cell reports 40: 1037–1045. https://doi.org/10.1007/s00299-020-02593-1.

Yu, Y., Z. Liu, L. Wang, S.G. Kim, P.J. Seo, Qiao et al. 2016. WRKY71 accelerates flowering via the direct activation of FLOWERING LOCUS T and LEAFY in Arabidopsis thaliana. The Plant Journal 85: 96–106.

CHAPTER 8

CRISPR/Cas9 Technology in Engineering Viral and Bacterial Immunity in Agricultural Crops

Murad Muhammad,[1,2,*] *Shumaila Batool,*[3] *Aqsa Batool,*[4] *Matoor Mohsin Gilani,*[5] *Ibad Ullah,*[1,2] *Abdul Wahab,*[2] *Ahmad M. Alqudah,*[6] *Samar G. Thabet*[7] and *Mona FA Dawood*[8]

1. Overview of CRISPR/Cas9 Technology

CRISPR/Cas9 technology has fundamentally revolutionized genetic engineering, providing a precise and efficient method for genome modification (Quétier 2016). It was initially recognized as a part of the bacteria's adaptive immune system, utilized to combat viral infections (Chandrasekaran et al. 2021). The mechanism acquires fragments of viral DNA and retains them in CRISPR arrays, enabling bacteria to identify and cleave the DNA of invading viruses with the aid of CRISPR-associated proteins (Zhang et al. 2013). The CRISPR/Cas9 system consists of two fundamental components: the Cas9 enzyme and the guide RNA. The Cas9 enzyme is guided to the specific genomic area designated for editing by the guide RNA, which is designed to be complementary to a particular target DNA sequence (Filippova et al. 2019). The gRNA generally has approximately 20 nucleotides that correspond to the target sequence, plus a scaffold region that interacts with Cas9 (Xu et al. 2014). The Cas9 enzyme is an endonuclease that creates Double-Stranded Breaks (DSBs) in DNA at the target location designated by the gRNA, necessitating

[1] State Key Laboratory of Desert and Oasis Ecology, Key Laboratory of Ecological Safety and Sustainable Development in Arid Lands, Xinjiang Institute of Ecology and Geography, Chinese Academy of Sciences, Urumqi, 830011, PR China.
[2] University of Chinese Academy of Sciences, Beijing 100049, PR China.
[3] Department of Zoology, Wildlife and Fisheries, University of Agriculture Faisalabad 38000, Pakistan.
[4] Department of Chemistry, Government College University Faisalabad, Pakistan.
[5] Department of Forestry and Range Management, Bahauddin Zakariya University, Multan.
[6] Shanghai Center for Plant Stress Biology, CAS Center for Excellence in Molecular Plant Sciences, Chinese Academy of Sciences, Shanghai 200032, China.
[7] Biological Science Program, Department of Biological and Environmental Sciences, College of Art and Science, Qatar University, Qatar.
[8] Department of Botany, Faculty of Science, Fayoum University, 63514 Fayoum, Egypt.
* Corresponding author: muradbotany1@uop.edu.pk

DOI: 10.1201/9781003518204-8

a short sequence termed the Protospacer Adjacent Motif (PAM) for identification and binding (Xue and Greene 2021).

CRISPR/Cas9 operates through a multi-step process. The guide RNA initially binds to the target DNA sequence, creating a complex with the Cas9 enzyme (Nasar et al. 2024a, b). This binding depends on the presence of the PAM sequence, which is crucial for Cas9 to identify and attach to the DNA (Hu et al. 2018). On formation of the gRNA-Cas9 complex, Cas9 activates the cell's DNA repair mechanisms by inducing a double-stranded break at the targeted site in the DNA. The cell can rectify the breakthrough through two principal pathways: Non-Homologous End Joining (NHEJ), which frequently results in insertions or deletions that may impair the target gene, effectively inactivating it, and Homology-Directed Repair (HDR), which facilitates accurate modifications when a donor DNA template is available (Park et al. 2019). CRISPR/Cas9 has various applications, including gene therapy, functional genomic screening, and the development of genetically modified species. Its capacity to alter numerous genes concurrently and its economic efficiency render it a valuable instrument in research and prospective clinical applications (Tan et al. 2018). The objectives of this chapter on CRISPR/Cas9 technology in engineering viral and bacterial immunity in crops are to provide a comprehensive overview of the technology, including its discovery and mechanism of action. The chapter will emphasize the urgent need for disease resistance in crops due to the economic impacts of viral and bacterial pathogens. It will explore how CRISPR/Cas9 can be tailored to target specific pathogens, supported by case studies of successful applications. A comparative analysis will highlight the advantages of CRISPR/Cas9 over traditional breeding methods. Additionally, ethical and regulatory considerations will be addressed, along with future perspectives on the technology's role in enhancing global food security. Finally, the chapter will summarize key findings, stressing the importance of ongoing research in CRISPR technology for sustainable agriculture.

1.1 Importance of Viral and Bacterial Resistance in Agriculture

The impact of viral and bacterial diseases on crop yields, food security, and economic stability cannot be overstated. These pathogens pose a significant threat to agriculture, potentially devastating entire crops and reducing plant vigor, leading to substantial losses in agricultural production (Verma et al. 2023). The agrarian losses brought on by illnesses and pests can range from 26 to 80%, depending on the crop and the severity of the infection (Singh et al. 2014). This stark reality underscores the urgent need for innovative solutions to combat these agricultural challenges. The economic implications of viral and bacterial diseases are far-reaching, affecting farmers' livelihoods and contributing to the broader economic instability of farming communities.

Reduced crop yields can lead to financial hardship for farmers as they struggle to maintain profitability in the face of diminished harvests (Naser et al. 2024a). Additionally, the ripple effects of these losses can be felt throughout the food supply chain, with increased prices and potential consumer shortages (Rosenstock et al. 2023). The need for effective disease management strategies has never been more critical. At the same time, the globe struggles to feed an expanding population in the face of global warming and other environmental pressures (Muhammad et al. 2023). Traditional methods of disease management, such as chemical pesticides and antibiotics, have limitations. While these treatments can be effective in the short term, they can also develop resistance among pathogens, rendering them less effective over time (Sundin et al. 2016). Furthermore, using these substances may harm human wellness and the environment, underscoring the need for more sustainable and targeted approaches to disease management (Muhammad et al. 2024).

Innovative strategies, such as resistant crop varieties, integrated pest management, and biotechnological advancements like CRISPR/Cas9, offer promising solutions for combating viral and bacterial diseases in agriculture (Kumar et al. 2021). These approaches aim to enhance crop resilience, reduce dependency on chemical treatments, and ultimately secure food production in the face of increasing agricultural challenges. Viral and bacterial resistance in agriculture is crucial

Table 1. Importance of viral and bacterial resistance in agriculture.

Aspect	**Viral Resistance**	**Bacterial Resistance**	**Reference**
Crop Health	Reduces crop losses due to viral infections.	Minimizes damage from bacterial diseases.	(Mann et al. 2021)
Yield Improvement	Increases overall crop yield by preventing viral damage.	Enhances yield stability by protecting against bacterial threats.	(Majeed et al. 2018)
Sustainable Practices	Promotes sustainable farming by reducing pesticide use.	Supports integrated pest management strategies.	(Miller et al. 2022)
Economic Benefits	Lower costs are associated with viral disease management.	Reduces economic losses from bacterial outbreaks.	(Svircev et al. 2018)
Genetic Diversity	Encourages the development of resistant varieties through breeding.	Stimulates research into bacterial resistance traits.	(Garrett et al. 2017)
Food Security	Contributes to food security by ensuring stable crop production.	Essential for maintaining a reliable food supply.	(McDonald and Stukenbrock 2016)
Environmental Impact	Decreases reliance on chemical treatments, benefiting ecosystems.	Reduces environmental pollution from antibiotics and pesticides.	(Buelow et al. 2021)
Adaptation to Climate Change	Helps crops adapt to changing conditions influenced by climate change.	Essential for developing resilient agricultural systems.	(Velásquez et al. 2018)

for enhancing crop health and yield, ensuring sustainable farming practices, as shown in Table 1. Reducing reliance on chemical treatments contributes to food security and environmental protection. Figure 1 presents the importance of viral and bacterial resistance in agriculture, highlighting benefits such as increased crop yield and reduced pesticide use. It also explores methods for achieving resistance, challenges faced, and future considerations for sustainable management and emerging technologies.

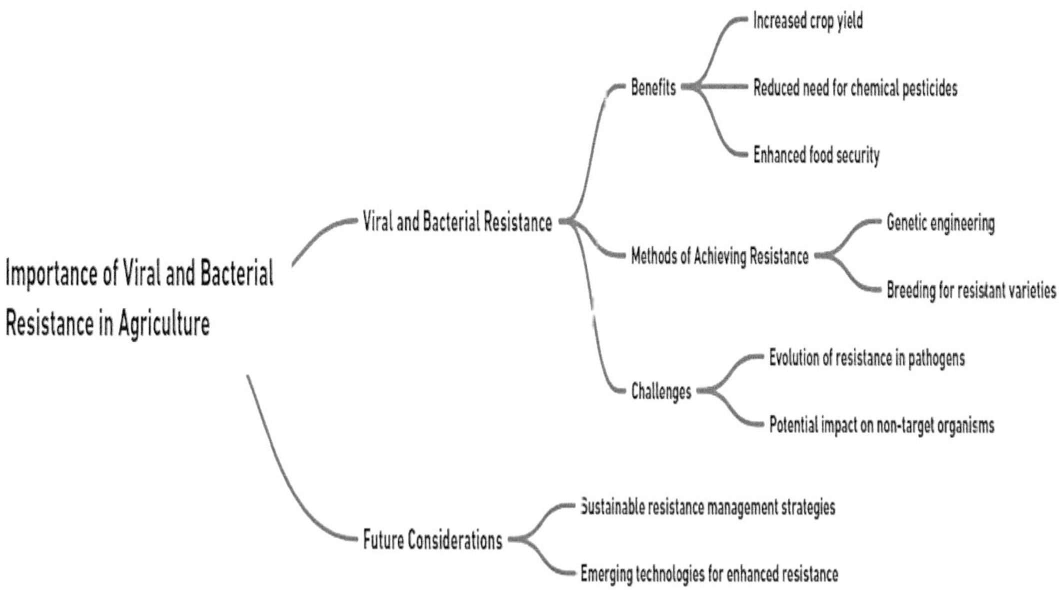

Figure 1. Importance of viral and bacterial resistance in agriculture.

1.2 Engineering CRISPR/Cas9-Based Resistance Against Viruses and Bacteria

CRISPR/Cas9 technology has revolutionized the field of agriculture, providing powerful tools for engineering viral and bacterial immunity in crops. This cutting-edge gene-editing system allows for precise modifications to the plant genome, enabling the development of crop varieties that can withstand a wide range of diseases, thereby enhancing global food security (Kumar et al. 2021). In terms of viral immunity, researchers are harnessing CRISPR/Cas9 to target specific genes associated with a plant's susceptibility to viral infections (Georges and Ray 2017). By editing these genes, scientists can effectively block the pathways that viruses exploit to infiltrate plant cells (Muhammad et al. 2024). This not only reduces crop losses but also helps minimize the reliance on chemical pesticides, promoting more sustainable agricultural practices that benefit both farmers and the environment (Elsharawy and Refat 2023). Similarly, CRISPR/Cas9 plays a crucial role in conferring resistance to bacterial and viral pathogens (Pandey and Vavilala 2024). By modifying genes involved in the plant's immune response, researchers can significantly enhance the plant's ability to fend off bacterial infections caused by organisms such as *Xanthomonas* and *Pseudomonas* (Tyagi et al. 2021). This genetic resilience can lead to healthier crops with improved yield, nutritional quality, and marketability. Despite the immense potential of CRISPR-engineered crops, several challenges remain, including off-target effects and regulatory hurdles that must be navigated (Kumar et al. 2020). Addressing these issues through rigorous research and extensive field trials is essential to ensure the safety, efficacy, and public acceptance of these innovations. Ultimately, the application of CRISPR/Cas9 technology in agriculture represents a significant advancement, offering innovative solutions to combat viral and bacterial threats and securing a more sustainable future for food production.

2. Mechanisms of CRISPR/Cas9

2.1 Structure and Function of CRISPR/Cas9

The CRISPR/Cas9 system is a groundbreaking genome editing tool characterized by its unique molecular structure and precise functionality. At its core, CRISPR (Clustered Regularly Interspaced Short Palindromic Repeats) consists of a guide RNA (gRNA) and the Cas9 protein, which together facilitate targeted DNA modification (Ghosh and Chatterjee 2024). The gRNA is typically synthesized as a single-guide RNA (sgRNA), which combines the functionalities of two RNA components: the CRISPR RNA (crRNA) that provides sequence specificity and the trans-activating CRISPR RNA (tracrRNA) that stabilizes the complex (Kelley et al. 2016). This sgRNA is designed to complement a specific target DNA sequence, allowing the Cas9 enzyme to locate and bind to the desired site within the genome (Burmistrz et al. 2020). Recognizing target DNA by the CRISPR/Cas9 system is a special process. The sgRNA guides Cas9 to the target DNA by pairing it with a complementary sequence. Still, this binding is contingent upon a Protospacer Adjacent Motif (PAM), a short sequence adjacent to the target DNA (Goldberg et al. 2021). PAM is crucial for Cas9's binding and cleavage activity, ensuring the enzyme interacts with the correct DNA sequence (Shalaby et al. 2020). The mechanism of double-strand break induction involves the coordinated action of these nuclease domains.

The HNH domain cleaves the DNA strand complementary to the sgRNA, while the RuvC domain cuts the non-target strand (Wang et al. 2023). This dual cleavage results in a double-stranded break in the DNA, which the cell tries to fix using Homology-Directed Repair (HDR) or Non-Homologous End Joining (NHEJ). NHEJ often leads to insertions or deletions that can disrupt gene function, while HDR can enable precise modifications if a repair template is provided (Yang et al. 2020). Recent studies have provided more profound insights into the molecular architecture of the CRISPR/Cas9 complex, revealing how the enzyme achieves specificity and efficiency in DNA cleavage (Song et al. 2021). For instance, cryo-electron microscopy has elucidated the conformational states of Cas9 during the DNA cleavage process, demonstrating how the HNH domain moves closer to the cleavage site to facilitate the cutting action (Jiang and Doudna 2017). These advancements in understanding the structural biology of CRISPR/Cas9 enhance our knowledge of its mechanisms and pave the way for improved applications in biotechnology and genetic research.

2.2 Targeting Mechanisms for Viral and Bacterial Genomes

CRISPR/Cas9 may be tailored to target specific viral and bacterial genomes by modifying the guide RNA (gRNA) to complement the intended target sequence (Quétier 2016). The pathogen's exact genomic region where editing is needed can be identified by the gRNA, which then directs the Cas9 enzyme there. Here are a few examples of the targeting of bacterial and viral genes using CRISPR/Cas9:

2.2.1 Targeting Viral Genes

Utilizing CRISPR technology to target viral genes in crops constitutes a groundbreaking method for enhancing plant resistance to viral infections (Shahriar et al. 2021). CRISPR (Clustered Regularly Interspaced Short Palindromic Repeats) is a very accurate gene-editing instrument that enables scientists to alter DNA with precision. This technology has achieved significant appeal owing to its efficiency and applicability across diverse applications, including agriculture (Manzoor et al. 2024).

In the realm of viral resistance, researchers may engineer guide RNAs (gRNAs) that precisely target sections within viral genomes. By incorporating the CRISPR-Cas9 system into plant cells, these gRNAs guide the Cas9 enzyme to the viral DNA or RNA, leading to precise cuts that hinder the replication and functionality of the virus. This technique efficiently impedes the virus's capacity to infect the plant, providing a strong defense against viral infections (Patil et al. 2024).

The advantages of employing CRISPR to target viral genes in crops are substantial. Firstly, it improves crop resistance, resulting in diminished losses from viral diseases. This is especially vital for essential crops such as tomatoes, potatoes, and rice, which are frequently impacted by diverse viral diseases (Shahriar et al. 2021; Zaidi et al. 2016). This strategy reduces dependence on chemical pesticides, fostering sustainable agriculture practices and aligning with worldwide initiatives to mitigate environmental impacts.

Nonetheless, this technology presents certain limitations. Ensuring specificity is crucial; off-target effects may unintentionally modify non-target genes, perhaps resulting in unforeseen consequences. Additionally, traversing regulatory frameworks for Genetically Modified Organisms (GMOs) can be intricate, and public opinion of genetically modified crops is a pivotal element in their acceptability (Carter 2020).

In conclusion, the targeting of viral genes by CRISPR presents a promising approach to enhancing agricultural output and food security. Ongoing research and development will be crucial in tackling problems and optimizing the potential of this new farming technology.

2.2.2 Targeting Bacterial Genes

Utilizing CRISPR technology to target bacterial genes in crops is a novel approach to improving plant resistance to bacterial infections. CRISPR (Clustered Regularly Interspaced Short Palindromic Repeats) has emerged as a potent gene-editing instrument that facilitates precise alterations in DNA, rendering it very efficacious in addressing plant diseases (Gosavi et al. 2020).

In this environment, researchers can formulate guide RNAs (gRNAs) that precisely target sections within bacterial genomes. The introduction of the CRISPR-Cas9 system into plant cells enables gRNAs to guide the Cas9 enzyme to bacterial DNA, effectuating precise cuts that impair the function and replication of the bacteria (Javed et al. 2023). This focused strategy significantly improves the plant's defense systems, allowing it to withstand bacterial infections that can drastically affect crop output and quality. The advantages of employing CRISPR for the targeting of bacterial genes are significant. Firstly, it enhances crop resilience, mitigating losses from bacterial illnesses such as bacterial blight and wilt. This is especially crucial for economically vital crops such as tomatoes, cotton, and rice. This strategy reduces reliance on chemical pesticides (Liu et al. 2020), thereby promoting sustainable agricultural practices vital for environmental preservation and public health.

Nonetheless, obstacles persist in the execution of this technology. Ensuring specificity is essential; off-target impacts may inadvertently modify non-target genes, resulting in possible adverse implications for the plant. Moreover, traversing the regulatory framework for genetically modified organisms (GMOs) can be intricate, and public endorsement of genetically modified crops is a crucial factor (Vento et al. 2019).

In summary, targeting bacterial genes using CRISPR technology is a viable avenue for enhancing crop resilience and bolstering food security. Continuous research and development are essential to tackle these problems and realize the full potential of this innovative technology in agriculture, facilitating healthier and more sustainable food production systems.

3. Engineering Immunity against Viral Infections

3.1 Overview of Viral Threats to Crops

Viral infections pose a severe danger to the agricultural sector, impacting various crops and leading to considerable economic losses. Among the most impactful viral infections are those caused by members of the begomoviruses, tospoviruses, and potyviruses, which are responsible for severe epidemics in various staple crops. For instance, the cassava mosaic disease, caused by begomoviruses, can yield losses of up to 100%, severely impacting food security in regions dependent on cassava as a staple food source (Chikoti and Tembo 2022).

Similarly, the maize lethal necrosis disease, caused by a combination of viruses, including the maize chlorotic mottle virus and the sugarcane mosaic virus, has led to devastating losses in maize production, a critical crop for food and feed globally. In legumes, significant viral threats include the bean typical mosaic virus (BCMV) and the cowpea severe mosaic virus (CPSMV). BCMV can cause yield losses ranging from 6 to 98%, depending on the strain and environmental conditions, while CPSMV can lead to losses between 13 and 87% in cowpea crops (Hema et al. 2014). These diseases not only diminish crop yields but also affect the quality of the produce, leading to economic instability for farmers and increased food prices for consumers (Chikoti and Tembo 2022). The prevalence of these viral diseases is exacerbated by climate change, which can alter the distribution and activity of viral vectors like aphids and whiteflies. Additionally, agricultural practices such as monoculture and the global movement of plant materials contribute to the spread of viruses, making it challenging to manage and control outbreaks effectively.

3.2 Case Studies: Successful Applications of CRISPR/Cas9

In a notable study, researchers engineered tomato plants to confer resistance against TYLCV, a danger to the global tomato yield. They utilized the CRISPR/Cas9 system to target the viral genome, specifically the Coat Protein (CP) and replicate genes. The engineered tomato plants reduced viral accumulation by 69 and 76% for the CP and replicate targets, respectively. This was achieved by introducing single-guide RNAs (sgRNAs) that directed Cas9 to cleave the viral DNA, resulting in indels that disrupted the viral genome (Sharma et al. 2021). Transgenic crops demonstrated considerably fewer diseases and viral loads than control plants, demonstrating the effectiveness of this approach in enhancing resistance to viral infections in crops.

Another successful application involved banana plants engineered to resist BSV, a virus that integrates into the banana genome and can lead to severe crop losses. Researchers targeted the integrated viral genome's Open Reading Frames (ORFs) using CRISPR/Cas9 (Tripathi et al. 2024). CRISPR/Cas9 was also employed to confer resistance against WDV in barley plants. In this case, researchers designed sgRNAs targeting overlapping regions of the WDV genome. Barley plants expressing Cas9 along with the sgRNAs showed significant resistance to WDV, with effective viral suppression observed at 112 days post-inoculation. This study highlights the ability of CRISPR/Cas9 to target viral genomes directly and provide durable resistance in economically important crops (Tripathi et al. 2024).

In *Arabidopsis thaliana*, researchers successfully targeted the coat protein region of CaMV using CRISPR/Cas9. The transgenic plants exhibited effective resistance to the virus, demonstrating that the CRISPR/Cas9 system can engineer resistance against DNA and RNA viruses (Zaidi et al. 2016). This study underscores the versatility of CRISPR/Cas9 in addressing various viral threats in different plant species. These case studies exemplify the successful application of CRISPR/Cas9 technology in engineering viral resistance in crops, showcasing its potential to enhance food security and agricultural sustainability in the face of increasing viral threats. The ability to precisely target viral genomes and plant susceptibility genes opens new avenues for developing resilient crop varieties capable of withstanding viral infections. Table 2 summarizes how CRISPR/Cas9 technology can be leveraged to improve plant immunity against both viral and bacterial threats, contributing to more resilient agricultural systems.

Table 2. The role of CRISPR/Cas9 in immunity against viruses and bacteria.

Application	Mechanism of Action	Importance in Agriculture	References
Dual Resistance in Crops	Editing genes to enhance resistance against both viruses and bacteria.	Increases crop resilience, reducing losses from multiple pathogens.	(Wahab et al. 2024)
Targeting Pathogen Genomes	Disrupting the genomes of both viral and bacterial pathogens.	Provides a broad-spectrum approach to disease management.	(Ali et al. 2015)
Gene Editing for Enhanced Defense	Modifying plant immune receptors to recognize a wider range of pathogens.	Promotes plant health and reduces dependency on chemical pesticides.	(Muhammad et al. 2024a)
Development of Transgenic Varieties	Creating crops with CRISPR-edited genes for dual resistance.	Supports sustainable agriculture by improving yield stability.	(Varanda et al. 2021)
Research and Development	Studying interactions of plants with both viral and bacterial pathogens.	Enhances understanding of plant immunity and pathogen strategies.	(Varanda et al. 2021)
Improving Soil Microbiomes	Engineering beneficial microbes to suppress both viral and bacterial diseases.	Enhances overall soil health and plant growth, fostering resilience.	(Muhammad et al. 2024b)

4. Engineering Immunity Against Bacterial Infections

Bacteria can rapidly evolve resistance mechanisms against CRISPR/Cas9 interventions. Research indicates that bacteria can acquire mutations that render them less susceptible to CRISPR-mediated targeting (Javed et al. 2023). These mutations may occur in the target sequences or the Cas genes, allowing bacteria to evade the CRISPR system's action. Additionally, the presence of anti-CRISPR (Acr) genes in bacterial genomes can inhibit the effectiveness of CRISPR/Cas systems, complicating efforts to engineer resistance. This adaptability highlights the need for continuous monitoring and potential modifications of CRISPR strategies to maintain their effectiveness over time (Serajian et al. 2021).

5. Efficiency and Precision of CRISPR/Cas9 vs. Traditional Breeding

CRISPR/Cas9 technology offers significant benefits over conventional breeding techniques in terms of accuracy and efficiency in achieving desired crop traits. Whereas traditional breeding depends on random genetic recombination and selection, CRISPR/Cas9 allows targeted modifications of specific genes or genomic regions (Ahmar et al. 2020). This precision enables researchers to introduce desired mutations or insert beneficial alleles at specific target sites, minimizing the risk of unintended genetic changes that can occur with traditional breeding approaches. By directly modifying the gene of interest, CRISPR/Cas9 can achieve the desired phenotype more efficiently than the lengthy backcrossing and selection process required in conventional breeding. In terms of efficiency, the CRISPR/Cas9 system is highly effective in producing desired mutations, with studies reporting mutagenesis frequencies ranging from 10 to 80% in various plant species, including wheat, maize, and barley.

In contrast, traditional breeding methods rely on natural genetic variation and recombination, which is less efficient in introducing specific traits (Breseghello and Coelho 2013). Furthermore, CRISPR/Cas9 can be used for multiplexed editing, allowing simultaneous modification of multiple genes or genomic regions, further enhancing efficiency. The timeframe for developing resistant crop varieties using CRISPR/Cas9 is significantly shorter than conventional breeding techniques.

Once the target gene is identified, the CRISPR/Cas9 technology can be designed and introduced into the plant, generating edited lines in months. These edited lines can then be evaluated for resistance and advanced through the breeding pipeline more rapidly than conventional methods. Researchers have successfully engineered tomato plants immune to bacterial speck disease caused by *P. syringae* in a single generation utilizing CRISPR/Cas9 technology. Similarly, CRISPR/Cas9 was used to create resistance to the banana streak virus in banana plants, illustrating the capacity for the fast generation of resistant cultivars. Despite the advantages of CRISPR/Cas9, societal attitudes toward genetic engineering and CRISPR technology include various concerns and misconceptions (Breseghello and Coelho 2013). Public perception of genetically modified organisms (GMOs) often extends to CRISPR-edited crops, leading to skepticism about their safety and environmental impact. Concerns about food safety, ecological consequences, and ethical implications can hinder the acceptance of CRISPR technology in agriculture (Sampath et al. 2023). Additionally, misinformation and a lack of understanding about the technology contribute to public apprehension. Therefore, effective communication and education about CRISPR technology's safety, benefits, and potential are essential for addressing these concerns and fostering acceptance among consumers, farmers, and policymakers.

6. Regulatory Frameworks for CRISPR/Cas9 in Crop Engineering

The regulatory framework overseeing the application of the CRISPR/Cas9 system in agriculture is intricate and differs markedly among various nations and areas. The legal status and approval processes of CRISPR-edited crops are essential for their commercialization and acceptance as they enter the market. The regulatory environment in the United States has developed to support the advancement of CRISPR-edited crops. The U.S. Department of Agriculture (USDA) has implemented a science-based and risk-proportionate strategy, permitting crops that may be produced via traditional breeding techniques to evade comprehensive regulatory examination (Jenkins et al. 2021). If a CRISPR-edited crop contains no foreign DNA and is indistinguishable from naturally occurring mutations, it may not be subject to the exact regulatory requirements of genetically modified organisms (GMOs) (Eriksson et al. 2019). This streamlined process aims to encourage innovation while ensuring safety.

Conversely, the European Union (EU) has taken a more cautious stance, treating CRISPR-edited crops similar to GMOs. In the EU, CRISPR-edited plants are subject to stringent regulations that require comprehensive risk assessments and approval processes before commercialization. This regulatory technique causes delays in the market entry of CRISPR-edited crops, as the legal status of these crops remains contentious and is under ongoing debate. For instance, if proposed legislation to classify certain gene-edited crops as non-GMO is not approved, they will continue to be regulated under the existing, more restrictive GMO framework (Rozas et al. 2022). Countries such as Argentina, Brazil, and Colombia have adopted more favorable regulatory environments for CRISPR-edited crops, facilitating their development and commercialization. These nations generally allow for a more flexible interpretation of regulations, promoting genome editing technologies to enhance agricultural productivity. In contrast, many developing countries have yet to establish comprehensive regulatory frameworks for assessing CRISPR-edited crops, which could hinder their ability to leverage this technology for agricultural improvement. The lack of clear guidelines may also lead to uncertainty among researchers and investors in these regions.

7. Future Perspectives

7.1 Innovations in CRISPR/Cas9 Technology

Current research and prospective improvements in CRISPR/Cas9 technology are poised to markedly augment its utilization in agriculture, tackling essential issues, including food security and climate resilience (Mafakheri and Kordrostami 2020). The refinement of the CRISPR/Cas9 system is an up-and-coming field of work aimed at enhancing precision and efficiency in genome editing. Innovations like base editing and prime editing are emerging as alternatives to typical CRISPR/Cas9, facilitating more precise alterations without causing double-strand breaks. These innovations can diminish off-target impacts, enhancing the safety and reliability of genetic changes for crop enhancement. Additionally, researchers are exploring CRISPR/Cas9 as a prospective instrument to enhance crop resilience against abiotic stressors such as salinity, drought, and elevated temperatures (Chennakesavulu et al. 2021; Villalobos-López et al. 2022). By pinpointing and focusing on specific genes associated with stress responses, researchers can cultivate crops that exhibit more resilience to the impacts of global warming. Research has demonstrated that CRISPR/Cas9 may modify genes associated with a plant's response to limited water availability (Sami et al. 2021), hence improving drought tolerance in crops like wheat and rice. Besides augmenting stress tolerance, CRISPR/Cas9 technology is employed to improve crop quality and output. The current study aims to enhance characteristics, including nutritional value, flavor, and longevity. Scientists are using CRISPR to develop low-gluten wheat types, benefiting those with gluten sensitivities and enhancing the overall quality of wheat for broader consumer markets (Gilissen and Smulders, 2021). The capacity for simultaneous exact modifications of several genes, referred to as multiplexing, expedites the advancement of crops with favorable characteristics. A critical domain of innovation pertains to enhancing the transport mechanisms for CRISPR components into plant cells. Enhancing delivery systems, such as viral vectors or nanoparticles, might improve gene editing efficiency and broaden its use among various plant species. This is especially significant for crops that are usually challenging to modify through conventional techniques. The regulatory framework governing CRISPR-edited crops is developing, with certain places adopting more permissive policies for gene-edited organisms. This transition may expedite the commercialization of CRISPR-engineered crops, enabling farmers to obtain enhanced varieties more rapidly.

7.2 Potential for Global Food Security

CRISPR/Cas9 technology has the potential to significantly impact global food security issues, especially with climate change and population expansion. A principal advantage of CRISPR/Cas9 is its capacity to enhance crop resilience to biotic and abiotic stressors (Rozas et al. 2022). Researchers can improve features like drought tolerance, pest resistance, and disease resistance in staple crops through precise genetic alterations. CRISPR has effectively been employed to modify rice and wheat types to enhance their resistance to bacterial blight and other diseases, which are becoming progressively problematic due to shifting climate conditions (Li et al. 2024). Such developments can result in increased yields and more reliable food supplies, even under harsh conditions, thereby enhancing food security.

Moreover, CRISPR/Cas9 can be utilized to biofortify crops, increasing their nutritional value to combat malnutrition (Kumar et al. 2022). This is especially crucial in areas where dietary deficits are common. By enhancing crops' essential vitamins and minerals, CRISPR technology can help address hidden hunger and improve overall public health (Sen et al. 2024). For example, researchers are exploring ways to increase the iron and zinc content in staple crops like rice and wheat, which can significantly impact the nutritional status of populations reliant on these foods. The efficiency and speed of CRISPR/Cas9 also allow for the rapid development and deployment of improved crop varieties (Cao et al. 2016). Traditional breeding methods can take years or even decades to achieve desired traits, whereas CRISPR can facilitate quicker advancements in crop improvement (Rao and Wang 2021). This rapid rate is crucial for addressing the urgent issues arising from food scarcity and climate change. Nonetheless, the application of CRISPR technology in agriculture must contend with regulatory obstacles and public image challenges. Despite an increasing comprehension of the advantages of CRISPR, apprehensions over genetic editing and its consequences for food safety and environmental effects persist (Zhang et al. 2020).

Consequently, ongoing research and clear communication regarding the safety and efficacy of CRISPR-engineered crops are essential for achieving public acceptability and securing regulatory approval (Abd-Elsalam and Ahmad 2023). Table 2 illustrates that CRISPR technology facilitates accurate gene editing in crops, permitting the creation of cultivars with enhanced characteristics such as yield, disease resistance, and nutritional value. This novel method significantly improves agricultural output and sustainability, tackling global food security issues as shown in Table 3.

Table 3. CRISPR technology in global food security.

Aspect	CRISPR Technology	Process	Impact	References
Gene Editing	The CRISPR-Cas9 system is used to make precise cuts in DNA.	Targeted DNA sequences are identified and modified.	Enables the development of crops with desired traits.	(Jiang and Doudna 2017)
Target Selection	Specific genes related to traits like yield or resistance are chosen.	Guide RNA (gRNA) is designed to match target genes.	Increases efficiency in crop improvement efforts.	(Bailey-Serres et al. 2019)
Delivery Mechanism	Various methods (e.g., *Agrobacterium, biolistics*) introduce CRISPR components into plant cells.	CRISPR components are delivered into the plant cells for editing.	Facilitates the transformation of crops with desired traits.	(Kuluev et al. 2019)
Screening and Validation	Edited plants are screened for successful modifications.	Molecular techniques confirm the presence of desired edits.	Ensures that new varieties meet agricultural and nutritional goals.	(Georges and Ray 2017)

8. Implications for Future Research and Development

Ongoing development in CRISPR technology is essential for progressing sustainable agriculture and improving food systems. As agricultural issues escalate due to climate change, population expansion, and heightened pest and disease pressures, CRISPR/Cas9 provides novel possibilities for creating resilient crop types and enhancing food security (Muha-Ud-Din et al. 2023). Investigations into CRISPR technology may facilitate the development of crops that are more resilient to diseases, pests, and environmental adversities. Targeted gene editing can enhance characteristics such as drought tolerance, nutrient utilization efficiency, and resistance to viral and bacterial diseases. This is particularly important as traditional breeding methods may not keep pace with the rapidly changing agricultural landscape.

Moreover, CRISPR technology can facilitate the creation of crops that require fewer chemical inputs, such as fertilizers and pesticides (Muha-Ud-Din et al. 2023). Farmers can reduce their reliance on chemical treatments by engineering plants to be more resilient, promoting more sustainable agricultural practices that are better for the environment. With over 10 billion people expected to live on the planet by 2050, increasing agricultural yields with CRISPR technology is essential for guaranteeing food safety (Rasheed et al. 2021). Improving the production of crucial crops via innovation can assist in fulfilling the world's rising food demand while minimizing the ecological footprint of agriculture. Additionally, CRISPR can be used to biofortify crops, enhancing their nutritional value and combating malnutrition in vulnerable populations.

The implications of CRISPR technology extend beyond individual crops; they can transform entire food systems. By enabling the rapid development of improved crop varieties, CRISPR can increase agricultural efficiency, leading to higher yields per acre and reducing the need for additional land conversion, thus helping to preserve natural ecosystems. As global warming alters growing conditions, CRISPR can facilitate the development of crops that thrive in new environments, ensuring that food production can adapt to changing climate patterns. Furthermore, by reducing crop losses due to disease and environmental stress, CRISPR-engineered crops can contribute to more stable food supplies and prices, benefiting producers and consumers. In conclusion, ongoing research in CRISPR technology is crucial for unlocking its capability in the sustainable food and agricultural sectors. By addressing the challenges of crop resilience, sustainability, and food security, CRISPR can significantly influence how agriculture develops and guarantees a steady food supply for the next generation.

9. Conclusion

In summary, CRISPR/Cas9 technology signifies revolutionary progress in agricultural biotechnology, especially in the enhancement of viral and bacterial resistance in crops. This chapter has emphasized the accuracy, efficacy, and swift advancement potential of CRISPR/Cas9, illustrating its capacity to tackle the urgent issues presented by plant diseases that jeopardize global food security. The provided case studies demonstrate the efficacy of CRISPR/Cas9 in improving crop resilience, resulting in increased yields and less dependence on chemical pesticides. CRISPR/Cas9 offers a precise and rapid strategy for crop enhancement by targeting specific genes linked to disease susceptibility, surpassing traditional breeding techniques in both speed and precision. Nonetheless, the implementation of CRISPR/Cas9 in agriculture presents some hurdles. Ethical considerations, public perception, and regulatory frameworks must be meticulously managed to ensure the realization of this technology's benefits while resolving social concerns. Interacting with stakeholders, such as farmers, consumers, and legislators, is crucial for promoting acceptance and comprehension of gene editing technology. The prospective impact of CRISPR/Cas9 on global food security is substantial. Ongoing research may advance CRISPR technology, facilitating the creation of crops capable of enduring fluctuating climates and novel diseases. The use of CRISPR/Cas9 in agricultural methods is expected to establish a more robust and sustainable food supply. Ongoing investment in research, collaboration among scientists and agricultural practitioners, and a dedication to responsible innovation will be essential for fully using CRISPR/Cas9 technology to benefit agriculture and society. In the context of addressing the simultaneous difficulties of nourishing an expanding population and responding to climate change, CRISPR/Cas9 emerges as a vital tool in the pursuit of sustainable agricultural solutions.

References

Abd-Elsalam, K.A. and A. Ahmad. 2023. Global Regulatory Outlook for CRISPRized Plants. Elsevier.

Ahmad, S., X. Wei, Z. Sheng, P. Hu and S. Tang. 2020. CRISPR/Cas9 for development of disease resistance in plants: recent progress, limitations and future prospects. Brief. Funct. Genomics 19(1): 26–39.

Ahmar, S., R.A. Gill, K.-H. Jung, A. Faheem, M.U. Qasim, M. Mubeen et al. 2020. Conventional and molecular techniques from simple breeding to speed breeding in crop plants: recent advances and future outlook. Int. J. Mol. Sci. 21(7): 2590.

Ali, Z., A. Abulfaraj, A. Idris, S. Ali, M. Tashkandi and M.M. Mahfouz. 2015 CRISPR/Cas9-mediated viral interference in plants. Genome Biol. 16: 1–11.

Bailey-Serres, J., J.E. Parker, E.A. Ainsworth, G.E. Oldroyd and J.I. Schroeder. 2019. Genetic strategies for improving crop yields. Nature 575(7781): 109–118.

Breseghello, F. and A.S.G. Coelho. 2013. Traditional and modern plant breeding methods with examples in rice (Oryza sativa L.). J. Agric. Food Chem. 61(35): 8277–8286.

Buelow, E., M.-C. Ploy and C. Dagot. 2021. Role of pollution on the selection of antibiotic resistance and bacterial pathogens in the environment. Curr. Opin. Microbiol. 64: 117–124.

Burmistrz, M., K. Krakowski and A. Krawczyk-Balska. 2020. RNA-targeting CRISPR–Cas systems and their applications. Int. J. Mol. Sci. 21(3): 1122.

Cao, H.X., W. Wang, H.T. Le and G.T. Vu. 2016. The power of CRISPR-Cas9-induced genome editing to speed up plant breeding. Int. J. Genomics 2016(1): 5078796.

Carter, M. 2020. Current Capabilities, Precision, and Risks of Genome Editing Techniques in Agricultural Systems.

Chandrasekaran, M., T. Boopathi and M. Paramasivan. 2021. A status-quo review on CRISPR-Cas9 gene editing applications in tomato. Int. J. Biol. Macromol. 190: 120–129.

Chennakesavulu, K., H. Singh, P.K. Trivedi, M. Jain and S.R. Yadav. 2021. State-of-the-art in CRISPR technology and engineering drought, salinity, and thermo-tolerant crop plants. Plant Cell Rep. 1–17.

Chikoti, P.C. and M. Tembo. 2022. Expansion and impact of cassava brown streak and cassava mosaic diseases in Africa: A review. Front. Sustain. Food Syst. 6: 1076364.

Elsharawy, H. and M. Refat. 2023. CRISPR/Cas9 genome editing in wheat: enhancing quality and productivity for global food security—a review. Funct. Integr. Genomics 23(3): 265.

Eriksson, D., D. Kershen, A. Nepomuceno, B.J. Pogson, H. Prieto, K. Purnhagen et al. 2019. A comparison of the EU regulatory approach to directed mutagenesis with that of other jurisdictions, consequences for international trade and potential steps forward. New Phytol. 222(4): 1673–1684.

Filippova, J., A. Matveeva, E. Zhuravlev and G. Stepanov. 2019. Guide RNA modification as a way to improve CRISPR/Cas9-based genome-editing systems. Biochimie 167: 49–60.

Garrett, K., K.F. Andersen, F. Asche, R. Bowden, G. Forbes, P. Kulakow et al. 2017. Resistance genes in global crop breeding networks. Phytopathology 107(10): 1268–1278.

Gendelman, H., P. Dash, B. Edagwa, S. Gorantla, R. Kaminski, R. Bella et al. 2018. D-110 Synergism between CRISPR/cas9 and Laser Art leads to elimination of Hiv-1 with no rebound in humanized mice. JAIDS J. Acquir. Immune Defic. Syndr. 77: 42.

Georges, F. and H. Ray. 2017. Genome editing of crops: a renewed opportunity for food security. GM Crops Food 8(1): 1–12.

Ghosh, S.K. and T. Chatterjee. 2024. Clustered Regularly Interspaced Short Palindromic Repeats (CRISPR)-Associated Proteins (Cas)[CRISPR–Cas]: An emerging technique in plant disease detection and management. pp. 589–645. *In*: Gene Editing in Plants: CRISPR-Cas and Its Applications. Springer.

Gilissen, L.J. and M.J. Smulders. 2021. Low gluten and coeliac-safe wheat through gene editing. Plant Biotechnol.: Exp. Future Prospect. 231–248.

Goldberg, G.W., J.M. Spencer, D.O. Giganti, B.R. Camellato, N. Agmon, D.M. Ichikawa et al. 2021. Engineered dual selection for directed evolution of SpCas9 PAM specificity. Nat. Commun. 12(1): 349.

Gosavi, G., F. Yan, B. Ren, Y. Kuang, D. Yan, X. Zhou et al. 2020. Applications of CRISPR technology in studying plant-pathogen interactions: overview and perspective. Phytopathol. Res. 2(1): 21.

Hema, M., P. Sreenivasulu, B.L. Patil, P.L. Kumar and D.V. Reddy. 2014. Tropical food legumes: virus diseases of economic importance and their control. Adv. Virus Res. 90: 431–505.

Hu, J.H., S.M. Miller, M.H. Geurts, W. Tang, L. Chen, N. Sun et al. 2018. Evolved Cas9 variants with broad PAM compatibility and high DNA specificity. Nature 556(7699): 57–63.

Javed, M.U., M.T. Hayat, H. Mukhtar and K. Imre. 2023. CRISPR-Cas9 system: A prospective pathway toward combatting antibiotic resistance. Antibiotics 12(6): 1075.

Jenkins, D., R. Dobert, A. Atanassova and C. Pavely. 2021. Impacts of the regulatory environment for gene editing on delivering beneficial products. *In Vitro* Cell. Dev. Biol.-Plant 57(4): 609–626.

Jiang, F. and J.A. Doudna. 2017. CRISPR–Cas9 structures and mechanisms. Annu. Rev. Biophys. 46(1): 505–529.

Jiang, L., H. Chen and C. Li. 2023. Advances in deciphering the interactions between viral proteins of influenza A virus and host cellular proteins. Cell Insight 2(2): 100079.

Kelley, M.L., Ž. Strezoska, K. He, A. Vermeulen and A. van Brabant Smith. 2016. Versatility of chemically synthesized guide RNAs for CRISPR-Cas9 genome editing. J. Biotechnol. 233: 74–83.

Kostyushev, D., S. Brezgin, A. Kostyusheva, D. Zarifyan, I. Goptar and V. Chulanov. 2019. Orthologous CRISPR/Cas9 systems for specific and efficient degradation of covalently closed circular DNA of hepatitis B virus. Cell. Mol. Life Sci. 76(9): 1779–1794.

Kuluev, B., G. Gumerova, E. Mikhaylova, G. Gerashchenkov, N. Rozhnova, Z. Vershinina et al. 2019. Delivery of CRISPR/Cas components into higher plant cells for genome editing. Russ. J. Plant Physiol. 66: 694–706.

Kumar, A., R. Kumar, D. Sengupta, S.N. Das, M.K. Pandey, A. Bohra et al. 2020. Deployment of genetic and genomic tools toward gaining a better understanding of rice-*Xanthomonas oryzae* pv. oryzae interactions for development of durable bacterial blight resistant rice. Front. Plant Sci. 11: 1152.

Kumar, A., T. Anju, S. Kumar, S.S. Chhapekar, S. Sreedharan, S. Singh et al. 2021. Integrating omics and gene editing tools for rapid improvement of traditional food plants for diversified and sustainable food security. Int. J. Mol. Sci. 22(15): 8093.

Kumar, D., A. Yadav, R. Ahmad, U.N. Dwivedi and K. Yadav. 2022. CRISPR-based genome editing for nutrient enrichment in crops: a promising approach toward global food security. Front. Genet. 13: 932859.

Kumar, J., R. Murali-Baskaran, S. Jain, P. Sivalingam, J. Mallikarjuna, V. Kumar et al. 2021. Emerging and re-emerging biotic stresses of agricultural crops in India and novel tools for their better management. Curr. Sci. 121: 26–36.

Li, C., B. Liu, H. Dong and B. Yang. 2024. Enhancing resistance to bacterial blight in rice using CRISPR-based base editing technology. Crop J.

Liu, Z., H. Dong, Y. Cui, L. Cong and D. Zhang. 2020. Application of different types of CRISPR/Cas-based systems in bacteria. Microb. Cell Fact. 19: 1–14.

Mafakheri, M. and M. Kordrostami. 2020. Role of molecular tools and biotechnology in climate-resilient agriculture. Plant Ecophysiol. Adapt. Clim. Change: Mech. Perspect. II: 491–529.

Majeed, A., Z. Muhammad and H. Ahmad. 2018. Plant growth promoting bacteria: role in soil improvement, abiotic and biotic stress management of crops. Plant Cell Rep. 37(12): 1599–1609.

Mann, A., K. Nehra, J. Rana and T. Dahiya. 2021. Antibiotic resistance in agriculture: Perspectives on upcoming strategies to overcome upsurge in resistance. Curr. Res. Microb. Sci. 2: 100030.

Manzoor, S., S.U. Nabi, T.R. Rather, G. Gani, Z.A. Mir, A.W. Wani et al. 2024. Advancing crop disease resistance through genome editing: a promising approach for enhancing agricultural production. Front. Genome Ed. 6: 1399051.

McDonald, B.A. and E.H. Stukenbrock. 2016. Rapid emergence of pathogens in agro-ecosystems: global threats to agricultural sustainability and food security. Philos. Trans. R. Soc. B: Biol. Sci. 371(1709): 20160026.

Miller, S.A., J.P. Ferreira and J.T. LeJeune. 2022. Antimicrobial use and resistance in plant agriculture: a one health perspective. Agriculture 12(2): 289.

Muhammad, M., A. Waheed, A. Wahab, M. Majeed, M. Nazim, Y.-H. Liu et al. 2023. Soil salinity and drought tolerance: An evaluation of plant growth, productivity, microbial diversity, and amelioration strategies. Plant Stress 100319.

Muhammad, M., M.W. Ahmad, A. Basit, S. Ullah, H.I. Mohamed, N. Nisar et al 2024a. Plant growth-promoting rhizobacteria and their applications and role in the management of soilborne diseases. pp. 59–82. *In*: Bacterial Secondary Metabolites. Elsevier.

Muhammad, M., S. Batool, V. Hivare, W.-J. Li, A. Waheed and D. Sinha. 2024b. Bioremediation techniques—classification, principles, advantages, limitations, and prospects. pp. 1–23. *In*: Microbiome-Assisted Bioremediation. Elsevier.

Muhammad, M., A. Wahab, A. Waheed, H.I. Mohamed, K.R. Hakeem, L. Li et al. 2024c. Harnessing bacterial endophytes for environmental resilience and agricultural sustainability. J. Environ. Manage. 368: 122201.

Muha-Ud-Din, G., F. Ali, A. Hameed, S.A.H. Naqvi, M.M. Nizamani, M. Jabran et al. 2023. CRISPR/Cas9-based genome editing: A revolutionary approach for crop improvement and global food security. Physiol. Mol. Plant Pathol. 102191.

Nasar, S., M. Muhammad, E.H. Siddiqi, M. Majeed, N. Ameen, S. Ullah et al. 2024a. OMICS-based knowledge for achieving food and nutritional security. OMICs-based Tech. Glob. Food Secur. 67–90.

Nasar, S., A. Rasool, W. Ikram, S. Shujat, M. Majeed, M. Muhammad et al. 2024b. Nano-osmics: a new approach to plant breeding. Sustain. Agric.: Nanotechnol. Biotechnol. Crop Prod. Prot. 129.

Oerke, E.-C. and H.-W. Dehne. 2004. Safeguarding production—losses in major crops and the role of crop protection. Crop Prot. 23(4): 275–285.

Pandey, P. and S.L. Vavilala. 2024. From gene editing to biofilm busting: CRISPR-CAS9 against antibiotic resistance—a review. Cell Biochem. Biophys. 1–12.

Park, J.S., S.T. Han, K.H. Lee, Y.M. Kim, H.J. Lee and J.Y. Han. 2019. Targeted genome modification mediated by homology-directed repair and non-homologous end joining and its application. J. Anim. Breed. Genet. 3(3).

Patil, A.M., S.G. Wagh, G.B. Janvale, B.D. Pawar and A.A. Daspute. 2024. Viral delivery of CRISPR/Cas9 genome editing for rapid crop improvement: A promising approach to enhance crop resilience against biotic and abiotic stresses. Int. J. Adv. Biochem. Res. 8: 782–796.

Quétier, F. 2016. The CRISPR-Cas9 technology: Closer to the ultimate toolkit for targeted genome editing. Plant Sci. 242: 65–76.

Rao, M.J. and L. Wang. 2021. CRISPR/Cas9 technology for improving agronomic traits and future prospective in agriculture. Planta 254: 1–16.

Rasheed, A., R.A. Gill, M.U. Hassan, A. Mahmood, S. Qari, Q.U. Zaman et al. 2021. A critical review: recent advancements in the use of CRISPR/Cas9 technology to enhance crops and alleviate global food crises. Curr. Issues Mol. Biol. 43(3): 1950–1976.

Rosenstock, T., F. CIAT, S. Ghosh-Jerath, H. Jordaan, O. Daley, W. Isaac et al. 2023. The COVID-19: Food System Frailties and Opportunities. 181.

Rozas, P., E.I. Kessi-Pérez and C. Martínez. 2022. Genetically modified organisms: adapting regulatory frameworks for evolving genome editing technologies. Biol. Res. 55.

Sami, A., Z. Xue, S. Tazein, A. Arshad, Z. He Zhu, Y. Ping Chen et al. 2021. CRISPR–Cas9-based genetic engineering for crop improvement under drought stress. Bioengineered 12(1): 5814–5829.

Sampath, V., N. Rangarajan, S. CH, M. Deori, M. Veeraragavan, B. Ghodake et al. 2023. Advancing crop improvement through CRISPR technology in precision agriculture trends—a review. Int. J. Environ. Clim. Change 13(11): 4683–4694.

Sen, H., A. Kumar and H.S. Janeja. 2024. Biofortification of major crops through conventional and modern biotechnological approaches to fight hidden hunger: an overview. J. Adv. Biol. Biotechnol. 27(7): 96–113.

Serajian, S., E. Ahmadpour, S.M.R. Oliveira, M.d.L. Pereira and S. Heidarzadeh. 2021. CRISPR-Cas technology: emerging applications in clinical microbiology and infectious diseases. Pharmaceuticals 14(11): 1171.

Shahriar, S.A., M.N. Islam, C.N.W. Chun, M.A. Rahim, N.C. Paul, J. Uddain et al. 2021. Control of plant viral diseases by CRISPR/Cas9: resistance mechanisms, strategies and challenges in food crops. Plants 10(7): 1264.

Shalaby, K., M. Aouida and O. El-Agnaf. 2020. Tissue-specific delivery of CRISPR therapeutics: Strategies and mechanisms of non-viral vectors. Int. J. Mol. Sci. 21(19): 7353.

Sharma, S.K., O.P. Gupta, N. Pathaw, D. Sharma, A. Maibam, P. Sharma et al. 2021. CRISPR-Cas-led revolution in diagnosis and management of emerging plant viruses: new avenues toward food and nutritional security. Front. Nutr. 8: 751512.

Singh, T., J. Satyanarayana and R. Peshin. 2014. Crop loss assessment in India-past experiences and future strategies. Integr. Pest Manag.: Pestic. Probl. 3: 227–243.

Song, B., S. Yang, G.-H. Hwang, J. Yu and S. Bae. 2021. Analysis of NHEJ-based DNA repair after CRISPR-mediated DNA cleavage. Int. J. Mol. Sci. 22(12): 6397.

Sundin, G.W., L.F. Castiblanco, X. Yuan, Q. Zeng and C.H. Yang. 2016. Bacterial disease management: challenges, experience, innovation and future prospects: challenges in bacterial molecular plant pathology. Mol. Plant Pathol. 17(9): 1506–1518.

Svircev, A., D. Roach and A. Castle. 2018. Framing the future with bacteriophages in agriculture. Viruses 10(5): 218.

Tan, O., R. Shrestha, M. Cunich and D. Schofield. 2018. Application of next-generation sequencing to improve cancer management: A review of the clinical effectiveness and cost-effectiveness. Clin. Genet. 93(3): 533–544.

Tripathi, L., V.O. Ntui and J.N. Tripathi. 2024. Application of CRISPR/Cas-based gene-editing for developing better banana. Front. Bioeng. Biotechnol. 12: 1395772.

Tyagi, S., R. Kumar, V. Kumar, S.Y. Won and P. Shukla. 2021. Engineering disease resistant plants through CRISPR-Cas9 technology. GM Crops Food 12(1): 125–144.

Varanda, C.M., M.d.R. Felix, M.D. Campos, M. Patanita and P. Materatski. 2021. Plant viruses: from targets to tools for CRISPR. Viruses 13(1): 141.

Velásquez, A.C., C.D.M. Castroverde and S.Y. He. 2018. Plant–pathogen warfare under changing climate conditions. Curr. Biol. 28(10): R619–R634.

Vento, J.M., N. Crook and C.L. Beisel. 2019. Barriers to genome editing with CRISPR in bacteria. J. Ind. Microbiol. Biotechnol. 46(9-10): 1327–1341.

Verma, R.C., D. Amulothu, P. Suhas, M. Nain, D. Kumar, D. Saikanth et al. 2023. A comprehensive overview of plant pathology understanding disease mechanisms and control. Int. J. Plant Soil Sci. 35(20): 1207–1214.

Villalobos-López, M.A., A. Arroyo-Becerra, A. Quintero-Jiménez and G. Iturriaga. 2022. Biotechnological advances to improve abiotic stress tolerance in crops. Int. J. Mol. Sci. 23(19): 12053.

Wahab, A., M. Muhammad, S. Ullah, G. Abdi, G.M. Shah, W. Zaman et al. 2024. Agriculture and environmental management through nanotechnology: Eco-friendly nanomaterial synthesis for soil-plant systems, food safety, and sustainability. Sci. Total Environ. 171862.

Wang, J., P.R. Arantes, M. Ahsan, S. Sinha, G.W. Kyro, F. Maschietto et al. 2023. Twisting and swiveling domain motions in Cas9 to recognize target DNA duplexes, make double-strand breaks, and release cleaved duplexes. Front. Mol. Biosci. 9: 1072733.

Xu, T., Y. Li, J.D. Van Nostrand, Z. He and J. Zhou. 2014. Cas9-based tools for targeted genome editing and transcriptional control. Appl. Environ. Microbiol. 80(5): 1544–1552.

Xue, C. and E.C. Greene. 2021. DNA repair pathway choices in CRISPR-Cas9-mediated genome editing. Trends Genet. 37(7): 639–656.

Yang, H., S. Ren, S. Yu, H. Pan, T. Li, S. Ge et al. 2020. Methods favoring homology-directed repair choice in response to CRISPR/Cas9 induced-double strand breaks. Int. J. Mol. Sci. 21(18): 6461.

Zaidi, S.S.-e.-A., M. Tashkandi, S. Mansoor and M.M. Mahfouz. 2016. Engineering plant immunity: using CRISPR/Cas9 to generate virus resistance. Front. Plant Sci. 7: 1673.

Zhang, D., A. Hussain, H. Manghwar, K. Xie, S. Xie, S. Zhao et al. 2020. Genome editing with the CRISPR-Cas system: an art, ethics and global regulatory perspective. Plant Biotechnol. J. 18(8): 1651–1669.

Zhang, Q., M. Rho, H. Tang, T.G. Doak and Y. Ye. 2013. CRISPR-Cas systems target a diverse collection of invasive mobile genetic elements in human microbiomes. Genome Biol. 14: 1–15.

Chapter 9

Defence Mechanisms Developed by CRISPR/Cas in Plants Against Abiotic and Biotic Stresses

Laiba Chaudhry,[1] *Pawan Kumar Verma,*[2] *Ravindra Solanki,*[3] *Yaqoob Sultan,*[4,5,*] *Murad Muhammad*[6] and *Heba I. Mohamed*[7]

1. Introduction

For humans and animals to survive and flourish, plants must provide food, fuel, and feed. By 2050, the global population is projected to exceed 9.5 billion, necessitating a 100–110% increase in food consumption. However, current agricultural systems indicate that yields of staple crops like rice, soybeans, corn, and wheat will only grow marginally, by an estimated 38–67% (Ray et al. 2013). During growth, plants face numerous challenges, both biotic, such as bacteria, viruses, fungi, and insects, and abiotic stresses (Borrelli et al. 2018; Rodriguez-Leal et al. 2017; Shaibu et al. 2020; Landa et al. 2022; Dong et al. 2019). Biotic stress alone is anticipated to cause over 40% yield losses, leading to a 15% reduction in the global food supply (Chen et al. 2019; Mushtaq et al. 2019; Tyagi et al. 2018). Abiotic stressors, exacerbated by climate change and human activities, further threaten crop production. Factors such as extreme temperatures, salinity, drought, waterlogging, and heavy metal contamination reduce global agricultural output by nearly 50% (Dracatos et al. 2018; Manghwar et al. 2019; Liu et al. 2022b). To address these challenges, pesticides, fertilizers, and other chemicals have been widely employed to enhance productivity, support plant health, and combat diseases. However, their extensive use has led to environmental concerns, including damage to soil, water, and plant ecosystems. Furthermore, they have an indirect impact on non-target plants,

[1] University of Sargodha, Pakistan.
[2] Department of Agronomy, Institute of Agricultural Sciences, Banaras Hindu University, Varanasi, India-221005.
[3] Department of Agricultural Biotechnology, B. A. College of Agriculture, Anand Agricultural University, Anand, Gujarat, India-388110.
[4] Department of Grass Breeding, Institute of Agriculture, Lithuanian Research Centre for Agriculture and Forestry, 58344 Kedainiu r., Lithuania.
[5] Floricultural Biotechnology Lab., Department of Horticulture, Graduate School, Kyungpook National University, 41566, Daegu, South Korea.
[6] State Key Laboratory of Desert and Oasis Ecology, Key Laboratory of Ecological Safety and Sustainable Development in Arid Lands, Xinjiang Institute of Ecology and Geography, Chinese Academy of Sciences, Urumqi, 830011, PR China.
[7] Biological and Geological Sciences Department, Faculty of Education, Ain Shams University, Cairo, Egypt.
* Corresponding author: yaqoob.sultan@lammc.lt

DOI: 10.1201/9781003518204-9

beneficial insects, fish, birds, and other animals (Tyagi et al. 2018; Tyagi et al. 2021; Vannier et al. 2019; Yin and Qiu 2019; Ahmad et al. 2020; Kaiser et al. 2020). To overcome these challenges and satisfy the expanding demands of humankind, it will be essential to be able to breed new species that are resistant to different pressures to minimize yield loss. The intricate interactions between pathogens and host plants in various biotic stress types depend on how susceptible or resistant crop plants are to various diseases. The most recent developments in molecular tools have shed light on a variety of pathogen infection mechanisms and how they interact with certain crop plants. Traditional plant breeding methods, physical (like gamma radiation), chemical (like ethyl methane sulfonate, EMS), and biological (like T-DNA and transposon insertion) methods have all been used to alter plant genomes, causing point mutations, deletions, and gene duplications. Even while these methods have improved crops, they are costly, time-consuming, and frequently encounter issues with traditional breeding. Additionally, they may result in unexpected changes to the genome. Enhancing the production of disease-free, abiotic stress-tolerant, high-yielding crops is therefore essential to prepare them for future challenges. The scientific community has long been dedicated to developing optimal crops in recognition of these difficulties. By using a trans-kingdom strategy to transfer advantageous genes to crops, Genetically Modified (GM) crops have been produced. Scientists have increasingly focused on developing Genome-Editing (GE) tools capable of modifying the genome without introducing transgenes (Tilman et al. 2011; Foley et al. 2011; Zhang et al. 2018c; Prado et al. 2014). Compared to methods such as crossbreeding, mutation breeding, and conventional transgenic breeding, GE offers a faster, more efficient, and precise approach to crop improvement. It enables the rapid enhancement of specific traits by directly altering target genes, regulatory sequences, or DNA/RNA bases in elite crop varieties.

The field of genome editing began with the development of Zinc Finger Nucleases (ZFNs) and Transcription Activator-Like Effector Nucleases (TALENs), but the introduction of the CRISPR/CRISPR-associated protein 9 (Cas9) system (CRISPR/Cas9) marked a significant breakthrough (Manghwar et al. 2019; Manghwar et al. 2020; Zhang et al. 2020a; Khalil 2020). This technology works by recognizing and cutting specific DNA sequences within the genome, a concept inspired by a bacterial immune mechanism that defends against viruses. The adaptive defense system of bacteria relies on CAS enzymes associated with CRISPR/Cas9 (Wang et al. 2022). CRISPR/Cas9 has not only advanced genome editing but also opened the door for the commercialization of edited crops and food products, helping to address public concerns about the presence of "foreign DNA." While countries like Argentina have embraced genome-edited crops, debates continue in other regions. Notably, Japan recently commercialized a CRISPR/Cas9-edited tomato with enhanced levels of gamma-aminobutyric acid (GABA), demonstrating the potential of this technology in agricultural innovation (Waltz 2022). This technology is anticipated to significantly facilitate the widespread acceptance of genome-edited crops in many countries. The CRISPR/Cas9 system has been utilized in various plant species, including model plants like *Nicotiana benthamiana* (Li et al. 2013), *Nicotiana tabacum* (Donovan et al. 2020; Kumar et al. 2022), and *Arabidopsis* (Li et al. 2013; Nekrasov et al. 2013). It has also been applied to important crop plants such as wheat (Shan et al. 2013), maize (Liang et al. 2014), rice (Miao et al. 2013; Shan et al. 2013), liverwort (Sugano et al. 2014), tomato (Brooks et al. 2014), potato (Wang et al. 2015), soybean (Jacobs et al. 2015), sweet orange (Jia and Nian 2014), banana (Tripathi et al. 2019a), pepper (Park et al. 2021), barley (Zeng et al. 2020; Garcia-Gimenez and Jobling 2021), peanut (Yuan et al. 2019; Han et al. 2023), foxtail millet (Lin et al. 2018), and sugarcane (Oz et al. 2021). Furthermore, CRISPR/Cas9 has been successfully employed for multiplex genome editing, targeting multiple genes within a single organism across crops like wheat (Wang et al. 2018b), rice (Miao et al. 2013), cotton (Gao et al. 2017), and maize (Char et al. 2017). This capability holds great promise for developing genome-edited crops that are resistant to multiple stresses by simultaneously modifying several stress-sensitive genes in high-yielding but stress-vulnerable cultivars. Additionally, CRISPR-mediated gene activation can be used to overexpress stress-tolerance genes (Zafar et al. 2020). Genes involved in stress-related regulatory networks, signal pathways, and metabolite biosynthesis can be targeted through CRISPR/Cas9 technology to create stress-tolerant crop varieties (Jain 2015).

Genome Editing (GE) has been applied to over 50 plant species, marking a significant leap forward in plant breeding innovation (Bisht et al. 2019; Gao 2021; Fauser et al. 2014; Schiml et al. 2014; Zhao et al. 2016; Ron et al. 2014; Ito et al. 2015; Pan et al. 2016; Shan et al. 2014; Liang et al. 2014; Zhu et al. 2016; Feng et al. 2016; Li et al. 2017a; Dong et al. 2018; Lawrenson et al. 2015). The introduction of the CRISPR/Cas9 system in 2013 transformed plant genetic engineering, breaking down barriers to precise genome editing (Schaeffer et al. 2016). This breakthrough has propelled genome editing beyond the capabilities of traditional breeding methods, driving substantial advancements in plant science (Chandrasekaran et al. 2016). Among various editing technologies, CRISPR/Cas9 has emerged as the most widely adopted due to its exceptional potential to address critical agricultural challenges, such as improving yields and enhancing resistance to biotic and abiotic stresses (Manghwar et al. 2020; Bortesi and Fischer 2015; Ma et al. 2016; Wang et al. 2020b). The type II CRISPR/Cas9 system, initially identified in Streptococcus pyogenes, has become the preferred choice for genome editing in plants, owing to its efficiency and versatility (Jinek et al. 2012; Bortesi and Fischer 2015; Pallarès Masmitjà et al. 2019). The emergence and adoption of genome-engineering technologies, particularly CRISPR/Cas9, have revolutionized modern plant development, cementing its status as one of the most powerful tools for gene editing (Zhang et al. 2014).

This chapter focuses on the advancements in plant genome editing using CRISPR/Cas systems to address various environmental stresses. As climate change exacerbates abiotic and biotic challenges like drought, salinity, extreme temperatures, and pathogen infestations, enhancing plant resilience has become a critical goal in agricultural research. The discussion highlights the molecular mechanisms underlying plant stress responses and how CRISPR/Cas technologies are being effectively employed to strengthen these responses. Through precise gene modifications, CRISPR/Cas facilitates the development of traits that enable plants to endure diverse environmental stresses, contributing to sustainable agricultural productivity.

2. Environmental Stresses and Genome Editing in Plants

2.1 Environmental Stresses Concerning Genome Editing in Plants

The global population is growing at an alarming rate, which is estimated to reach around 10 billion by 2050 (Clarke and Zhang 2013; Rasheed et al. 2021), thereby posing a great challenge to global food and nutrition security, which massively interferes with the agricultural system (Chaudhary and Sidhu 2022). Hence, there is a need to accelerate food production by 25–70% more than that of the present status (Hunter et al. 2017). Understanding the present condition, there is a need to tackle this challenge for sustainable agricultural development (El-Mounadi et al. 2020). To cope with this, there has been the need to use many techniques along the timeline, ranging from molecular markers, molecular breeding, and conventional breeding approaches to modern-day biotechnological tools such as genetic engineering and genome editing. Using a genetic engineering approach, it uses transgenes having certain benefit traits, but the biggest demerit in this approach is the risk of environmental concerns regarding legal restrictions and regulations of Genetically Modified Organisms, which interfere in its normal consumer acceptance (GMOs) (Herman et al. 2019; Tripathi et al. 2022, 2023). To overcome this, scientists and researchers discovered a new robust, simple, efficient, cost-effective, highly-specific method, which does not require any requirement of the transgene, genome editing that involves precise sequence-specific degradation of DNA which involves four types namely: Mega Nucleases (MN's), Zinc-Finger Nucleases (ZFN's), Transcription-Activator Like Effector Nucleases (TALENS) and Clustered Regularly Interspaced Short Palindromic Repeats (CRISPR) (Gao 2021; Gaj et al. 2013; Khalil 2020; Manghwar et al. 2019, 2020; Puchta and Fauser 2014; Zhang et al. 2020c) and all these techniques work on the principle of Site-Directed Nucleases (SDN, which is categorized into SDN1, SDN2 and SDN3)

(Abdallah et al. 2021; Ahmad et al. 2021; Matres et al. 2021). Amongst these, CRISPR is the most actively used technique isolated from the bacterium *Streptococcus pyogenes* (Bortesi and Fischer 2015; Jinek et al. 2012; Pallarès Masmitjà et al. 2019; Wright et al. 2016), along with elaborated categories, the most common to date is CRISPR Cas9 (CRISPR-associated protein 9). This protein uses a DNA-RNA interaction (Garneau et al. 2010) to disrupt the DNA of a specific sequence using a two-component system: a chimeric single guide RNA (a small stretch of 20 nucleotides) (Cong et al. 2013; Jiang and Doudna 2017; Zetsche et al. 2015; Zhong et al. 2024) formed by CRISPR RNA (crRNA) and trans-activator RNA (tracrRNA) (Anders et al. 2014; Anzalone et al. 2020; Jiang and Doudna 2017) and a Cas 9 enzyme (Erdoğan et al. 2023) consisting of an entire ribonucleoprotein complex (RNP). There is yet another important sequence, PAM (Protospacer Adjacent Motif), which is a generally NGG sequence located just adjacent to the cleavage site that ensures efficient cleavage of target-specific DNA hence proving to be a precise means of genetic modifications (Baltes et al. 2014; Endo et al. 2016; Fauser et al. 2014; Li et al. 2016a; Negi et al. 2022; Sun et al. 2016a; Sun et al. 2016b). After cleavage of DNA, they form blunt ends of Double-Stranded Breaks (DSBs), which are naturally or artificially repaired by two repair pathways (Hiom 2010) either Homology Directed Repair (HDR) or Non-Homologous End Joining (NHEJ) (Choudry et al. 2024; Feng et al. 2013; Gupta et al. 2024; Liu et al. 2018; Voytas and Gao 2014; Zhang and Showalter 2020c) without the requirement of a template, which causes small nucleotide InDels (insertions-deletions) resulting in mutations. The process of CRISPR-Cas9 relies upon the fact that prokaryotes restore a memory of viral infections by integrating a small piece of DNA or plasmid DNA, which is often referred to as a Protospacer, into its CRISPR loci, which is known as the "Adaptation" or "Acquisition" phase. Further, this protospacer is then transcribed and processed to form a mature CRISPR RNA (crRNA) under the "Expression and Maturation" phase. Finally, this crRNA is ready to degrade/disrupt the viral or plasmid DNA upon the second encounter by binding to its complementary sequence, which is known "Interference" phase (Mohammed et al. 2024; Negi et al. 2022; Rajput et al. 2021). At present, it is the most widely accepted system that has revolutionized the field of agriculture as it does not require the integration of transgenes, so it is a non-transgenic method of crop improvement, a term referred to as transgene-free genome editing, which provides a great advantage as it bypasses the regulatory concerns applicable for genetically modified organisms and has no negative impact from environmental concerns (Agapito-Tenfen et al. 2018; Erdoğan et al. 2023; Gu et al. 2021; Jenkins et al. 2021; Kumar et al. 2023). Also, it is feasible to use a single Cas9 to edit multiple traits of plants, known as multiplexing of genome editing (Gu et al. 2021; Kumar et al. 2023), which has been taken into account regarding the successful implementation of genome editing in agriculture (Bortesi et al. 2016; Najera et al. 2019; Zhang et al. 2018b; Zhu et al. 2017).

On the contrary, there are many stress conditions in which plant growth and development are severely hampered, which are broadly classified into biotic and abiotic stress, which are subcategorized into subgroups. For elucidation of their genetic basis, the complex molecular response of both biotic and abiotic stress involves numerous networks of genes (Abdelrahman et al. 2018) that are detailed below.

2.1.1 Abiotic Stresses (Drought, Salinity, Heat, Cold, Heavy Metal Toxicity)

It includes environmental factors responsible for the overall decrease in crop productivity (Boscaiu and Fita 2020; Khan et al. 2015), which include drought (water stress), salinity (salt stress), heat and cold stress (temperature stress), and heavy metals (He et al. 2018; Sharma et al. 2020; Zafar et al. 2020), which affect the overall metabolism of plants (Abobatta 2020; Khan et al. 2010; Zulfiqar 2021). With the ongoing climate change, it has become more challenging to minimize the losses, so CRISPR proves worthy in the production of abiotic-stress resilient crops (Biswas et al. 2021; Kaur et al. 2022; Pramanik et al. 2021).

Drought and salinity stress are two main abiotic stresses that negatively impact plant growth and productivity (Billah et al. 2021; Ma et al. 2020; Singh et al. 2018). The study of any abiotic stress involves multiple genes, signals, and transduction molecules is an elaborate and complex pathway involving multiple cascades of genes (Dong et al. 2022). In plants, various Transcription Factors (TF's) of different families, e.g., *NAC*, *DREB*, *bZIP*, *MYB*, *TCP*, and *WRKY*, have been identified in imparting abiotic-stress tolerance (Erpen et al. 2018). Several studies in different crops reveal that CRISPR enhances abiotic stress tolerance; for instance, the first example of improving drought tolerance using CRISPR/Cas9 was reported in maize, where *AGROS8* (AUXIN REGULATED GENE INVOLVED IN ORGAN SIZE 8) was targeted for improved drought tolerance (Ghosh and Dey 2022). In crops like rice (*Oryza sativa*), a transcription factor (TF) *OsRR22* is involved in salinity stress; when knocked out, the crop's performance was enhanced (Zhang et al. 2019a). Another gene, *stress/ABA–activated protein kinase 2* (*SAPK2*) gene, which is a primary mediator in abscisic acid (ABA) signaling, was knocked out, which resulted in decreased performance under drought and oxidative stress (Lou et al. 2017). Moreover, the negative regulators for drought and salinity tolerance, such as SINA protein 1 (*OsDIS1*), drought and salt-tolerant protein 1 (*OsDST*), and ring finger protein 1 (*OsSRFP1*), whose expression was deliberately minimized for drought resistance (Santosh Kumar et al. 2020). Other examples of edited genes in rice include phytoene desaturase (*OsPDS*), mitogen-activated protein kinase (*OsMPK2*), and betaine aldehyde dehydrogenase (*OsBADH2*), the consequence of which is an improved abiotic stress tolerance (Ashokkumar et al. 2020; Kaur et al. 2018; Kumar et al. 2023; Shan et al. 2013). Ogata et al. (2020) also reported that CRISPR/Cas9-mediated mutagenesis of *OsERA1* resulted in a higher extent of drought stress tolerance in rice. Also, in crops like tomato (*Solanum lycopersicum* L.), upregulation and downregulation of many genes play a key role in drought resistance or tolerance, like overexpression of *AREB1* is responsible for drought resistance and vice versa (Singh and Laxmi 2015). In another study, it was reported that the knockout of *SlLBD40* (lateral organ boundaries domain transcription factor) and *SlMAPK3* aided in drought tolerance (Liu et al. 2020a, b). Similarly, when *Auxin Responsive Factor* (*SlARF4*) in tomato was knocked out using CRISPR-Cas, its performance improved under drought conditions (Chen et al. 2021).

Salinity stress is encountered in plants due to fluctuation in the osmoregulation of ions, adversely affecting plant growth and development (Dawood et al. 2022). In crops like *Arabidopsis*, when *AtWRKY3* and *AtWRKY4* were knocked out resulted in increased ion escape and decreased antioxidant actions, creating saline sensitivity (Chen et al. 2021), while acquired genes for osmotolerance resulted in salinity tolerance in plants (Kim et al. 2021). In rice loss of *Slender Rice 1* (*SLR1*) functioned to promote mesocotyl and root growth, specifically under dark and salt stress conditions (Mo et al. 2020). Additionally, the mutants of the target gene PARAQUAT TOLERANCE 3 (*OsPQT3*) developed using CRISPR Cas9 resulted in enhanced salinity tolerance in rice (Alfatih et al. 2020). Many other studies in rice indicate that mutants developed from CRISPR that resist salinity stress, such as *OsNAC45* (Zhang et al. 2020b), *AGO2* (ARGONAUTE2) (Yin et al. 2020), and *OsBBS1* (bilateral blade senescence) (Zeng et al. 2018). In tomato, the deletion of negative response domains (s) like *SlHyPRP1* increased the salinity tolerance at both vegetative and germination stages (Tran et al. 2021), and *SlARF4,* which effectively improves the crop's ability to resist the salinity stress (Bouzroud et al. 2020).

Heat stress negatively impacts the entire life cycle of crops, in specific stages like early establishment, flowering, and gamete formation (Jagadish et al. 2021). These kinds of situations are countered by endogenous plant molecular mechanisms by a group of genes known as heat shock proteins or heat stress genes (Awasthi et al. 2015). Like in other cases, CRISPR/Cas9 proves to be a useful technology here also (Duan et al. 2016). In tomato, when the wild type was compared with *SlMAPK3* mutant generated by CRISPR/Cas9 maintained ROS homeostasis by regulation of antioxidant enzyme expression and *HSPs/HSFs* genes, thereby increasing heat stress tolerance (Yu et al. 2019). Also, in the apoplast region of the tomato, the production of Reactive Oxygen Species (ROS) by positive regulator *BRZ1* which also contributes to heat stress tolerance, and this confirmation was done when CRISPR/Cas9 knockout *BRZ1* mutants and reduced ROS production and heat tolerance were reported as measured by Respiratory Burst Oxidase Homolog 1 (*RBOH1*) (Yin et al. 2018). Another example of heat stress: when wild type was compared with CRISPR-Cas9 knockout mutants, heat-stress-sensitive albino 1 (*HSA1*) exhibited higher heat stress sensitivity (Qiu et al. 2018). In rice, triple knockout rice was edited using CRISPR/Cas9 using *pyl1/4/6* mutants, which exhibited higher yield, enhanced temperature tolerance, and less germination before harvest compared to the wild variety (Miao et al. 2018). Furthermore, other proteins and transcription factors, such as protein kinase *SAPK6* and *OsbZIP46CA1,* are also responsible for enhanced response to heat stress (Chang et al. 2017). In other crops like maize, CRISPR/Cas9 mutants of the thermosensitive genic male sterile 5 (*TMS5*) gene and, as a result, enhanced thermos-sensitivity were achieved (Li et al. 2017b, c).

Cold stress interferes with plant metabolic activity, resulting in injury to the plant tissues (Jha and Mohamed 2023). It is due to chilling (< 20°C) and freezing (< 0°C), which directly influences osmotic stress, oxidative stress, and others (Li et al. 2022; Yadav et al. 2023). This stress was combated by CRISPR/Cas9 for improvement in cold resistance, where negative regulators are directly eliminated; e.g., a transcription factor (TF), *OsMYB30,* when bound to the promoter of the β-amylase gene, had a detrimental effect on cold tolerance (Mohamed et al. 2024). In rice, when three genes, *OsPIN5b*, *GS3*, and *OsMYB30,* by CRISPR underwent simultaneous mutations by CRISPR, they exhibited an increase in yield and cold stress resistance (Zeng et al. 2020) which has also been accomplished by CRISPR/Cas-mediated deletion of *OsPRP1* that encodes for a proline-rich protein (Shakiba et al. 2017). Moreover, cold stress resistance in rice was also achieved by the CRISPR/Cas technique when the alteration of genes related to G-complex, viz., *OsRGA1*, *OsGS3*, *OsDEP1*, and *OsPXLG4* (Cui et al. 2020). In other crops like tomato, it has been reported that C-repeat Binding Factor 1 (CBF1) was knocked out using CRISPR/Cas9, and the *CBF1* mutants remained protected from cold stress and decreased electrolyte leakage (Li et al. 2018).

Heavy metals seriously interfere with plant growth and development (Alengebawy et al. 2021), which includes metals like Cs, As, Cd, Pb, etc. (Hamdam et al. 2022; Yadav et al. 2023). This is because they disturb the overall homeostasis of plants (Hoque et al. 2021), and the accumulation of these metals poses a great threat to human health (Kaur et al. 2021). In crops like rice, the CRISPR/Cas9 knockdown of *OsNramp5* and *OsLCT1* resulted in a significant reduction of cadmium (Cd) accumulation (Songmei et al. 2019). Also, the CRISPR/Cas9 knockout of *OsNRAMP1* resulted in reduced levels of cadmium (Cd) and lead (Pb) in rice grains (Chang et al. 2020; Chu et al. 2022). Another metal, arsenic (As) whose absorption and transport is inhibited when eliminated by CRISPR/Cas9 after the target R2R3 MYB transcription factor *OsARM1* was knocked out (Wang et al. 2017a). Likewise, CRISPR/Cas9 was also used to inactivate K^+ transporter *OsHAK1* to reduce the cesium (Cs) levels in rice (Nieves-Cordones et al. 2017).

2.1.2 Biotic Stresses (Pathogens, Pests, Fungi, Bacteria, Viruses)

It includes factors such as bacteria, fungi, viruses, and other pathogens which are responsible for significant economic losses that depend on the extent of attack (Das et al. 2023; Li et al. 2022; Wang et al. 2022). In response, there are elucidated defense mechanisms involving Resistance (R) or Susceptibility (S) genes against pathogens (Erdoğan et al. 2023). So, to improve disease resistance, the CRISPR/Cas system has been used extensively for a broad spectrum of pathogens (Arora and Narula 2017; Bisht et al. 2019).

For fungal resistance, many crops have improved the traits of fungal resistance by knocking out negative regulators that hinder the resistance or tolerance process by using the CRISPR/Cas system (Borrelli et al. 2018). In wheat (*Triticum aestivum* L.), all three alleles of TaMLO were knocked out using CRISPR-Cas for enhanced resistance for powdery mildew (Wang et al. 2014). Another S gene to powdery mildew, *Edr1* (*Enhanced disease resistance 1*), which encodes a Raf-like mitogen-activated protein, when targeted via CRISPR-Cas9, showed a significant reduction of powdery mildew in wheat (Zhang et al. 2017). In crops like rice, rice blast is a major problem but was countered by CRISPR-Cas targeting a key factor, *OsERF922,* which is a negative regulator for blast resistance (Zaidi et al. 2020) and *OsSEC3A* (Ma et al. 2018). Moreover, in crops like tomato (*Solanum lycopersicum* L.), many studies regarding CRISPR-Cas targeting of important genes have been done; for instance, "*Tomelo*" is a tomato variety (targeted gene is *SlMLO1*) resistant to powdery mildew (Nekrasov et al. 2017). Other than that, mutant lines in tomato, when *reduced mycorrhizal colonization* (*rmc*) that consist of chromosomal deletion affecting five genes that resulted in susceptibility to Fusarium wilt, and this was countered by knockout of one of the five genes, *Solyc08g075770*, which enhances tolerance to fusarium wilt (Prihatna et al. 2018).

Bacterial pathogens differ widely and multiply much more rapidly, so it becomes a great challenge to check them. Now, to confer bacterial resistance to different crops, one of the important sugar transporter genes in rice has been extensively studied: *SWEET* genes, *OsSWEET13*, a Susceptibility (S) gene that codes for a sucrose transporter that has a key role in plant-pathogen interaction. Here, *Xanthomonas oryzae* (*X. oryzae*) produces an effector protein, *PthXo2*, which in response triggers expression of *OsSWEET13* in the plant host, making the plant susceptible, which is rectified by knockout of *OsSWEET13* promoter (Zhou et al. 2015). This makes CRISPR a successful example of the introduction of bacterial blight resistance. The rice SWEET clade III family consists of the genes *OsSWEET11*, *OsSWEET13*, and *OsSWEET14*, which encode sugar transporters that mediate glucose and sucrose export (Zaidi et al. 2020). Similarly, when *OsNRAMP1* mutants generated from CRISPR-Cas resulted in the production of Reactive Oxygen Species (ROS) content and decreased antioxidant enzymatic activity, they showed broad-spectrum resistance against bacteria and fungi (Chu et al. 2022). For tomatoes, CRISPR/Cas9-induced mutations of the susceptible gene, *SlDMR6-1,* resulted in resistance to a wide variety of pathogens, including bacteria, fungi, and oomycetes (Thomazella et al. 2021). The downy mildew in tomatoes is caused by a negative regulator, *downy mildew resistance 6* (*DMR6*), by CRISPR-Cas-mediated knockout of *SlDMR6-1* (tomato ortholog) to increase resistance against *Phytophthora capsici* without having any adverse effect on the plant (Langner et al. 2018; Zeilmaker et al. 2015). Additionally, when the *Jasmonate ZIM-domain-2* ortholog (*SlJAZ2*) gene was mutated by CRISPR-Cas technology in tomato, it resulted in bacterial speck disease (Ortigosa et al. 2019).

Viruses are obligatory parasites that hamper crop productivity at considerable levels (Das et al. 2023; Shahriar et al. 2021), and their replication and multiplication depend on plant machinery. So, plant virus resistance can be achieved either by host susceptibility modification or by targeting pathogenicity-related genes (Tyagi et al. 2021). Using the CRISPR-Cas approach, plant resistance can be conferred directly by targeting DNA or RNA viruses, splitting their genomes (Mushtaq et al. 2020; Rath et al. 2015). The nature of viral genomes is categorized into six major groups: single-stranded DNA (ssDNA), double-stranded DNA (dsDNA) viruses with no plant viruses in this group, double-stranded RNA (dsRNA), reverse-transcribing viruses, positive-sense single-stranded RNA (ssRNA+) viruses, and negative-sense single-stranded RNA (ssRNA) (Roossinck et al. 2015). Amongst them, most plant viruses consist of RNA genomes that cannot be directly targeted by CRISPR-Cas as they require a double-stranded template (preferentially double-stranded DNA) for their action (Das et al. 2023). But the major focus for virus resistance in the many studies in CRISPR-Cas-edited plants was the ssDNA geminivirus genome (Baltes et al. 2015). From an economic perspective, the most significant genus of geminiviruses is begomovirus, having a bipartite genome that is mainly associated with the phloem of infected plants and spreads to dicots via the sweet potato, tobacco, and silverleaf whitefly (*Bemisia tabaci*) (Gilbertson et al. 2015). Tremendous work has been done in the case of crops like tomatoes, tobacco, and many more. For instance, in tomato, *SlDCL2b,* when edited by CRISPR/Cas9, gave the highest expression in four *DCL2* subfamilies (consists of *SlDCL2a-d*) and as a result, resistance to tomato mosaic virus with the simultaneous editing of *SlDCL2a* and *SlDCL2b* resulted in enhanced Potato Virus X (PVX) and Tobacco Mosaic Virus (TMV) resistance in tomato (Wang et al. 2018a, c). Also, the CRISPR-Cas9 technology has been successful in imparting viral resistance in some other crops, such as turnip mosaic virus in *Arabidopsis* (Pyott et al. 2016), Rice Tungro Spherical Virus (RTSV) (Macovei et al. 2018), and banana streak virus (eBSV) (Tripathi et al. 2019b).

For insect resistance, it is important to study the genome at the nucleotide and protein level and thereby enhance insect resistance using the CRISPR/Cas9 technique. When *Arabidopsis* was knocked in with the potato protease inhibitor II (*pinII*) gene, it provided resistance against pathogen attack by Lepidoptera (Ghosh and Dey 2022). When *Slabd-A*-specific single guide RNA (sgRNA) and Cas9 mRNA were injected into the lepidopteran pest *Spodoptera litura,* it resulted in an abnormal larval developmental stage (Bi et al. 2016). Also, resistance provided in transgenic peanuts against *Spodoptera litura* under the promoter *PR1* (a prominent promoter for transgenics that is considered for aphid resistance) (War et al. 2012) and expressing transgene *Cry1Ac*, adapted from *Bacillus thuringiensis* (Yu et al. 2022). Another *cry* protein, *Cry1Ab* plays a key role in resistance for broccoli under PR-1a promoter against the insect *Plutella xylostella* (Marium et al. 2025).

2.2 Plant Defense Mechanisms Against Environmental Stresses and the Role of Genome Editing

Numerous environmental cues that plants encounter result in significant reductions in crop productivity. Plants struggle to reach their full genetic potential for growth and reproduction because of changing environmental conditions. The frequent attacks on plants by microbial diseases and herbivores are one example of this environmental situation. Plants have evolved a wide variety of intricate defence mechanisms to fend off such attacks. There are two types of defence mechanisms: inducible, which activates defence upon detecting an attack, and preformed, which stores poisonous secondary metabolites. Plants detect biotic stressors, trigger transcriptional or regulatory processes, and ultimately produce the proper reaction. The interaction and influence of many signals to produce defence responses against biotic stress are still unclear, even though plant defence against pathogen attack is well understood. One such area to understand is how light and dark signals affect the biotic stress response. Changes in light and dark not only control defense systems that affect plant biochemistry and development, but they also confer resistance to invasive infections. A signaling cascade is triggered by the interplay between plant defense and the dark/light environment. The perception of biotic stress, the dark/light environment, and the production of a suitable physiological or biochemical response are all connected by this signaling cascade (Iqbal et al. 2021; Nejat and Mantri 2017; Saijo and Loo 2020; Zhang et al. 2017). A summary of crop species that have been conferred with biotic stress resistance is presented in Table 2, adapted from the studies by Wang et al. (2022) and Erdoğan et al. (2023). The modification of *S* genes of host or pathogen has made it possible to acquire disease resistance with CRISPR-Cas9 playing a crucial role in it (Karmakar et al. 2022), which is accomplished by silencing the host component to survive a pathogen attack (Hussain et al. 2024; Tripathi et al. 2022).

Abiotic stress arises in plants due to individual or collective factors of different subtypes of these occurring in the plant and is detrimental to plant growth and development (Angon et al. 2023). The detailed study of all the regulatory elements, key genes, and Transcription Factors (TFs) provides us the insights into where the action of improvement in the pathway is needed to endure abiotic stress resistance (Khokhar Voytas et al. 2023). CRISPR-Cas knocks out the negative regulator, promoter, or any other sequence that hinders the resistance of the plants by up-regulating and down-regulating certain genes or their associated factors, thereby tackling the adversaries of the plant under stressed conditions. In this manner, either from a specific or broader perspective of any kind of stress, CRISPR always serves as a potent, useful tool in achieving resistance and tolerance of both abiotic and biotic stresses (Gupta et al. 2024).

3. CRISPR/Cas System and Its Mechanism in Plant Stresses

3.1 CRISPR/Cas System and its Mechanism in Plants Stresses (Including a Detailed Figure Reflecting the Whole Pathway/Mechanism)

CRISPR-Cas systems comprise of highly sophisticated, diverse, and adaptive immune systems in microbes as a defense mechanism from invading viruses or plasmids in many archaea (90%) and bacteria (40%) (Gajardo et al. 2023). This plays an important role in the recognition of incoming 'foreign' DNA from 'self' DNA and is used as an alternative to transgene-free genome editing (Manghwar et al. 2019). There is a specific loci/locus for CRISPR present on the host genome consisting of a leader sequence, *cas* genes, and an array of protospacer sequences present so that a memory for the viral DNA is sustained for the attack by the virus in the next infection.

CRISPR-Cas9 requires a complete genome sequence for sequence-specific degradation of a DNA sequence, which induces Double-Stranded Breaks (DSBs). When DSBs are introduced, two DNA repair pathways are stimulated: Non-Homologous EndJoining (NHEJ), which is an error-prone pathway for inducing mutation for gene disruption, and the Homology-Directed Repair (HDR) pathway, which induces precise replacement for mutations. These detailed pathways, therefore, alter the final product/genotype/phenotype of the crop (Yang et al. 2020). For achieving this process, there are different obligate components of CRISPR-Cas, namely, single guide RNA (sgRNA), Protospacer Adjacent Motif (PAM) site, CRISPR RNA (crRNA), and trans activating RNA (tracrRNA). Single guide RNA (sgRNA) is a synthetic oligonucleotide of about 20 bp that binds to its complementary DNA sequence. This guide RNA consists of crRNA and tracrRNA that bind to the DNA for the degradation process (Erdoğan et al. 2023). Another very important component is the Protospacer Adjacent Motif (PAM) site, which is an NGG sequence (N is any of the 4 nucleotides, i.e., A, T, G, and C) located just adjacent to the site where sgRNA binds to its target sequence so it is suggested that the PAM sequence is essential for binding of guide RNA for gene disruption (Chennakesavulu et al. 2021). The entire procedure of CRISPR-Cas9 is very economical, more accurate, and less time-consuming when compared to the other genome editing technologies (Rath et al. 2015; Jinak et al. 2012; Gasiunas et al. 2012). In this manner, CRISPR-Cas9 is used for regulating gene expression, the study of which has been done in stress conditions of both biotic and abiotic points of concern, in crops like rice (Le et al. 2022). A list of crops that have been conferred with abiotic stress resistance is presented in Table 1 (adapted from Kumar et al. 2023; Wang et al. 2022).

The complete mechanism of CRISPR-Cas9 is explained in three stages:

- **Acquisition/Adaptation phase:** Protospacer (a unique sequence of viral DNA) gets integrated into the host genome, which acts as a memory for encountering the same viral DNA infection next time, which plays an important role in the degradation of invading viral DNA (Mohamed et al. 2024).
- **Expression or Maturation phase:** Following the integration of invading viral DNA in CRISPR loci, the CRISPR locus is then transcribed into pre-CRISPR RNA (pre-crRNA). The transcribed pre-crRNAs are then processed to yield mature crRNAs (Mohamed et al. 2024).
- **Interference phase:** Here, the *Cas* proteins form a ribonucleoprotein (RNP) complex with crRNA in which crRNA forms base pairing with the invading protospacer DNA. At the end , the invading DNA is degraded by crRNA-directed cleavage by complementary base pairing between the protospacer and crRNA spacer sequence (Mohamed et al. 2024).

Under stress conditions, the pathway that operates is different from those that operate under normal conditions, so it becomes important to employ modern-day cutting-edge technologies such as CRISPR-Cas9 to enhance stress (biotic and abiotic) tolerance. For this to be achieved, the metabolic biochemical pathways are extensively studied to employ the optimum combination of sgRNA, promoters, and optimization of *Cas* proteins. The result of which has been reported, like, for instance, a derivative, CRISPR/*Cpf1* (CRISPR from Prevotella and Francisella 1), is derived from the CRISPR/Cas9 system that recognizes a T-rich PAM sequence, enables editing of AT-rich regions, and has a lower rate of off-target edits in comparison to CRISPR/Cas9 (Zetsche et al. 2015; Scheben et al. 2017). In most cases, for the desired gene sequence, the respective guide RNAs are designed, which are then used for sequence-specific gene knockout after transformation. This is

Table 1. List of crops conferred abiotic stress resistance tolerance (Adopted from Kumar et al. 2023; Wang et al. 2022).

Drought stress tolerance			
Plant species	**Target Gene**	**Method of Delivery**	**Reference**
Arabidopsis thaliana	*AtOST2*	Agrobacterium-Mediated	(Osakabe et al. 2016)
A. thaliana	*AtAREB1*	Agrobacterium-mediated	(Roca Paixão et al. 2019)
A. thaliana	*AtAVP1*	Agrobacterium-Mediated	(Park et al. 2017)
A. thaliana	*AtmiR169a*	Agrobacterium-Mediated	(Zhao et al. 2016)
Brassica napus	*BnaA6.RGA*	Agrobacterium-Mediated	(Wu et al. 2020)
Cicer arietinum	*At4CL, AtRVE7*	PEG-mediated	(Badhan et al. 2021)
Glycine max	*GmMYB118*	Agrobacterium-Mediated	(Du et al. 2018)
Oryza sativa	*OsERA1*	Agrobacterium-Mediated	(Ogata et al. 2020)
O. sativa	*OsSAPK2*	Agrobacterium-mediated	(Lou et al. 2017)
O. sativa	*OsSRL1, OsSRL2*	Agrobacterium-mediated	(Liao et al. 2019)
O. sativa	*OsDST*	Agrobacterium-Mediated	(Santosh Kumar et al. 2020)
O. sativa	*OsNAC14*	Agrobacterium-Mediated	(Shim et al. 2018)
O. sativa	*OsPUB67*	Agrobacterium-Mediated	(Qin et al. 2020)
O. sativa	*OsmiR535*	Agrobacterium-Mediated	(Yue et al. 2020)
Solanum lycopersicum	*SlNPR1*	Agrobacterium-Mediated	(Li et al. 2019)
S. lycopersicum	*SlMAPK3*	Agrobacterium-Mediated	(Wang et al. 2017b)
S. lycopersicum	*SlLBD40*	Agrobacterium-Mediated	(Liu et al. 2020)
S. lycopersicum	*SlARF4*	Agrobacterium-Mediated	(Chen et al. 2021)
Triticum aestivum	*TaDREB2, TaDREB3, TaERF3*	PEG-mediated	(Kim et al. 2018)
T. aestivum	*TaDREB2, TaERF3*	Agrobacterium-Mediated	(Kim et al. 2018)
Zea mays	*ZmARGOS8*	Agrobacterium-Mediated	(Shi et al. 2017)
Salt stress tolerance			
Plant species	**Target Gene**	**Method of Delivery**	**Reference**
A. thaliana	*AtWRKY, AtWRKY4*	Agrobacterium-Mediated	(Li et al. 2021b)
A. thaliana	*AtACQOS*	Agrobacterium-Mediated	(Kim et al. 2021)
A. thaliana	*UGT79-B2, and B3*	Agrobacterium-Mediated	(Li et al. 2017c)
Glycine max	*GmDrb2a, GmDrb2b*	Agrobacterium-Mediated	(Curtin et al. 2018)
G. max	*GmAITR*	Agrobacterium-Mediated	(Wang et al. 2021b)
Medicago truncatula	*MtHEN1*	Agrobacterium-Mediated	(Curtin et al. 2018)
O. sativa	*OsDST*	Agrobacterium-Mediated	(Santosh Kumar et al. 2020)
O. sativa	*OsRAV2*	Agrobacterium-Mediated	(Duan et al. 2016)
O. sativa	*OsRR22*	Agrobacterium-Mediated	(Zhang et al. 2019a)
O. sativa	*OsNAC45*	Agrobacterium-Mediated	(Zhang et al. 2020b)
O. sativa	*OsBBS1*	Agrobacterium-Mediated	(Zeng et al. 2018)
O. sativa	*OsAGO2*	Agrobacterium-Mediated	(Yin et al. 2020)
O. sativa	*OsVDE*	Agrobacterium-Mediated	(Xu et al. 2023)
O. sativa	*OsNAC041*	Agrobacterium-Mediated	(Bo et al. 2019)
O. sativa	*OsSAPK2*	Agrobacterium-Mediated	(Lou et al. 2017)
O. sativa	*OsPQT3*	Agrobacterium-Mediated	(Alfatih et al. 2020)

Table 1 contd. ...

...Table 1 contd.

Plant species	Target Gene	Method of Delivery	Reference
O. sativa	*OsPIL14*	Agrobacterium-Mediated	(Mo et al. 2020)
O. sativa	*OsBGE3*	Agrobacterium-Mediated	(Yin et al. 2017)
O. sativa	*OsSPL10*	Agrobacterium-Mediated	(Lan et al. 2019)
O. sativa	*OsDOF15*	Agrobacterium-Mediated	(Qin et al. 2019)
O. sativa	*OsFLN2*	Agrobacterium-Mediated	(Chen et al. 2020)
O. sativa	*OsbHLH024*	Agrobacterium-Mediated	Alam et al. 2022
O. sativa	*OsGTγ-2*	Agrobacterium-Mediated	(Liu et al. 20202a)
O. sativa	OsmiR535	Agrobacterium-Mediated	(Yue et al. 2020)
S. tuberosum	*StCoilin*	PEG-mediated	(Makhotenko et al. 2019)
S. lycopersicum	*SlHyPRP1*	PEG-mediated	(Tran et al. 2021)
S. lycopersicum	*SlARF4*	Agrobacterium-Mediated	Bouzroud et al. 2020
T. aestivum	*TaHAG1*	Agrobacterium-Mediated	(Zheng et al. 2021)
Hordeum vulgare	*ITPK*	Agrobacterium-Mediated	(Golfam et al. 2021)
Heat stress tolerance			
Gossypium hirsutum	*GhPGF, GhCLA1*	Agrobacterium-Mediated	(Li et al. 2021a)
Lactuca sativa	*LsNCED4*	Agrobacterium-Mediated	(Bertier et al. 2018)
O. sativa	*OsPDS*	Gene gun	(Nandy et al. 2019)
O. sativa	*OsHSA1*	Agrobacterium-Mediated	(Qiu et al. 2018)
O. sativa	*OsNAC006*	PEG-mediated	(Wang et al. 2020a)
O. sativa	*OsPYL1/4/6*	Agrobacterium-Mediated	(Miao et al. 2018)
S. lycopersicum	*SlAGL6*	Agrobacterium-Mediated	(Klap et al. 2017)
S. lycopersicum	*SlCPK28*	Agrobacterium-Mediated	(Hu et al. 2021)
S. lycopersicum	*SlMAPK3*	Agrobacterium-Mediated	(Yu et al. 2019)
S. lycopersicum	*SlBZR1*	Agrobacterium-Mediated	(Yin et al. 2018)
Z. mays	*ZmTMS5gene*	Particle bombardment	(Li et al. 2017b)
Cold stress tolerance			
A. thaliana	*AtCBF1, AtCBF2*	Agrobacterium-Mediated	(Chen et al. 2010)
A. thaliana	*UGT79-B2, and B3*	Agrobacterium-Mediated	(Li et al. 2017c)
O. sativa	*OsAnn3*	Agrobacterium-Mediated	(Shen et al. 2017)
O. sativa	*OsPIN5b, GS3, OsMYB30*	Agrobacterium-Mediated	(Zeng et al. 2020)
O. sativa	*OsAnn5*	Agrobacterium-Mediated	(Shen et al. 2017)
O. sativa	*OsPRP1*	Agrobacterium-Mediated	(Nawaz et al. 2019)
O. sativa	*OsMYB30*	Agrobacterium-Mediated	(Liao et al. 2019)
S. lycopersicum	*SlCBF1*	Agrobacterium-Mediated	(Li et al. 2018)
Metal stress tolerance			
A. thaliana	*Atoxp1*	Agrobacterium-mediated	(Baeg et al. 2021)
O. sativa	*OsARM1*	Agrobacterium-mediated	(Wang et al. 2017a)
O. sativa	*OsNRAMP5*	Agrobacterium-mediated	(Tang et al. 2017b)
O. sativa	*OsNRAMP1*	Agrobacterium-mediated	(Chang et al. 2020; Chu et al. 2022)
O. sativa	*OsLCT1*	Agrobacterium-mediated	(Lu et al. 2017)

Table 1 contd. ...

...Table 1 contd.

Plant species	Target Gene	Method of Delivery	Reference
O. sativa	*OsHAK1*	Agrobacterium-Mediated	(Nieves-Cordones et al. 2017)
O. sativa	*OsPRX2*	Agrobacterium-Mediated	(Mao et al. 2019)
O. sativa	*OsATX1*	Agrobacterium-Mediated	(Zhang et al. 2021b)
Herbicide resistance			
B. napus	*BnALS*	Agrobacterium-Mediated	(Wu et al. 2020)
Manihot esculenta	*MeEPSPS*	Agrobacterium-Mediated	(Hummel et al. 2018)
O. sativa	*OsALS*	Agrobacterium-Mediated	(Zhang et al. 2021a)
O. sativa	*OsALS*	Agrobacterium-Mediated	(Wang et al. 2021a)
O. sativa	*ALS-FTIP1e*	Agrobacterium-Mediated	(Shimatani et al. 2017)
O. sativa	*OsTB1*	Agrobacterium-Mediated	(Butt et al. 2020)
O. sativa	*OsPUT1/2/3*	Agrobacterium-Mediated	(Lyu et al. 2022)
O. sativa	*OsACC*	Agrobacterium-Mediated	(Liu et al. 2018)
O. sativa	*OsEPSPS*	PEG-mediated	(Li et al. 2016b)
O. sativa	*OsEPSPS*	Biolistic gene transfer	(Li et al. 2016b)
O. sativa	*OsALS-1, OsALS-2, OsALS-3, OsALS-4*	Agrobacterium-Mediated	(Endo et al. 2016)
O. sativa	*PPO1 and HPPD*	PEG-mediated	(Lu et al. 2021)
O. sativa	*OsTubA2*	Agrobacterium-Mediated	(Liu et al. 2021)
Saccharum officinarum	*SoALS*	Biolistic gene transfer	(Oz et al. 2021)
S. lycopersicum	*SlEPSPS*	Agrobacterium-Mediated	(Yang et al. 2022)
S. lycopersicum	*SlALS1, SlALS2*	Agrobacterium-Mediated	(Yang et al. 2022)
S. lycopersicum	*Slpds1*	Agrobacterium-Mediated	(Yang et al. 2022)
T. aestivum	*TaALS-P174*	Biolistic-mediated	(Zhang et al. 2019b)
Z. mays	*ZmALS1, ZmALS2*	Agrobacterium-Mediated	(Svitashev et al. 2015)
Z. mays	*ZmALS1, ZmALS2*	PEG-mediated protoplast transformation	(Li et al. 2020b)
Z. mays	*MS26*	Biolistic-mediated	(Svitashev et al. 2015)
Z. mays	*ARGOS8*	Agrobacterium-Mediated	(Shi et al. 2017)
Citrullus lanatus	*ALS*	Agrobacterium-Mediated	(Tian et al. 2018)

then confirmed by designing primers of the respective guide RNAs and analyzing them in PCR (Yadav et al. 2023).

3.2 Targeted Genes for Stress Resistance

Crops often face challenges, right from the emergence of multiple stresses, that are negative influencers of plant growth and development, and these include biotic (Erdoğan et al. 2023) and abiotic (Yadav et al. 2023) factors. In biotic factors, the growth is affected mainly by viruses, fungi, and bacteria, through the defense mechanism employed, such as the Hypersensitive Response (HR) (Erdoğan et al. 2023). In abiotic factors, plant growth is affected by factors like drought, temperature (hot/cold), heavy metals, herbicides, etc. (Biswas et al. 2021). Studies reveal that more than 20 crops have been modified in response to stress, be it horticultural (Liu et al. 2022a) or field crops (Mathivanan et al. 2023). At present, the study of various crops to withstand abiotic stress plays a crucial role in climate-resilient crops. Drought conditions (an abiotic stress) negatively affect

Table 2. List of crops conferred biotic stress resistance (Adapted from Wang et al. 2022; Erdoğan et al. 2023).

Bacterial resistance				
Crop	**Disease name**	**Target genes**	**Method of Delivery**	**References**
S. lycopersicum	Bacterial speck disease	*JAZ2*	Agrobacterium-Mediated	(Ortigosa et al. 2019)
O. sativa	Bacterial blight (*Xanthomonas oryzae pv. oryzae*)	*SWEET13*	Agrobacterium-Mediated	(Zhou et al. 2015)
O. sativa	Bacterial blight (*Xanthomonas oryzae pv. oryzae*)	*OsSWEET11* and *OsSWEET14*	Agrobacterium-Mediated	(Jiang et al. 2013)
O. sativa	Bacterial blight (*Xanthomonas oryzae pv. oryzae*)	*SWEET11*, *SWEET13*, and *SWEET14*	Agrobacterium-Mediated	(Oliva et al. 2019)
O. sativa	Bacterial blight (*Xanthomonas oryzae pv. oryzae*)	*Xa13 promoter*	Agrobacterium-Mediated	(Li et al. 2020a)
Citrus sinensis	Citrus canker (*Xanthomonas citri* subsp. *Citri*)	*LOB1*	Agrobacterium-Mediated	(Jia et al. 2016)
Citrus sinensis	Citrus canker (*Xanthomonas citri* subsp. *Citri*)	*CsWRKY22*	Agrobacterium-Mediated	(Wang et al. 2019)
Citrus sinensis	Citrus canker (*Xanthomonas citri* subsp. *Citri*)	*CsLOB1*/promoter	Agrobacterium-Mediated	(Peng et al. 2017)
Fungal resistance				
S. lycopersicum	Powdery mildew	*PMR4*	Agrobacterium-Mediated	(Santillán Martínez et al. 2020)
S. lycopersicum	Powdery mildew	*SlMlo1*	Agrobacterium-Mediated	(Nekrasov et al. 2017)
S. lycopersicum	Fungal disease	*PL*	Agrobacterium-Mediated	(Silva et al. 2021)
S. lycopersicum	Fungal disease	*ACET1a* and *ACET1b*	Agrobacterium-Mediated	(Jeon et al. 2020)
O. sativa	Rice blast	*SEC3A*	Protoplast transformation with Cas9/gRNA expression binary	(Ma et al. 2018)
O. sativa	Rice blast	*OSERF922*	Agrobacterium-Mediated	(Wang et al. 2016)
Triticum aestivum	Powdery mildew	*TaMlo1*	Agrobacterium-Mediated	(Wang et al. 2014)
Triticum aestivum	Powdery mildew	*TaEDR1*	Agrobacterium-Mediated	(Zhang et al. 2017)
Vitis vinifera	Gray mold (*Botrytis cinerea*)	*WRKY52*	Agrobacterium-Mediated	(Wang et al. 2018b)
V. vinifera	Powdery mildew (*Erysiphe necator*)	*MLO-7*	PEG-mediated	(Malnoy et al. 2016)
V. vinifera	Powdery mildew (*Erysiphe necator*)	*VvMLO3*	Agrobacterium-mediated	(Wan et al. 2020)

Table 2 contd. ...

...Table 2 contd.

Crop	Disease name	Target genes	Method of Delivery	References
Malus domestica	Fire blight (*Erwinia amylovora*)	*DIPM-1, DIPM-2, DIPM-4*	PEG-mediated	(Malnoy et al. 2016)
Carica papaya	Root, stem, and fruit rot (*Phytophthora. palmivora*)	*alEPIC8*	Agrobacterium-mediated	(Gumtow et al. 2018)
Theobroma cacao	Black pod rot (*Phytophthora tropicalis*)	*TcNPR3*	Agrobacterium-mediated	(Fister et al. 2018)
Viral resistance				
Arabidopsis thaliana	Turnip mosaic virus (TuMV)	*eIF(iso)4E*	Agrobacterium-mediated	(Pyott et al. 2016)
Arabidopsis thaliana	Clover yellow vein virus (ClYVV)	*eIF4E1*	Agrobacterium-mediated	(Bastet et al. 2019)
O. sativa	Rice tungro spherical virus (RTSV)	*eIF4G*	Agrobacterium-mediated	(Macovei et al. 2018)
Solanum tuberosum	Potato virus Y (PVY)	*P3, CI, NIb* and *CP*	Agrobacterium-mediated	(Zhan et al. 2019)
Solanum tuberosum	Potato virus Y (PVY)	*eIF4E*	Agrobacterium-mediated	(Noureen et al. 2022)
Solanum tuberosum	Potato virus Y (PVY)	*eIF4E1*	Protoplast transformation	(Lucioli et al. 2022)
S. lycopersicum	Tomato yellow leaf curl virus (TYLCV)	*SlPelo* and *SlMlo1*	Agrobacterium-mediated	(Pramanik et al. 2021)
Hordeum vulgare	Wheat dwarf virus (WDV)	*Rep*, *MP*, and *LIR*	Agrobacterium-mediated	(Kis et al. 2019)
Nicotiana benthamiana	Tomato Yellow Leaf Curl Virus (TYLCV) and Beet Curly Top Virus (BCTV)	*CP, Rep, and IR*	Agrobacterium-mediated	(Ali et al. 2015)
N. benthamiana	Bean Yellow Dwarf Virus (BeYDV)	*LIR and Rep/RepA*	Agrobacterium-mediated	(Baltes et al. 2015)
N. benthamiana	Turnip mosaic virus (TuMV)	*GFP1, GFP2, HC-Pro, and CP*	Agrobacterium-mediated	(Aman et al. 2018)
N. benthamiana and *A. thaliana*	Cucumber mosaic virus (CMV) and Tobacco mosaic virus (TMV)	*ORF1,2,3, CP and 30UTR*	Agrobacterium-mediated	(Zhang et al. 2018a)
N. benthamiana and *A. thaliana*	Bean Yellow Dwarf Virus (BeYDV)	*CP, Rep, and IR*	Agrobacterium-mediated	(Ji et al. 2015)

plant growth, and the early examples of genome-edited *AGROS8* have been reported in maize (Li et al. 2022). Other stress, like osmotic stress and drought, stimulates ABA (abscisic acid)-dependent pathways, and important genes identified are *LOS6*/*ABA*, which are involved in the ABA biosynthesis pathway and are activated by salinity-induced osmotic stress (Angon et al. 2023). Moreover, a tremendous number of studies and works have been done in crops like rice, tomato, and maize for conferring all the subclasses of abiotic stress tolerance (Bhat et al. 2021; Bashir et al. 2021; Kaur et al. 2022; Rajput et al. 2021; Zaidi et al. 2020). CRISPR/Cas9 enables up-regulation or down-regulation of certain genes or its elements, such as negative regulators, promoters, other regulatory elements, and transcription factors (TFs), thereby playing an instrumental role in conferring stress (biotic and abiotic) resistance which is one of the defining objectives of crop improvement (Mohamed et al. 2024; Praveen et al. 2023).

3.3 Developing Multi-Stress Resistance using CRISPR-Cas

When a plant encounters stress conditions, it endogenously initiates defense pathways that are very complex and diverse (Dong et al. 2022). But when plants encounter multiple stresses simultaneously, it is important to respond in a suitable manner, which can potentially be achieved by CRISPR-Cas9, which targets multiple genes using multiple guide RNAs (gRNA) and Cas9 enzymes that help in achieving desired changes at various nucleotide levels in the target genome, which makes this technique a very worthy, versatile, and highly powerful tool which improves genome-editing technology at a very high magnitude, and this is known as multiplex genome editing (Jacobs et al. 2017; Kumar et al. 2023). This enables multiple genes to be modified for any crop, which results in the introduction of various traits into new plant varieties by precise modification at numerous loci (Komor et al. 2017; Meng et al. 2017). The major advantage of multiplex genome editing is that it covers various aspects of plant improvements, such as metabolic engineering and hormone biosynthesis and perception, and molecular farming, capturing > 100 concurrent targeting events (Najera et al. 2019). So, multiple genome-editing technologies will encourage the creation and utilization of novel genetic variation at a much faster rate in a short time (Kumar et al. 2023).

4. Challenges and Limitations

4.1 Technical Challenges in Applying CRISPR/Cas Technology in Plants

There are undoubtedly some challenges to overcome when attempting to use CRISPR/Cas9 technology in plants, as it is a relatively new toolkit for genome editing. To allow for precise editing, the gene that is specifically responsible for the desired function must be identified before any genome manipulation can take place. The function of genes that have been sequenced within the genomes of many relevant plant species is still poorly understood, which hinders efforts to use precision editing to achieve desired results (Haque et al. 2018; Adhikari and Poudel 2020). Fortunately, gene functions can be accurately predicted by using Genome-Wide Association Studies (GWAS), which can lead to additional research on the necessary plant manipulations.

For example, the genes OsC1 and OsRb, which are involved in controlling anthocyanins in rice leaves, were identified by Zheng et al. (2019). This made it possible for Hu et al. (2020) to manipulate the levels of anthocyanins in rice using CRISPR/Cas9 technology. The RDP1 gene of *A. thaliana* was also studied using a similar methodology (Tsuchimatsu et al. 2020). Similar to how GWAS can advance CRISPR/Cas9 plant editing, CRISPR/Cas9 technology is also employed as a substitute technique for cross-population validation (Alseekh et al. 2021). For example, it has been used to validate GWAS results in maize (*Zea mays*) (Liu et al. 2020) and rice (*Oryza sativa*) (Lu et al. 2017; Meng et al. 2017).

This sheds light on how crucial it is to identify the causal links and gene interactions that can propel the advancement of CRISPR/Cas9 technology in plants (Yin et al. 2017). Despite the CRISPR/Cas systems' strong potential for crop genetic improvement, there are still certain obstacles to be addressed in this area (Razzaq et al. 2022). Because SpCas9 can only recognize NGG PAM sites, its effectiveness is limited, and needs a 5′-NGG-3CPAM right next to the 20 nt DNA target sequence. Even though this restriction is essential, the objective is always to use selective mutagenesis to shut down the gene (Fig. 1). Thus, the primary effort to create Cas9-like systems is in progress, altering PAM sequences or causing *Streptococcus pyogenes*' single CRISPR/Cas9 to detect other PAMs.

To detect a variety of PAM sequences with GAT, NG, and GAA in mammalian cells, for instance, SpCas9 is modified to become xCas9 (Niu et al. 2020). utilizing an xCas9 and Cas9-NG

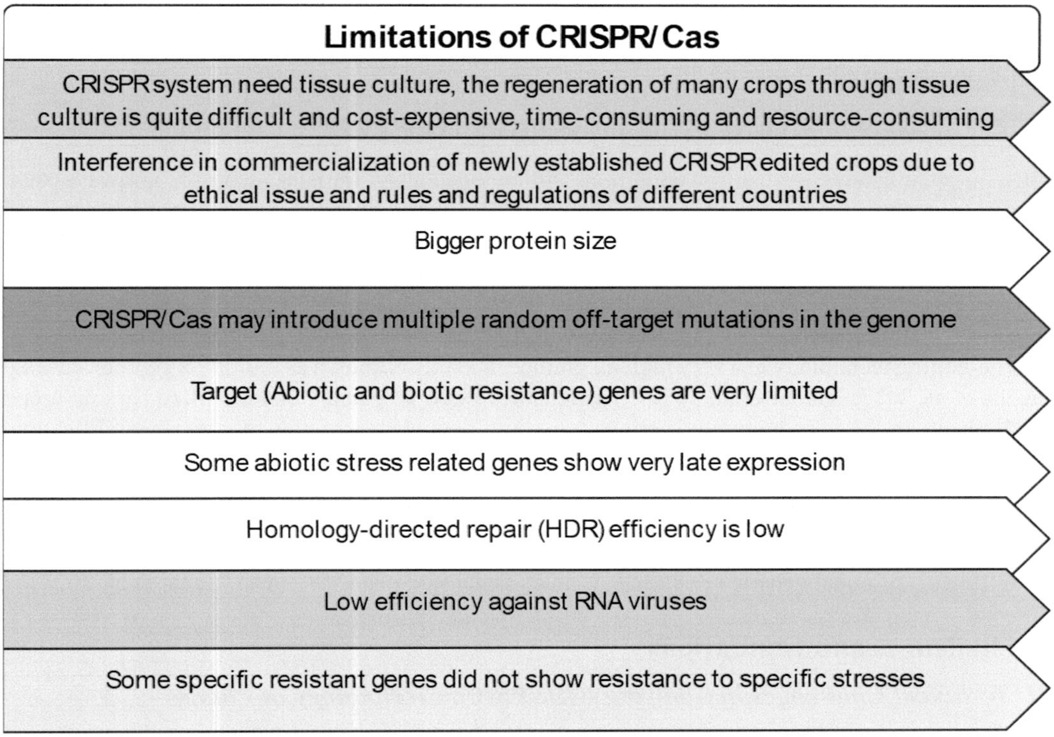

Figure 1. Current CRISPR/Cas system limitations.

hybrid to broaden the application of CRISPR/Cas9-mediated GE in plants (Walton et al. 2020). An expanded set of NGN PAMs can be targeted by a recently developed SpCas9 variant. This enzyme was optimized to create SpRY (NRN > NYN PAMs), a near-PAMless SpCas9 variant. Nearly all PAMs can be targeted by SpRY nuclease and base-editor variants (Cao et al. 2022).

Agrobacterium tumefaciens-mediated transformation is necessary for the CRISPR/Cas system to be used in plants, but this process is time-consuming, expensive, and labor-intensive (Fig. 1). There is very little choice in target gene selection. On the one hand, resistance genes serve a redundant purpose, and resistance cannot be attained by merely knocking down a gene. On the other hand, PAM limits the knockout of resistance genes, so sequences near PAM need to be chosen. Random off-target mutations could be introduced into the plant genome by CRISPR/Cas. Since it takes several generations to fully isolate Cas proteins and produce transgene-free crops, the commercialization of CRISPR-edited crops has been hampered. The efficiency of homologous donor sequence transformation into plant cells is currently low, and the homologous recombination pathway (knock-in/gene replacement) is less effective. As a result, knock-in is not as difficult or as effective. Hence effective gene knock-in using CRISPR/Cas-mediated homologous recombination in plants is still a way off.

4.2 Challenge: Transforming Edited Plants into Commercial Crops

In fact, all major crops and some minor crops have benefited from the successful application of CRISPR/Cas genome-editing technology. Translating this scientific breakthrough into a meaningful socio-economic contribution is the next challenge in this context. Technical limitations also slow down the quick expansion of the advantages of the CRISPR/Cas system in agriculture, even though political tensions like trade barriers may restrict the use of edited crops globally. Original articles and reviews frequently ignore these limitations in favor of concentrating on the appealing potential uses of this groundbreaking technique. These limitations, some of which also exist in the production of commercial transgenic crops, must be recognized and addressed in order to optimize the advantages of gene editing.

4.3 The Unexpected Delay in the Use of Genetic Modifications: A Lesson Learned from Transgenic

When several research teams reported successfully establishing Agrobacterium-mediated plant transformation in 1983, the race to create the first transgenic plant came to an end (Zambryski 2013). Plant research and breeding through genetic modifications continue to be based on the effective delivery of nucleic acids into the plant cell and the regeneration of plants from modified cells.

Nevertheless, only two transgenic traits—herbicide tolerance and insect resistance—have been extensively accepted by farmers inspite of the apparent promise (Tyczewska et al. 2023). Since the introduction of the first transgenic crops, glyphosate-tolerant soybean, and insect-resistant cotton, in 1996, the commercial use of transgenic traits has remained largely stagnant, despite the emergence of some notable exceptions, such as crops with increased provitamin A content like Golden Rice and those with drought stress tolerance like HB4 Wheat.

4.4 Concern for Delay in Massive Use of Edited Crops

When combined with molecular tools (like the modular cloning system), transformation protocols (like PEG-mediated protoplast transformation), sequencing technology (like NGS technology), genomic data (like the target crop's entire sequence), and bioinformatics platforms (like Geneious software), CRISPR/Cas technology can be utilized in ways that are far superior to those employed by the early 1980s pioneers of plant biotechnology. In a similar vein, during the past forty years, there has been a significant shift in our knowledge of plant physiology, the availability of better methods, and the regulatory frameworks for the management of genetic modifications in plants. Indeed, a growing number of nations have legal frameworks that permit the use of gene-edited crops that do not contain foreign DNA as conventional crops (Lewi and Vici´en 2020; Mallapaty 2022). Nonetheless, there has been a comparable delay in the commercialization of transgenic and edited events.

For instance, the first CRISPR/Cas-edited crop (GABA-enriched tomato) was introduced to the market nearly ten years after the first CRISPR/Cas-edited plant was created in 2013 (Shan et al. 2013; Waltz 2022). Few genome-edited varieties and edited traits have received preliminary approval for widespread use globally to date, despite numerous reports detailing the CRISPR/Cas system's ability to edit crops (Pixley et al. 2022; Nidhi et al. 2021). Consequently, it is critical to create plans to boost the production of commercially modified crops.

4.5 Ethical and Regulatory Concerns

Although the CRISPR-Cas system enables us to efficiently modify genomic DNA (Lyon 2017), several moral, ethical, political, and scientific issues could result in a moratorium that would prevent further research of this kind. To articulate an ideal ethical and legal framework, it is crucial to identify the issues, clarify concepts, and encourage dialogue among all relevant stakeholders, including societies, universities, and science and technology systems (de Lecuona et al. 2017; Lyon 2017).

Accordingly, at a 2015 meeting in Napa Valley, a group of CRISPR-Cas9 developers, scientists, and ethicists discussed the legal, ethical, and biomedical aspects of the technology (Baltimore et al. 2015). Widespread conversations were also urged following this meeting. To discuss how, when, and where to apply this technology in humans, the Chinese Academy of Sciences, and the Royal Society of the United Kingdom (UK) were invited to attend an International Summit on Human Gene Editing by the United States (US) National Academies of Sciences, Engineering, and Medicine (NASEM) (Brokowski and Adli 2019). The Nuffield Council on Bioethics, shortly after this meeting, requested evidence regarding the ethical concerns arising from these biological techniques that enable us to precisely modify the genes of living cells and their applications in industrial biotechnology, ecology, plants and animals, biomedicine, and reproduction (Mulvihill et al. 2017; Nuffield Council on Bioethics 2016).

The rapidly developing genome editing technology known as CRISPR/Cas9 is revolutionizing the field of molecular biology by allowing researchers to alter the DNA of a wide range of organisms as they see fit. Due to its simplicity and ease of use, it was swiftly adopted after its 2012 introduction. Applications ranging from clinical therapeutics to agriculture, including human germline modifications to treat genetic diseases, are being explored (Carroll and Charo 2015).

CRISPR/Cas9-mediated genome editing has given the ethical debate over human genetic modifications a new edge, even though it is not new. A thorough discussion of the ethical and societal ramifications of this technology is required due to the unpredictable and far-reaching consequences that come with its alluring applications. The ethical issues surrounding this new technology are covered in the current review, with particular attention paid to its potential and concerns regarding human germline modifications. It calls for a thorough discussion between scientists, ethnic groups, businesspeople, and decision-makers. The opinions of various societal segments, such as the general public and religious scholars, are crucial.

4.6 Major Concerns

4.6.1 Off-target Mutations

One of the main issues with CRISPR/Cas9 genome engineering is off-target mutations, which could have harmful consequences for both people and the environment. Studies have shown that humans have a comparatively greater number of off-target mutations than zebrafish. Bigger genomes can contain a lot of identical sites for cleavage, allowing CRISPR/Cas9 to cut unwanted sequences. Cell death and an aberrant transformation may result from such mutations. In addition, CRISPR/Cas9-based therapies are challenging to implement in hard-to-infect cells. Accurate alteration using CRISPR/Cas9 will be crucial to the technology's adoption. Both the advantages and the risks need to be assessed (Cong et al. 2013; Hwang et al. 2013; Ma et al. 2014; Rodriguez 2016; Yang et al. 2013).

Furthermore, the integrity of the environment may be threatened by such off-target mutations. Gene drive is an ongoing process that has the potential to transfer genes between organisms. The integrity of the environment may be compromised in such a pattern since the off-target mutations may also spread to other organisms.

4.6.2 Regulatory Issues

Because CRISPR/Cas9 is so easy to use to create genetic changes, it is also challenging to identify the altered organism outside of a laboratory, which raises questions about its regulation. We can anticipate a growing market for genetically modified organisms, which will raise concerns about their regulations, if CRISPR/Cas9 continues to spread around the globe. The patenting problems also need to be fixed. The patenting of genetically altered organisms for medical purposes is a point of contention among scientific communities. The CRISPR/Cas9 system is the focus of numerous economic interests. Patenting CRISPR/Cas9 techniques is thought to grant the relevant companies a great deal of power (Ledford 2015b; Rodriguez 2016).

4.6.3 Next Generation CRISPR Technologies (cpf1, Base Editing and Prime Editing)

CRISPR-Cpf1

CRISPR from Prevotella and Francisella, formerly known as Cas12a, is represented by the Cpf1 endonuclease. As a more sophisticated and effective genome-editing tool and a superior replacement for CRISPR/Cas9, the CRISPR/Cpf1 system has recently grown in popularity (Bin Moon et al. 2018).

Compared to Cas9, the Cpf1 endonuclease is smaller and needs shorter CRISPR RNA (crRNA) to function correctly (Liu et al. 2017). According to Zetsche et al. (2015), Cpf1 is an effector nuclease protein that is guided by a single RNA. It binds upstream of the protospacer adjacent motif (PAM) and introduces five-base-pair (bp) staggered cuts into the DNA at the proximal end of the PAM, which is far from the seed region.

Re engineering the desired DNA as the target and maintaining the PAM sequence (5′-TTTN-3′) are the primary benefits of a CRISPR/Cpf1-mediated genome-editing tool. The ortholog's origin may cause variations in the PAM sequence. The Cpf1-database1 is an online tool that can identify AsCpf1 and LbCpf1 nucleases using DNA recognition sequences and provides a straightforward method for identifying a possible target within the genome and designing gRNA (Park and Bae 2017).

However, Cpf1 is different from Cas9 in the following ways: (Dong et al. 2016; Zetsche 2015)

1) T-rich PAM sequences, like 50-TTTN-30, are recognized by Cpf1;
2) The PAM sequence is situated upstream of a protospacer sequence at the 50-end of a target DNA sequence.
3) Cpf1 is a ribonuclease that processes precursor crRNAs; (4) Cpf1 is guided by a single crRNA and does not require trans-acting crRNA9 (Fonfara et al. 2016). AsCpf1 from Acidaminococcus sp. BV3L6 and LbCpf1 from the Lachnospiraceae bacterium ND 2006 are two of the Cpf1 family's proteins that function better in human cells than their orthologues (Kim et al. 2016).

4.6.4 Base Editing

Gene base conversion is not possible with the CRISPR/Cas9 system. They work best with gene knock-ins or knock-outs. These constraints look for a stable and accurate method of crop genome editing essential. Base editing has been considered a more effective and alternative method (Veillet et al. 2019). For nucleotide conversions without the creation of DNA DSBs, it is an easy and accurate method (Fig. 1; Yin et al. 2017; Molla et al. 2021). By using a tethered deaminase domain or nickase Cas9 for base conversion from A > G or C > T, it gets around some of the drawbacks of CRISPR/Cas9.

This method has been applied in recent research to induce single and multiple nucleotide changes in cells (Wang et al. 2020). Thus, this technology enables high-throughput automatic plant genetic engineering.

5. Applications for Base Editors in Plants

In the longer breeding process known as "mutation breeding," scientists use physical and chemical mutagenesis to produce random mutations across the plant genome. To create crop varieties with increased yields and improved nutritional quality, transgenic breeding techniques enable the direct introduction of beneficial genes specific to a crop or genes from other species (Arak et al. 2015). Nevertheless, this breeding technique is tightly regulated since it necessitates the incorporation of exogenous genes into the plant genome. The three breeding techniques mentioned above are effectively supplemented by BE technology. To swiftly achieve the desired variety, BEs enable targeted modification of the plant genome without the introduction of exogenous genes (Hufford et al. 2021).

5.1 Prime Editing

Prime Editors (PEs) were created recently to produce precise editing without the need for donor DNA or DSBs (Anzalone et al. 2020). PE is primarily caused by the fusion of a Cas9 nickase (nCas9) and a Reverse Transcriptase (RT) from the Moloney murine leukemia virus (M-MLV). nCas9 was engineered to mediate site-specific nicking by a prime-editing guide RNA (pegRNA), which subsequently functioned as a template for RT to introduce mutagenesis that could be customized. In human cells, PEs could effectively create all potential base conversions and minor insertions in a larger target range with few byproducts (Anzalone et al. 2020).

6. Contribution of CRISPR to Sustainable Agriculture and Climate-Resilient Crops

6.1 Enhancing Stress Tolerance

Abiotic or biotic stress tolerance is one of the CRISPR targets associated with agricultural applications (Bigini et al. 2021). Nonetheless, the development of CRISPR may entail straightforward or intricate mutations or the integration of specific genes in the target genome, which could improve breeding efficiency and lead to the development of hitherto unattainable plant traits (Ahmad et al. 2016).

The CRISPR/Cas9 technique has been used in numerous studies to target multiple genes (Xie and Yang 2013). The high concentration of salt in the soil severely restricts rice productivity by impeding plant growth and development, which in turn affects yield. Furthermore, diseases brought on by biotic agents—such as bacteria, fungi, viruses, and insects—also have a detrimental impact on yield loss and result in subpar products (Mishra et al. 2018).

6.2 High Yield

The primary drivers of sustainable crop yield and food quality improvement are the effects of climate change and the need to meet dietary needs brought on by the world's population growth. Crop yield is a complicated characteristic that is strongly impacted by environmental factors and polygenic regulation. For instance, several characteristics, including the number of panicles per plant, the number of grains per panicle, and the size and weight of the grains, affect rice yield. Numerous studies have demonstrated that some genes' expression has a negative impact on yield, and that knocking them out using CRISPR/Cas9 may have beneficial effects. To date, CRISPR/Cas9 technology has been used to silence several genes in rice, including GN1a, DEP1, GS3, GW2, GW5, and TGW6, resulting in a notable increase in grain weight and size, grain number, and dense and erect panicles (Xu et al. 2016).

6.3 Quality Improvement

Given that 2 billion people worldwide suffer from malnutrition, especially in developing nations, genetic improvement aims to increase the nutritional value of food. One of the main causes of health issues in children and pregnant women is vitamin A deficiency. In order to biofortify rice and tomatoes with carotenoid, CRISPR/Cas9 genome editing has been used. This has produced marker-free mutants and an accumulation of lycopene, which can be used to treat chronic illnesses and reduce the risk of cancer and cardiovascular disease (Waltz 2016).

7. Prospects of CRISPR Technology in Developing Multi-Stress–Resistant Plants

The future of genome editing technology is represented by CRISPR/Cas, which can be applied to a variety of fields under both biotic and abiotic agricultural stressors (Fig. 2). CRISPR/Cas is being used in a wider range of applications as Genome Editing (GE) technology advances and new GE tools are created. Now, CRISPR/Cas can replace a single base to alter amino acids, knock out,

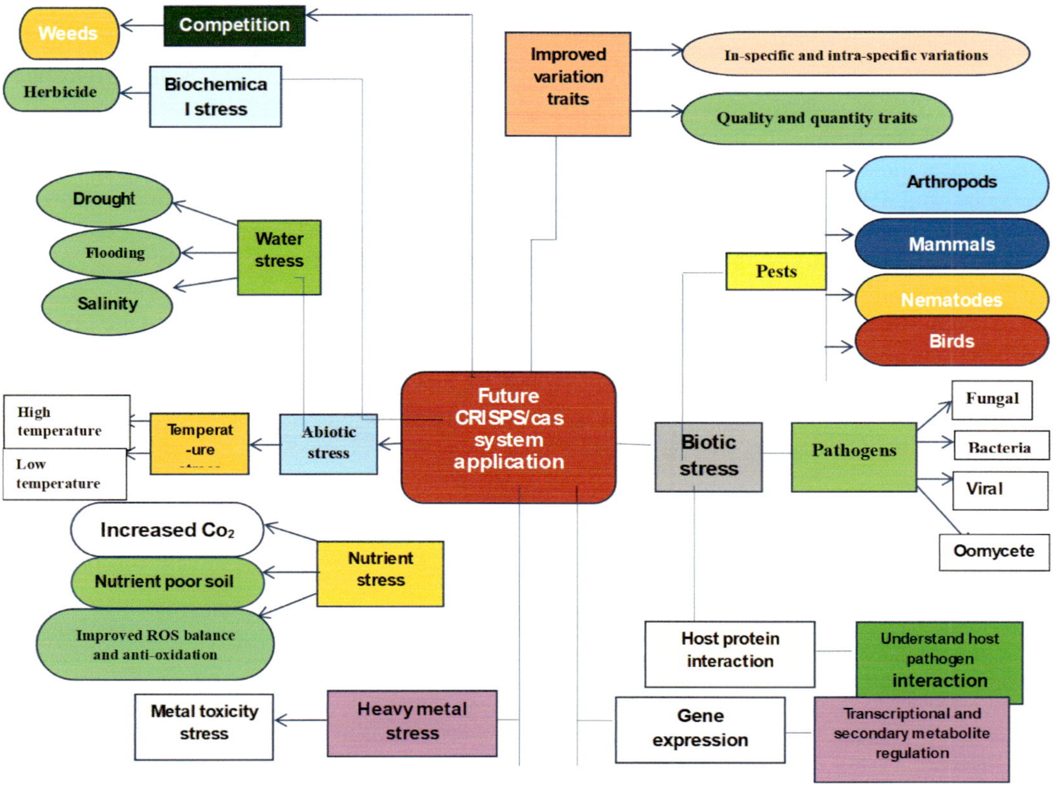

Figure 2. Prospective uses of CRISPR/Cas in plants to combat biotic and abiotic stressors.

knock in, and knock up genes in plants, among other things. In order to improve crop growth and development and help them cope with different environmental stresses, CRISPR/Cas can be used to modify important genes of biotic and abiotic stresses. This will increase the amount of germplasm resources that can be used to meet human needs.

Conclusion

CRISPR/Cas is a precise, efficient, and cost-effective genome editing technology in plant biology and biotechnology that provides plant resistance against environmental stresses, i.e., biotic and abiotic stresses. This technology works by editing the genes of specified crops without inserting the DNA, like in GMOs, and modifying the genome without introducing transgenes. Multiplex genome editing further enhances its potential by enabling simultaneous modifications of multiple genes, thus enabling crops to possess broad-spectrum resistance traits.

Abiotic stress, like drought, salinity, heat, cold, and heavy metal toxicity, affects the overall metabolism of plants, the life cycle of crops, and plant growth and productivity. Addressing abiotic stress involves multiple genes, signals, and transduction molecules, which form an elaborate and complex pathway involving multiple cascades of genes. Similarly, to address biotic stress challenges, CRISPR/Cas in plants provides mechanisms against pathogen and pest attacks by inserting genes against pathogens, fungal and bacterial attacks are reduced by knocking out negative regulators, and virus attacks are overcome by host susceptibility modification or by targeting pathogenicity-related genes. There is a need to tackle this challenge for sustainable agricultural development against biotic and abiotic stresses. Numerous environmental cues that plants encounter result in significant reductions in crop productivity, and frequent attacks on plants by microbial diseases and herbivores where defense mechanisms encounter these barriers. The detailed study of all the regulatory elements, key genes, and Transcription Factors (TFs) provides us the insights into where the action of improvement in the pathway is needed to endure abiotic stress resistance.

This chapter highlights the successful application of CRISPR/Cas in various crops such as rice, tomato, maize, and wheat for traits like drought tolerance, salinity resistance, heat and cold resilience, and defense against fungal, bacterial, viral, and insect pests. These findings affirm the potential of CRISPR/Cas as a powerful tool in developing climate-smart and disease-resistant crops.

Despite its success, several challenges remain because the CRISPR system needs tissue culture, which is quite difficult and expensive; large protein size; Interference in the commercialization of newly established CRISPR-edited crops; and multiple random off-target mutations in the genome. Besides CRISPR/Cpf1, which is derived from the CRISPR/Cas9 system and SpRY (NRN > NYN PAMs), the scope of genome editing continues to expand, and combining CRISPR/Cas with genomic and transcriptomic data continuous struggle towards sustainable crop production and addressing the growing food demands under climate change scenarios.

References

Abdallah, N.A., A. Hamwieh, K. Radwan, N. Fouad and C. Prakash. 2021. Genome editing techniques in plants: a comprehensive review and future prospects toward zero hunger. GM Crops Food. Dec 31; 12(2): 601–615.

Abdelrahman, M., A.M. Al-Sadi, A. Pour-Aboughadareh, D.J. Burritt and L.S.P. Tran. 2018. Genome editing using CRISPR/Cas9–targeted mutagenesis: An opportunity for yield improvements of crop plants grown under environmental stresses. Plant Physiol. Biochem. 131: 31–36.

Abobatta, W.F. 2020. Plant responses and tolerance to combined salt and drought stress. Salt and Drought Stress Tolerance in Plants: Signaling Networks and Adaptive Mechanisms 17–52.

Adhikari, P. and M. Poudel. 2020. CRISPR-Cas9 in agriculture: Approaches, applications, future perspectives, and associated challenges. Malays. J. Halal Res. 3(1): 6–16. 10.2478/mjhr-2020-0002 [DOI] [Google Scholar][Ref list].

Agapito-Tenfen, S.Z., A.S. Okoli, M.J. Bernstein, O.G. Wikmark and A.I. Myhr. 2018. Revisiting risk governance of GM plants: the need to consider new and emerging gene-editing techniques. Front. Plant Sci. 9: 1874.

Ahmad, A., N. Munawar, Z. Khan, A.T. Qusmani, S.H. Khan, A. Jamil et al. 2021. An outlook on global regulatory landscape for genome-edited crops. Int. J. Mol. Sci. 22(21): 11753.

Ahmad, P., A.A.H. Abdel Latef, S. Rasool, N.A. Akram, M. Ashraf and S. Gucel. 2016. Role of proteomics in crop stress tolerance. Front. Plant Sci. 7: 1–14.

Ahmad, S., X. Wei, Z. Sheng, P. Hu and S. Tang. 2020. CRISPR/Cas9 for development of disease resistance in plants: recent progress, limitations and future prospects. Brief. Funct. Genomics 19(1): 26–39.

Alam, M.S., J. Kong, R. Tao, T. Ahmed, M. Alamin, S.S. Alotaibi et al. 2022. CRISPR/Cas9 mediated knockout of the OsbHLH024 transcription factor improves salt stress resistance in rice (*Oryza sativa* L.). Plants 11(9): 1184.

Alengebawy, A., S.T. Abdelkhalek, S.R. Qureshi and M.Q. Wang. 2021. Heavy metals and pesticides toxicity in agricultural soil and plants: Ecological risks and human health implications. Toxics 9(3): 42.

Alfatih, A., J. Wu, S.U. Jan, Z.S. Zhang, J.Q. Xia and C.B. Xiang. 2020. Loss of rice PARAQUAT TOLERANCE 3 confers enhanced resistance to abiotic stresses and increases grain yield in field. Plant Cell Environ. 43(11): 2743–2754.

Ali, Z., A. Abulfaraj, A. Idris, S. Ali, M. Tashkandi and M.M. Mahfouz. 2015. CRISPR/Cas9-mediated viral interference in plants. Genome Biol. 16: 1–11.

Alseekh, S., D. Kostova, M. Bulut and A.R. Fernie. 2021. Genome-wide association studies: assessing trait characteristics in model and crop plants. Cell. Mol. Life Sci. 78(15): 5743–5754. 10.1007/S00018-021-03868-W [DOI] [PMC free article] [PubMed] [Google Scholar][Ref list].

Aman, R., Z. Ali, H. Butt, A. Mahas, F. Aljedaani, M.Z. Khan et al. 2018. RNA virus interference via CRISPR/Cas13a system in plants. Genome Biol. 19: 1–9.

Anders, C., O. Niewoehner, A. Duerst and M. Jinek. 2014. Structural basis of PAM-dependent target DNA recognition by the Cas9 endonuclease. Nature 513(7519): 569–573.

Angon, P.B., S. Mondal, S. Akter, M.A. Sakil and M.A. Jalil. 2023. Roles of CRISPR to mitigate drought and salinity stresses on plants. Plant Stress 8: 100169.

Anzalone, A.V., L.W. Koblan and D.R. Liu. 2020. Genome editing with CRISPR–Cas nucleases, base editors, transposases and prime editors. Nat. Biotechnol. 38(7): 824–844.

Araki, M. and T. Ishii. 2015. Towards social acceptance of plant breeding by genome editing. Trends Plant Sci. 20: 145–149.

Arora, L. and A. Narula. 2017. Gene editing and crop improvement using CRISPR-Cas9 system. Front. Plant Sci. 8: 1932.

Ashokkumar, S., D. Jaganathan, V. Ramanathan, H. Rahman, R. Palaniswamy, R. Kambale et al. 2020. Creation of novel alleles of fragrance gene OsBADH2 in rice through CRISPR/Cas9 mediated gene editing. PloS ONE 15(8): e0237018.

Awasthi, R., K. Bhandari and H. Nayyar. 2015. Temperature stress and redox homeostasis in agricultural crops. Front. Environ. Sci. 3: 11.

Badhan, S., A.S. Ball and N. Mantri. 2021. First report of CRISPR/Cas9 mediated DNA-free editing of 4CL and RVE7 genes in chickpea protoplasts. Int. J. Mol. Sci. 22: 396.

Baeg, G.J., S.H. Kim, D.M. Choi, S. Tripathi, Y.J. Han and J. Kim II. 2021. CRISPR/Cas9-mediated mutation of 5-oxoprolinase gene confers resistance to sulfonamide compounds in Arabidopsis. Plant Biotechnol. Rep. 15: 753–764.

Baltes, N.J., J. Gil-Humanes, T. Cermak, P.A. Atkins and D.F. Voytas. 2014. DNA replicons for plant genome engineering. Plant Cell 26(1): 151–163.

Baltes, N.J., A.W. Hummel, E. Konecna, R. Cegan, A.N. Bruns, D.M. Bisaro et al. 2015. Conferring resistance to geminiviruses with the CRISPR–Cas prokaryotic immune system. Nat. Plants 1(10): 1–4.

Baltimore, D., P. Berg, M. Botchan, D. Carroll, R.A. Charo, G. Church et al. 2015. A prudent path forward for genomic engineering and germline gene modification. Science 348: 36–38

Bashir, S.S., A. Hussain, S.J. Hussain, O.A. Wani, S. Zahid Nabi, N.A. Dar et al. 2021. Plant drought stress tolerance: Understanding its physiological, biochemical and molecular mechanisms. Biotechnol. Biotechnol. Equip. 35(1): 1912–1925.

Bastet, A., D. Zafirov, N. Giovinazzo, A. Guyon-Debast, F. Nogué, C. Robaglia et al. 2019. Mimicking natural polymorphism in eIF4E by CRISPR-Cas9 base editing is associated with resistance to potyviruses. Plant Biotechnol. J. 17(9): 1736–1750.

Bertier, L.D., M. Ron, H. Huo, K.J. Bradford, A.B. Britt and R.W. Michelmore. 2018. High-resolution analysis of the efficiency, heritability, and editing outcomes of CRISPR/Cas9-induced modifications of NCED4 in lettuce (Lactuca sativa). G3: Genes Genomes Genet. 8: 1513–1521.

Bhat, M.A., R.A. Mir, V. Kumar, A.A. Shah, S.M. Zargar, S. Rahman et al. 2021. Mechanistic insights of CRISPR/Cas-mediated genome editing towards enhancing abiotic stress tolerance in plants. Physiol. Plant. 172(2): 1255–1268.

Bi, H.L., J. Xu, A.J. Tan and Y.P. Huang. 2016. CRISPR/Cas9-mediated targeted gene mutagenesis in Spodoptera litura. Insect Sci. 23(3): 469–477.

Bigini, V., F. Camerlengo, E. Botticella, F. Sestili and D.V. Savatin. 2021. Biotechnological resources to increase disease-resistance by improving plant immunity: A sustainable approach to save cereal crop production. Plants 10(6): 1146.

Billah, M., S. Aktar, M. Brestic, M. Zivcak, A.B.M. Khaldun, M.S. Uddin et al. 2021. Progressive genomic approaches to explore drought-and salt-induced oxidative stress responses in plants under changing climate. Plants 10(9): 1910.

Bin Moon, S., J.M. Lee, J.G. Kang, N.E. Lee, D.I. Ha, D.Y. Kim et al. 2018. Highly efficient genome editing by CRISPR-Cpf1 using CRISPR RNA with a uridinylate-rich 3'-overhang. Nature Communications 9(1): 3651.

Bisht, D.S., V. Bhatia and R. Bhattacharya. 2019. Improving plant-resistance to insect-pests and pathogens: The new opportunities through targeted genome editing. Semin. Cell Dev. Biol. 96: 65–76.

Biswas, D., S.C. Saha and A. Dey. 2021. CRISPR-Cas genome-editing tool in plant abiotic stress-tolerance. Plant Gene 26: 100286.

Bo, W., Z. Zhaohui, Z. Huanhuan, W. Xia, L. Binglin, Y. Lijia et al. 2019. Targeted mutagenesis of NAC transcription factor gene, OsNAC041, leading to salt sensitivity in rice. Rice Sci. 26: 98–108.

Bonifacio, J. 2019. Transformation of *Arabidopsis thaliana* with CRISPR/Cas9-modified AtHMA4 (Heavy-metal ATPase-4) using floral dip to increase Zn^{2+} tolerance.

Borrelli, V.M.G., V. Brambilla, P. Rogowsky, A. Marocco and A. Lanubile. 2018. The enhancement of plant disease resistance using CRISPR/Cas9 technology. Front. Plant Sci. 9: 1245.

Bortesi, L. and R. Fischer. 2015. The CRISPR/Cas9 system for plant genome editing and beyond. Biotechnol. Adv. 33(1): 41–52.

Bortesi, L., C. Zhu, J. Zischewski, L. Perez, L. Bassié, R. Nadi and S. Schillberg. 2016. Patterns of CRISPR/Cas9 activity in plants, animals and microbes. Plant Biotechnol. J. 14(12): 2203–2216.

Boscaiu, M. and A. Fita. 2020. Physiological and molecular characterization of crop resistance to abiotic stresses. Agronomy 10(9): 1308.

Bouzroud, S., K. Gasparini, G. Hu, M.A.M. Barbosa, B.L. Rosa, M. Fahr et al. 2020. Down regulation and loss of auxin response factor 4 function using CRISPR/Cas9 alters plant growth, stomatal function and improves tomato tolerance to salinity and osmotic stress. Genes 11(3): 272.

Brokowski, C. and M. Adli. 2019. CRISPR ethics: moral considerations for applications of a powerful tool. J. Mol. Biol. 431: 88–101.

Brooks, C., V. Nekrasov, Z.B. Lippman and J. Van Eck. 2014. Efficient gene editing in tomato in the first generation using the clustered regularly interspaced short palindromic repeats/CRISPR-associated9 system. Plant Physiol. 166(3): 1292–1297.

Butt, H., A. Eid, A.A. Momin, J. Bazin, M. Crespi, S.T. Arold et al. 2019. CRISPR directed evolution of the spliceosome for resistance to splicing inhibitors. Genome Biol. 20(1): 1–9.

Butt, H., G.S. Rao, K. Sedeek, R. Aman, R. Kamel and M. Mahfouz. 2020. Engineering herbicide resistance via prime editing in rice. Plant Biotechnol. J. 18: 2370.

Cao, X., J. Guo, S. Huang, W. Yu, G. Li, L. An et al. 2022. Engineering of near-PAMless adenine base editor with enhanced editing activity and reduced off-target. Molecular Therapy Nucleic Acids 28: 732–742.

Carroll, D. and R.A. Charo. 2015. The societal opportunities and challenges of genome editing. Genome Biol. 16: 242.

Chandrasekaran, J., M. Brumin, D. Wolf, D. Leibman, C. Klap, M. Pearlsman et al. 2016. Development of broad virus resistance in non-transgenic cucumber using CRISPR/Cas9 technology. Mol. Plant Pathol. 17(7): 1140–1153.

Chang, J.D., S. Huang, N. Yamaji, W. Zhang, J.F. Ma and F.J. Zhao. 2020. *OsNRAMP1* transporter contributes to cadmium and manganese uptake in rice. Plant Cell Environ. 43(10): 2476–2491.

Chang, Y., B.H. Nguyen, Y. Xie, B. Xiao, N. Tang, W. Zhu et al. 2017. Co-overexpression of the constitutively active form of OsbZIP46 and ABA-activated protein kinase SAPK6 improves drought and temperature stress resistance in rice. Front. Plant Sci. 8: 1102.

Char, S.N., A.K. Neelakandan, H. Nahampun, B. Frame, M. Main, M.H. Spalding et al. 2017. An Agrobacterium-delivered CRISPR/Cas9 system for high-frequency targeted mutagenesis in maize. Plant Biotechnol. J. 15(2): 257–268.

Chaudhry, S. and G.P.S. Sidhu. 2022. Climate change regulated abiotic stress mechanisms in plants: A comprehensive review. Plant Cell Rep. 41(1): 1–31.

Chen, G., J. Hu, L. Dong, D. Zeng, L. Guo, G. Zhang et al. 2020. The tolerance of salinity in rice requires the presence of a functional copy of FLN2. Biomol. 10: 17.

Chen, H.J., C.T. Su, C.H. Lin, G.J. Huang and Y.H. Lin. 2010. Expression of sweet potato cysteine protease SPCP2 altered developmental characteristics and stress responses in transgenic Arabidopsis plants. J. Plant Physiol. 167: 838–847.

Chen, K., Y. Wang, R. Zhang, H. Zhang and C. Gao. 2019. CRISPR/Cas genome editing and precision plant breeding in agriculture. Annu. Rev. Plant Biol. 70(1): 667–697.

Chen, S., N. Zhang, G. Zhou, S. Hussain, S. Ahmed, H. Tian et al. 2021. Knockout of the entire family of AITR genes in Arabidopsis leads to enhanced drought and salinity tolerance without fitness costs. BMC Plant Biol. 21: 1–15.

Chen, Y., Z. Wang, H. Ni, Y. Xu, Q. Chen and L. Jiang. 2017. CRISPR/Cas9-mediated base-editing system efficiently generates gain-of-function mutations in Arabidopsis. Sci. China. Life. Sci. 60(5): 520–523.

Chennakesavulu, K., H. Singh, P.K. Trivedi, M. Jain and S.R. Yadav. 2021. State-of-the-art in CRISPR technology and engineering drought, salinity, and thermo-tolerant crop plants. Plant. Cell. Rep. 1–17.

Choudry, M.W., R. Riaz, P. Nawaz, M. Ashraf, B. Ijaz and A. Bakhsh. 2024. CRISPR-Cas9 mediated understanding of plants' abiotic stress-responsive genes to combat changing climatic patterns. Funct. Integr. Genomics 24(4): 132.

Chu, C., R. Huang, L. Liu, G. Tang, J. Xiao, H. Yoo et al. 2022. The rice heavy-metal transporter OsNRAMP1 regulates disease resistance by modulating ROS homoeostasis. Plant Cell Environ. 45(4): 1109–1126.

Clarke, J.L. and P. Zhang. 2013. Plant biotechnology for food security and bioeconomy. Plant Mol. Biol. 83: 1–3.

Cong, L., F.A. Ran, D. Cox, S. Lin, R. Barretto, N. Habib et al. 2013. Multiplex genome engineering using CRISPR/Cas systems. Science 339(6121): 819–823.

Cong, L., F.A. Ran, D. Cox, S. Lin, R. Barreto, N. Habib et al. 2013. Multiplex genome engineering using CRISPR/Cas systems. Science 339: 819–823. http://dx.doi.org/10.1126/science.1231143.

Cui, Y., N. Jiang, Z. Xu and Q. Xu. 2020. Heterotrimeric G protein are involved in the regulation of multiple agronomic traits and stress tolerance in rice. BMC Plant Biol. 20: 1–13.

Curtin, S.J., Y. Xiong, J.M. Michno, B.W. Campbell, A.O. Stec, C T.˘ermák et al. 2018. CRISPR/Cas9 and TALENs generate heritable mutations for genes involved in small RNA processing of glycine max and *Medicago truncatula*. Plant Biotechnol. J. 16: 1125–1137.

Das, K., B.Y. Ayim, N. Borodynko-Filas, S.C. Das and F.M. Aminuzzaman. 2023. Genome editing (CRISPR/Cas9) in plant disease management: challenges and future prospects. J. Plant Prot. Res. 159–172.

Dawood, M.F., M.R. Sofy, H.I. Mohamed, A.R. Sofy and H.A. Abdel-Kader. 2022. Hydrogen sulfide modulates salinity stress in common bean plants by maintaining osmolytes and regulating nitric oxide levels and antioxidant enzyme expression. Journal of Soil Science and Plant Nutrition 22(3): 3708–3726.

de Lecuona, I., M. Casado, G. Marfany, M.L. Baroni and M. Escarrabill. 2017. Focus: genome editing: gene editing in humans: towards a global and inclusive debate for responsible research. Yale J. Biol. Med. 90: 673

Dong, D. et al. 2016. The crystal structure of Cpf1 in complex with CRISPR RNA. Nature 532: 522–526.

Dong, L., L. Li, C. Liu, C. Liu, S. Geng, X. Li et al. 2018. Genome editing and double-fluorescence proteins enable robust maternal haploid induction and identification in maize. Mol. Plant 11(9): 1214–1217.

Dong, O.X. and P.C. Ronald. 2019. Genetic engineering for disease resistance in plants: recent progress and future perspectives. Plant Physiol. 180(1): 26–38.

Dong, Q., L. Wallrad, B.O. Almutairi and J. Kudla. 2022. Ca^{2+} signaling in plant responses to abiotic stresses. J. Integr. Plant Biol. 64(2): 287–300.

Donovan, S., Y. Mao, D.J. Orr, E. Carmo-Silva and A.J. McCormick. 2020. CRISPR-Cas9-mediated mutagenesis of the Rubisco small subunit family in *Nicotiana tabacum*. Front. Genome Ed. 2: 605614.

Dracatos, P.M., R. Haghdoust, D. Singh, and R.F. Park. 2018. Exploring and exploiting the boundaries of host specificity using cereal rust and mildew models. New Phytologist 218(2): 453–462.

Du, Y.T., M.J. Zhao, C.T. Wang, Y. Gao, Y.X. Wang, Y.W. Liu et al. 2018. Identification and characterization of GmMYB118 responses to drought and salt stress. BMC Plant Biol. 18: 1–18.

Duan, Y.B., J. Li, R.Y. Qin, R.F. Xu, H. Li, Y.C. Yang et al. 2016. Identification of a regulatory element responsible for salt induction of rice OsRAV2 through *ex situ* and *in situ* promoter analysis. Plant Mol. Biol. 90: 49–62.

El-Mounadi, K., M.L. Morales-Floriano and H. Garcia-Ruiz. 2020. Principles, applications, and biosafety of plant genome editing using CRISPR-Cas9. Front. Plant Sci. 11: 56.

Endo, M., M. Mikami and S. Toki. 2016. Biallelic gene targeting in rice. Plant Physiol. 170(2): 667–677.

Erdoğan, İ., B. Cevher-Keskin, Ö. Bilir, Y. Hong and M. Tör. 2023. Recent developments in CRISPR/Cas9 genome-editing technology related to plant disease resistance and abiotic stress tolerance. Biology 12(7): 1037.

Erpen, L., H.S. Devi, J.W. Grosser and M. Dutt. 2018. Potential use of the DREB/ERF, MYB, NAC and WRKY transcription factors to improve abiotic and biotic stress in transgenic plants. Plant Cell, Tissue and Organ Culture (PCTOC) 132: 1–25.

Fauser, F., S. Schiml and H. Puchta. 2014. Both CRISPR/C as-based nucleases and nickases can be used efficiently for genome engineering in *Arabidopsis thaliana*. Plant J. 79(2): 348–359.

Feng, C., J. Yuan, R. Wang, Y. Liu, J.A. Birchler and F. Han. 2016. Efficient targeted genome modification in maize using CRISPR/Cas9 system. J. Genet. Genomics 43(1): 37–43.

Feng, Z., B. Zhang, W. Ding, X. Liu, D.L. Yang, P. Wei et al. 2013. Efficient genome editing in plants using a CRISPR/Cas system. Cell Res. 23(10): 1229–1232.

Fister, A.S., L. Landherr, S.N. Maximova and M.J. Guiltinan. 2018. Transient expression of CRISPR/Cas9 machinery targeting TcNPR3 enhances defense response in Theobroma cacao. Front. Plant Sci. 9: 268.

Foley, J.A., N. Ramankutty, K.A. Brauman, E.S. Cassidy, J.S. Gerber, M. Johnston et al. 2011. Solutions for a cultivated planet. Nature 478(7369): 337–342.

Fonfara, I., H. Richter, M. Bratovic ˇ, A. Le Rhun and E. Charpentier. 2016. The CRISPR-associated DNA-cleaving enzyme Cpf1 also processes precursor CRISPR RNA. Nature 532: 517–521.

Gaj, T., C.A. Gersbach and C.F. Barbas. 2013. ZFN, TALEN, and CRISPR/Cas-based methods for genome engineering. Trends Biotechnol. 31(7): 397–405.

Gajardo, H.A., O. Gómez-Espinoza, P. Boscariol Ferreira, H. Carrer and L.A. Bravo. 2023. The potential of CRISPR/Cas technology to enhance crop performance on adverse soil conditions. Plants 12(9): 1892.

Gao, C. 2021. Genome engineering for crop improvement and future agriculture. Cell 184(6): 1621–1635.

Gao, W., L. Long, X. Tian, F. Xu, J. Liu, P.K. Singh et al. 2017. Genome editing in cotton with the CRISPR/Cas9 system. Front. Plant Sci. 8: 1–12.

Garcia-Gimenez, G. and S.A. Jobling. 2021. Gene editing for barley grain quality improvement. J. Cereal Sci. 103394.

Garneau, J.E., M.È. Dupuis, M. Villion, D.A. Romero, R. Barrangou, P. Boyaval et al. 2010. The CRISPR/Cas bacterial immune system cleaves bacteriophage and plasmid DNA. Nature 468(7320): 67–71.

Gasiunas, G., R. Barrangou, P. Horvath and V. Siksnys. 2012. Cas9 crRNA ribonucleoprotein complex mediates specific DNA cleavage for adaptive immunity in bacteria. Proc. Nat. Acad. Sci. 109(39): 2579–2586.

Ghosh, S. and G. Dey. 2022. Biotic and abiotic stress tolerance through CRISPR-Cas mediated genome editing. J. Plant Biochem. Biotechnol. 31(2): 227–238.

Ghosh, S. and G. Dey. 2022. Biotic and abiotic stress tolerance through CRISPR-Cas mediated genome editing. Journal of Plant Biochemistry and Biotechnology 31(2): 227–238.

Gilbertson, R.L., O. Batuman, C.G. Webster and S. Adkins. 2015. Role of the insect supervectors *Bemisia tabaci* and *Frankliniella occidentalis* in the emergence and global spread of plant viruses. Annu. Review Virol. 2(1): 67–93.

Golfam, R., K. Kiarostami, T. Lohrasebi, K. Esfahani and H. Ghazvini. 2021. Application of molecular farming based on barley plant (Hordeum vulgare L.) platform: A review. Research on Crops 22(2): 410–424.

Gu, X., L. Liu and H. Zhang. 2021. Transgene-free genome editing in plants. Front. Genome Ed. 3: 805317.

Gumtow, R., D. Wu, J. Uchida and M. Tian. 2018. A phytophthora palmivora extracellular cystatin-like protease inhibitor targets papain to contribute to virulence on papaya. Molecular Plant-Microbe Interactions 31(3): 363–373.

Gupta, D., A. Saini, C. van der Vyver and S.K. Panda. 2024. Gene editing: paving the way for enhancing plant tolerance to abiotic stresses-mechanisms, breakthroughs, and future prospects. J. Plant Growth Regul. 1–17.

Hamdan, M.F., C.K.S. Karlson, E.Y. Teoh, S.E. Lau and B.C. Tan. 2022. Genome editing for sustainable crop improvement and mitigation of biotic and abiotic stresses. Plants 11(19): 2625.

Han, H.W., S.T. Yu, Z.W. Wang, Z. Yang, C.J. Jiang, X.Z. Wang et al. 2023. In planta genetic transformation to produce CRISPRed high-oleic peanut. Plant Growth Regul. 101(2): 443–451.

Haque, E., H. Taniguchi, M.M. Hassan, P. Bhowmik, M.R. Karim, M. Śmiech et al. 2018. Application of CRISPR/Cas9 genome editing technology for the improvement of crops cultivated in tropical climates: recent progress, prospects, and challenges. Front Plant Sci. 9: 617.

He, M., C.Q. He and N.Z. Ding. 2018. Abiotic stresses: general defenses of land plants and chances for engineering multistress tolerance. Front. Plant Sci. 9: 1771.

Herman, R.A., M. Fedorova and N.P. Storer. 2019. Will following the regulatory script for GMOs promote public acceptance of gene-edited crops? Trends Biotechnol. 37(12): 1272–1273.

Hiom, K. 2010. Coping with DNA double strand breaks. DNA Repair 9(12): 1256–1263.

Hoque, M.N., M. Tahjib-Ul-Arif, A. Hannan, N. Sultana, S. Akhter, M. Hasanuzzaman et al. 2021. Melatonin modulates plant tolerance to heavy metal stress: morphological responses to molecular mechanisms. Int. J. Mol. Sci. 22(21): 11445.

Hu, J.H., S.M. Miller, M.H. Geurts, W. Tang, L. Chen, N. Sun et al. 2018. Evolved Cas9 variants with broad PAM compatibility and high DNA specificity. Nature 556(7699): 57–63.

Hu, W., T. Zhou, Z. Han, C. Tan and Y. Xing. 2020. Interesting complementary interaction between OsC1 and two tightly linked genes, Rb1 and Rb2, controls the purple leaf sheath in rice. Theore Appl. Gene 133: 2555–2566.

Hu, Z., J. Li, S. Ding, F. Cheng, X. Li, Y. Jiang et al. 2021. The protein kinase CPK28 phosphorylates ascorbate peroxidase and enhances thermotolerance in tomato. Plant Physiol. 186: 1302–1317.

Hufford, M.B., A.S. Seetharam, M.R. Woodhouse, K.M. Chougule, S. Ou, J. Liu et al. 2021. *De novo* assembly, annotation, and comparative analysis of 26 diverse maize genomes. Science 373(6555): 655–662.

Hummel, A.W., R.D. Chauhan, T. Cermak, A.M. Mutka, A. Vijayaraghavan, A. Boyher et al. 2018. Allele exchange at the EPSPS locus confers glyphosate tolerance in cassava. Plant Biotechnol. J. 16: 1275–1282.

Hunter, M.C., R.G. Smith, M.E. Schipanski, L.W. Atwood and D.A. Mortensen. 2017. Agriculture in 2050: recalibrating targets for sustainable intensification. Bioscience 67(4): 386–391.

Hussain, A., M. Munir, B.H. Aloufi, M. Ali, M. Amanullah, Q. Ali et al. 2024. Engineering biotic stress tolerance via CRISPR-Cas mediated genome editing in crop plants. Plant Stress 100650.

Hwang, W.Y., Y. Fu, D. Reyon, M.L. Maeder, S.Q. Tsai, J.D. Sander et al. 2013. Efcient genome editing in zebrafsh using a CRISPR-Cas system. Nat. Biotechnol. 31: 227–229. htp://dx.doi.org/10.1038/nbt.2501.

Iqbal, Z., M.S. Iqbal, A. Hashem, E.F. Abd_Allah and M.I. Ansari. 2021. Plant defense responses to biotic stress and its interplay with fluctuating dark/light conditions. Front. Plant Sci. 12: 631810.

Ito, Y., A. Nishizawa-Yokoi, M. Endo, M. Mikami and S. Toki. 2015. CRISPR/Cas9-mediated mutagenesis of the RIN locus that regulates tomato fruit ripening. Biochem. Biophys. Res. Commun. 467(1): 76–82.

Jacobs, T.B., P.R. LaFayette, R.J. Schmitz and W.A. Parrott. 2015. Targeted genome modifications in soybean with CRISPR/Cas9. BMC Biotechnol. 15: 1–10.

Jacobs, T.B., N. Zhang, D. Patel and G.B. Martin. 2017. Generation of a collection of mutant tomato lines using pooled CRISPR libraries. Plant Physiol. 174(4): 2023–2037.

Jagadish, S.K., D.A. Way and T.D. Sharkey. 2021. Plant heat stress: Concepts directing future research. Plant Cell Environ. 44(7): 1992–2005.

Jain, M. 2015. Function genomics of abiotic stress tolerance in plants: A CRISP approach. Front. Plant Sci. 6: e93806.

Jenkins, D., R. Dobert, A. Atanassova and C. Pavely. 2021. Impacts of the regulatory environment for gene editing on delivering beneficial products. *In Vitro* Cellular and Developmental Biology-Plant 57(4): 609–626.

Jeon, J.E., J.G. Kim, C.R. Fischer, N. Mehta, C. Dufour-Schroif, K. Wemmer et al. 2020. A pathogen-responsive gene cluster for highly modified fatty acids in tomato. Cell 180(1): 176–187.

Jha, Y. and H.I. Mohamed. 2023. Inoculation with Lysinibacillus fusiformis strain YJ4 and Lysinibacillus sphaericus strain YJ5 alleviates the effects of cold stress in maize plants. Gesunde Pflanzen 75(1): 77–95.

Ji, X., H. Zhang, Y. Zhang, Y. Wang and C. Gao. 2015. Establishing a CRISPR–Cas-like immune system conferring DNA virus resistance in plants. Nat. Plants 1(10): 1–4.

Jia, H. and W. Nian. 2014. Targeted genome editing of sweet orange using Cas9/sgRNA. PloS ONE 9: e93806.

Jia, H., V. Orbovic, J.B. Jones and N. Wang. 2016. Modification of the PthA4 effector binding elements in Type I CsLOB1 promoter using Cas9/sg RNA to produce transgenic Duncan grapefruit alleviating XccΔpthA4: dCs LOB 1.3 infection. Plant Biotechnol. J. 14(5): 1291–1301.

Jiang, F. and J.A. Doudna. 2017. CRISPR–Cas9 structures and mechanisms. Annu. Rev. Biophys. 46(1): 505–529.

Jiang, W., H. Zhou, H. Bi, M. Fromm, B. Yang and D.P. Weeks. 2013. Demonstration of CRISPR/Cas9/sgRNA-mediated targeted gene modification in Arabidopsis, tobacco, sorghum and rice. Nucleic Acids Res. 41(20): e188–e188.

Jinek, M., K. Chylinski, I. Fonfara, M. Hauer, J.A. Doudna and E. Charpentier. 2012. A programmable dual-RNA–guided DNA endonuclease in adaptive bacterial immunity. Science 337(6096): 816–821.

Kaiser, N., D. Douches, A. Dhingra, K.C. Glenn, P.R. Herzig, E.C. Stowe et al. 2020. The role of conventional plant breeding in ensuring safe levels of naturally occurring toxins in food crops. Trends in Food Science and Technology 100: 51–66.

Karmakar, S., P. Das, D. Panda, K. Xie, M.J. Baig and K.A. Molla. 2022. A detailed landscape of CRISPR-Cas-mediated plant disease and pest management. Plant Sci. 323: 111376.

Kaur, N., A. Alok, N. Shivani, N. Kaur, P. Pandey, P. Awasthi et al. 2018. CRISPR/Cas9-mediated efficient editing in phytoene desaturase (PDS) demonstrates precise manipulation in banana cv. Rasthali genome. Functional and Integrative Genomics 18: 89–99.

Kaur, N., S. Sharma, M. Hasanuzzaman and P.K. Pati. 2022. Genome editing: a promising approach for achieving abiotic stress tolerance in plants. Int. J. Genomics 2022(1): 5547231.

Kaur, R., S. Das, S. Bansal, G. Singh, S. Sardar, H. Dhar et al. 2021. Heavy metal stress in rice: Uptake, transport, signaling, and tolerance mechanisms. Physiol. Plant. 173(1): 430–448.

Khalil, A.M. 2020. The genome editing revolution. J. Genet. Eng. Biotechnol. 18(1): 68.

Khan, M.I.R., M. Fatma, T.S. Per, N.A. Anjum and N.A. Khan. 2015. Salicylic acid-induced abiotic stress tolerance and underlying mechanisms in plants. Front. Plant Sci. 6: 462.

Khan, N., S. Syeed, A. Masood, R. Nazar and N. Iqbal. 2010. Application of salicylic acid increases contents of nutrients and antioxidative metabolism in mungbean and alleviates adverse effects of salinity stress. Int. J. Plant Biol. 1(1): e1–e1.

Khokhar Voytas, A., M. Shahbaz, M.F. Maqsood, U. Zulfiqar, N. Naz, U.Z. Iqbal et al. 2023. Genetic modification strategies for enhancing plant resilience to abiotic stresses in the context of climate change. Funct. Integr. Genomics 23(3): 283.

Kim, D., J. Kim, J.K. Hur, K.W. Been, S.H. Yoon and J.S. Kim. 2016. Genome-wide analysis reveals specificities of Cpf1 endonucleases in human cells. Nat Biotechnol. 34(8): 863–868.

Kim, D., B. Alptekin and H. Budak. 2018. CRISPR/Cas9 genome editing in wheat. Funct. Integr. Genomics 18: 31–41.

Kim, S.T., M. Choi, S.J. Bae and J.S. Kim. 2021. The functional association of acqos/victr with salt stress resistance in *Arabidopsis thaliana* was confirmed by crispr mediated mutagenesis. Int. J. Mol. Sci. 22: 11389.

Kim, S., D. Kim, S.W. Cho, J. Kim and J.S. Kim. 2014. Highly efficient RNA guided genome editing in human cells via delivery of purified Cas9 ribonucleoproteins. Genome Res. 24: 1012–1019.

Kis, A., É. Hamar, G. Tholt, R. Bán and Z. Havelda. 2019. Creating highly efficient resistance against wheat dwarf virus in barley by employing CRISPR/Cas9 system. Plant Biotechnol. J. 17(6): 1004.

Klap, C., E. Yeshayahou, A.M. Bolger, T. Arazi, S.K. Gupta, S. Shabtai et al. 2017. Tomato facultative parthenocarpy results from SlAGAMOUS-LIKE 6 loss of function. Plant Biotechnol. J. 15: 634–647.

Kleinstiver, B.P., S.Q. Tsai, M.S. Prew, N.T. Nguyen, M.M. Welch, J.M. Lopez et al. 2016. Genome-wide specificities of CRISPR-Cas Cpf1 nucleases in human cells. Nat Biotechnol. 34(8): 869–74.

Komor, A.C., K.T. Zhao, M.S. Packer, N.M. Gaudelli, A.L. Waterbury, L.W. Koblan et al. 2017. Improved base excision repair inhibition and bacteriophage Mu Gam protein yields C: G-to-T: A base editor with higher efficiency and product purity. Sci. Adv. 3(8): eaao4774.

Kumar, M., D. Ayzenshtat, A. Marko and S. Bocobza. 2022. Optimization of TDNA configuration with UBIQUITIN10 promoters and tRNA–sgRNA complexes promotes highly efficient genome editing in allotetraploid tobacco. Plant Cell Rep. 41: 1–20.

Kumar, M., M.R. Prusty, M.K. Pandey, P.K. Singh, A. Bohra, B. Guo et al. 2023. Application of CRISPR/Cas9-mediated gene editing for abiotic stress management in crop plants. Front. Plant Sci. 14: 1157678.

Lan, T., Y. Zheng, Z. Su, S. Yu, H. Song, X. Zheng et al. 2019. OsSPL10, a SBPbox gene, plays a dual role in salt tolerance and trichome formation in rice (*Oryza sativa* l.). G3: Genes Genomes Genet. 9: 4107–4114.

Landa, B.B., M. Saponari, O.R. Feitosa-Junior, A. Giampetruzzi, F.J. Vieira, E. Mor et al. 2022. Xylella fastidiosa's relationships: The bacterium, the host plants, and the plant microbiome. New Phytologist 234(5): 1598–1605.

Langner, T., S. Kamoun and K. Belhaj. 2018. CRISPR crops: plant genome editing toward disease resistance. Annu. Rev. Phytopathol. 56(1): 479–512.

Lawrenson, T., O. Shorinola, N. Stacey, C. Li, L. Østergaard, N. Patron et al. 2015. Induction of targeted, heritable mutations in barley and *Brassica oleracea* using RNA-guided Cas9 nuclease. Genome Biol. 16: 1–13.

Le, V.T., M.S. Kim, Y.J. Jung, K.K. Kang and Y.G. Cho. 2022. Research trends and challenges of using CRISPR/Cas9 for improving rice productivity. Agronomy 12(1): 164.

Ledford, H. 2015a. Biologists create more precise molecular scissors for genome editing. Nature. Available at: www.nature.com/news/biologists-create-more-precise-molecular-scissors-for-genome-editing-1.18932.

Lewi, C. and Vici´en. 2020. Argentina's local crop biotechnology developments: why have they not reached the market yet? Front. Bioeng. Biotechnol. 8.

Li, B., S. Liang, M. Alariqi, F. Wang, G. Wang, Q. Wang et al. 2021a. The application of temperature sensitivity CRISPR/LbCpf1 (LbCas12a) mediated genome editing in allotetraploid cotton (*G. hirsutum*) and creation of non-transgenic, gossypol free cotton. Plant Biotechnol. J. 19: 221.

Li, C., W. Li, Z. Zhou, H. Chen, C. Xie and Y. Lin. 2020a. A new rice breeding method: CRISPR/Cas9 system editing of the Xa13 promoter to cultivate transgene-free bacterial blight-resistant rice. Plant Biotechnol. J. 18(2): 313.

Li, C., C. Liu, X. Qi, Y. Wu, X. Fei, L. Mao et al. 2017a. RNA-guided Cas9 as an *in vivo* desired-target mutator in maize. Plant Biotechnol. J. 15(12): 1566–1576.

Li, J.F., J.E. Norville, J. Aach, M. McCormack, D. Zhang, J. Bush et al. 2013. Multiplex and homologous recombination-mediated genome editing in arabidopsis and *Nicotiana benthamiana* using guide RNA and Cas9. Nat. Biotechnol. 31: 688–691.

Li, J., X. Meng, Y. Zong, K. Chen, H. Zhang, J. Liu et al. 2016a. Gene replacements and insertions in rice by intron targeting using CRISPR–Cas9. Nat. Plants 2(10): 1–6.

Li, J., H. Zhang, X. Si, Y. Tian, K. Chen, J. Liu et al. 2017b. Generation of thermosensitive male-sterile maize by targeted knockout of the ZmTMS5 gene. J. Genet. Genomics= Yi chuan xue bao 44(9): 465–468.

Li, P., X. Li and M. Jiang. 2021b. CRISPR/Cas9-mediated mutagenesis of WRKY3 and WRKY4 function decreases salt and me-JA stress tolerance in *Arabidopsis thaliana*. Mol. Biol. Rep. 48: 5821–5832.

Li, P., Y.J. Li, F.J. Zhang, G.Z. Zhang, X.Y. Jiang, H.M. Yu et al. 2017c. The Arabidopsis UDP-glycosyltransferases UGT79B2 and UGT79B3, contribute to cold, salt and drought stress tolerance via modulating anthocyanin accumulation. Plant J. 89(1): 85–103.

Li, R., C. Liu, R. Zhao, L. Wang, L. Chen, W. Yu et al. 2019. CRISPR/Cas9-mediated SlNPR1 mutagenesis reduces tomato plant drought tolerance. BMC Plant Biol. 19: 1–13.

Li, R., L. Zhang, L. Wang, L. Chen, R. Zhao, J. Sheng et al. 2018. Reduction of tomato-plant chilling tolerance by CRISPR-Cas9-Mediated SlCBF1 mutagenesis. J. Agric. Food Chem. 66: 9042–9051.

Li, T., B. Liu, C.Y. Chen and B. Yang. 2016b. TALEN-mediated homologous recombination produces site-directed DNA base change and herbicide-resistant rice. J. Genet. Genomics 43(5): 297–305.

Li, Y., X. Wu, Y. Zhang and Q. Zhang. 2022. CRISPR/Cas genome editing improves abiotic and biotic stress tolerance of crops. Front. Genome Ed. 4: 987817.

Li, Y., J. Zhu, H. Wu, C. Liu, C. Huang, J. Lan et al. 2020b. Precise base editing of non-allelic acetolactate synthase genes confers sulfonylurea herbicide resistance in maize. Crop J. 8(3): 449–456.

Liang, Z., K. Zhang, K. Chen and C. Gao. 2014. Targeted mutagenesis in *Zea mays* using TALENs and the CRISPR/Cas system. J. Genet. Genomics 41: 63–68.

Liao, S., X. Qin, L. Luo, Y. Han, X. Wang, B. Usman et al. 2019. CRISPR/Cas9-induced mutagenesis of semi-rolled Leaf1,2 confers curled leaf phenotype and drought tolerance by influencing protein expression patterns and ROS scavenging in rice (*Oryza sativa* L.). Agronomy 9: 728.

Lin, C.S., C.T. Hsu, L.H. Yang, L.Y. Lee, J.Y. Fu, Q.W. Cheng et al. 2018. Application of protoplast technology to CRISPR/Cas9 mutagenesis: from single-cell mutation detection to mutant plant regeneration. Plant Biotechnol. J. 16: 1295–1310.

Liu, H., M. Hu, Q. Wang, L. Cheng and Z. Zhang. 2018. Role of papain-like cysteine proteases in plant development. Front. Plant Sci. 9: 1717.

Liu, H., K. Wang, Z. Jia, Q. Gong, Z. Lin, L. Du et al. 2020. Efficient induction of haploid plants in wheat by editing of TaMTL using an optimized Agrobacterium-mediated CRISPR system. Journal of Experimental Botany 71(4): 1337–1349.

Liu, L., Y. Kuang, F. Yan, S. Li, B. Ren, G. Gosavi et al. 2021. Developing a novel artificial rice germplasm for dinitroaniline herbicide resistance by base editing of *OsTubA2*. Plant Biotechnol. J. 19(1): 5.

Liu, L., J. Zhang, J. Xu, Y. Li, L. Guo, Z. Wang et al. 2020a. CRISPR/Cas9 targeted mutagenesis of *SlLBD40*, a lateral organ boundaries domain transcription factor, enhances drought tolerance in tomato. Plant Sci. 301: 110683.

Liu, X., D. Wu, T. Shan, S. Xu, R. Qin, H. Li et al. 2020b. The trihelix transcription factor *OsGTγ-2* is involved adaption to salt stress in rice. Plant Mol. Biol. 103: 545–560.

Liu, Y., C. Zhang, X. Wang, X. Li and C. You. 2022a. CRISPR/Cas9 technology and its application in horticultural crops. Hortic. Plant. J. 8(4): 395–407.

Liu, Z., C. Ma, L. Hou, X. Wu, D. Wang, L. Zhang et al. 2022b. Exogenous SA affects rice seed germination under salt stress by regulating Na(+)/K(+) balance and endogenous GAs and ABA homeostasis. Int. J. Mol. Sci. 23(6): 3293.

Lou, D., H. Wang, G. Liang and D. Yu. 2017. OsSAPK2 confers abscisic acid sensitivity and tolerance to drought stress in rice. Front. Plant Sci. 8: 993.

Lu, H.P., S.M. Liu, S.L. Xu, W.Y. Chen, X. Zhou, Y.Y. Tan et al. 2017. CRISPR-s: an active interference element for a rapid and inexpensive selection of genome-edited, transgene-free rice plants. Plant Biotechnol. J. 15: 1371.

Lu, Y., J. Wang, B. Chen, S. Mo, L. Lian, Y. Luo and L. Jiang. 2021. A donor-DNA-free CRISPR/Cas-based approach to gene knock-up in rice. Nat. Plants 7(11): 1445–1452.

Lu, Y., X. Ye, R. Guo, J. Huang, W. Wang, J. Tang et al. 2017. Genome-wide targeted mutagenesis in rice using the CRISPR/Cas9 system. Molecular Plant 10(9): 1242–1245.

Lucioli, A., R. Tavazza, S. Baima, K. Fatyol, J. Burgyan and M. Tavazza. 2022. CRISPR-Cas9 targeting of the eIF4E1 gene extends the potato virus y resistance spectrum of the Solanum tuberosum L. cv. desirée. Front. Microbiol. 13: 873930.

Lyon, J. 2017. Bioethics panels open door slightly to germline gene editing. JAMA 318: 1639–1640

Lyu, Y.-S., L.-M. Cao, W.-Q. Huang, J.-X. Liu and H.-P. Lu. 2022. Disruption of three polyamine uptake transporter genes in rice byCRISPR/Cas9 gene editing confers tolerance to herbicide paraquat. aBIOTECH. 3(2): 140–145.

Ma, J., J. Chen, M. Wang, Y. Ren, S. Wang, C. Lei et al. 2018. Disruption of OsSEC3A increases the content of salicylic acid and induces plant defense responses in rice. J. Exp. Bot. 69(5): 1051–1064.

Ma, X., Q. Zhu, Y. Chen and Y.G. Liu. 2016. CRISPR/Cas9 platforms for genome editing in plants: developments and applications. Mol. Plant 9(7): 961–974.

Ma, Y., M.C. Dias and H. Freitas. 2020. Drought and salinity stress responses and microbe-induced tolerance in plants. Front. Plant Sci. 11: 591911.

Ma, Y., L. Zhang and X. Huang. 2014. Genome modifcation by CRISPR/Cas9. FEBS. J. 281: 5186–5193. http://dx.doi.org/10.1111/febs.13110.

Macovei, A., N.R. Sevilla, C. Cantos, G.B. Jonson, I. Slamet-Loedin, T. Čermák et al. 2018. Novel alleles of rice eIF4G generated by CRISPR/Cas9-targeted mutagenesis confer resistance to Rice tungro spherical virus. Plant Biotechnol. J. 16(11): 1918–1927.

Makhotenko, A.V., A.V. Khromov, E.A. Snigir, S.S. Makarova, V.V. Makarov, T.P. Suprunova et al. 2019. Functional analysis of coilin in virus resistance and stress tolerance of potato Solanum tuberosum using CRISPR-Cas9 editing. Dokl Biochem. Biophys. 484: 88–91.

Mallapaty, S. 2022. China's approval of gene-edited crops energizes researchers. Nature 602: 559–560.

Malnoy, M., R. Viola, M.H. Jung, O.J. Koo, S. Kim, J.S. Kim et al. 2016. DNA-free genetically edited grapevine and apple protoplast using CRISPR/Cas9 ribonucleoproteins. Front. Plant Sci. 7: 1904.

Manghwar, H., B. Li, X. Ding, A. Hussain, K. Lindsey, X. Zhang et al. 2020. CRISPR/Cas systems in genome editing: methodologies and tools for sgRNA design, off-target evaluation, and strategies to mitigate off-target effects. Adv. Sci. 7(6): 1902312.

Manghwar, H., K. Lindsey, X.L. Zhang and S.X. Jin. 2019. CRISPR/Cas system: recent advances and future prospects for genome editing. Trends Plant Sci. 24: 1102–1125.

Mao, X., Y. Zheng, K. Xiao, Y. Wei, Y. Zhu, Q. Cai et al. 2018. OsPRX2 contributes to stomatal closure and improves potassium deficiency tolerance in rice. Biochem. Biophys. Res. Commun. 495(1): 461–467.

Mao, Y., J.R. Botella, Y. Liu and J.K. Zhu. 2019. Gene editing in plants: Progress and challenges. Natl. Sci. Rev. 6: 421–437.

Marium, A., Tariq, S., Arshad, H., Khalid, M.S., Murtaza, B., Aslam, H. et al. 2025. CRISPR/CAS genome editing to enhance resistance leading to enhanced productivity in crops. Kashf Journal of Multidisciplinary Research 2(02): 53–89.

Mathivanan, S., S. Sundaresan and C. Arumugam. 2023. Genome editing for enhancing abiotic stress tolerance in crop plants. Cornous Biology 1(1): 7–17.

Matres, J.M., Hilscher, J., Datta, A., Armario-Nájera, V., Baysal, C., He, W. et al. 2021. Genome editing in cereal crops: an overview. Transgenic Research 30: 461–498.

Meng, X., Yu, H., Zhang, Y., Zhuang, F., Song, X., Gao, S. et al. 2017. Construction of a genome-wide mutant library in rice using CRISPR/Cas9. Mol. Plant 10(9): 1238–1241.

Miao, C., Xiao, L., Hua, K., Zou, C., Zhao, Y., Bressan, R.A. et al. 2018. Mutations in a subfamily of abscisic acid receptor genes promote rice growth and productivity. Proceedings of the National Academy of Sciences 115(23): 6058–6063.

Miao, J., Guo, D., Zhang, J., Huang, Q., Qin, G., Zhang, X. et al. 2013. Targeted mutagenesis in rice using CRISPR-cas system. Cell Res. 23: 1233–1236.

Mishra, R., R.K. Joshi and K. Zhao. 2018. Genome editing in rice: recent advances, challenges, and future implications. Frontiers in Plant Science 9: 1361.

Mo, W., W. Tang, Y. Du, Y. Jing, Q. Bu and R. Lin. 2020. *PHYTOCHROME-INTERACTINGFACTOR-LIKE14* and *SLENDER RICE1* interaction controls seedling growth under salt stress. Plant Physiol. 184(1): 506–517.

Mohamed, H.I., A. Khan and A. Basit. 2024. CRISPR-Cas9 system mediated genome editing technology: an ultimate tool to enhance abiotic stress in crop plants. J. Soil Sci. Plant Nutrit. 24(2): 1799–822.

Molla, K.A., Sretenovic, S., Bansal, K.C., & Qi, Y. 2021. Precise plant genome editing using base editors and prime editors. Nature Plants 7(9): 1166–1187.

Mulvihill, J.J., Capps, B., Joly, Y., Lysaght, T., Zwart, H.A. and Chadwick, R. 2017. Ethical issues of CRISPR technology and gene editing through the lens of solidarity. Br. Med. Bull. 122: 17–29

Mushtaq, M., Mukhtar, S., Sakina, A., Dar, A.A., Bhat, R., Deshmukh, R. et al. 2020. Tweaking genome-editing approaches for virus interference in crop plants. Plant Physiology and Biochemistry 147: 242–250.

Mushtaq, M., Sakina, A., Wani, S.H., Shikari, A.B., Tripathi, P., Zaid, A. et al. 2019. Harnessing genome editing techniques to engineer disease resistance in plants. Front. Plant Sci. 10: 550.

Najera, V.A., Twyman, R.M., Christou, P. and Zhu, C. 2019. Applications of multiplex genome editing in higher plants. Curr. Opin. Biotechnol. 59: 93–102.

Nandy, S., Pathak, B., Zhao, S. and Srivastava, V. 2019. Heat-shock-inducible CRISPR/Cas9 system generates heritable mutations in rice. Plant Direct 3(5): e00145.

Nawaz, G., Han, Y., Usman, B., Liu, F., Qin, B. and Li, R. 2019. Knockout of *OsPRP1*, a gene encoding proline-rich protein, confers enhanced cold sensitivity in rice (*Oryza sativa* l.) at the seedling stage. 3 Biotech. 9: 1–18.

Negi, C., Vasistha, N.K., Singh, D., Vyas, P. and Dhaliwal, H.S. 2022. Application of CRISPR-mediated gene editing for crop improvement. Mol. Biotechnol. 64(11): 1198–1217.

Nejat, N. and N. Mantri. 2017. Plant immune system: crosstalk between responses to biotic and abiotic stresses the missing link in understanding plant defence. Curr. Issues Mol. Biol. 23(1): 1–16.

Nekrasov, V., Staskawicz, B., Weigel, D., Jones, J.D. and Kamoun, S. 2013. Targeted mutagenesis in the model plant *Nicotiana benthamiana* using Cas9 RNA-guided endonuclease. Nat. Biotechnol. 31(8): 691–693.

Nekrasov, V., Wang, C., Win, J., Lanz, C., Weigel, D. and Kamoun, S. 2017. Rapid generation of a transgene-free powdery mildew resistant tomato by genome deletion. Sci. Rep. 7(1): 482.

Nieves-Cordones, M., Mohamed, S., Tanoi, K., Kobayashi, N.I., Takagi, K., Vernet, A. et al. 2017. Production of low-Cs^+ rice plants by inactivation of the K^+ transporter *OsHAK1* with the CRISPR-Cas system. Plant. J. 92(1): 43–56.

Niu, Q., Wu, S., Li, Y., Yang, X., Liu, P., Xu, Y. et al. 2020. Expanding the scope of CRISPR/Cas9-mediated genome editing in plants using an xCas9 and Cas9-NG hybrid. Journal of Integrative Plant Biology 62(4): 398–402.

Nidhi, S., Anand, U., Oleksak, P., Tripathi, P., Lal, J.A., Thomas, G. et al. 2021. Novel CRISPR–Cas systems: an updated review of the current achievements, applications, and future research perspectives. International Journal of Molecular Sciences 22(7): 3327.

Noureen, A., Khan, M.Z., Amin, I., Zainab, T. and Mansoor, S. 2022. CRISPR/Cas9-mediated targeting of susceptibility factor *eIF4E*-enhanced resistance against potato virus Y. Front. Genet. 13: 922019.

Nuffield Council on Bioethics. 2016. Genome Editing: An Ethical Review. http://nuffieldbioethics.org/wp-content/uploads/Genome-editing-an-ethical-review.pdf (7 April 2019, last assessed).

Ogata, T., Ishizaki, T., Fujita, M. and Fujita, Y. 2020. CRISPR/Cas9-targeted mutagenesis of *OsERA1* confers enhanced responses to abscisic acid and drought stress and increased primary root growth under nonstressed conditions in rice. PLoS ONE 15: e0243376.

Oliva, R., Ji, C., Atienza-Grande, G., Huguet-Tapia, J.C., Perez-Quintero, A., Li, T. et al. 2019. Broad-spectrum resistance to bacterial blight in rice using genome editing. Nat. Biotechnol. 37(11): 1344–1350.

Ortigosa, A., Gimenez-Ibanez, S., Leonhardt, N. and Solano, R. 2019. Design of a bacterial speck resistant tomato by CRISPR/Cas9-mediated editing of SlJAZ2. Plant Biotechnol. J., 17(3), 665-673.

Osakabe, Y., Watanabe, T., Sugano, S.S., Ueta, R., Ishihara, R., Shinozaki, K. et al. 2016. Optimization of CRISPR/Cas9 genome editing to modify abiotic stress responses in plants. Sci. Rep. 6: 26685.

Oz, M.T., A. Altpeter, R. Karan, A. Merotto and F. Altpeter. 2021. CRISPR/ Cas9-mediated multi-allelic gene targeting in sugarcane confers herbicide tolerance. Front. Genome Ed. 3.

Paix˜ao, J.F.R., F. Gillet, T.P. Ribeiro, C. Bournaud, I.T. Lourenço-tess, D.D. Noriega et al. 2019. Improved drought stress tolerance in Arabidopsis by CRISPR/dCas9 fusion with a histone Acetyl Transferase. Sci. Rep. 1–9.

Pallarès Masmitjà, M., N. Knödlseder and M. Güell. 2019. CRISPR-gRNA design. CRISPR Gene Editing: Methods and Protocols 3–11.

Pan, C., L. Ye, L. Qin, X. Liu, Y. He, J. Wang et al. 2016. CRISPR/Cas9-mediated efficient and heritable targeted mutagenesis in tomato plants in the first and later generations. Sci. Rep. 6(1): 24765.

Park, J.J., E. Dempewolf, W. Zhang and Z.Y. Wang. 2017. RNA-Guided transcriptional activation via CRISPR/dCas9 mimics overexpression phenotypes in Arabidopsis. PloS ONE 12: e0179410.

Park, J. and S. Bae. 2017. Cpf1-database: web-based genome-wide guide RNA library design for gene knockout screens using CRISPR-Cpf1. Bioinformatics 34: 1077–1079. doi: 10.1093/bioinformatics/btx695.

Park, S.II, H.B. Kim, H.J. Jeon and H. Kim. 2021. Agrobacterium-mediated capsicum annuum gene editing in two cultivars, hot pepper CM334 and bell pepper dempsey. Int. J. Mol. Sci. 22: 3921.

Peng, A., S. Chen, T. Lei, L. Xu, Y. He, L. Wu et al. 2017. Engineering canker-resistant plants through CRISPR/Cas9-targeted editing of the susceptibility gene *CsLOB1* promoter in citrus. Plant Biotechnol. J. 15(12): 1509–1519.

Pixley, J.B. Falck-Zepeda, R.L. Paarlberg, P.W.B. Phillips, I.H. Slamet-Loedin, K.S. Dhugga et al. 2022. Genome-edited crops for improved food security of smallholder farmers, Nat. Genet. 54: 364–367.

Prado, J.R., G. Segers, T. Voelker, D. Carson, R. Dobert, J. Phillips et al. 2014. Genetically engineered crops: from idea to product. Annu. Rev. Plant Biol. 65(1): 769–790.

Pramanik, D., R.M. Shelake, J. Park, M.J. Kim, I. Hwang, Y. Park et al. 2021. CRISPR/Cas9-mediated generation of pathogen-resistant tomato against tomato yellow leaf curl virus and powdery mildew. Int. J. Mol. Sci. 22(4): 1878.

Praveen, A., S. Dubey, S. Singh and V.K. Sharma. 2023. Abiotic stress tolerance in plants: a fascinating action of defense mechanisms. 3 Biotech 13(3): 102.

Prihatna, C., M.J. Barbetti and S.J. Barker. 2018. A novel tomato fusarium wilt tolerance gene. Front Microbiol. 9: 1226.

Puchta, H. and F. Fauser. 2014. Synthetic nucleases for genome engineering in plants: prospects for a bright future. Plant J. 78(5): 727–741.

Pyott, D.E., E. Sheehan and A. Molnar. 2016. Engineering of CRISPR/Cas9-mediated potyvirus resistance in transgene-free *Arabidopsis* plants. Mol. Plant Pathol. 17(8): 1276–1288.

Qin, H., J. Wang, X. Chen, F. Wang, P. Peng, Y. Zhou et al. 2019. Rice *OsDOF15* contributes to ethylene-inhibited primary root elongation under salt stress. New Phytol. 223: 798–813.

Qin, Q., Y. Wang, L. Huang, F. Du, X. Zhao, Z. Li et al. 2020. A U-box E3 ubiquitin ligase *OsPUB67* is positively involved in drought tolerance in rice. Plant. Mol. Biol. 102: 89–107.

Qiu, Z., S. Kang, L. He, J. Zhao, S. Zhang, J. Hu et al. 2018. The newly identified heat-stress sensitive albino 1 gene affects chloroplast development in rice. Plant Sci. 267: 168–179.

Rajput, M., K. Choudhary, M. Kumar, V. Vivekanand, A. Chawade, R. Ortiz et al. 2021. RNA interference and CRISPR/Cas gene editing for crop improvement: Paradigm shift towards sustainable agriculture. Plants 10(9): 1914.

Randolph, P.B. and J.R. Davis. 2019. Search-and-replace genome editing without double-strand breaks or donor DNA. Nature 576: 149–157.

Rasheed, A., R.A. Gill, M.U. Hassan, A. Mahmood, S. Qari, Q.U. Zaman et al. 2021. A critical review: recent advancements in the use of CRISPR/Cas9 technology to enhance crops and alleviate global food crises. Curr. Issues Mol. Biol. 43(3): 1950–1976.

Rath, D., L. Amlinger, A. Rath and M. Lundgren. 2015. The CRISPR-Cas immune system: biology, mechanisms and applications. Biochimie 117: 119–128.

Ray, D.K., N.D. Mueller, P.C. West and J.A. Foley. 2013. Yield trends are insufficient to double global crop production by 2050. PloS ONE 8(6): e66428.

Razzaq, M.K., M. Akhter, R.M. Ahmad, K.L. Cheema, A. Hina, B. Karikari et al. 2022. CRISPR-Cas9 based stress tolerance: New hope for abiotic stress tolerance in chickpea (*Cicer arietinum*). Mol. Biol. Rep. 49: 8977–8985.

Richter, J., J.M. Watson, P. Stasnik, M. Borowska, J. Neuhold, M. Berger et al. 2018. Multiplex mutagenesis of four clustered *CrRLK1L* with CRISPR/Cas9 exposes their growth regulatory roles in response to metal ions. Sci. Rep. 8(1): 12182.

Roca Paixão, J.F., F.X. Gillet, T.P. Ribeiro, C. Bournaud, I.T. Lourenço-Tessutti, D.D. Noriega et al. 2019. Improved drought stress tolerance in arabidopsis by CRISPR/dCas9 fusion with a histone AcetylTransferase. Sci. Rep. 9: 8080.

Rodriguez, E. 2016. Ethical issues in genome editing using Crispr/Cas9 System. J. Clin. Res. Bioethics 2016.

Rodríguez-Leal, D., Z.H. Lemmon, J. Man, M.E. Bartlett and Z.B. Lippman. 2017. Engineering quantitative trait variation for crop improvement by genome editing. Cell 171(2): 470–480.

Ron, M., K. Kajala, G. Pauluzzi, D. Wang, M.A. Reynoso, K. Zumstein et al. 2014. Hairy root transformation using Agrobacterium rhizogenes as a tool for exploring cell type-specific gene expression and function using tomato as a model. Plant Physiol. 166(2): 455–469.

Roossinck, M.J., D.P. Martin and P. Roumagnac. 2015. Plant virus metagenomics: advances in virus discovery. Phytopathology 105(6): 716–727.

Safari, F., Zare, K., Negahdaripour, M., Barekati-Mowahed, M., and Ghasemi, Y. 2019. CRISPRCpf1proteins: structure, function and implications for genomeediting. Cell Biosci. 9: 1–21.

Saijo, Y. and Loo, E.P.I. 2020. Plant immunity in signal integration between biotic and abiotic stress responses. New Phytologist 225(1): 87–104.

Sandhu, K.S., Patil, S.S., Aoun, M. and Carter, A.H. 2022. Multi-trait multi-environment genomic prediction for end-use quality traits in winter wheat. Front. Genet. 13.

Santillán Martínez, M.I., Bracuto, V., Koseoglou, E., Appiano, M., Jacobsen, E., Visser, R.G. and Bai, Y. 2020. CRISPR/Cas9-targeted mutagenesis of the tomato susceptibility gene PMR4 for resistance against powdery mildew. BMC Plant Biol. 20: 1–13.

Santosh Kumar, V.V., Verma, R.K., Yadav, S.K., Yadav, P., Watts, A., Rao, M.V. and Chinnusamy, V. 2020. CRISPR-Cas9 mediated genome editing of drought and salt tolerance (OsDST) gene in indica mega rice cultivar MTU1010. Physiol. Mol. Biol. Plants. 26: 1099–1110.

Schaeffer, S.M. and P.A. Nakata. 2016. The expanding footprint of CRISPR/Cas9 in the plant sciences. Plant Cell Rep. 35: 1451–1468.

Scheben, A., Wolter, F., Batley, J., Puchta, H. and Edwards, D. 2017. Towards CRISPR/Cas crops - bringing together genomics and genome editing. New Phytol. 216: 682–698.

Schiml, S., Fauser, F., and Puchta, H. 2014. The CRISPR/C as system can be used as nuclease for in planta gene targeting and as paired nickases for directed mutagenesis in Arabidopsis resulting in heritable progeny. Plant J. 80(6): 1139–1150.

Shahriar, S.A., Islam, M.N., Chun, C.N.W., Rahim, M.A., Paul, N.C., Uddain, J. et al. 2021. Control of plant viral diseases by CRISPR/Cas9: resistance mechanisms, strategies and challenges in food crops. Plants 10(7): 1264.

Shaibu, A.S., Li, B., Zhang, S., and Sun, J. 2020. Soybean cyst nematode-resistance: gene identification and breeding strategies. Crop J. 8(6): 892–904.

Shakiba, E., Edwards, J.D., Jodari, F., Duke, S.E., Baldo, A.M., Korniliev, P. et al. 2017. Genetic architecture of cold tolerance in rice (*Oryza sativa*) determined through high resolution genome-wide analysis. PloS ONE 12(3): e0172133.

Shan, Q., Wang, Y., Li, J., Zhang, Y., Chen, K., Liang, Z. et al. 2013. Targeted genome modification of crop plants using a CRISPR-Cas system. Nat. Biotechnol. 31(8): 686–688.

Shan, Q.W., Y.P. Wang, J. Li and C.X. Gao. 2014. Genome editing in rice and wheat using the CRISPR/Cas system. Nat. Protoc. 9: 2395–2410.

Shan, Y. Wang, J. Li, Y. Zhang, K. Chen, Z. Liang et al. 2013. Targeted genome modification of crop plants using a CRISPR-Cas system, Nat. Biotechnol. 31: 686–688.

Sharma, A., Soares, C., Sousa, B., Martins, M., Kumar, V., Shahzad, B. et al. 2020. Nitric oxide-mediated regulation of oxidative stress in plants under metal stress: a review on molecular and biochemical aspects. Physiol. Plant. 168(2): 318–344.

Shen, C., Que, Z., Xia, Y., Tang, N., Li, D., He, R. et al. 2017. Knock out of the annexin gene OsAnn3 via CRISPR/Cas9-mediated genome editing decreased cold tolerance in rice. J. Plant Biol. 60: 539–547.

Sheng, J., Zhang, S., Li, R. and Shen, L. 2019. Knockout of *SlMAPK3* enhances tolerance to heat stress involving ROS homeostasis in tomato plants. BMC Plant Biol., 19, 1-13.

Shi, J., Gao, H., Wang, H., Lafitte, H.R., Archibald, R.L., Yang, M. et al. 2017. *ARGOS8* variants generated by CRISPR-Cas9 improve maize grain yield under field drought stress conditions. Plant Biotechnol. J. 15: 207–216.

Shim, J.S., Oh, N., Chung, P.J., Kim, Y.S., Choi, Y.D. and Kim, J.K. 2018. Overexpression of *OsNAC14* improves drought tolerance in Rice. Front. Plant Sci. 9: 310.

Shimatani, Z., Kashojiya, S., Takayama, M., Terada, R., Arazoe, T., Ishii, H. et al. 2017. Targeted base editing in rice and tomato using a CRISPR-Cas9 cytidine deaminase fusion. Nat. Biotechnol. 35(5): 441–443.

Silva, C.J., van den Abeele, C., Ortega-Salazar, I., Papin, V., Adaskaveg, J.A., Wang, D. et al. 2021. Host susceptibility factors render ripe tomato fruit vulnerable to fungal disease despite active immune responses. J. Exp. Bot. 72(7): 2696–2709.

Singh, D. and A. Laxmi. 2015. Transcriptional regulation of drought response: a tortuous network of transcriptional factors. Front. Plant Sci. 6: 895.

Singh, S., Kumar, V., Chauhan, A., Datta, S., Wani, A.B., Singh, N. et al. 2018. Toxicity, degradation and analysis of the herbicide atrazine. Environmental Chemistry Letters 16: 211–237.

Songmei, L., Jie, J., Yang, L., Jun, M., Shouling, X., Yuanyuan, T. et al. 2019. Characterization and evaluation of *OsLCT1* and *OsNramp5* mutants generated through CRISPR/Cas9-mediated mutagenesis for breeding low Cd rice. Rice Sci. 26(2): 88–97.

Sugano, S.S., M. Shirakawa, J. Takagi, Y. Matsuda, T. Shimada, I. Hara-Nishimura et al. 2014. CRISPR/Cas9-mediated targeted mutagenesis in the liverwort marchantia polymorpha l. Plant Cell Physiol. 55: 475–481.

Sun, Y., J. Li and L. Xia. 2016a. Precise genome modification via sequence-specific nucleases-mediated gene targeting for crop improvement. Front. Plant Sci. 7: 1928.

Sun, Y., X. Zhang, C. Wu, Y. He, Y. Ma, H. Hou et al. 2016b. Engineering herbicide-resistant rice plants through CRISPR/Cas9-mediated homologous recombination of acetolactate synthase. Mol. Plant 9(4): 628–631.

Svitashev, S., J.K. Young, C. Schwartz, H. Gao, S.C. Falco and A.M. Cigan. 2015. Targeted mutagenesis, precise gene editing, and site-specific gene insertion in maize using Cas9 and guide RNA. Plant Physiol. 169(2): 931–945.

Tang, L., B. Mao, Y. Li, Q. Lv, L. Zhang, C. Chen et al. 2017a. Knockout of *OsNramp5* using the CRISPR/Cas9 system produces low Cd-accumulating indica rice without compromising yield. Sci. Rep. 7(1): 14438.

Tang, X., L.G. Lowder, T. Zhang, A.A. Malzahn, X. Zheng, D.F. Voytas et al. 2017b. A CRISPR-Cpf1 system for efficient genome editing and transcriptional repression in plants. Nat. Plants 3: 1–5.

Thomazella, D.P.D.T., K. Seong, R. Mackelprang, D. Dahlbeck, Y. Geng, U.S. Gill et al. 2021. Loss of function of a *DMR6* ortholog in tomato confers broad-spectrum disease resistance. Proceedings of the National Academy of Sciences 118(27): e2026152118.

Tian, S., L. Jiang, X. Cui, J. Zhang, S. Guo, M. Li et al. 2018. Engineering herbicide-resistant watermelon variety through CRISPR/Cas9-mediated base-editing. Plant Cell Rep. 37: 1353–1356.

Tilman, D., C. Balzer, J. Hill and B.L. Befort. 2011. Global food demand and the sustainable intensification of agriculture. Proc. Nat. Acad. Sci. 108(50): 20260–20264.

Tran, M.T., D.T.H. Doan, J. Kim, Y.J. Song, Y.W. Sung, S. Das et al. 2021. CRISPR/Cas9-based precise excision of *SlHyPRP1* domain(s) to obtain salt stress tolerant tomato. Plant Cell Rep. 40: 999–1011.

Tripathi, J.N., V.O. Ntui, M. Ron, S.K. Muiruri, A. Britt and L. Tripathi. 2019a. CRISPR/Cas9 editing of endogenous banana streak virus in the b genome of musa spp. overcomes a major challenge in banana breeding. Commun. Biol. 2: 46.

Tripathi, L., V.O. Ntui and J.N. Tripathi. 2019b. Application of genetic modification and genome editing for developing climate-smart banana. Food Energy Secur. 8(4): e00168.

Tripathi, M.K., N. Tripathi, S. Tiwari, N. Mishra, A. Sharma, S. Tiwari et al. 2023. Identification of Indian soybean (*Glycine max* [L.] Merr.) genotypes for drought tolerance and genetic diversity analysis using SSR markers. Scientist 3(3): 31–46.

Tripathi, N., M.K. Tripathi, S. Tiwari and D.K. Payasi. 2022. Molecular breeding to overcome biotic stresses in soybean: update. Plants 11(15): 1967.

Tsuchimatsu, T., H. Kakui, M. Yamazaki, C. Marona, H. Tsutsui, A. Hedhly et al. 2020. Adaptive reduction of male gamete number in the selfing plant Arabidopsis thaliana. Nature Communications 11(1): 2885.

Tyagi, S., R. Kumar, V. Kumar, S.Y. Won and P. Shukla. 2021. Engineering disease resistant plants through CRISPR-Cas9 technology. GM Crops and Food 12(1): 125–144.

Tyagi, S., S.I. Mulla, K.J. Lee, J.C. Chae and P. Shukla 2018. VOCs-mediated hormonal signaling and crosstalk with plant growth promoting microbes. Crit. Rev. Biotechnol. 38(8): 1277–1296.

Tyczewska, T. Twardowski and E. Wo´zniak-Gientka. 2023. Agricultural biotechnology for sustainable food security. Trends Biotechnol. 41: 331–341

Vannier, N., M. Agler and S. Hacquard. 2019. Microbiota-mediated disease resistance in plants. PLoS Pathog. 15(6): e1007740.

Veillet, F., L. Perrot, L. Chauvin, M.P. Kermarrec, A. Guyon-Debast, J.E. Chauvin et al. 2019. Transgene-free genome editing in tomato and potato plants using agrobacterium-mediated delivery of a CRISPR/Cas9 cytidine base editor. Int. J. Mol. Sci. 20: 402.

Vlcko, T. and L. Ohnoutkova. 2020. Allelic variants of CRISPR/Cas9 induced mutation in an inositol trisphosphate 5/6 kinase gene manifest different phenotypes in barley. Plants 9(2): 195.

Voytas, D.F. and C. Gao. 2014. Precision genome engineering and agriculture: opportunities and regulatory challenges. PLoS Biol. 12(6): e1001877.

Walton, R.T., K.A. Christie, M.N. Whittaker and B.P. Kleinstiver. 2020. Unconstrained genome targeting with near-PAMless engineered CRISPR-Cas9 variants. Science 368(6488): 290–296.

Waltz, E. 2016. Gene-edited CRISPR mushroom escapes US regulation. Nature 532(7599): 293.

Waltz, E. 2022. GABA-enriched tomato is first CRISPR-edited food to enter market. Nat. Biotechnol. 40(1): 9–11.

Wan, D.Y., Y. Guo, Y. Cheng, Y. Hu, S. Xiao, Y. Wang et al. 2020. CRISPR/Cas9-mediated mutagenesis of VvMLO3 results in enhanced resistance to powdery mildew in grapevine (*Vitis vinifera*). Hortic. Res. 7.

Wang, B., Z. Zhong, X. Wang, X. Han, D. Yu, C. Wang et al. 2020a. Knockout of the *OsNAC006* transcription factor causes drought and heat sensitivity in rice. Int. J. Mol. Sci. 21: 2288.

Wang, D., F. Zhang and G. Gao. 2020. CRISPR-based therapeutic genome editing: strategies and *in vivo* delivery by AAV vectors. Cell 181: 136–150.

Wang, F.Z., M.X. Chen, L.J. Yu, L.J. Xie, L.B. Yuan, H. Qi et al. 2017a. *OsARM1*, an *R2R3 MYB* transcription factor, is involved in regulation of the response to arsenic stress in rice. Front. Plant. Sci. 8: 1868.

Wang, F., Wang, C., Liu, P., Lei, C., Hao, W., Gao, Y. et al. 2016. Enhanced rice blast resistance by CRISPR/Cas9-targeted mutagenesis of the ERF transcription factor gene *OsERF922*. PloS ONE 11(4): e0154027.

Wang, F., Xu, Y., Li, W., Chen, Z., Wang, J., Fan, F. et al. 2021a. Creating a novel herbicide-tolerance OsALS allele using CRISPR/Cas9-mediated gene editing. Crop J. 9: 305–312.

Wang, J.Y., P. Pausch and J.A. Doudna. 2022. Structural biology of CRISPR–Cas immunity and genome editing enzymes. Nature Reviews Microbiology 20(11): 641–656.

Wang, L., Chen, L., Li, R., Zhao, R., Yang, M. and Sheng, J. 2017b. Reduced drought tolerance by CRISPR/Cas9-mediated *SlMAPK3* mutagenesis in tomato plants. J. Agric. Food. Chem. 65: 8674–8682.

Wang, L., Chen, S., Peng, A., Xie, Z., He, Y. and Zou, X. 2019. CRISPR/Cas9-mediated editing of *CsWRKY22* reduces susceptibility to *Xanthomonas citri subsp. citri* in Wanjincheng orange (*Citrus sinensis* (L.) Osbeck). Plant Biotechnol. Rep. 13: 501–510.

Wang, S., Yang, Y., Guo, M., Zhong, C., Yan, C. and Sun, S. 2020b. Targeted mutagenesis of amino acid transporter genes for rice quality improvement using the CRISPR/Cas9 system. The Crop Journal 8(3): 457–464.

Wang, S., Zhang, S., Wang, W., Xiong, X., Meng, F., and Cui, X. 2015. Efficient targeted mutagenesis in potato by the CRISPR/Cas9 system. Plant Cell Rep. 34: 1473–1476.

Wang, T., Deng, Z., Zhang, X., Wang, H., Wang, Y., Liu, X. et al. 2018a. Tomato DCL2b is required for the biosynthesis of 22-nt small RNAs, the resulting secondary siRNAs, and the host defense against ToMV. Hortic. Res. 5: 62.

Wang, T., Xun, H., Wang, W., Ding, X., Tian, H., Hussain, S. et al. 2021b. Mutation of *GmAITR* genes by CRISPR/Cas9 genome editing results in enhanced salinity stress tolerance in soybean. Front. Plant Sci. 12.

Wang, W., Pan, Q., He, F., Akhunova, A., Chao, S., Trick, H. et al. 2018b. Transgenerational CRISPR-Cas9 activity facilitates multiplex gene editing in allopolyploid wheat. CRISPR J. 1(1): 65–74.

Wang, Y.P., Cheng, X., Shan, Q.W., Zhang, Y., Liu, J.X., Gao, C.X. et al. 2014. Simultaneous editing of three homoeoalleles in hexaploid bread wheat confers heritable resistance to powdery mildew. Nat. Biotechnol. 32(9): 947–951.

Wang, Y., Zafar, N., Ali, Q., Manghwar, H., Wang, G., Yu, L. et al. 2022. CRISPR/Cas genome editing technologies for plant improvement against biotic and abiotic stresses: advances, limitations, and future perspectives. Cells 11(23): 3928.

Wang, Z., Hardcastle, T.J., Pastor, A.C., Yip, W.H., Tang, S. and Baulcombe, D.C. 2018c. A novel *DCL2*-dependent miRNA pathway in tomato affects susceptibility to RNA viruses. Genes and Development 32(17-18): 1155–1160.

War, A.R., Paulraj, M.G., Ahmad, T., Buhroo, A.A., Hussain, B., Ignacimuthu, S. and Sharma, H.C. 2012. Mechanisms of plant defense against insect herbivores. Plant Signal. Behav. 7(10): 1306–1320.

Wright, A.V., Nuñez, J.K. and Doudna, J.A. 2016. Biology and applications of CRISPR systems: harnessing nature's toolbox for genome engineering. Cell 164(1): 29–44.

Wu, J., Yan, G., Duan, Z., Wang, Z., Kang, C., Guo, L. et al. 2020. Roles of the brassica napus DELLA protein BnaA6.RGA, in modulating drought tolerance by interacting with the ABA signaling component BnaA10.ABF2. Front. Plant Sci. 11.

Xie, K, and Yang, Y. 2013. RNA-guided genome editing in plants using a CRISPR–Cas system. Mol Plant 6(6): 1975–1983.

Xu, H., Chang, Q., Huang, L., Wei, P., Song, Y., Guo, Z. et al. 2023. An Agrobacterium-mediated transient expression method for functional assay of genes promoting disease in monocots. Inter J. Mol. Sci. 24(8): 7636.

Xu, R., Li, H., Qin, R., Wang, L., Li, L., Wei, P. et al. 2014. Gene targeting using the Agrobacterium tumefaciens-mediated CRISPR-Cas system in rice. Rice 7: 1–4.

Xu, R., Yang, Y., Qin, R., Li, H., Qiu, C., Li, L. et al. 2016. Rapid improvement of grain weight via highly efficient CRISPR/Cas9-mediated multiplex genome editing in rice. Journal of Genetics and Genomics = Yi chuan xue bao 43(8): 529–532.

Yadav, R.K., Tripathi, M.K., Tiwari, S., Tripathi, N., Asati, R., Chauhan, S. et al. 2023. Genome editing and improvement of abiotic stress tolerance in crop plants. Life 13(7): 1456.

Yamano, T., Nishimasu, H., Zetsche, B., Hirano, H., Ian, M., Li, Y., et al. (2016). EIGER detector: application in macromolecular crystallography. Acta Crystallogr. Sect. D Struct. Biol. 165: 949–962.

Yang, H., Ren, S., Yu, S., Pan, H., Li, T., Ge, S. et al. 2020. Methods favoring homology-directed repair choice in response to CRISPR/Cas9 induced-double strand breaks. Int. J. Mol. Sci. 21(18): 6461.

Yang, H., Wang, H., Shivalila, C.S., Cheng, A.W., Shi, L., and Jaenisch, R. 2013. One-step generation of mice carrying reporter and conditional alleles by CRISPR/Cas-mediated genome engineering. Cell 154: 1370–1379.

Yang, S.H., Kim, E., Park, H., and Koo, Y. 2022. Selection of the high efficient sgRNA for CRISPR-Cas9 to edit herbicide related genes, PDS, ALS, And EPSPS in tomato. Appl. Biol. Chem. 65: 13.

Yin, K., and Qiu, J.L. 2019. Genome editing for plant disease resistance: applications and perspectives. Philosophical Transactions of the Royal Society B 374(1767): 20180322.

Yin, K., Gao, C., and Qiu, J.L. 2017. Progress and prospects in plant genome editing. Nature Plants 3(8): 1–6.

Yin, K., C. Gao and J.L. Qiu. 2017. Progress and prospects in plant genome editing. Nat. Plants 3: 17107.

Yin, W., Y. Xiao, M. Niu, W. Meng, L. Li, X. Zhang et al. 2020. ARGONAUTE2 enhances grain length and salt tolerance by activating *BIG GRAIN3* to modulate cytokinin distribution in rice. The Plant Cell 32(7): 2292–2306.

Yin, Y., K. Qin, X. Song, Q. Zhang, Y. Zhou, X. Xia et al. 2018. *BZR1* transcription factor regulates heat stress tolerance through FERONIA receptor-like kinase-mediated reactive oxygen species signaling in tomato. Plant Cell Physiol. 59: 2239–2254.

Yu, S., C. Wang, K. Li, Y. Yang, Y.Z. He and Y. Wu. 2022. Transcriptional analysis of cotton bollworm strains with different genetic mechanisms of resistance and their response to *Bacillus thuringiensis* Cry1Ac toxin. Toxins 14(6): 366.

Yu, W., L. Wang, R. Zhao, J. Sheng, S. Zhang, R. Li et al. 2019. Knockout of SlMAPK3 enhances tolerance to heat stress involving ROS homeostasis in tomato plants. BMC Plant Biol. 19: 1–13.

Yuan, M., J. Zhu, L. Gong, L. He, C. Lee, S. Han et al. 2019. Mutagenesis of FAD2 genes in peanut with CRISPR/Cas9 based gene editing. BMC Biotech. 19: 1–7.

Yue, E., H. Cao and B. Liu. 2020. *OsmiR535*, a potential genetic editing target for drought and salinity stress tolerance in *Oryza sativa*. Plants 9(10): 1337.

Zafar, S.A., S.S.E.A. Zaidi, Y. Gaba, S.L. Singla-Pareek, O.P. Dhankher, X. Li et al. 2020. Engineering abiotic stress tolerance via CRISPR/Cas-mediated genome editing. J. Exp. Bot. 71: 470–9.

Zaidi, S.S.E.A., A. Mahas, H. Vanderschuren and M.M. Mahfouz. 2020. Engineering crops of the future: CRISPR approaches to develop climate-resilient and disease-resistant plants. Genome Biol. 21(1): 289.

Zambryski, P. 2013. Fundamental discoveries and simple recombination between circular plasmid DNAs led to widespread use of Agrobacterium tumefaciens as a generalized vector for plant genetic engineering, Int. J. Dev. Biol. 57: 449–452.

Zeilmaker, T., N.R. Ludwig, J. Elberse, M.F. Seidl, L. Berke, A. Van Doorn et al. 2015. *DOWNY MILDEW RESISTANT 6* and *DMR 6-LIKE OXYGENASE 1* are partially redundant but distinct suppressors of immunity in *Arabidopsis*. Plant J. 81(2): 210–222.

Zeng, D.D., C.C. Yang, R. Qin, M. Alamin, E.K. Yue, X.L. Jin et al. 2018. A guanine insert in *OsBBS1* leads to early leaf senescence and salt stress sensitivity in rice (*Oryza sativa* L.). Plant Cell Rep. 37: 933–946.

Zeng, Y., J. Wen, W. Zhao, Q. Wang and W. Huang. 2020. Rational improvement of rice yield and cold tolerance by editing the three genes *OsPIN5b*, *GS3*, and *OsMYB30* with the CRISPR–Cas9 system. Front. Plant Sci. 10: 933–946.

Zetsche, B., J.S. Gootenberg, O.O. Abudayyeh, I.M. Slaymaker, K.S. Makarova, P. Essletzbichler et al. 2015. Cpf1 is a single RNA-guided endonuclease of a class 2 CRISPR-Cas system. Cell 163(3): 759–771.

Zetsche, B., J. Strecker, O.O. Abudayyeh, J.S. Gootenberg, D.A. Scott and F. Zhang. 2019. A survey of genome editing activity for 16Cas 12 aorthologs. Keio J. Med. 6: 1–5.

Zhan, X., F. Zhang, Z. Zhong, R. Chen, Y. Wang, L. Chang et al. 2019. Generation of virus-resistant potato plants by RNA genome targeting. Plant Biotechnology Journal 17(9): 1814–1822.

Zhang, A., Y. Liu, F. Wang, T. Li, Z. Chen, D. Kong et al. 2019a. Enhanced rice salinity tolerance via CRISPR/Cas9-targeted mutagenesis of the OsRR22 gene. Mol. Breed. 39: 1–10.

Zhang, D., A. Hussain, H. Manghwar, K. Xie, S. Xie, S. Zhao et al. 2020a. Genome editing with the CRISPR-Cas system: an art, ethics and global regulatory perspective. Plant Biotechnol. J. 18(8): 1651–1669.

Zhang, F., Y. Wen and X. Guo. 2014. CRISPR/Cas9 for genome editing: progress, implications and challenges. Hum. Mol. Genet. 23(R1): R40–R46.

Zhang, R., S. Chen, X. Meng, Z. Chai, D. Wang, Y. Yuan et al. 2021a. Generating broad-spectrum tolerance to ALS-inhibiting herbicides in rice by base editing. Sci. China Life Sci. 64: 1624–1633.

Zhang, R., J. Liu, Z. Chai, S. Chen, Y. Bai, Y. Zong et al. 2019b. Generation of herbicide tolerance traits and a new selectable marker in wheat using base editing. Nature Plants 5(5): 480–485.

Zhang, T., Q. Zheng, X. Yi, H. An, Y. Zhao, S. Ma et al. 2018a. Establishing RNA virus resistance in plants by harnessing CRISPR immune system. Plant Biotechnol. J. 16(8): 1415–1423.

Zhang, X., Y. Long, J. Huang and J. Xia. 2020b. OsNAC45 is involved in ABA response and salt tolerance in rice. Rice 13: 1–13.

Zhang, Y. and A.M. Showalter. 2020c. CRISPR/Cas9 genome editing technology: a valuable tool for understanding plant cell wall biosynthesis and function. Front. Plant Sci. 11: 589517.

Zhang, Y.W., Y. Bai, G.H. Wu, S.H. Zou, Y.F. Chen, C.X. Gao et al. 2017. Simultaneous modification of three homoeologs of *TaEDR1* by genome editing enhances powdery mildew resistance in wheat. Plant J. 91: 714–724.

Zhang, Y., B. Iaffaldano, and Y. Qi. 2021b. CRISPR ribonucleoprotein-mediated genetic engineering in plants. Plant Commun. 2: 100168.

Zhang, Y., D. Li, D. Zhang, X. Zhao, X. Cao, L. Dong et al. 2018b. Analysis of the functions of *TaGW2* homoeologs in wheat grain weight and protein content traits. Plant J. 94(5): 857–866.

Zhang, Y., K. Massel, I.D. Godwin and C. Gao. 2018c. Applications and potential of genome editing in crop improvement. Genome Biol. 19(1): 210.

Zhao, Y., C. Zhang, W. Liu, W. Gao, C. Liu, G. Song et al. 2016. An alternative strategy for targeted gene replacement in plants using a dual-sgRNA/Cas9 design. Sci. Rep. 6: 1–11.

Zheng, J., H. Wu, H. Zhu, C. Huang, C. Liu, Y. Chang et al. 2019. Determining factors, regulation system, and domestication of anthocyanin biosynthesis in rice leaves. New Phytologist 223(2): 705–721.
Zheng, M., J. Lin, X. Liu, W. Chu, J. Li, Y. Gao et al. 2021. Histone acetyltransferase TaHAG1 acts as a crucial regulator to strengthen salt tolerance of hexaploid wheat. Plant Physiol. 186: 1951–1969.
Zhong, X., L. Hu and G. Tang. 2024. The application of genome editing technologies in soybean (Glycine max L.) for abiotic stress tolerance. Plant Genome Editing Technologies: Speed Breeding, Crop Improvement and Sustainable Agriculture 221–237.
Zhou, J., Z. Peng, J. Long, D. Sosso, B.O. Liu, J.S. Eom et al. 2015. Gene targeting by the TAL effector PthXo2 reveals cryptic resistance gene for bacterial blight of rice. Plant J. 82(4): 632–643.
Zhu, C., L. Bortesi, C. Baysal, R.M. Twyman, R. Fischer, T. Capell et al. 2017. Characteristics of genome editing mutations in cereal crops. Trends Plant Sci. 22(1): 38–52.
Zhu, J., N. Song, S. Sun, W. Yang, H. Zhao, W. Song et al. 2016. Efficiency and inheritance of targeted mutagenesis in maize using CRISPR-Cas9. J. Genet. Genomics 43(1): 25–36.
Zulfiqar, F. 2021. Effect of seed priming on horticultural crops. Sci. Hortic. 286: 110197.

CHAPTER 10

Trends in CRISPR/Cas9 Editing Technology as an Approach for Climate-Smart Agriculture

Samina Yasmin,[1,*] *Amjad Ali,*[2] *Muhammad Tatar*[2] and *Jaweria Iqbal*[3]

1. Introduction

By addressing important issues related to climate change, CRISPR/Cas9 has shown its revolutionary potential in the context of climate-smart agriculture. For instance, the method has been used to improve salt tolerance by altering the genes for the High Affinity Potassium Transporter (*HKT*) and increase agricultural resilience to drought by targeting genes like *DREB2* (Drought-Responsive Element-Binding Protein). In a similar vein, genes implicated in the Heat Shock Protein (HSP) pathways have been altered to provide characteristics like heat tolerance (Chen et al. 2023). These innovations make it possible to create crop varieties that are climate resilient, meaning they can adjust to harsh conditions and maintain steady production even when confronted with fluctuations (Li et al. 2022).

The combination of CRISPR/Cas9 technology with both conventional and cutting-edge breeding techniques is another noteworthy development. In addition to Genome-Wide Association Studies (GWAS) and genomic selection, this method expedites the creation of climate-resilient crops by effectively identifying and editing key genes (Altaf et al. 2024). Additionally, crops with enhanced Nitrogen Use Efficiency (*NUE*) are being engineered using CRISPR/Cas9, which lessens reliance on synthetic fertilizers and reduces greenhouse gas emissions, which are an essential part of sustainable agricultural systems (Karunarathne et al. 2022).

In climate-smart agriculture, CRISPR/Cas9 has a potential in carbon sequestration, whereby root architecture can be modified to increase soil carbon storage and aid in climate mitigation. Furthermore, a possible way to boost agricultural yields under climate stress is to employ CRISPR/Cas9 to alter genes linked to the Calvin cycle, improving photosynthetic efficiency. Multiple genes can now be targeted simultaneously thanks to advancements in multiplex genome editing, which improves complicated qualities like yield stability and environmental adaptability (Zheng et al. 2021). The CRISPR/Cas9 technology is a key component of sustainable development

[1] Department of Horticulture, Faculty of Agriculture, University of Cukurova, 01339 Bakcalı Adana, Türkiye.
[2] Department of Plant Protection, Faculty of Agricultural Science and Technologies, Sivas University of Science and Technology, Sivas, Türkiye.
[3] Department of Plant Breeding and Genetics, Bahauddin Zakariya University, Multan Pakistan.
* Corresponding author: samyasmin@ymail.com

DOI: 10.1201/9781003518204-10

as the agriculture industry faces previously unresolved climate concerns. However, resolving legal issues, enhancing public acceptance, and developing delivery methods for modifying a variety of crop species are necessary for this strategy to continue to be successful. CRISPR/Cas9 has the potential to transform agriculture by bringing sustainability and productivity in line with the needs of a changing global economy through sustained innovation and judicious application (Ghulam Muha-Ud-Din et al. 2024). The purpose of this chapter is to explore the role of CRISPR/Cas9 genome editing technology in promoting climate-smart agriculture.

1.1 Climate-Smart Agriculture

The intricate interplay between population growth, climate change, and food shortage presents significant challenges to the global community. Meeting the escalating dietary demands of the expanding global population requires inevitable advancement in technologies and researchers persistently confront this challenge (Lawrence et al. 2023). A significant portion of agricultural research is directed toward alleviating the effects of environmental challenges and bolstering resilience against the variability and extremes of climate. A key emphasis lies in advancing and expanding climate-smart agriculture practices. The core component of these practices are illustrated in Fig. 1. Current trends in plant breeding leverage new technical innovations to address the pressing need for ensuring global food security under changing climatic conditions. Climate-smart agriculture encompasses a diverse range of agricultural technologies and practices, such as abiotic and biotic stresses, conservation agriculture, agroforestry, soil, and water (Zheng et al. 2024). However, after identifying the updated technologies and management practices, the ensuing challenges lie in scaling

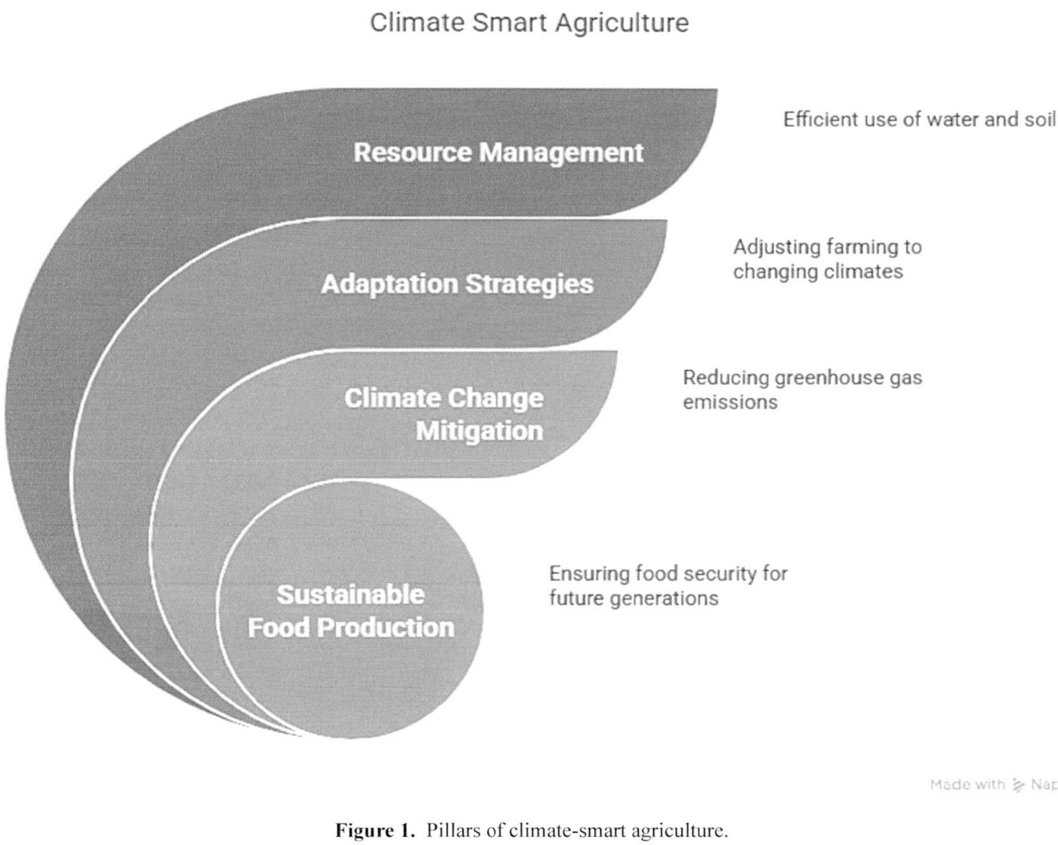

Figure 1. Pillars of climate-smart agriculture.

them up through extensive dissemination among the designated target populations. Over the past few decades, scientists have demonstrated through the number of studies on Climate-Smart Agriculture (CSA) that, adoption of CSA technologies can lead to heightened productivity, increased income, and improved food security across diverse regions (Sissoko et al. 2023).

2. CRISPR/CAS9 Technology - An Overview

The CRISPR-Cas system functions as an acquired immune system in prokaryotes, specifically engineered to counteract invading genetic material. The CRISPR-Cas system consists of (i) a CRISPR array and (ii) a group of Cas genes that encode Cas proteins, which begins with an AT-rich sequence (Marraffini 2015). These sequences are structured in an alternating manner. Homogeneous CRISPR array sequences exhibit a high level of conservation, whereas sequences in different CRISPR arrays vary in both their sequence and structure. These arrays are surrounded by various Cas genes. The Cas genes code for unique Cas proteins that have diverse functional domains. These include endonuclease domains, RNA and DNA binding domains, and domains involved in transcriptional control (Goldberg et al. 2024).

The CRISPR system's immunological process has essential stages: adaptation, processing, and interference. In the adaptation stage, new spacers are obtained from the invading genome and quickly incorporated into the CRISPR array. These spacers are positioned directly after the leader sequences. This mechanism requires a stable complex, two Cas1 dimers, and one Cas2 dimer. Some CRISPR-Cas systems may engage additional auxiliary proteins, like as Cas4 and reverse transcriptase (Wang et al. 2021).

In the processing step, the CRISPR RNA (crRNA) undergoes maturation. The complete CRISPR array is decoded into precursor crRNA (pre-crRNA) with the help of promoter regions present in the leader sequence. The ribonucleases involved in the maturation of pre-crRNA are Cas nucleases (Cas6, Cas5d, and Cas12) and non-Cas nucleases (RNase III, RNase E, and PNPase). Afterwards, the fully developed crRNA attaches to multi-subunit effector complexes, creating a ribonucleoprotein complex (RNP). During the last phase of target interference, the ribonucleoprotein specifically recognizes and cuts the foreign genome by forming base pairs between the crRNA and the target genome sequence (Wang et al. 2021). Figure 2 illustrates the application of

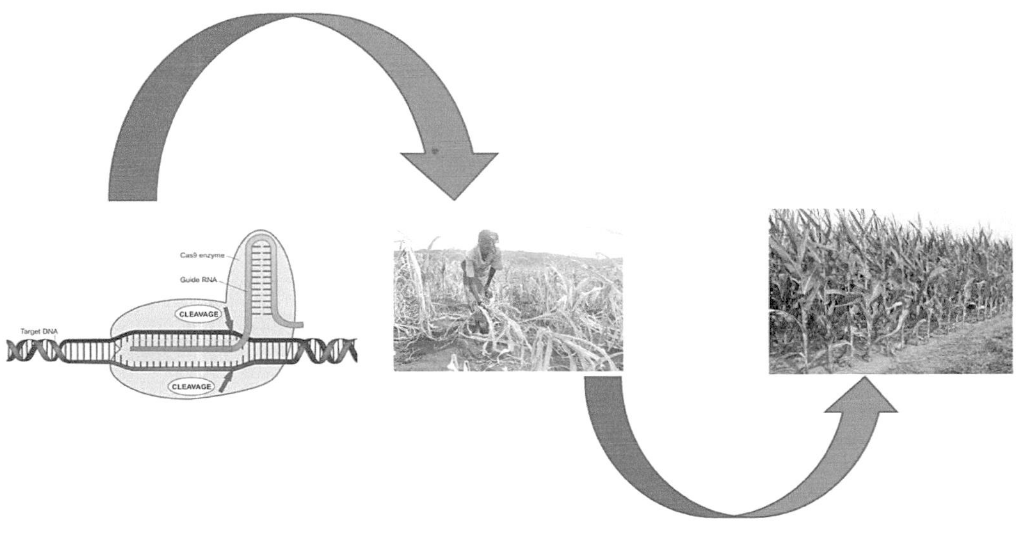

Figure 2. CRISPR/Cas9 genome editing application for climate resilient agriculture. The process involves targeting gene modification (left) to develop crop varieties resilient to climate induced stresses, resulting in improved performance and yield under adverse conditions (Ndudzo et al. 2024).

CRISPR/Cas9 for climate resilient agriculture, highlighting the process of targeted gene modification to develop crop varieties with enhanced tolerance to climate induced stresses, ultimately improving productivity under adverse environmental conditions.

2.1 Climate Challenges in Agriculture

Agriculture encounters complex and distinct challenges regarding climate change. Primarily, the agriculture sector exhibits heightened vulnerability due to its dependence on weather patterns and climatic conditions. High temperatures, variable rainfall, invasive pests, and heightened frequency of extreme weather events (Habib-ur-Rehman et al. 2022).

2.2 Temperature and Precipitation Impacts

Research indicates that an increase in temperature typically leads to a decrease in agricultural yield, while an increase in precipitation has the potential to counteract or mitigate the adverse effects of increasing temperature. In the case of Iran, where climatic variables exert influence, crop productivity is contingent upon factors such as the adaptive capacity of crops, the specific crop type, prevailing climatic scenario, and the fertilization effects of CO_2. In Cameroon, a decrease in rainfall or a temperature rise is associated with a significant decline in the income earned by farmers. Statistical data indicates the correlation between temperature and coffee production in Mexico. Furthermore, there are indications that coffee production may confront economic feasibility challenges for producers in the foreseeable future, with a projected 34% reduction (Javadi et al. 2023).

2.3 Regional Impact of Climate Change

The influence of climate change on crop production varies depending on the geographical region and irrigation practices (Kang et al. 2009). The aggregate production of wheat, maize, and rice could decline if both temperate and tropical regions experience 2°C warming (Challinor et al. 2014). The effects of climate change are more noticeable in tropical areas because tropical crops are closer to their high-temperature optima and are, therefore, more vulnerable to high-temperature stress during high temperatures. Furthermore, the occurrence of biotic stress is more widespread in places characterized by high humidity and warmer temperatures. Since the 1900s, the Netherlands has witnessed a heightened frequency of extreme weather events, significantly impacting the yield of wheat (Heikonen et al. 2025).

3. Case Studies from Developing Nations

In contrast, previous research suggests that the impact of climate change on the global food supply is restricted, although emerging nations are expected to experience significant adverse effects. In India, it is forecasted that the temperature will increase by a range of 2.33°C to 4.87°C. This will be accompanied by a doubling of CO_2 concentration and longer heat waves, which will have harmful consequences on the agriculture sector (Kumar and Gautam 2014). According to Shakoor et al. (2011), farmers in the dry region of Rawalpindi, Pakistan, are projected to have a yearly loss of 4180 INR per acre by the year 2100 due to a 1°C rise in temperature. Projections suggest that the yield in cereal crops will increase by 10 to 25% with a 1°C rise in the world average surface temperature. Climate change in sub-Saharan Africa is projected to cause a significant decline, ranging from 6 to 24%, in the average crop yield (Alimagham et al. 2024).

3.1 Plant Responses to Climate Change

The relationships between plants and water are extremely sensitive to changes in temperature and precipitation. Extreme alterations in these characteristics are predicted to have a more significant impact on physiological changes compared to changes in mean climate. Plants exhibit varied responses to climate change contingent upon specific plant species and developmental stage. Distinct species-specific thresholds exist, influencing responses such as root elongation and yield reduction across different plant species (Raza et al. 2024).

3.2 Nutritional Quality under Elevated CO_2

Elevated atmospheric CO_2 levels have been associated with reduced transpiration in plants, resulting in an increase in air temperature by 0.42 ± 0.02 K. This indirect physiological effect of heightened CO_2, coupled with a direct radioactive effect, has the potential to amplify land surface warming by 3.33 ± 0.03 K. While the rise in CO_2 levels is expected to enhance the harvestable produce of crops, the developmental stages depend on the type of crop. C3 crops are expected to produce more, however, both C3 and C4 crops are expected to have reduced water requirements in the absence of stressful conditions. The adverse consequences of climate change are anticipated to be substantial in the tropical regions of developing countries, with a magnitude of impact largely contingent on the specific climate scenario prevailing in each region (Zhang et al. 2024).

Elevated levels of carbon dioxide (CO_2) lead to a reduction in zinc and iron levels in C3 grain crops and legumes, resulting in negative impacts on human health. Additionally, C3 plants and legumes exhibit a reduction in protein concentration under elevated CO_2 conditions, whereas C4 plants remain unaffected. Climate-induced changes also affect microbial community and their enzymatic activities (Coronado-Lira 2024).

3.3 Pests, Diseases, and Crop Vulnerability

The microbial population in a temperature-gradient tunnel with a temperature that was 4–5°C high was much more than in field conditions. More precisely, the number of bacteria and fungi that can fix nitrogen and solubilize phosphorus, as well as the activities of enzymes, showed a considerable increase across a wide range of temperatures. The highest values were found at or very close to the ideal temperature. The expected changes in climate patterns are expected to have the greatest impact on the growth and survival of diseases. Changes in the climate or weather patterns of a place are anticipated to heighten a crop's vulnerability to certain pests, diseases, and weeds. Projections suggest that countries located at high and mid-latitudes will see higher crop yields, whereas regions at lower latitudes are expected to face a decrease in yields (Jiang et al. 2023). Projections indicate that a one-degree increase in temperature could lead to a 10–25% surge in losses caused by insect pest infestations (Shrestha 2019). The possible consequences of climate change on insect population dynamics and their movement pose a substantial risk to agricultural productivity and sustainability. The proliferation of pests is mostly influenced by abiotic variables, specifically humidity and temperature. The infection of coffee nematodes and leaf miners in Brazil is projected to rise due to an increase in the number of generations per month, as compared to the prevailing climatic conditions between 1961 and 1990. As a result, the presence of pests has led to significant expenses for controlling them, mostly through the purchase of pesticides. Empirical data suggests that the rise in rainfall and temperatures has led to a surge in pesticide expenses for commodities such as corn, potatoes, and soybeans in the United States (Burc et al. 2025). However, there has been a notable decrease in pesticide prices for wheat, as observed in the study conducted by Chen and McCarl in 2001. In the HadCM3-high 2050 scenario, it is projected that the proportion of arable area impacted by the European corn borer and the Colorado potato beetle will rise by 43–48% for the second generation. Furthermore, the high-latitude regions in central Europe that were formerly uninhabited are now recognized as susceptible to these pests as a result of rising temperatures. According to current projections, the acceptable habitat for the wheat aphid (*Schizaphis graminum*) is expected to expand towards higher latitudes in the northern hemisphere by 2030 due to global warming. However, this expansion is expected to be offset by a decrease in suitable habitat in the northern hemisphere as a whole. Due to rising temperatures, the frequency of insect outbreaks involving

30 pest species is projected to increase in Sweden. This will result in the expansion of affected areas and have a significant impact on the country's forestry sector (Lehmann et al. 2020).

Increasing temperatures are expected to impose constraints on the life cycle of pathogens, such as *Puccinia striiformis f.* sp. *tritici.* On the other hand, a rise in the concentration of CO_2 in the atmosphere is anticipated to create advantageous conditions for *Fusarium pseudograminearum*. Climate change affects the geographic distribution and growth rates of people, increasing the number of generations. This shift in climate trends has the potential to extend the duration of pests' growth season and disrupt the timing of crop and pest cycles, increasing the likelihood of invasive pests (Zhang et al. 2025). The effectiveness of plant defense strategies is expected to decrease due to climate change. Climate change has the potential to cause unforeseeable linkages between the weather patterns of farming systems and the dynamics of pests. Insects, especially in temperate climates, may undergo alterations in their development and metabolic rates as a result of climate-induced changes. There are three prevalent bug species found in Africa. *Tuta absoluta*, *Ceratitis cosyra*, and *Bactrocera invadens* exhibit a growing level of suitability for their habitats throughout the entire continent, especially in regions near their most favorable habitats. Moreover, the elevated levels of carbon dioxide (CO_2) and escalating temperatures are amplifying the vulnerability linked to late blight in potatoes, blast disease, and sheath blight in rice, thus presenting a significant peril to worldwide food security (Shen et al. 2023).

3.4 Impact on Weeds

Climate change affects the growth of weeds in crops. C3 weeds exhibit a significant response to higher levels of carbon dioxide (CO_2), increasing both leaf surface area and biomass. C3 weeds present a notable obstacle in C4 plants, whereas C4 weeds in C3 plants exhibit reduced competitiveness. Weeds engage in competition with plants for water and nutrients, as weeds possess greater nutrient demands in comparison to agricultural plants (Malhi et al. 2020). Climate change has a considerable impact on the effectiveness of herbicides by altering the way they work, in addition to promoting the growth of weeds. Shifting weather patterns and climatic conditions are leading to the emergence of new geographical areas for weeds. To effectively manage this, it is necessary to design new methods that consider the impact of climate change. Predictions indicate that climate change would exacerbate the problem of pest infestation, since the increased warmth and humidity will provide favorable circumstances for the rapid growth of pests. However, the extent of this impact will vary depending on the locality and the adaptability of pests to climate change (Bajwa et al. 2020).

The recent advancements in CRISPR-based genome-editing technologies have proven instrumental in the pursuit of enhanced disease resistance across various biological threats, including bacteria, fungi, viruses, and other pathogens. In the subsequent section, we delve into contemporary trends about disease resistance achieved through the CRISPR/Cas system in distinct pathogenic contexts.

4. Novel Approaches to Enhance Crop Resilience

The ability of crops to withstand and recover from climate challenges is crucial for maintaining food security. Novel solutions are crucial in tackling these difficulties. CRISPR-Cas is a cutting-edge gene-editing technique that enables the accurate and precise change of genes. There are several ways in which CRISPR technology contributes to enhancing crop resilience (Mishra et al. 2018).

4.1 Use of CRISPR-Cas for Enhancing Crop Traits for Ever-Changing Climatic Conditions

CRISPR-Cas9 has gained widespread importance due to many advantages, such as efficacy, accuracy, cost-effectiveness, remarkable adaptability, and the ability to target multiple genes simultaneously. These features greatly enhance the breeding of various polyploidy crop species, which are challenging to improve using conventional methods (Mascher et al. 2017). Additionally, CRISPR technology facilitates the generation of both monoallelic and biallelic mutants, along with a diverse array of phenotypic outcomes. Such achievements are typically not possible through conventional breeding approaches, making CRISPR-Cas9 a more versatile and efficient tool for genome editing (Villette et al. 2024). The CRISPR-Cas9 system has been widely employed in diverse plant species to address multiple issues encountered in plant breeding. The implementation of this technology has resulted in substantial enhancements in monocot and dicot crops, encompassing a wide range of characteristics such as increased yield, improved quality, enhanced disease resistance, and greater resilience to climate fluctuations. The CRISPR-Cas9 technology has been utilized in various crops (Uranga et al. 2024).

5. Application of CRISPR-Cas9 for Enhancing Agricultural Productivity and Improving Crop Characteristics

5.1 Enhancing Agricultural Productivity

This novel innovation offers a multitude of applications aimed at enhancing both the yield and quality of crops. CRISPR-Cas9 enables the development of cultivars with high nutritional value and resistance traits, marking a significant milestone in genome editing technology. In recent developments, CRISPR-Cas9 has been instrumental in obtaining commercial oil from the seeds of *Camelina sativa*, a crop valued for its high fatty acid composition (Wang et al. 2021). In addition, genes such as JAZI (jasmonate-zim-domain protein), BRII (brassinosteroid insensitive 1), and GA-insensitive genes have been effectively manipulated, leading to mutation rates ranging from 26 to 84%. The improvements underscore the extraordinary capacity of CRISPR-Cas9 to transform contemporary agriculture. Manipulating a squamosal promoter binding protein and flowering locus using Cas9 in the same species led to plants with a mutation frequency of over 90% during the flowering stage. In addition, a complex CRISPR-Cas9 system, in conjunction with other techniques, was developed and utilized to modify six modified PYL genes of ABO receptors. This resulted in mutation frequencies ranging from 13–93% in the initial generation. In addition, the green fluorescent gene in *Nicotiana benthamiana* was altered using RNA-facilitated endonuclease (Zhang et al. 2020). Later, the instructions to the plant gene were modified, and the transcriptional characteristics were manipulated using the tobacco rattle virus.

The researchers used CRISPR-Cas9 to alter the progeny of rice species that had the same genetic makeup, resulting in the removal of a gene cluster on the chromosome and the introduction of inheritable differences in the genetic composition. The CRISPR-Cas system was used to induce mutations in three members of the aldehyde oxidase gene family in rice, namely *OsAox1b*, *OsAox1a*, and *OsAox1c*, as well as the *OsBEL* gene. The genetic mutation was discovered, and following examinations were conducted to explore the intrinsic alteration of genes in consecutive progenies. Moreover, the activation of genes related to grain dormancy was attributed to the presence of a protein called *HvPM19*, which is produced by the barley gene and stimulated by ABA. The use of CRISPR-Cas to induce mutations in *HvPM19* resulted in the creation of mutations with a frequency of 10%. Researchers have utilized CRISPR-Cas9 technology to specifically modify important genes in rice, including *Gn1a, GS3, OsSPL16, WAXY, MIR 396f, MIR396e, GL3.2, Os03g0568400, Os03g0603100, OsBADH2, OsAAP6*, and OsAAP10. This targeted gene editing has led to the development of rice mutants that exhibit improved yield, cooking quality, and aroma (Chen et al. 2024). Furthermore, research has shown that CRISPR-Cas9 can create rice plants with excellent yields by specifically altering the amino acid transporter gene (Wang et al. 2021; Usman et al. 2020). Targeted mutagenesis of *TaSBEIIa* using CRISPR-Cas9 in wheat has resulted in the formation of high amylose and increased resistant starch content. CRISPR-Cas9 has been employed in *Brassica napus* to enhance the synthesis of high-quality oleic acid. In addition, targeted mutagenesis using CRISPR-Cas9 has been employed to change the activity of genes such as *ALCATRAZ INDEHSCENT* and *JAGGED*, thereby improving resistance to pod shattering in *Camellina sativa*. CRISPR-Cas9 has been used to do targeted mutagenesis on genes such as INDEHSCENT and ALCATRAZ in Brassica species. This has resulted in enhanced fruit dehiscence. Furthermore, the manipulation of the *JAGGED* and *ALCATRAZ* genes in *Brassica napus* through genome editing plays a crucial role in enhancing resistance to pod shattering. Targeted mutation of the *SP5G* gene in tomatoes led to accelerated flowering and enhanced crop productivity (Li et al. 2021). Furthermore, there have been reports of targeted disruption of genes including *ZmIPK1A*, *ZmIPK*, and *ZmMRP4* in maize, which are involved in the manufacture of phytic acid. The CRISPR-Cas9 system has been employed to modify the apple protoplast to improve its flavor and overall quality. The CRISPR-Cas9 technique has been used to modify the characteristics of potatoes, specifically by manipulating the *St16Dox* gene to alter steroidal glycoalkaloid synthesis. This has led to the creation of potato lines that are free of steroidal glycoalkaloids (SGAs) (Toda et al. 2019). The CRISPR-Cas9 system has been employed to improve the quality and productivity of various essential and fruit-bearing plants, including citrus, grapes, and tomatoes. The CRISPR-Cas9 technique has been used to modify the *CsPDS* genes in sweet orange leaves. This was done using the agro-infusion approach using *Xanthomonas citri*. The efficacy of CRISPR-Cas9 assures that the modified fruits are safe to eat, as it does not introduce foreign genes into the genome. Previous research has also demonstrated the applicability of the Cas9 techniques in editing genes in apple protoplast, which can be extended to other crops as well (Espley et al. 2019).

5.2 Breeding of Disease-Resistant Cultivars

Viruses, fungi, nematodes, insects, and bacteria are responsible for plant diseases that have a substantial impact on agricultural productivity and quality (Yin and Qiu 2019). CRISPR-Cas9 is a new method that is being used to create crops that are resistant to diseases (Khan et al. 2021). The desire for wheat-based products that are safe for those with celiac disease has grown in wheat farming. This is to reduce the danger of chronic problems, a characteristic that is difficult to achieve using standard breeding techniques. CRISPR-Cas9 has recently been employed to specifically target glutenin genes. This has resulted in the development of a celiac-safe cultivar with silenced and deleted gliadins. There is a potential reduction in the presence of Coeliac Disorder (CD) epitopes, therefore minimizing the risk of exposure to CD (Jouanin et al. 2020). Previous research findings revealed the significance of CRISPR-Cas9 in developing gluten-free wheat (Verma et al. 2021). These findings hold promise for developing coeliac-safe rice as well, for example target mutation of *Elf4g* gene in rice has shown increased resistance to *Tungro spherical virus*. These advancements underscore the potential of CRISPR-Cas9 in enhancing crop resilience and addressing critical challenges in agriculture. In rice, bacterial leaf blight poses a significant threat to yield losses, targeted mutation of *OsSWEET14* gene by CRISPR-Cas9 induced resistance against the pathogen (Sam et al. 2021). Therefore, the targeted mutagenesis technique has the potential to boost plant resistance against many infections by changing genes that make the plant susceptible. As an illustration, the use of Cas9 to disable the *OsERF922* gene led to a heightened ability to withstand blast disease induced by *Magnaporthe oryzae*, as demonstrated by Wang et al. in 2016. The CRISPR-Cas9 method has demonstrated efficacy in comprehending the interaction between plants and pathogens, as well as in the creation of disease-resistant cultivars by targeted mutagenesis. An instance of Cas9-targeted mutagenesis was employed to boost resistance to citrus canker by targeting the promoter region of the *CsLOBI* gene. The mutant lines DLoB10 and DLoB9 had elevated mutation frequencies, resulting in enhanced resistance to *Xanthomonas citri*. Furthermore, the deliberate alteration of the effector binding motif in the promoter region of *CsLOBI* resulted in improved citrus resistance to *Xanthomonas*. Targeted mutagenesis of the *TaMLO* gene in wheat protoplasts led to an increase in resistance to powdery mildew. Similarly, targeting EDRI homologs using Cas9 led to the development of powdery mildew-resistant wheat varieties. These examples underscore the versatility and effectiveness of CRISPR-Cas9 in bolstering plants' resistance against various pathogens, offering promising prospects for crop protection and improvement. The targeting efficiency of the genome has been greatly improved by DNA viral amplicons. Geminivirus replicon was employed in hexaploid wheat to achieve transitory Cas9 expression for combating the dwarf virus. This led to a significant increase of roughly 12 times in gene expression. The CRISPR-Cas9 system has successfully targeted the genome of geminiviruses, thereby inhibiting their proliferation and providing a promising approach to controlling viral infections. The utilization of Cs9-mediated genome editing in viral DNA has emerged as a viable strategy to mitigate the impact of diseases. This approach involves leveraging the virus promoter to regulate the expression of sgRNA cassata. A novel Cas9 ortholog, discovered in *Francisella novicida*, has been employed for genome editing of RNA-based viruses. The FnCas9 gene inhibits the reproduction process of TMV and cucumber virus, hence providing resistance against both viruses. The manipulation of the *SIPeLo* and *SIMIO1* genes in tomatoes using Cas9 led to the development of immunity against leaf curl virus and powdery mildew (Pramanik et al. 2021). Nevertheless, the application of Cas-mediated mutagenesis on the *OsPETI* gene in rice resulted in enhanced resistance against rice sheath blight. Novel disease resistance paralogs have been created in soybean using Cas9, offering enhanced resistance (Nagy et al. 2021). Tobacco is an important cash crop and is severely affected by PVY, knockout of the susceptible gene *Ntab0942120*, resulting in transgene-free homozygous edited plants resistant to potato virus Y. In Brassica, 5 genes were knocked out to confer resistance against Phytophthora in potatoes. Hence, Cas9 is a versatile and efficient approach in enhancing resistance against various diseases as shown in Table 1 (Ruyi et al. 2021).

Table 1. Disease-resistant crops by using CRISPR/Cas9.

Crops	**Gene**	**Trait**	**References**
Tomato	*SIPeLo and SIMIO1*	Resistance to leaf curl virus	Pramanik et al. 2021
Rice	*OsSWEET14*	Bacterial leaf blight resistance	Sam et al. 2021
Citrus		Resistance to citrus canker	Jia et al. 2014
Tobacco	*Ntab0942120*	Resistance to potato virus Y	Ruyi et al. 2021
Grape	*Mlo-7*	Mildew Resistance	Malnoy et al. 2016
Papaya	*PpalEPIC8*	Resistance to phytophthora	Gumtow et al. 2018

6. Climate Smart Crops Using CRISPR-CAS9

Cas9 has been essential in mitigating several abiotic stresses in key agricultural crops. The creation of CRISPR-Cas9 has revolutionized plant breeding by enabling the development of climate-smart and stress-resistant crops. This novel strategy allows for the modification of any sequence, enabling the introduction of desirable traits into horticultural crops.

Molecular breeders have identified many genes associated with abiotic stress and engineered them into crop genomes. As a prime example, the use of CRISPR-Cas9 technology to modify the genome of tomato plants has resulted in enhanced defense mechanisms against drought stress, namely in the slmapk3 protein gene. The Cas9 system was used to specifically target two genes, *TaDREB3* and *TaDREB2*, which are associated with abiotic stress in wheat protoplasts. The T7 endonuclease assay revealed mutant gene expression in around 70% of transfected wheat protoplasts, resulting in enhanced tolerance in all mutant plants as compared to all wild-type plants (Lv et al. 2025). As a prime example, the use of CRISPR-Cas9 technology to modify the genome of tomato plants has resulted in enhanced defense mechanisms against drought stress, namely in the slmapk3 protein gene. The Cas9 system was used to specifically target two genes, *TaDREB3* and *TaDREB2*, which are associated with abiotic stress in wheat protoplasts (Kim et al. 2018). Lou et al. (2017) conducted a study where they altered the *SAPK2* gene in rice to investigate its role in the stress response to

Table 2. Applying CRISPR/Cas9 to develop climate-resilient crops.

Crops	**Genes**	**Traits**	**References**
Rice	*OsPRPI*	Cold sensitive	Nawaz et al. 2019
	OsMYB30	Cold tolerance, enlarged grain size	Zeng et al. 2019
	OsERA1	Drought tolerance through stomatal regulation	Ogata et al. 2020
	OsDST	Salinity tolerance, enhanced leaf area	Santosh et al. 2020
	OcNAC041	Salinity tolerance, enhanced MDA content	Bo et al. 2019
	OsTB1	High grain yield	Wang et al. 2020
	OsAAP10	Good cooking quality	
	P450	High grain yield	Usman et al. 2020
	IPA	Enhanced yield	Li et al. 2016
Tomato	*SlIAA9*	Parthenocarpy	Ueta et al. 2017
	SICPK28	Heat sensitive, accumulation of ROS	Ueta et al. 2017
	SINPRI	Reduced drought tolerance, increased stomatal aperture	Li et al. 2019
	SILBD40	Enhanced drought tolerance and reduced stomatal conductance	
	SlARF4	Enhanced drought tolerance and stem thickness	Chen et al. 2021
	SLIAA9	Seedless tomato	Ueta et al. 2017
	IncRNA1459	Prolonged shelf life	
Arabidopsis	*AtAITR*	Enhanced drought and salt tolerance, reduced ABA sensitivity	Chen et al. 2021
	AREBI	Enhanced drought tolerance and chlorophyll content	
	AtAITR	Enhanced salinity tolerance, reduced ABA sensitivity	
	ACQOS	Enhanced salinity tolerance and chlorophyll content	
Soyabean	*GmMYB118*	Reduced drought and salinity tolerance	
	GmAITR	Enhanced salinity tolerance, more sensitivity to ABA	Wang et al. 2021
	GmFATB1	Low saturated fatty acid	
	GmFT2a	Delay in flowering time	
Maize	*CLE*	Grain yield	Liu et al. 2021
Rapeseed	*BnITPK*	Low phytic acid	Sashidhar et al. 2020
	Bna MAXI	Yield increase	
Cabbage	*Bolc. GA4.a*	Dwarf plant	Lawrenson et al. 2015
RVE7	*Chickpea*	Grain quality	Badhan et al. 2021
Sorghum	*Wholek1gene*	Increased lysine content	Massel et al. 2023
Wheat	*TaSBEIIa*	High amylose content	Li et al. 2021
Wheat	*GASR7*	Grain weight	Zhang et al. 2016
Watermelon	*CIPDS*	Albino phenotype	Tian et al. 2017
Banana	*PDS*	Albino phenotype	Kaur et al. 2018
Camelina	*CsFAD2*	Oleic acid	Jiang et al. 2017

drought and salinity. The genes Leaf 1, 2 were modified using Cas9 to enhance drought tolerance by influencing Reactive Oxygen Species (ROS). Zhang et al. (2019) improved the ability of rice to tolerate high salinity by specifically modifying the *OsRR22* gene. The *ARGOS8* gene in maize was specifically manipulated to enhance drought tolerance, resulting in a novel and significant variation. Chickpea, a crucial legume crop, is greatly impacted by drought conditions. The Cas9 enzyme was used to delete the *4CL* and *RVE7* genes to improve the drought tolerance of chickpea. Zhang et al. 2019 enhanced the salt tolerance of rice by targeted modification of the *OsRR22* gene. The *ARGOS8* gene in maize was deliberately modified to improve its ability to withstand drought, leading to the emergence of a new and noteworthy variant. Drought conditions have a significant impact on chickpea, which is an essential legume crop. To enhance the drought tolerance of chickpea, the Cas9 enzyme was employed to eliminate the *4CL* and *RVE7* genes. The study focused on investigating the potential of the mitogen-activated protein kinase gene to alleviate salt stress by safeguarding membranes against oxidative stress and regulating gene transcription. The function of *SlMPAK3* in tomato's response to drought stress has been investigated by employing Cas9-mediated deletion, resulting in improved drought tolerance (Table 2) (Wang et al. 2021).

7. Future Prospects of CRISPR-CAS9 in Climate-Smart Agriculture

CRISPR-Cas9 technology has a very bright future in climate-smart agriculture, providing innovative solutions to the problems of resource scarcity, food insecurity, and climate change. CRISPR-Cas9 has the potential to completely transform sustainable farming methods due to its unparalleled precision and adaptability.

7.1 Development of Climate-Resilient Crops

The capacity to precisely modify genes linked to stress tolerance will help create crops that can flourish in harsh environmental conditions, including salinity, drought, and temperature swings. It will be possible to produce robust crops that guarantee steady yields in spite of unpredictable weather patterns by focusing on genes like *DREB* for drought tolerance or *HKT1* for salinity tolerance (Li et al. 2022).

7.2 Improved Nutritional Quality

To combat malnutrition, crops can be biofortified using CRISPR-Cas9, increasing their nutritional value. For example, nutrient-rich crops can be produced by modifying genes involved in the manufacture of vital vitamins, minerals, and proteins, thereby promoting global food security (Kumar et al. 2022).

7.3 Enhanced Resource Efficiency

By engineering crops with improved nitrogen and water use efficiency, the method has the potential to lessen reliance on chemical fertilizers and irrigation. As a result, farming will become more sustainable and consume less agricultural inputs and greenhouse gas emissions (Ali et al. 2025).

7.4 Carbon Sequestration and Climate Mitigation

By modifying root architecture and promoting the deposition of organic matter in soils, CRISPR-Cas9 can improve carbon sequestration. This strategy can improve soil health and reduce atmospheric CO_2 levels, which will help slow down climate change (Zhang et al. 2024).

7.5 Targeting Photosynthetic Efficiency

CRISPR-Cas9-assisted improvements in photosynthetic efficiency can result in more productive crops grown in less-than-ideal circumstances. Even in marginal conditions, yield potential could be greatly increased by editing genes involved in the Calvin cycle or C4 photosynthetic pathways (Paul 2021).

7.6 Multiplex Genome Editing

By targeting many genes at once, complex features like yield stability and stressor tolerance can be developed more quickly. This will make it possible to quickly develop crops that are appropriate for specific ecological and climatic circumstances (Abdelrahman et al. 2021).

7.7 Integration with Emerging Technologies

The discovery and editing of important genes will be made easier by CRISPR-Cas9's integration with other technologies, including machine learning, artificial intelligence, and high-throughput phenotyping. This technological convergence will increase the scalability of CRISPR-Cas9 applications and optimize breeding processes (Chen et al. 2024).

7.8 Gene Drive Systems for Pest Control

By changing the genetic composition of pest populations or vectors, CRISPR-Cas9 gene drive systems can control agricultural diseases and pests, lowering the need for pesticides and minimizing crop losses (Yadav et al. 2023).

7.9 Regeneration of Marginal Lands

By using technology to create crops that can be grown on marginal or degraded soil, arable land availability can be increased and strain on already fertile lands can be lessened (Assem et al. 2023).

7.10 Global Collaboration and Open-Source Platforms

Open-source platforms and global collaboration can democratize CRISPR-Cas9 technology and provide fair access to its advantages, especially in impoverished nations where climate impacts are most acute (Assem et al. 2023).

8. Challenges and Opportunities

Although CRISPR-Cas9 has revolutionary promise, ethical issues, public acceptance, and regulatory barriers that must be addressed before it can be widely used in climate-smart agriculture. Its application will be further expanded by ongoing research to enhance delivery systems for a variety of crops, especially non-model, and orphan crops.

To sum up, CRISPR-Cas9 is a key technology in the development of climate-smart, resilient, and sustainable agricultural systems. It holds the key to solving the twin problems of feeding a growing population and reducing the effects of climate change by permitting precise and targeted genetic alterations (Verdezoto-Prado et al. 2025).

9. Conclusion

The necessity to provide food security for an expanding population and the increased unpredictability of global climate patterns necessitate creative and flexible agricultural practices. CRISPR/Cas9 has become a potent and revolutionary instrument that can solve these issues. Because of its accuracy, effectiveness, and adaptability, CRISPR/Cas9 promotes the creation of climate-resilient crops, increases resource efficiency, boosts nutritional quality, and aids in the reduction of the effects of climate change through sustainable practises.

As discussed in this chapter, CRISPR/Cas9 can help climate-smart agriculture by facilitating targeted genetic alteration for enhanced carbon sequestration, photosynthetic optimization, and stress tolerance. Furthermore, its incorporation with cutting-edge technologies like machine learning and artificial intelligence presents a fresh approach to speed up decision-making and crop improvements. The use of open-access platforms, gene-driven systems, and marginal land regeneration further expands its applicability in addressing the ecological and socio-economic aspects of agriculture. In conclusion, CRISPR/Cas9 stands at the forefront of a new era in agriculture and plant sciences. By harnessing its full potential, we can pave the way toward more resilient, productive, and sustainable agriculture systems that are well equipped to thrive in the face of climate change.

References

Abdelrahman, M., Z. Wei, J.S. Rohila and K. Zhao. 2021. Multiplex genome-editing technologies for revolutionizing plant biology and crop improvement. Front. Plant Sci. 12: 721203.

Ali, A., N. Jabeen, R. Farruhbek, Z. Chachar, A.A. Laghari, S. Chachar et al. 2025. Enhancing nitrogen use efficiency in agriculture by integrating agronomic practices and genetic advances. Front. Plant Sci. 16: 1543714.

Alimagham, S., M.P. van Loon, J. Ramirez-Villegas, S. Adjei-Nsiah, F. Baijukya, A. Bala et al. 2024. Climate change impact and adaptation of rainfed cereal crops in sub-Saharan Africa. Eur. J. Agron. 155: 127137.

Altaf, M.T., M. Tatar, A. Ali, W. Liaqat, P. Mortazvi, C. Kayihan et al. 2024. Advancements in QTL mapping and GWAS application in plant improvement. Turk. J. Bot. 48(7): 376–426.

Assem, S.K., M.A. Basry, T.A. Taha, M.H.A. El-Aziz, T. Alwa and W.M. Fouad. 2023. Development of an *in vitro* regeneration system from immature inflorescences and CRISPR/Cas9-mediated gene editing in sudangrass. J. Genet. Eng. Biotechnol. 21: 58.

Badhan, S., A.S. Ball and N. Mantri. 2021. First report of CRISPR/Cas9 mediated DNA-free editing of 4CL and RVE7 genes in chickpea protoplasts. Int. J. Mol. Sci. 22: 396.

Bajwa, A.A., M. Farooq, A.M. Al-Sadi, A. Nawaz, K. Jabran and K.H.M. Siddique. 2020. Impact of climate change on biology and management of wheat pests. Crop Prot. 137: 105304.

Biber-Freudenberger, L., J. Ziemacki, H.E.Z. Tonnang and C. Borgemeister. 2016. Future risks of pest species under changing climatic conditions. PLoS ONE 11: e0153237.

Bo, W., Z. Zhaohui, Z. Huanhuan, W. Xia, L. Binglin, Y. Lijia et al. 2019. Targeted mutagenesis of NAC transcription factor gene, OsNAC041, leading to salt sensitivity in rice. Rice Sci. 26: 98–108.

Burc, E., C. Girard-Tercieux, M. Metz, E. Cazaux, J. Baur, M. Koppic et al. 2025. Life-history adaptation under climate warming magnifies the agricultural footprint of a cosmopolitan insect pest. Nat. Commun. 16: 827.

Challinor, A.J., J. Watson, D.B. Lobell, S.M. Howden, D.R. Smith and N. Chhetri. 2014. A meta-analysis of crop yield under climate change and adaptation. Nat. Clim. Chang. 4: 287–291.

Chen, C.C. and B.A. McCarl. 2001. An investigation of the relationship between pesticide usage and climate change. Clim. Chang. 50: 475–487.

Chen, F., L. Chen, Z. Yan, J. Xu, L. Feng, N. He et al. 2024. Recent advances of CRISPR-based genome editing for enhancing staple crops. Front. Plant Sci. 15: 1478398.

Chen, L., G. Liu and T. Zhang. 2024. Integrating machine learning and genome editing for crop improvement. aBIOTECH. 5: 262–277.

Chen, Y., T. Du, J. Zhang, S. Chen, J. Fu, H. Li et al. 2023. Genes and pathways correlated with heat stress responses and heat tolerance in maize kernels. Front. Plant Sci. 14: 1228213. doi: 10.3389/fpls.2023.1228213.

Chen, S., N. Zhang, G. Zhou, S. Hussain, S. Ahmed, H. Tian, et al. 2021. Knockout of the entire family of AITR genes in Arabidopsis leads to enhanced drought and salinity tolerance without fitness costs. BMC Plant Biol. 21: 137.

Coronado Lira, Z. 2024. Effect of elevated CO_2 on nutritional quality of legume crops: A meta-analysis. Rev. Latinoam. Papa. 28(2): 38–49. ISSN 1853-4961.

Espley, R.V., D. Leif, B. Plunkett, T. McGhie, R. Henry-Kirk, Hall et al. 2019. Red to brown: An elevated anthocyanic response in apple drives ethylene to advance maturity and fruit flesh browning. Front. Plant Sci. 10: 1248.

Gao, J., G. Wang, S. Ma, X. Xie, X. Wu, X. Zhang et al. 2015. CRISPR/Cas9-mediated targeted mutagenesis in *Nicotiana tabacum*. Plant Mol. Biol. 87: 99–110.

Ghulam Muha-Ud-Din, Ali F., A. Hameed, S. Naqvi, M.M. Nizamani, M. Jabran, S. Sarfraz et al. 2024. CRISPR/Cas9-based genome editing: A revolutionary approach for crop improvement and global food security. Physiol. Mol. Plant Pathol. 129: 102191.

Goldberg, G.W., M. Kogenaru, S. Keegan, M.A.B. Haase, L. Kagermazova and M. Arias. 2024. Engineered transcription-associated Cas9 targeting in eukaryotic cells. Nat. Commun. 15: 10287.

Gray, S.B. and S.M. Brady. 2016. Plant developmental responses to climate change. Dev. Biol. 419: 64–77.

Gumtow, R., D. Wu, J. Uchida and M. Tian. 2018. A *Phytophthora palmivora* extracellular cystatin-like protease inhibitor targets papain to contribute to virulence on papaya. Mol. Plant-Microbe Interact. 31: 363–373.

Habib-ur-Rahman, M., A. Ahmad, A. Raza, M.U. Hasnain, H.F. Alharby, Y.M. Alzahrani et al. 2022. Impact of climate change on agricultural production; issues, challenges, and opportunities in Asia. Front. Plant Sci. 13: 925548. doi: 10.3389/fpls.2022.925548.

Heikonen, S., M. Heino, M. Jalava, S. Siebert, D. Viviroli, M. Kummu et al. 2025. Climate change threatens crop diversity at low latitudes. Nat. Food. 6: 331–342.

Jansen, R., J.D. Embden, W. Gaastra and L.M. Schouls. 2002. Identification of genes that are associated with DNA repeats in prokaryotes. Mol. Microbiol. 43: 1565–1575.

Javadi, A., M. Ghahremanzadeh, M. Sassi, O. Javanbakht and B. Hayati. 2023. Impact of climate variables change on the yield of wheat and rice crops in Iran (application of stochastic model based on Monte Carlo simulation). Comput. Econ. 2023: 1–18.

Jia, H. and N. Wang. 2014. Targeted genome editing of sweet orange using Cas9/sgRNA. PLoS One 9.

Jiang, M., Y. Tian, R. Guo, S. Li, J. Guo and T. Zhang. 2023. Effects of warming and nitrogen addition on soil fungal and bacterial community structures in a temperate meadow. Front. Microbiol. 14: 1231442.

Jiang, W.Z., I.M. Henry, P.G. Lynagh, L. Comai, E.B. Cahoon and D.P. Weeks. 2017. Significant enhancement of fatty acid composition in seeds of the allohexaploid, *Camelina sativa*, using CRISPR/Cas9 gene editing. Plant Biotech. J. 15: 648–657.

Jouanin, A., L.J. Gilissen, J.G. Schaart, F.J. Leigh, J. Cockram, Wallington et al. 2020. CRISPR/Cas9 gene editing of gluten in wheat to reduce gluten content and exposure—reviewing methods to screen for coeliac safety. Front. Nutr. 7: 51.

Kang, Y., S. Khan and X. Ma. 2009. Climate change impacts on crop yield, crop water productivity and food security—A review. Prog. Nat. Sci. 19: 1665–1674.

Karunarathne, S.D., Y. Han, X.Q. Zhang and C. Li. 2022. CRISPR/Cas9 gene editing and natural variation analysis demonstrate the potential for HvARE1 in improvement of nitrogen use efficiency in barley. J. Integr. Plant Biol. 64: 756–770.

Kaur, N., A. Alok, N. Kaur, P. Pandey, P. Awasthi and S. Tiwari. 2018. CRISPR/Cas9-mediated efficient editing in phytoene desaturase (PDS) demonstrates precise manipulation in banana cv. Rasthali genome. Funct. Integr. Genom. 18: 89–99.

Khan, Z., T. Saboor, M. Ashfaq, A. Saddique and P. Khanum. 2021. Disease resistance in crops through CRISPR/Cas. pp. 151–175. *In*: CRISPR Crops. Springer, Berlin/Heidelberg, Germany.

Kim, D., B. Alptekin and H. Budak. 2018. CRISPR/Cas9 genome editing in wheat. Funct. Integr. Genomics 18: 31–41. https://doi.org/10.1007/s10142-017-0572-x.

Korres, N.E., J.K. Norsworthy, P. Tehranchian, T.K. Gitsopoulos, D.A. Loka, D.M. Oosterhuis et al. 2016. Cultivars to face climate change on crops and weeds: A review. Agron. Sustain. Dev. 36: 12.

Kumar, D., A. Yadav, R. Ahmad, U.N. Dwivedi and K. Yadav. 2022. CRISPR-based genome editing for nutrient enrichment in crops: A promising approach toward global food security. Front. Genet. 13: 932859.

Kumar, R. and H.R. Gautam. 2014. Climate change and its impact on agricultural productivity in India. J. Climatol. Weather Forecast. 2: 1–3.

Lawrence, T.J., J.M. Vilbig, G. Kangogo, E.M. Fèvre, S.L. Deem, I. Gluecks et al. 2023. Spatial changes to climatic suitability and availability of agropastoral farming systems across Kenya (1980–2020). Outlook Agric. 52: 117–131.

Lawrenson, T., O. Shorinola, N. Stacey, C. Li, L. Østergaard, N. Patron et al. 2015. Induction of targeted, heritable mutations in barley and *Brassica oleracea* using RNA-guided Cas9 nuclease. Genome Biol. 16: 258.

Lehmann, P., T. Ammunét, M. Barton, A. Battisti, S.D. Eigenbrode, J.U. Jepsen et al. 2020. Complex responses of global insect pests to climate warming. Front. Ecol. Environ. 18: 141–150.

Li, A., S. Jia, A. Yobi, Z. Ge, S.J. Sato, C. Zhang et al. 2021. Editing of starch branching enzyme IIa in wheat generates high-amylose wheat with altered starch structure and properties. Plant Biotechnol. J. 19: 937–947.

Li, J., G. Jiao, Y. Sun, J. Chen, Y. Zhong and L. Yan. 2021. Modification of starch composition, structure and properties through editing of TaSBEIIa in both winter and spring wheat varieties by CRISPR/Cas9. Plant Biotechnol. J. 19: 937–951.

Li, M., X. Li, Z. Zhou, P. Wu, M. Fang, X. Pan et al. 2016. Reassessment of the four yield-related genes Gn1a, DEP1, GS3, and IPA1 in rice using a CRISPR/Cas9 system. Front. Plant Sci. 7: 377.

Li, R., C. Liu, R. Zhao, L. Wang, L. Chen, W. Yu et al. 2019. CRISPR/Cas9-mediated SlNPR1 mutagenesis reduces tomato plant drought tolerance. BMC Plant Biol. 19: 38. doi: 10.1186/s12870-018-1627-4.

Li, X., S. Xu, M.B. Fuhrmann-Aoyagi, S. Yuan, T. Iwama and M. Kobayashi. 2022. CRISPR/Cas9 technique for temperature, drought, and salinity stress responses. Curr. Issues Mol. Biol. 44: 2664–2682.

Li, Y., X. Wu, Y. Zhang and Q. Zhang. 2022. CRISPR/Cas genome editing improves abiotic and biotic stress tolerance of crops. Front. Genome Ed. 4: 987817.

Liu, L., J. Gallagher, E.D. Arevalo, R. Chen, T. Skopelitis, Q. Wu et al. 2021. Enhancing grain-yield-related traits by CRISPR–Cas9 promoter editing of maize CLE genes. Nat. Plants 7: 287–294.

Liu, L., J. Zhang, J. Xu, Y. Li, L. Guo, Z. Wang et al. 2020. CRISPR/Cas9 targeted mutagenesis of SlLBD40, a lateral organ boundaries domain transcription factor, enhances drought tolerance in tomato. Plant Sci. 301: 110683.

Lou, D., H. Wang, G. Liang and D. Yu. 2017. OsSAPK2 confers abscisic acid sensitivity and tolerance to drought stress in rice. Front. Plant Sci. 8: 993.

Lv, H., X. Wang, X. Dong, M. Gao, D. Dong, C. Li et al. 2025. CRISPR/Cas9 edited SlGT30 improved both drought resistance and fruit yield through endoreduplication. Plant Cell Environ. 48: 2581–2595.

Malhi, G.S., M.C. Rana, S.S. Rana and P. Kaushik. 2020. Effect of individual or combined application of herbicide imazethapyr on nutrient uptake by blackgram (*Vigna mungo* L.). J. Exp. Biol. Agric. Sci. 8: 441–446.

Malnoy, M., R. Viola, M.-H. Jung, O.-J. Koo, S. Kim, Kim et al. 2016. DNA-free genetically edited grapevine and apple protoplast using CRISPR/Cas9 ribonucleoproteins. Front. Plant Sci. 7: 1904.

Marraffini, L. 2015. CRISPR-Cas immunity in prokaryotes. Nature 526: 55–61. https://doi.org/10.1038/nature15386.

Mascher, M., N. Stein, A. Himmelbach and C. Jung. 2017. CRISPR-Cas9 targeted mutagenesis leads to simultaneous modification of different homoeologous gene copies in polyploid oilseed rape (*Brassica napus*). Plant Physiol. 174: 935–942.

Massel, K., K.J. Vining, Y. Tao, Z.H. Cheah, B. Chong, Y. Xu et al. 2023. Large and stable genome edits at the sorghum alpha kafirin locus result in changes in chromatin accessibility and globally increased expression of genes encoding lysine enrichment. Front. Plant Sci. 14: 1116886.

Mishra, R., R.K. Joshi and K. Zhao. 2018. Genome editing in rice: Recent advances, challenges, and future implications. Front. Plant Sci. 9: 1361. doi: 10.3389/fpls.2018.01361.

Nagy, E.D., J.L. Stevens, N. Yu, C.S. Hubmeier, N. LaFaver, M. Gillespie et al. 2021. Novel disease resistance gene paralogs created by CRISPR/Cas9 in soy. Plant Cell Rep. 40: 1047–1058.

Nawaz, G., Y. Han, B. Usman, F. Liu, B. Qin and R. Li. 2019. Knockout of *OsPRP1*, a gene encoding proline-rich protein, confers enhanced cold sensitivity in rice (*Oryza sativa* L.) at the seedling stage. 3 Biotech 9: 254.

Ndudzo, A., A.S. Makuvise, S. Moyo and E.D. Bobo. 2024. CRISPR-Cas9 genome editing in crop breeding for climate change resilience: Implications for smallholder farmers in Africa. J. Agric. Food Res. 16: 101132.

Nishizawa-Yokoi, A., M. Endo, S. Kojima, H. Saika and S. Toki. 2016. Precision genome editing in plants via gene targeting and piggyBac-mediated marker excision. Plant J. 85: 173–183.

Ogata, T., T. Ishizaki, M. Fujita and Y. Fujita. 2020. CRISPR/Cas9-targeted mutagenesis of OsERA1 confers enhanced responses to abscisic acid and drought stress and increased primary root growth under nonstressed conditions in rice. PLoS One 15: e0243376. doi: 10.1371/journal.pone.0243376.

Patil, G.B., N. Lakhssassi, S. Mansoor and C.N. Stewart, Jr. 2022. CRISPR/Cas9-mediated genome editing in crops: Recent advances, applications and future prospects. Plant Cell Rep. 41: 1855–1887.

Paul, M.J. 2021. Improving photosynthetic metabolism for crop yields: What is going to work? Front. Plant Sci. 12: 743862.

Pauwels, L., G.F. Barbero, J. Geerinck, S. Tilleman, W. Grunewald, A.C. Pérez et al. 2010. NINJA connects the co-repressor TOPLESS to jasmonate signalling. Nature 464: 788–791.

Pramanik, D., R.M. Shelake, M.J. Kim and J.Y. Kim. 2021. CRISPR-mediated engineering across the central dogma in plant biology for basic research and crop improvement. Mol. Plant 14: 127–150.

Qi, Y., Y. Zhang, F. Zhang, J.A. Baller, S.C. Cleland, Y. Ryu et al. 2013. Increasing frequencies of site-specific mutagenesis and gene targeting in Arabidopsis by manipulating DNA repair pathways. Genome Res. 23: 547–554.

Qin, R., J. Li, J. Zhao, L. Xiao, Y. Liu, C. Wu et al. 2021. Stable transformation of WRKY45 RNAi construct and knockout of WRKY45 gene significantly compromise resistance to Xoo in rice. Rice 14: 1–12.

Raza, A., S. Bashir, T. Khare, B. Karikari, R.G.R. Copeland, M. Jamla et al. 2024. Temperature-smart plants: A new horizon with omics-driven plant breeding. Physiol. Plant. 176: e14188.

Razzaq, A., F. Saleem, S.H. Wani, R. Sah, M. Mushtaq, N.B. Singh et al. 2019. Precision breeding for sustainable agriculture: Genome editing in plants. Biotechnol. 9: 1–25.

Rodríguez-Leal, D., Z.H. Lemmon, J. Man, M.E. Bartlett and Z.B. Lippman. 2017. Engineering quantitative trait variation for crop improvement by genome editing. Cell 171: 470–480.e8.

Roy, S., R. Kapoor and P. Mathur. 2024. Revisiting changes in growth, physiology and stress responses of plants under the effect of enhanced CO_2 and temperature. Plant Cell Physiol. 65: 4–19.

Ruyi, R., Z. Qiang, N. Futai, J. Qiu, W. Xiuqing and W. Jicheng. 2021. Breeding for PVY resistance in tobacco LJ911 using CRISPR/Cas9 technology. Crop Breed. Appl. Biotechnol. 21.

Sabar, M., M. Arif, M.A. Nadeem, M. Ali, S.S. Alotaibi, T. Mahmood. et al. 2022. CRISPR/Cas mediated editing of yield-related traits to develop climate-resilient crops: A way forward to mitigate food insecurity under changing climate. Plants 11: 3169.

Saidi, M.N., A. Mahas and M.M. Mahfouz. 2020. CRISPR/Cas9-based multiplex genome editing in plants. Methods Mol. Biol. 2172: 159–177.

Santosh, K.V.V., R.K. Verma, S.K. Yadav, P. Yadav, A. Watts, M.V. Rao et al. 2020. CRISPR/Cas9-mediated genome editing of drought and salt tolerance (OsDST) gene in indica mega rice cultivar MTU1010. Physiol. Mol. Biol. Plants 26: 1099–1110. doi: 10.1007/s12298-020-00819-w.

Sam, V.H., P.T. Van, N.T. Ha, N.T.T. Ha, P.T.T. Huong, P.X. Hoi et al. 2021. Design and transfer of OsSWEET14-editing T-DNA construct to Bac Thom 7 rice cultivar. Acad. J. Biol. 43.

Sarkar, T., R. Thankappan, A. Kumar, G.P. Mishra, J.R. Dobaria, D. Singh et al. 2022. Genome editing for crop improvement: A perspective from India. 3 Biotech 12: 5.

Sashidhar, N., H.J. Harloff, L. Potgieter and C. Jung. 2020. Gene editing of three BnITPK genes in tetraploid oilseed rape leads to significant reduction of phytic acid in seeds. Plant Biotechnol. J. 18: 2241–2250.

Scheben, A., F. Wolter, J. Batley, H. Puchta and D. Edwards. 2017. Towards CRISPR/Cas crops—Bringing together genomics and genome editing. New Phytol. 216: 682–698.

Shakoor, U., A. Saboor, I. Ali and A.Q. Mohsin. 2011. Impact of climate change on agriculture: Empirical evidence from arid region. Pak. J. Agri. Sci. 48: 327–333.

Shen, M., C. Cai, L. Song, J. Qiu, C. Ma, D. Wang et al. 2023. Elevated CO_2 and temperature under future climate change increase severity of rice sheath blight. Front. Plant Sci. 14: 1115614.

Shrestha, S. 2019. Effects of climate change in agricultural insect pest. Acta Sci. Agric. 3(12): 74–80.

Shimatani, Z., S. Kashojiya, M. Takayama, R. Terada, T. Arazoe, H. Ishii et al. 2017. Targeted base editing in rice and tomato using a CRISPR-Cas9 cytidine deaminase fusion. Nat. Biotechnol. 35: 441–443.

Singh, B., T. Gautam, H. Vishwakarma, S. Kunwar, K. Verma, A.K. Pal. et al. 2023. Genome editing for climate-resilient agriculture: Challenges and perspectives. pp. 317–331. *In*: New Frontiers in Stress Management for Durable Agriculture. Springer, Singapore.

Singh, M., H. Chauhan, J.P. Khurana and P. Khurana. 2020. Wheat biotechnology: Present status and future developments. pp. 1–40. *In*: Al-Khayri, J.M., S.M. Jain and D.V. Johnson (eds.). Advances in Plant Breeding Strategies: Cereals. Springer, Cham.

Sissoko, P., S.S. Guindo, S. Togola, B.D. Dembélé, L.K. Grimsby and J.B. Aune. 2023. Effect of adoption of climate-smart-agriculture technologies on cereal production, food security and food diversity in Central Mali. Agriculture 13: 1196.

Sita, K., A. Sehgal, J. Kumar, S. Kumar, S. Singh, K.H.M. Siddique et al. 2017. Identification of genetic variability for synergistic traits related to heat tolerance in chickpea. Field Crops Res. 210: 73–84.

Tamang, B.G. and T. Fukao. 2015. Plant adaptation to multiple stresses during submergence and following desubmergence. Int. J. Mol. Sci. 16: 30164–30180.

Tang, X., G. Liu, J. Zhou, Q. Ren, Q. You, L. Tian et al. 2018. A large-scale whole-genome sequencing analysis reveals highly specific genome editing by both Cas9 and Cpf1 (Cas12a) nucleases in rice. Genome Biol. 19: 84.

Tian, S., L. Jiang, Q. Gao, J. Zhang, M. Zong, H. Zhang et al. 2017. Efficient CRISPR/Cas9-based gene knockout in watermelon. Plant Cell Rep. 36: 399–406.

Tripathi, P., D. Singh, A.K. Singh, A. Chaudhury, P.K. Singh, S. Sharma et al. 2022. Emerging role of CRISPR/Cas for crop improvement in changing climatic scenario. Physiol. Mol. Biol. Plants 28: 1207–1221.

Tripathi, S., P. Yadava and S. Tiwari. 2021. CRISPR/Cas9: A revolutionary approach for targeted genome editing in crops. pp. 1–18. *In*: Gosal, S.S. and S.H. Wani (eds.). Accelerated Plant Breeding, Volume 1: Biotechnological Advancements. Springer, Cham.

Ueta, R., C. Abe, T. Watanabe, S.S. Sugano, R. Ishihara, H. Ezura et al. 2017. Rapid breeding of parthenocarpic tomato plants using CRISPR/Cas9. Sci. Rep. 7: 507.

Uranga, M., A.M. Martín-Hernández, N. De Storme and F. Pasin. 2024. CRISPR–Cas systems and applications for crop bioengineering. Front. Bioeng. Biotechnol. 12: 1483857.

Usman, M., M. Farooq, A. Wakeel, A. Nawaz, H. Rehman, S. Peng et al. 2020. Nanotechnology in agriculture: Current status, challenges and future opportunities. Sci. Total Environ. 727: 138855.

Verdezoto-Prado, J., C. Chicaiza-Ortiz, A.B. Mejía-Pérez, C. Freire-Torres, M. Viteri-Yanez, L. Deng et al. 2025. Advances in environmental biotechnology with CRISPR/Cas9: bibliometric review and cutting-edge applications. Discov. Appl. Sci. 7: 167.

Verma, A.K., S. Mandal, A. Tiwari, C. Monachesi, G.N. Catassi, A. Srivastava et al. 2021. Current status and perspectives on the application of CRISPR/Cas9 gene-editing system to develop a low-gluten, non-transgenic wheat variety. Foods 10: 2351.

Villette, J., F. Lecourieux, E. Bastiancig, M.C. Héloir and B. Poinssot. 2024. New improvements in grapevine genome editing: high efficiency biallelic homozygous knock-out from regenerated plantlets by using an optimized zCas9i. Plant Methods 20: 45.

Wang, F., C. Wang, P. Liu, C. Lei, W. Hao, Y. Gao et al. 2016. Enhanced rice blast resistance by CRISPR/Cas9-targeted mutagenesis of the ERF transcription factor gene *OsERF922*. PLoS ONE 11: e0154027.

Wang, B., Z. Zhong, X. Wang, X. Han, D. Yu, C. Wang. et al. 2020. Knockout of the OsNAC006 transcription factor causes drought and heat sensitivity in rice. Int. J. Mol. Sci. 21: 2288. doi: 10.3390/ijms21072288.

Wang, J.Y., C.M. Hoel, B. Al-Shayeb, J.F. Banfield, S.G. Brohawn, J.A. Dounda et al. 2021. Structural coordination between active sites of a CRISPR reverse transcriptase-integrase complex. Nat. Commun. 12: 2571.

Wang, M., Y. Mao, Y. Lu, X. Tao and J.K. Zhu. 2017. Multiplex gene editing in rice using the CRISPR-Cpf1 system. Mol. Plant 10: 1011–1013.

Wang, W., Q. Pan, F. He, A. Akhunova, S. Chao, H. Trick et al. 2018. Transgenerational CRISPR-Cas9 activity facilitates multiplex gene editing in allopolyploid wheat. CRISPR J. 1: 65–74.

Wang, Y., X. Cheng, Q. Shan, Y. Zhang, J. Liu, C. Gao et al. 2014. Simultaneous editing of three homoeoalleles in hexaploid bread wheat confers heritable resistance to powdery mildew. Nat. Biotechnol. 32: 947–951.

Xing, H.L., L. Dong, Z.P. Wang, H.Y. Zhang, C.Y. Han, B. Liu et al. 2014. A CRISPR/Cas9 toolkit for multiplex genome editing in plants. BMC Plant Biol. 14: 327.

Xu, Z., S. Zhong, X. Li, W. Li, S.J. Rothstein and S. Zhang. 2020. Genome editing for horticultural crop improvement. Hortic. Res. 7: 1–20.

Yadav, A.K., C. Butler, A. Yamamoto, A.A. Patil, A.L. Lloyd and M.J. Scott. 2023. CRISPR/Cas9-based split homing gene drive targeting doublesex for population suppression of the global fruit pest Drosophila suzukii. Proc. Natl. Acad. Sci. U.S.A. 120: e2301525120.

Yadav, M., P. Pandey, A. Verma, A.K. Singh and P.K. Singh. 2021. Genome editing by CRISPR/Cas9: A potential tool for enhancing abiotic stress tolerance in crop plants. pp. 137–164. *In*: Gosal, S.S. and S.H. Wani (eds.). Accelerated Plant Breeding, Volume 1: Biotechnological Advancements. Springer, Cham.

Yadav, P., M. Biswal, D. Das, K. Thirugnanasambantham, B. Bhatt, N. Tripathi et al. 2021. A comprehensive overview on CRISPR/Cas9 gene editing applications in improving abiotic stress tolerance of cereal crops. Int. J. Mol. Sci. 22: 12968.

Yin, K. and J.L. Qiu. 2019. Genome editing for plant disease resistance: applications and perspectives. Philos. Trans. R. Soc. Lond. B Biol. Sci. 374(1767): 20180322. doi: 10.1098/rstb.2018.0322.

Zhang, D., H. Zhang, T. Li, K. Chen, J.L. Qiu and C. Gao. 2020. Perfectly matched 20-nucleotide guide RNA sequences enable robust genome editing using high-fidelity SpCas9 nucleases. Genome Biol. 21: 1–7.

Zhang, J., Y. Li, D. Liu, J. Zang, Z. Zhang, Q. Yao et al. 2025. Re-delimiting oversummering regions of *Puccinia striiformis* f. sp. tritici in China. Phytopathol. Res. 7: 13.

Zhang, P., N.K. Lenka and Y. Song. 2024. Editorial: Sustaining CO_2 fertilization gains under water and nutrient stress in crop plants. Front. Plant Sci. 15: 1375175.

Zhang, Y., A.A. Malzahn, S. Sretenovic and Y. Qi. 2019. The emerging and uncultivated potential of CRISPR technology in plant science. Nat. Plants 5: 778–794.

Zhang, Y., F. Zhang, X. Li, J.A. Baller, Y. Qi, C.G. Starker et al. 2013. Transcription activator-like effector nucleases enable efficient plant genome engineering. Plant Physiol. 161: 20–27.

Zheng, H., W. Ma and Q. He. 2024. Climate-smart agricultural practices for enhanced farm productivity, income, resilience, and greenhouse gas mitigation: a comprehensive review. Mitig. Adapt. Strateg. Glob. Change 29: 28.

Zheng, S., C. Ye, J. Lu, J. Liufu, L. Lin, Z. Dong et al. 2021. Improving the rice photosynthetic efficiency and yield by editing OsHXK1 via CRISPR/Cas9 system. Int. J. Mol. Sci. 22(17): 9554. doi: 10.3390/ijms22179554.

Zhou, H., B. Liu, D.P. Weeks, M.H. Spalding and B. Yang. 2014. Large chromosomal deletions and heritable small genetic changes induced by CRISPR/Cas9 in rice. Nucleic Acids Res. 42: 10903–10914.

CHAPTER 11

CRISPR/Cas9

An Innovative Biotechnological Approach for the Improvement of Oilseeds

Muhammad Tanveer Altaf,[1,*] *Waqas Liaqat,*[1] *Amna Jamil,*[2] *Muhammad Faheem Jan,*[3] *Asia Maqbool,*[4] *Naveed Abbas*[5] and *Faheem Shehzad Baloch*[6,7]

1. Introduction

The world population is growing, dietary tastes are changing, and consumption rates are rising, leading to a significant increase in food and energy requirements (Altaf et al. 2023b; Villanueva-Mejia and Alvarez 2017). According to estimates, the present global agricultural production must be expanded by at least 60–100% by 2050 to fulfil the demands. However, there is very limited or no possibility to expand the area of arable land. To satisfy the growing world population's need for oil, it is necessary to make substantial advancements in the output of oil from agricultural seeds (Subedi et al. 2020a). However, attaining these yield enhancements will be challenging due to the adverse environmental impacts caused by climate change on oil seed production. Additionally, there will be an increase in pest infestations and disease attack, which are expected to pose greater difficulties in the future (Jaradat 2016; Raman et al. 2019). There are several potential approaches to improve oilseed yields, either individually or in combination. These include increasing the amount of oil in the seeds, boosting seed production, and minimizing losses in oilseed yield caused by various factors (Attia et al. 2021; Ali et al. 2021).

Oilseeds play an essential role in the daily calorie consumption both in human and animal diets. Additionally, plant-based oils are utilized as primary ingredients in various industrial goods,

[1] Department of Field Crops, Faculty of Agriculture, Recep Tayyip Erdoğan University, Rize/Pazar, Türkiye.
[2] Department of Horticultural Sciences, The Islamia University of Bahawalpur, 63100-Pakistan.
[3] Guangxi Key Laboratory of Agro-environment and Agro-products Safety, College of Agriculture, Guangxi University, Nanning, China.
[4] College of Horticulture, Hebei Agricultural University, Baoding, 071001, Hebei, China.
[5] Department of Genetics, Plant Breeding and Seed Science, University of Agriculture in Krakow, Poland.
[6] Department of Biotechnology, Faculty of Science, Mersin University, 33343, Yenişehir, Mersin, Türkiye.
[7] Department of Plant Resources and Environment, Jeju National University, Jeju 63243, Korea.
* Corresponding author: muhammadtanveer.altaf@erdogan.edu.tr

DOI: 10.1201/9781003518204-11

including plastics, detergents, surfactants, cosmetics, plasticizers, emulsifiers, paints, and waxes (Farooq and Nazim 2024). Furthermore, they serve as a sustainable energy source and a supplement for animal feed (Rahman and de Jiménez 2016). Among oilseeds, palm oil plant, accounts for over 40% of the world's vegetable oil production (Subedi et al. 2020a). Rapeseed/canola and soybean collectively account for around 40% of the world's vegetable oil production (Altaf et al. 2023a). The remaining majority is comprised of peanut, sunflower, coconut, cotton, and olive (Abrol 2024; Yildiz et al. 2022; Ali et al. 2021). Rapeseed or canola, Indian mustard, sunflower, and soybean dominate the global oilseed market due to a growing need for high-quality food oil and sustainable industrial oils (Gupta et al. 2021). *Brassica juncea* and *Brassica napus* are important oilseed crops that are cultivated mostly in India, Bangladesh, Pakistan, Canada, Russia, Australia, and China for their economic and industrial value (Gupta et al. 2021).

Extensive research has focused on improving the quantity and quality of seed oil by classical or modern breeding methods, as well as precise genetic alteration (Subedi et al. 2020b). Although some progress has been made utilizing all these breeding methods, each strategy has its own benefits and limitations. Traditional breeding is complex and time-consuming due to various challenges. The obstacles encompass linkage drag, the polyploid character of most oilseed plants, and the reality that oil content (and occasionally fatty acid composition) represents a quantitative characteristic (Weselake et al. 2009; Yang et al. 2017). To fulfill the future need for seed oil, it would be advantageous to employ advanced molecular breeding techniques as additional instruments in breeding (Altaf et al. 2024; Baloch et al. 2023a). Although transgenic methods have been effectively used in various oilseed crops to enhance different characteristics (Ali et al. 2022; Kumar et al. 2020b), only a few of these plants have been commercialized. This is primarily due to negative public perception and the high expenses and lengthy regulatory procedures involved (Mall et al. 2018).

Genome Editing (GE) technologies are well recognized as efficient techniques for carrying out knockouts, chromosome Recombination, and targeted insertion/substitution at specific places within genes and chromosomal regions (Zhang et al. 2017). GE is a technique that involves altering DNA using specific enzymes called Site-Directed Nucleases (SDNs), including Transcriptional Activator-Like Effector Nucleases (TALENs), Zinc Finger Nucleases (ZFNs), and CRISPR/Cas, has recently been employed in crop enhancement to modify genes precisely and selectively (Nadeem et al. 2022; Gupta et al. 2019) (Fig. 1). Zinc finger nucleases (ZFNs) and transcription activator-like effector nucleases (TALENs) are two systems that use sequence-specific endonucleases. These systems consist of a particular DNA-binding domain and a cleavage domain. They have been effectively employed in rice (Shan et al. 2013), maize (Shan et al. 2013), wheat (Wang et al. 2014), and tomato (Fahad et al. 2023; Čermák et al. 2015). Nevertheless, the widespread use of these techniques in plants is restricted due to their reliance on an intricate construction process (Gaj et al. 2013; Liu et al. 2020a). The CRISPR/Cas systems are more efficient than other SDNs for GE because they have a high degree of editing specificity, which is dictated by the genetic homology between the RNA that serves as the guide and a particular sequence (Lino et al. 2018). The utilization of CRISPR/Cas techniques in the field of agricultural development can speed up determining gene function and producing novel cultivars. The CRISPR/Cas approaches have been effectively employed in more than 50 plant species (Metje-Sprink et al. 2018). Over the past few years, there have been notable improvements in boosting agronomic traits in rapeseed through GE by utilizing CRISPR/Cas9. Several studies have investigated the efficacy of CRISPR/Cas9 technology in enhancing the efficiency of brassica plant (Jedličková et al. 2022; Liu et al. 2022). GE technology has been extensively applied in oilseed crops, demonstrating substantial advancement and widespread utilization in the development of germplasm resources and genetic enhancement of oilseeds (Subedi et al. 2020a, b; Lu and Tian 2022; Mabuza et al. 2023). The CRISPR/Cas9 technology is an essential tool for investigating gene activity and advancing molecular processes in oilseed crops. This chapter provides insights into the current applications of CRISPR/Cas9 in improving oilseed quality, yield, and resilience against stresses. By highlighting recent advancements and outlining future directions,

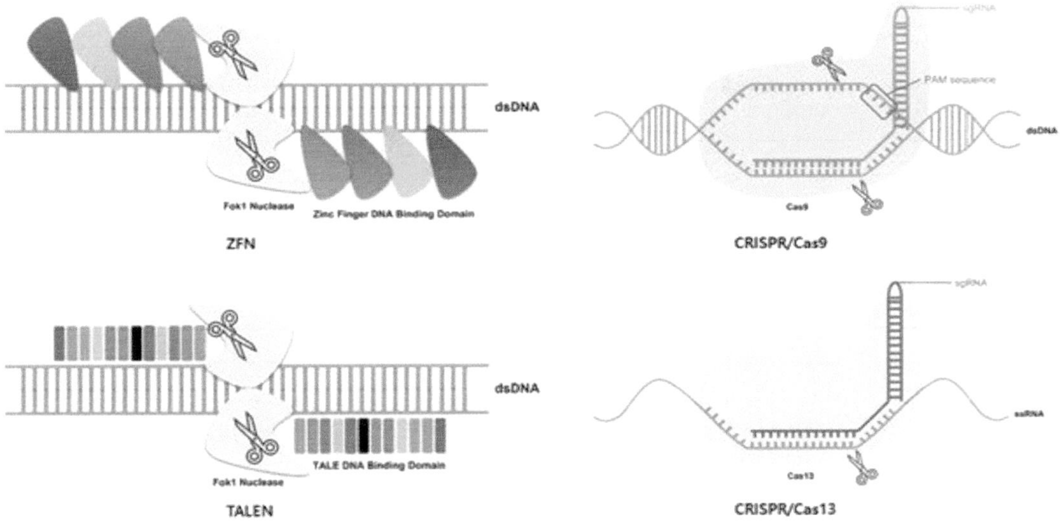

Figure 1. Representative tools for genome editing, including Zinc Finger Nucleases (ZFNs), Artificially Linked Endonucleases (ALENs), and CRISPR/Cas systems (Kim and Lee 2022).

the chapter underscores the transformative potential of GE in fostering sustainable agriculture and addressing the growing global demand for high-quality vegetable oils.

2. A Breakthrough in Plant Genome Editing: CRISPR/Cas9

ZFNs and TALENs are two types of SDNs that have been effectively utilized for GE in plants. ZFNs possess the ability to induce Double-Stranded Breaks (DSBs) at certain genomic loci (Farooq et al. 2018), their design and construction in plants is not straightforward. Furthermore, ZFNs are expensive and susceptible to inaccurate identification of DNA sequences (Ramirez et al. 2008; Sanjana et al. 2012). Compared to other technologies, TALENs reduce off-target effects and increase target binding specificity (Joung et al. 2013). Nevertheless, the process of constructing TALENs is arduous and intricate because of the intricate tandem repeat characteristic found in the DNA binding domains of the TALEN protein (Zhang et al. 2017). Consequently, the utilization of both ZFNs and TALENs has been restricted because of intricacy, decreased accuracy, and reduced efficiency in cutting targets. There is a pressing need for a straightforward, dependable, effective, and cost-effective technique to achieve accurate mutation of plant genomes (Gupta et al. 2019).

The CRISPR/Cas9 GE technique has been discovered as a defense mechanism in prokaryotic organisms to protect against invading genetic elements such as plasmids and invasive phages (Altenbuchner 2016). The utilization of the CRISPR-associated (CRISPR/Cas) platform for GE has just set in recently in plant breeding methods. This technique is especially beneficial in polyploid species since it has the capability to induce specific bi-allelic or homozygous mutations in several alleles. CRISPR/Cas is a genome-editing system that utilizes the nuclease enzyme Cas9 to induce double-stranded breaks (DSBs) in DNA. The system includes a single-guide RNA (sgRNA), which consists of a user-defined sequence of approximately 20 nucleotides (Fig. 2). This sequence guides Cas9 to a specific chromosomal location that is proximal to a Protospacer-Adjacent Motif (PAM), a short-conserved sequence required for target recognition. Various Cas enzymes, including Cas9, and cas12a have been effectively used in plants (Ming et al. 2020; Raitskin et al. 2019). Among these, the most employed Cas nuclease is Cas9 from *Streptococcus pyogenes* (PAM ¼ 50 –NGG–30). The double-strand breaks (DSBs) caused by these Cas enzymes are primarily fixed by the plant's innate repairing process of Non Homologous End-Joining (NHEJ). This repair mechanism often results in

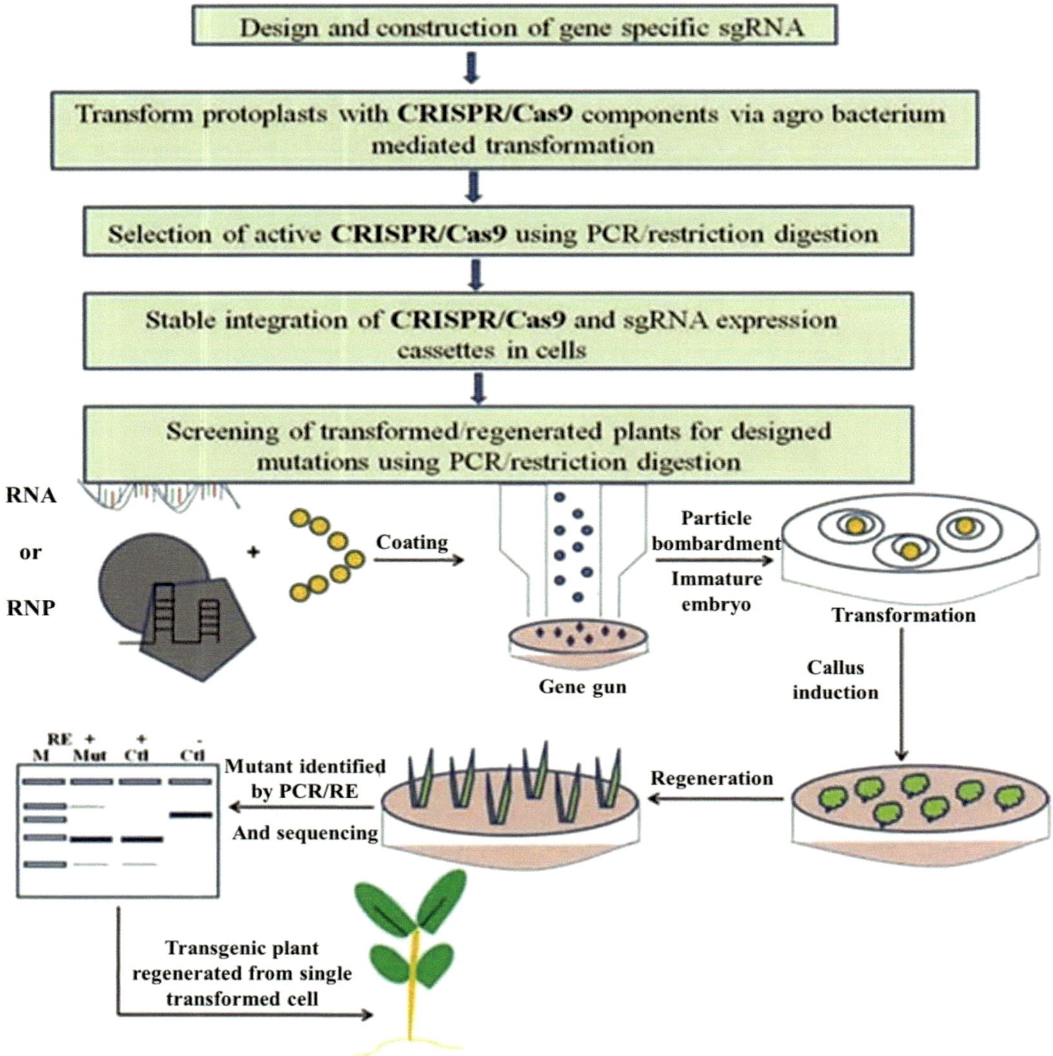

Figure 2. Procedure for CRISPR/Cas9-mediated genome modification in plants (Mushtaq et al. 2021).

a little alteration, like as a tiny insertion or deletion (indel), at the specific location of the break. This results in a genetic alteration that is difficult to differentiate from naturally occurring changes or changes generated by mutagenesis. It usually disables or reduces the activity of the genes undergoing editing (Subedi et al. 2020a). HDR (Homology-Directed Repair) can be utilized to specifically insert transgenes and replace alleles in plants (Shi et al. 2017). The crisper system can boost the precision of editing without the need for intricate protein engineering of DNA-recognition domains for every DNA target site that needs to be edited (Hu et al. 2016). The use of several sgRNAs allows for multiplex editing, which greatly simplifies and enhances the efficiency of introducing mutations at specific DNA target locations using CRISPR/Cas9 technology compared to ZFNs and TALENs (Ma et al. 2015b). To date, this novel technology is utilizing speedily in various plant species as tomato, wheat, rice, coffee, dandelion lotus, lily, rose, soybean, rapeseed, sunflower, cotton, and alfalfa (Agnès et al. 2017; Liu et al. 2021). CRISPR/Cas9 is making a substantial contribution to the field of plant functional genomics research and plants improvement.

3. CRISPR Mediated Improvement of Oilseed Quality

The primary objective of oilseed crops breeding has been to enhance the oil content of seeds and optimize the composition of fatty acids. Triacylglycerol is the principal component of vegetable oils and functions as the main storage substance in the fruits and seeds of most oil-producing plants (Subedi et al. 2020a). The increasing need for vegetable oil, which mostly consist of the storage lipid triacylglycerol, is driving extensive research to enhance oil content and optimize fatty acid composition in oilseeds. Advancements in this field have been made through the utilization of both traditional breeding methods and transgenic techniques. However, the higher ploidy level of many oilseed crops and time constraints makes it difficult to make additional improvements using traditional breeding methods. The public's opinion and the high cost of regulatory processes also make it hard to sell modified oilseed crops. CRISPR/Cas gene modification is emerging as a transformative breeding technique capable of addressing the rising demand while minimizing regulatory limitations.

3.1 Reducing Saturated and Elevating Monounsaturated Fatty Acids in Oilseeds

From a health perspective, there is a tendency to favour oils that have a lower amount of Saturated Fatty Acids (SFS). SFS is linked to elevated total cholesterol, LDL, and chronic heart disease and type II diabetes. Reductions in Saturated Fatty Acids (SFA) generally correspond with elevations in monounsaturated fatty acids (MUFA) and polyunsaturated fatty acids (PUFA), including 18:1, 18:2, and 18:3. These alterations reduce heart disease risk (Belide et al. 2012). The regulation of genes encoding fatty acid thioesterase B (FATB), a specific type of enzyme that hydrolyses the 16:0-Acyl Carrier Protein (ACP) to release free 16:0 fatty acids, can affect the accumulation of palmitic acid (16:0) and therefore the quantity of SFS in seed oil. However, the elimination of this gene in Arabidopsis resulted in impaired growth (Bonaventure et al. 2003). This effect can restrict the potential application of this technology in lowering SFA contents by GE, as the ability to specifically modify tissues in a heritable manner is not achievable without a transgene. A different approach to decrease SFA in seed oils is to adjust the activity of Stearoyl-ACP Desaturase (SAD) to increase the conversion of 18:0-ACP to 18:1-ACP. A study demonstrated that conventional GE methods, such as knock-out or knock-down, may not be efficient. However, the study revealed that employing directed evolution techniques, specifically by modifying a single amino acid in the cyanobacterial $\Delta 9$ desaturase, increased the desaturation capacity in yeast by 25 times (Bai et al. 2016). If the plant SAD enzyme can be modified using base or prime editing approaches, it may reduce SFA levels in plants. However, the potential impact on plant vegetative tissues would require further evaluation.

Conversely, oils that have high quantities of PUFA, provide numerous health advantages. However, oils with more MUFA are fewer oxidative and rancid (Belide et al. 2012). Thus, vegetable oils with a high content of MUFA and a low content of PUFA can be advantageous for many dietary uses. Researchers have successfully acquired seed oils enhanced with a ratio of 18:1 from different oilseed species including sunflower and rapeseed using traditional breeding methods (Singer et al. 2014) as well as using transgenic techniques (Chen et al. 2015). Canola has experienced notable yet meaningful enhancements in 18:1 level by using RNA interference to decrease the activity of phosphatidylcholine: diacylglycerol cholinephosphotransferase (PDCT) homologs. These homologs play a role in the conversion of phosphatidylcholine (PC) and diacylglycerol (DAG) into triacylglycerol (TAG), thereby influencing the flow of PUFAs into TAG (Bai et al. 2020b). However, most studies aiming to do this have usually focused on reducing or eliminating the activity of *FAD2*, a gene that produces an enzyme responsible for converting 18:1 to 18:2. This leads to an increase in the concentration of 18:1 at the cost of 18:2 and 18:3 (Subedi et al. 2020b).

3.2 Seed Oil Composition Improvement Using Genome Editing

The production of high-quality oil from oilseeds is critical for agricultural and industrial applications, and GE technologies have emerged as transformative tools for improving both the content and composition of oilseed-derived oils. Triacylglycerol (TAG), the primary storage lipid in oilseeds, serves as a focus for enhancing both the quantity of oil and the characteristics of its fatty acid profile.

The properties of TAG are dictated by the specific fatty acids linked to the glycerol backbone, affecting melting point, sensory attributes, stability under oxidation, and chemical utility. Fatty acids are categorized as SFA, MUFA, or polyunsaturated (PUFA) according to the quantity of double-bonded molecules in the chains of hydrocarbons (Dolowy and Pyka 2015). The ratios of these fatty acids vary across oilseed species and genotypes, defining the ultimate use of the oil (Villanueva-Mejia and Alvarez 2017; McVetty and Scarth 2002). Consequently, improving fatty acid composition has been a longstanding goal in oilseed breeding, targeting healthier profiles or properties suited to industrial applications.

Beyond oil composition, breeding efforts also aim to enhance oil production per unit area by increasing both seed yield and the amount of oil stored in seeds. TAG synthesis and accumulation involve intricate genetic networks that remain only partially understood. However, advances in molecular biology have identified key regulatory genes, such as PEPC1 and PEPC2, which negatively affect lipid production. Suppression of these genes has been shown to enhance oil content in cotton seeds by 7.3 to 16.7% (Zhao et al. 2018; Xu et al. 2016). Similarly, the DGAT-1 gene in maize has been modified to significantly increase oil content (Zheng et al. 2008). These genes, along with transcription factors like MYB, WRKY6, and TT8, present promising targets for GE (Kumar et al. 2020a; Zafar et al. 2019).

By employing GE technologies such as CRISPR/Cas9, researchers have begun to modulate these genetic factors with unprecedented precision. This includes tailoring fatty acid composition to reduce saturated fats or increase monounsaturated and polyunsaturated fats for health and stability benefits. Simultaneously, modifications in TAG-related genes aim to boost oil yields, providing dual benefits in addressing the global demand for sustainable, high-quality vegetable oils. Together, these advances underscore the potential of GE to revolutionize oilseed crop development for both nutritional and economic purposes.

4. CRISPR Cas9 and Resistance Against Biotic Stresses in Oilseeds

Pathogens such as viruses, fungi, and bacteria cause biotic stresses that result in substantial losses of 20 to 40% in worldwide agricultural production (Bhoi et al. 2022). In addition, as climate change continues, it is expected that existing phytopathogens will become more widespread and more harmful species will arise and spread in the near future (Wu et al. 2020a; Chattopadhyay et al. 2019). To address the difficulties caused by the food crisis, the introduction of pathogen resistance into host plants can help alleviate the negative impact of diseases on agricultural output (Borrelli et al. 2018). Despite the use of traditional breeding strategies to create cultivars with improved resistance to specific diseases, there are still considerable difficulties in properly managing several biotic stress factors in different crop species. Therefore, the potential to breed new oilseed varieties that have a high level of resistance to pests and diseases will be extremely important for oilseed breeders. This will allow them to fulfil the increasing demand for oilseeds and potentially decrease the need for fungicides and pesticides. Most breeding endeavours aimed at enhancing disease resistance in crop plants have mostly concentrated on incorporating or amplifying Resistance (R) genes using gene introgression or transgenic techniques (Ercolano et al. 2012). These strategies have often led to a reduction in the severity and occurrence of diseases (Qi et al. 2019; Wang et al. 2019; Xun et al. 2019).

Fungal diseases pose a substantial risk to the productivity of oilseeds and are primarily addressed through cultural techniques, host plant resistance, and by the application of chemical fungicides (Ali et al. 2021). Nevertheless, the utilization of fungicides can incur significant expenses and provide health risks for cultivators, while achieving an ideal time for application can be a difficult task. In addition, the rise of fungicide resistance is becoming more and more disturbing (Derbyshire and Denton-Giles 2016; Carter et al. 2014). In addition, numerous crop species do not have cultivars that offer full resistance to specific diseases (Derbyshire and Denton-Giles 2016). As a result of these considerations, there has been a considerable increase in the use of CRISPR/Cas-based strategies to enhance fungal disease resistance. Various approaches have shown promising results in this field (Wang et al. 2016; Wang et al. 2018). Researchers have effectively utilized the CRISPR/Cas9 knockout technology to create plants that possess substantial resistance against viral, bacterial, and fungal infections (Chen et al. 2019). Oilseed crops can experience substantial harm from Powdery Mildew (PM) when exposed to moderate temperatures and high humidity, resulting in yield reductions of 20–30% (Dunn and Gaynor 2020; Uloth et al. 2018). The CRISPR/Cas-based modification of the *S* gene *powdery-mildew-resistance locus o* (MLO) has been used successfully to increase resistance to PM in wheat and soybean (Shen et al. 2012; Wang et al. 2014). Currently, there is limited information available on MLO homologs in oilseed species. However, it is worth noting that these genes are largely preserved across different species of plants (Pessina et al. 2016). Arabidopsis mlo mutants are more resistant to Powdery Mildew (PM) (Acevedo-Garcia et al. 2017; Consonni 2010; Frye 2001). Based on our findings, targeting MLO homologs with CRISPR/Cas9 may benefit oilseed crops. According to Zhang et al. (2018b), deleting two Gh14-3-3d homologs in cotton using CRISPR/Cas9 via non-homologous end joining (NHEJ) increases resistance to Verticillium dahliae. The members of 14-3-3 protein family play diverse roles in many biological processes of plants. They achieve these functions by binding to multiple other proteins, thereby controlling their degradation, activity, or location within the cell (Paul et al. 2005). The proteins mentioned are strongly preserved among different plant species, such as rapeseed and soybean. Some of these proteins are believed to be involved in signalling networks and stress responses. Although it is possible to achieve similar outcomes in other oilseed species using CRISPR/Cas.

Verticillium longisporum (Vl43) is a type of fungus that infects rapeseed plants. This infection causes a decrease in crop yield by hindering plant growth and causing premature ageing (Lopisso et al. 2017). Pröbsting et al. (2020) found that silencing AtCRT1a, a gene responsible for increasing host susceptibility, led to a marked reduction in the sensitivity of *Arabidopsis* to Vl43. By introducing the CRISPR/Cas9 system, the gene *CRT1a* was effectively disabled in rapeseed leading to the production of materials that are resistant to *Vl43*. *CRT1a*, showed reduced susceptibility to *Vl43*, with the *BnaC09*. *CRT1a* mutants displaying a less pronounced decrease in susceptibility compared to *BnaA09.CRT1a* (Pröbsting et al. 2020). Similarly, the fungus *Sclerotinia sclerotiorum*, which causes sclerotinia stem rot in rapeseed, often leads to significant reductions in crop yield, ranging from 5 to 100% (Saharan et al. 2008). Prior research has shown that *WRKY11* and *WRKY70* have a role in controlling the Salicylic Acid (SA) and Jasmonic Acid (JA) pathways, which in turn affect the plant's ability to withstand diseases caused by pathogens in Arabidopsis (Jiang et al. 2016; Shim et al. 2013; Journot-Catalino et al. 2006; Li et al. 2004). CRISPR/Cas9 technology has also been successfully employed in soybean breeding endeavours aimed at enhancing resistance against biotic stresses. The devastating Soybean Mosaic Virus (SMV) has inflicted significant damage on soybean yields (Liu et al. 2016). In order to combat this, soybean root hairs and plants were subjected to CRISPR/Cas9 treatment, targeting three genes involved in isoflavonoid synthesis simultaneously (*GmFNSII-1*, *GmF3H2*, and *GmF3H1*). This led to an elevation in the concentration of isoflavones in both the seeds and leaves of mutant soybean plants (Luo et al. 2021). Soybean plants with elevated isoflavone levels exhibited notable resistance to the SMV strain SC7 (Luo et al. 2021).

5. CRISPR and Abiotic Stress Tolerance in Oilseeds

Abiotic stressors detrimentally impact overall plant development, resulting in significant agricultural loss on a global scale (Liaqat et al. 2024; Liaqat et al. 2023a). Factors such as drought, waterlogging, salinity, and extreme temperatures play a critical role in limiting the growth, quality, and productivity of oilseed crops (Liaqat et al. 2023b; Elferjani and Soolanayakanahally 2018; Singer et al. 2016; Purty et al. 2008; Boem et al. 1996). For instance, when soybean plants face a substantial lack of water in the soil throughout the stages of flowering and pod setting, they showed a decrease in the total area of leaves per plant by 61%, a reduction in the amount of above-ground biomass by 67%, and a loss of seed output up to 82% (Wei et al. 2018). The occurrence and intensity of these pressures are increasing in several parts of the globe because of climate change (AghaKouchak et al. 2020). As a result, it is anticipated that the negative impact of abiotic stressors on oilseed output would worsen in the next years. Drought is the most significant stress, affecting almost 40% of the world's land area, and it has a severe influence on agricultural growth and production due to global warming. Salinity is the second main abiotic stress factor that limits agricultural productivity, after drought (Liaqat et al. 2023c). It impacts around 7% of the Earth's surface area, which is equivalent to covering an area of 60 and 10.5 million square kilometres (Chikkaputtaiah et al. 2017). Abiotic stressors cause an increase in the concentration of calcium (Ca^{2+}) in the soluble part of the plant cells. This results in an increase in the production of Reactive Oxygen Species (ROS), which, when surpassing a certain threshold can inflict damage to nucleic acids (DNA and RNA), proteins, and lipids in plants. The buildup of Ca^{2+} and ROS subsequently initiates many signalling pathways. These systems, together with phytohormones like abscisic acid (ABA), work together to organize an adaptive reaction with the goal of maximizing lifespan (Mohanta et al. 2018). This reaction involves triggering many processes, such as the production of protein structures or defensive proteins, the accumulation of osmoprotectants, and the enhancement of antioxidative capacity. So far, a significant number of plants have been found to have increased tolerance to different abiotic stresses. This is achieved through the over-expression of miRNAs, genes encoding transcription factors, antioxidant enzymes, proteins associated with the production of molecular antioxidants or osmoprotectants, and proteins that play a role in phytohormone signalling pathways (Arshad et al. 2017; Ho et al. 2019; Hao et al. 2011; Zhu et al. 2018; Saxena et al. 2020; Kim et al. 2019; Sun et al. 2019; Sahni et al. 2016). However, there is a scarcity of research on improving tolerance to abiotic stress factors using CRISPR/Cas-mediated methods.

Although other phytohormones are believed to contribute to the plant stress response, ABA is particularly renowned for its crucial function in enabling plants to endure abiotic stresses. During drought, plants increase the production of ABA by *de novo* biosynthesis. This leads to changes in the transcription of several genes and promotes the closing of stomata, which helps to reduce water loss caused by transpiration (Nakashima and Yamaguchi-Shinozaki 2013). RACK1, a WD40-repeat protein, inhibits ABA responses in plants (Zhang et al. 2013). In soybean, RNA interference-induced downregulation of a RACK1 homolog enhances salt and drought tolerance compared to wild-type plants (Zheng et al. 2019). Given the worsening climatic conditions, it is becoming increasingly necessary to produce crop cultivars that are resistant to these conditions to improve food security and maintain agricultural sustainability. This can be achieved by employing either genetic engineering or the selection of crossbred lines, with the assistance of a comprehensive understanding of the molecular pathways that govern plant stress responses. Furthermore, the progress in developing research fields such as miRNA and CRISPR/Cas9 GE has greatly expedited the comprehension of plant reactions to various abiotic stress factors (Chikkaputtaiah et al. 2017). Researchers have discovered that in *rapeseed,* the genes *BnFTA* and *BnERA1* are down-regulated in response to drought, which helps protect the yield during flowering without affecting growth when there is sufficient irrigation (Wang et al. 2005; Wang et al. 2009). However, the effectiveness of this approach may depend on the specific down-regulation of these genes during drought conditions. It has been observed that mutating both genes in Arabidopsis leads to multiple defects,

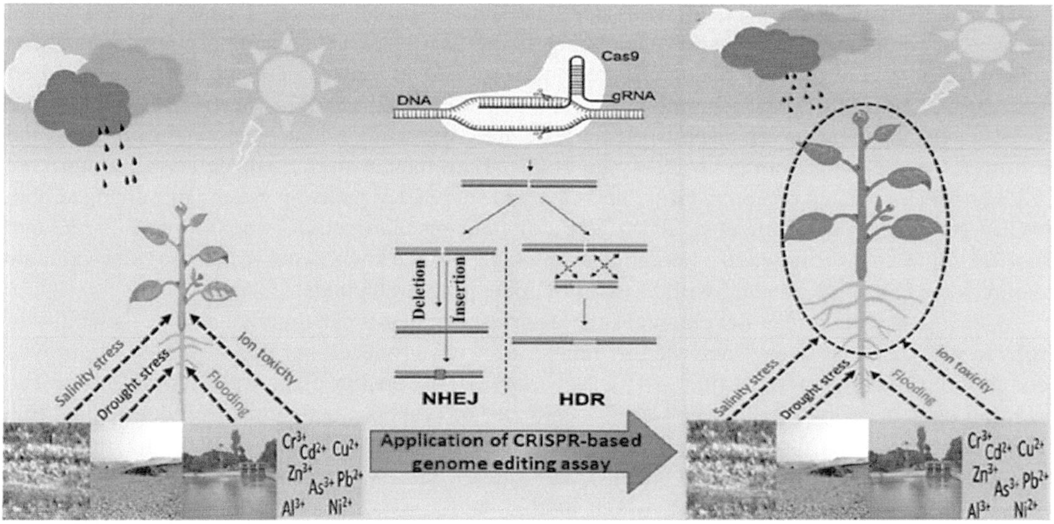

Figure 3. Illustration showcasing CRISPR-mediated enhancement of plant resistance to abiotic stresses (Adapted from Rahman et al. 2022).

such as increased vulnerability to pathogens and reduced ability to tolerate moderate and prolonged heat stress. These negative effects would likely restrict the practical usefulness of this approach in agriculture. It is yet to be determined if the use of transgene-free variants of the CRISPR/Cas system would provide advantages in achieving inducible expression. Rapeseed has high sensitivity to water scarcity throughout its whole growth and development cycle (Zhu et al. 2016). Hence, it is crucial to examine the molecular and physiological mechanisms by which rapeseed responds to drought stress to enhance drought resistance through breeding. DELLA proteins have been shown to enhance drought tolerance (Nir et al. 2017; Wang et al. 2020a). The researchers utilized CRISPR/Cas9 technology to modify the *BnaRGA* genes responsible for encoding DELLA proteins (Yang et al. 2017). This resulted in the creation of a mutant *BnaA6.rga-D* variety that exhibited enhanced drought tolerance (Wu et al. 2020c). Consequently, this mutant variety has become a valuable resource for developing drought-resistant rapeseed strains. Figure 3 present the graphical representation of the CRISPR-based GE to improve resistance to abiotic stresses.

6. CRISPR/Cas9 and Improvement of Oilseed Yield

The primary objective of oilseed breeding has consistently been to enhance crop productivity. The overall yield of oilseed crops is greatly influenced by both plant shape and the primary yield components. Zhai et al. (2019) conducted path analysis on important traits related to crop yield. The analysis revealed that the main factors influencing yield were the number of pods per plant, number of seeds per pod, size, and weight of the seeds, as well as plant morphological traits such as plant height and top branch angle. Seed size and number are the main factors that determine the oilseed yield per unit area. Both qualities are intricate, and although there is an increasing amount of genetic and genomic evidence on how they are controlled, the exact molecular mechanisms that govern them are unclear (Ali and Zhang 2023). Furthermore, silique length is a crucial factor affecting the overall yield of rapeseed, alongside these two intricate characteristics. ENHANCER3 OF DA1 (EOD3/CYP78A6) has been identified as a regulator of silique length and seed size in the model plant (Arabidopsis) and wheat (Fang et al. 2012; Ma et al. 2015a). The crisper has been utilized to eliminate *BnaEOD3* (all four homologous copies) in rapeseed, with the aim of enhancing crop productivity. The quadruple mutants demonstrated a mean elevation of 13.9% in seed weight per plant in comparison to the wild type (Ma et al. 2015a). The CLV signalling pathway has been recognized as a controller of features linked to crop yield in maize, rice, and tomato (Bommert et al. 2005; Chu et al. 2006; Bommert et al. 2013; Xu et al. 2015). The *CLAVATA3* (*CLV3*) and *CLAVATA1* (*CLV1*) genes are responsible for regulating the multilocular phenotype in *B. rapa* and *B. juncea*, respectively (Fan et al. 2014; Xu et al. 2017). Researchers used the crisper to turn off CLV3 and its associated receptors CLV1 and CLV2 to study CLV pathway genes. This changed both BnCLV genes in J9707 (Yang et al. 2018). The simultaneous elimination of both *BnaA04.CLV3* and *BnaC04.CLV3* genes resulted in a heritable multilocular silique phenotype, which provides valuable resources for rapeseed breeding with the goal of generating high yields.

Altering the structure or development characteristics of plants can greatly enhance the ability of crops to withstand stress and increase the number of seeds produced per pod, leading to improved seed output (Teichmann and Muhr 2015). Nevertheless, enhancing plant architecture in rapeseed breeding presents a substantial obstacle. The number of branches is a significant component that affects the structure of rapeseed plants and is strongly correlated with crop output. Rapeseed semi-dwarf germplasm was created using CRISPR/Cas9. This was done by altering two homologs of more axillary growth 1 (MAX1), which inhibits axillary bud vegetative growth (Zheng et al. 2020). This study is the initial application of the CRISPR/Cas9 technology to enhance the structure of rapeseed plants, offering substantial genetic resources for the breeding of high-yield rapeseed. The Arabidopsis plant species harbours a gene called *BREVIPEDICELLUS* (*BP*), which codes for a knotted1-like homeobox gene. This gene is responsible for controlling the bending of the pedicel and the shape of the leaves. This information was discovered by Lincoln et al. (1994) and further supported by Venglat et al. (2002). In a similar vein, Fan et al. (2021) discovered that by suppressing the activity of two closely related counterparts of BP, specifically *BnaA03.BP* and *BnaC03.BP*, plants exhibited a more condensed growth pattern due to a decrease in branch angle. CRISPR/Cas9 vectors were designed to disrupt these genes. As expected, plants with homozygous mutations in the *BnaA03.BP* gene displayed a semi-dwarf phenotype, with slightly drooping seed pods and upright side shoots. Similarly, the *bnac03.bp* homozygous mutants exhibited phenotypic characteristics comparable to those of *bnaa03.bp* mutants, however they had a reduced height. Significantly, plants with homozygous mutations in both *BnaBP* genes exhibited severe stunted growth, narrower branch angles, and highly drooping, shortened siliques, ultimately resulting in infertility.

Pod dehiscence at the point of ripeness in rapeseed frequently results in reduced crop production, causing roughly 20% of the overall yield to be lost. This percentage can escalate to 50% under unfavourable weather conditions (Hu et al. 2017). To reduce production losses, it is common practice to harvest rapeseed before it reaches complete maturity. Nevertheless, this untimely harvest leads to the extraction of oil from underdeveloped seeds, resulting in the contamination of chlorophyll and a decrease in the quality of the oil (Hu et al. 2015). Furthermore, pod shatter has a considerable impact on the process of automated harvesting in rapeseed. Increasing tolerance to pod dehiscence can extend the period during which mechanical harvesting can take place and reduce the amount of crop loss. At present, the primary emphasis of CRISPR/Cas9 gene editing technology is on enhancing features associated with yield, such as resistance to shattering and the creation of multiple chambers in pods. Nevertheless, the advancement in studying genes linked to grain weight, oil content, and other characteristics has been somewhat slow because of their intricate nature. Premature rupture of the silique before harvest frequently leads to substantial reductions in rapeseed production. Currently, gene editing technology is being used to alter resistance genes linked to silique shattering. The objective is to impede the segregation layer (*BnaALC*), lignified cells, and the separation layer (*BnaIND*) that prevent pod dehiscence (Zhai et al. 2020). In their study, Braatz et al. (2017) modified *BnaALC* by editing four alleles in two homologous genes located on chromosomes A and C, resulting in mutant plants. When the pod length was 5–6 cm, these mutants showed greater tolerance to dehiscence compared to the wild-type. Zhai et al. (2020) modified both *BnaIND* and *BnaALC*, observing that *BnaIND* has a greater capacity to withstand

Table 1. Genes edited by using CRISPR/Cas9 technology in rapeseed and soybean.

Target genes	Associated traits	References
BnaEOD3	Number of seeds and seed size	Khan et al. (2021)
BnALC	Shattering resistant pod	Zhai et al. (2019); Braatz et al. (2017)
BnCLV3	Multilocular silique	Yang et al. (2018)
BnFAD2	Enhancement of oleic acid content	Okuzaki et al. (2018); Huang et al. (2020)
BnaMAX1	Plant morphology and yield	Zheng et al. (2020)
BnWRKY70	Sclerotinia-resistant	Sun et al. (2018)
BnWRKY11	The mutant strain shows resistance to Sclerotinia that is not significantly different from that of the wildtype.	Sun et al. (2018)
BnCRT1a	*V. longisporum*-resistant	Pröbsting et al. (2020)
BnaBP	Plant architecture	Fan et al. (2021)
BnaC04EPSPS	Glyphosate-tolerant	Wang et al. (2021)
BnWRKY70	Increased resistance to Sclerotinia	Sun et al. (2018)
BnALC	Shattering resistant pod	Braatz et al. (2017)
BnA10.LMI1	In the HY backdrop, unlobed leaves are produced, but no discernible alterations occur in the J9709 background.	Hu et al. (2018)
BnIND	Shattering resistant pod	Zhai et al. (2019)
BnaSDG8	Early flowering attributes	Jiang et al. (2018)
BnJAG, BnSHP1, BnSHP2	Shattering resistant pod	Zaman et al. (2021); Zaman et al. (2019)
BnLPAT2/5	Seeds obtained from the mutants had shrivelled appearance, enlarged oil bodies, disrupted protein body distribution, and increased starch accumulation in mature seeds	Zhang et al. (2019)
BnLPAT2 BnLPAT5	Oil content	Zhang et al. (2019)
BnTT8	Oil content	Zhai et al. (2020)
BnaMAX1	Semi-dwarf stature and enhanced branching, along with increased silique formation	Zheng et al. (2020)
BnTT2	Oil content	Xie et al. (2020)
BnITPK	Phytic acid	Sashidhar et al. (2020)
BnGTR1, BnGTR2	Glucosinolate	Li et al. (2021)
BnaA9.WRKY47	Produced phenotypes more sensitive to low boron contents	Feng et al. (2020)
BnaTFL1	Flowering time	Sriboon et al. (2020)
BnS6-Smi2	Promotes self-incompatibility	Dou et al. (2021)
BnaSDG8	Floral transition	Jiang et al. (2018)
BnaEOD3	Smaller siliques with slighter seeds, and high seed count per silique	Khan et al. (2021)
BnaRGA	Drought tolerant	Yang et al. (2017)
BnAP2	Flower development	Zhang et al. (2018a)
Bn.YCO	Cotyledons yellow in colour and chlorotic true leaves	Liu et al. (2021)
BnaZEP	Flower colour	Liu et al. (2020b)
BnALS	Herbicide-resistant	Wu et al. (2020b)
BnaLEC1	Decreased seed oil content and C18:1, while increased C18:2	Yan et al. (2021)

Table 1 contd. ...

...Table 1 contd.

Target genes	**Associated traits**	**References**
GmFT2a	Late flowering	Cai et al. (2018)
GmPDS11/18	Dwarf buds with Albino phenotype	Du et al. (2016)
GmFATB1	Significant decrease in seed saturated fatty acids	Ma et al. (2021)
GmPRR37	Early flowering	Wang et al. (2020b)
GmFT5a	Extremely delayed flowering phenotypes under long-day conditions	Cai et al. (2020)
GmSPL9	Higher-order mutant crop plants with more branching and nodulation capacity on the main stem	Bao et al. (2019)
GmRIC1/2	Increased nodulation	Bai et al. (2020a)
GmPPD1/2	Trifoliate and dome shaped leaves with extremely twisted pods having few seeds	Kanazashi et al. (2018)

diagonal fruit breaking compared to *BnaALC*. Furthermore, they noted that the impact of *BnaA03.IND* on the two sub genomes was more significant compared to that of *BnaC03.IND*. The genes that have been modified using CRISPR/Cas technology in certain oilseed crops such as rapeseed and soybean are outlined in Table 1.

7. Conclusion and Future Prospects

Traditional plant breeding methods, while foundational in crop improvement, often face limitations due to their non-specific nature and the unintended introduction of undesirable traits during genetic introgression. These techniques remain time-intensive, particularly in segregating unwanted traits through backcrossing. In contrast, CRISPR/Cas9 offers a precise, efficient, and predictable alternative for GE, expediting the development of improved cultivars. This technology holds immense potential to address critical challenges in oilseed breeding, including increasing yield, reducing losses from seed cracking, and enhancing resistance to environmental stresses and weed competition. Despite its promise, the application of CRISPR/Cas9 to oilseed crops remains in its infancy, with limited research focused on enhancing specific traits. Challenges such as reliance on *Agrobacterium tumefaciens*-mediated transformations and arbitrary gene insertions hinder precise GE. Recent advancements, such as optimized protoplast regeneration protocols, provide promising avenues for overcoming these obstacles and guiding future research in oilseed improvement.

Further exploration into enhancing traits such as photosynthetic capacity, disease resistance, and response to agronomic inputs through CRISPR-mediated approaches is essential. Employing techniques like base or prime editing to refine protein functions and enhance gene expression, rather than solely disrupting gene function, represents a promising direction for advancing oilseed breeding. Additionally, targeting domestication-associated genes using CRISPR/Cas9 can facilitate the cultivation of underutilized or wild oilseed species, broadening the genetic base for sustainable oilseed production. As CRISPR/Cas technologies evolve, their application to agriculturally significant oilseed crops are expected to accelerate, contributing to improvements in yield, stress tolerance, and productivity. These advancements are critical for meeting the global demand for sustainable oilseed-derived products and ensuring food security in the face of climate change and population growth.

References

Abrol, D.P. 2024. Pollination Biology of Cultivated Oil Seeds and Pulse Crops. CRC Press.

Acevedo-Garcia, J., K. Gruner, A. Reinstädler, A. Kemen, E. Kemen, L. Cao et al. 2017. The powdery mildew-resistant Arabidopsis mlo2 mlo6 mlo12 triple mutant displays altered infection phenotypes with diverse types of phytopathogens. Sci. Rep. 7: 9319.

AghaKouchak, A., F. Chiang, L.S. Huning, C.A. Love, I. Mallakpour, O. Mazdiyasni et al. 2020. Climate extremes and compound hazards in a warming world. Annu. Rev. Earth Planet. Sci. 48: 519–530.

Agnès, R., C. Pauline and W, H.J. 2017. Use of CRISPR systems in plant genome editing: Toward new opportunities in agriculture. Emerg. Top. Life Sci. 1: 169–182.

Al Amin, N., N. Ahmad, N. Wu, X. Pu, T. Ma, Y. Du et al. 2019. CRISPR-Cas9 mediated targeted disruption of FAD2-2 microsomal omega-6 desaturase in soybean (*Glycine max* L). BMC Biotechnol. 19: 9.

Alexandratos, N. and J. Bruinsma. 2012. World Agriculture Towards 2030/2050: The 2012 Revision. ESA Working Paper No. 12-03.

Ali, A., M.A. Zeshan, M. Mehtab, S. Khursheed, M. Mudasir, M. Abid et al. 2021. A comprehensive note on Trichoderma as a potential biocontrol agent against soil borne fungal pathogens: a review. Plant Prot. 5(3): 171–196.

Ali, A., M.T. Altaf, M.A. Nadeem, T. Karaköy, A.N. Shah, H. Azeem et al. 2022. Recent advancement in OMICS approaches to enhance abiotic stress tolerance in legumes. Front. Plant Sci. 13: 952759.

Ali, E. and K. Zhang. 2023. CRISPR-mediated technology for seed oil improvement in rapeseed: challenges and future perspectives. Front Plant Sci. 14: 1086847.

Ali, F., M.A. Nadeem, E. Habyarimana, M.T. Altaf, M. Barut, C. Kurt et al. 2021. Identification of genetic basis associated with agronomic traits in a global safflower panel using genome-wide association study. Turk. J. Agric. For. 45(6): 834–849.

Altaf, M.T., W. Liaqat, J. Iqbal, M.M.A. Baig, A. Ali, M.A. Nadeem et al. 2023a. Exploring omics approaches to enhance stress tolerance in soybean for sustainable bioenergy production. pp. 123–154. *In*: Biotechnology and Omics Approaches for Bioenergy Crops. Singapore: Springer Nature Singapore.

Altaf, M.T., W. Liaqat, M.A. Nadeem and F.S. Baloch. 2023b. Recent trends and applications of omics-based knowledge to end global food hunger. pp. 381–397. *In*: Sustainable Agriculture in the Era of the OMICs Revolution. Cham: Springer International Publishing.

Altaf, M.T., W. Liaqat, A. Ali, A. Jamil, M. Bedir, M.A. Nadeem et al. 2024. Conventional and biotechnological approaches for the improvement of industrial crops. pp. 1–48. *In*: Industrial Crop Plants. Singapore: Springer Nature Singapore.

Altenbuchner, J. 2016. Editing of the *Bacillus subtilis* genome by the CRISPR-Cas9 system. Appl. Environ. Microbiol. 82: 5421–5427.

Arshad, M., B.A. Feyissa, L. Amyot, B. Aung and A. Hannoufa. 2017. MicroRNA156 improves drought stress tolerance in alfalfa (Medicago sativa) by silencing SPL13. Plant Sci. 258: 122–136.

Attia, Z., C.S. Pogoda, S. Reinert, N.C. Kane and B.S. Hulke. 2021. Breeding for sustainable oilseed crop yield and quality in a changing climate. Theor. Appl. Genet. 134(6): 1817–1827.

Bai, M., J. Yuan, H. Kuang, P. Gong, S. Li, Z. Zhang et al. 2020a. Generation of a multiplex mutagenesis population via pooled CRISPR-Cas9 in soybean. Plant Biotechnol. J. 18: 721–731.

Bai, S., J.G. Wallis, P. Denolf, S. Engelen, J.D. Bengtsson, M. Van Thournout et al. 2020b. The biochemistry of headgroup exchange during triacylglycerol synthesis in canola. Plant J. https://doi.org/10.1111/tpj.14709.

Bai, S., J.G. Wallis, P. Denolf and J. Browse. 2016. Directed evolution increases desaturation of a cyanobacterial fatty acid desaturase in eukaryotic expression systems. Biotechnol Bioeng. 113(7): 1522–1530.

Baloch, F.S., M.T. Altaf, W. Liaqat, M. Bedir, V. Çatalkaya, G. Cömertpay et al. 2023a. Genomics and phenomics for sustainable agriculture with few examples. PAG Australia 2023.

Baloch, F.S., M.T. Altaf, W. Liaqat, M. Bedir, M.A. Nadeem, G. Cömertpay et al. 2023b. Recent advancements in the breeding of sorghum crop: current status and future strategies for marker-assisted breeding. Front. Genet. 14: 1150616.

Bao, A., H. Chen, L. Chen, S. Chen, Q. Hao, W. Guo et al. 2019. CRISPR/Cas9-mediated targeted mutagenesis of GmSPL9 genes alters plant architecture in soybean. BMC Plant Biol. 19: 1–12.

Belide, S., J.R. Petrie, P. Shrestha and S.P. Singh. 2012. Modification of seed oil composition in Arabidopsis by artificial microRNA mediated gene silencing. Front. Plant Sci. 3: 168.

Bhoi, A., B. Yadu, J. Chandra and S. Keshavkant. 2022. Mutagenesis: A coherent technique to develop biotic stress resistant plants. Plant Stress 3: 100053.

Boem, F.H.G., R.S. Lavado and C.A. Porcelli. 1996. Note on the effects of winter and spring waterlogging on growth, chemical composition and yield of rapeseed. Field Crop Res. 47: 175–179.

Bommert, P., C. Lunde, J. Nardmann, E. Vollbrecht, M. Running, D. Jackson et al. 2005. Thick tassel dwarf1 encodes a putative maize ortholog of the Arabidopsis CLAVATA1 leucine-rich repeat receptor-like kinase. Development 132: 1235–1245.

Bommert, P., N.S. Nagasawa and D. Jackson. 2013. Quantitative variation in maize kernel row number is controlled by the FASCIATED EAR2 locus. Nat. Genet. 45: 334–337.

Bonaventure, G., J.J. Salas, M.R. Pollard and J.B. Ohlrogge. 2003. Disruption of the FATB gene in Arabidopsis demonstrates an essential role of saturated fatty acids in plant growth. Plant Cell 15: 1020–1033.

Borrelli, V.M.G., V. Brambilla, P. Rogowsky, A. Marocco and A. Lanubile. 2018. The enhancement of plant disease resistance using CRISPR/Cas9 technology. Front. Plant Sci. 9: 1245.

Braatz, J., H.-J. Harloff, M. Mascher, N. Stein, A. Himmelbach and C. Jung. 2017. CRISPR-Cas9 targeted mutagenesis leads to simultaneous modification of different homoeologous gene copies in polyploid oilseed rape (Brassica napus). Plant Physiol. 174(2): 935–942.

Cai, Y., L. Chen, X. Liu, C. Guo, S. Sun, C. Wu et al. 2018. CRISPR/Cas9-mediated targeted mutagenesis of GmFT2a delays flowering time in soybean. Plant Biotechnol. J. 16: 176–185.

Cai, Y., L. Wang, L. Chen, T. Wu, L. Liu, S. Sun et al. 2020. Mutagenesis of GmFT2a and GmFT5a mediated by CRISPR/Cas9 contributes to expanding the regional adaptability of soybean. Plant Biotechnol. J. 18: 298–309.

Carter, H.E., B.A. Fraaije, J.S. West, S.L. Kelly, A. Mehl, M.W. Shaw et al. 2014. Alterations in the predicted regulatory and coding regions of the sterol 14α-demethylase gene (CYP51) confer decreased azole sensitivity in the oilseed rape pathogen Pyrenopeziza brassicae. Mol. Plant Pathol. 15: 513–522.

Čermák, T., N.J. Baltes, R. Čegan, Y. Zhang and D.F. Voytas. 2015. High-frequency, precise modification of the tomato genome. Genome Biol. 16: 1–15.

Chattopadhyay, C., A. Birah and B.L. Jalali. 2019. Climate change: impact on biotic stresses afflicting crop plants. pp. 133–146. *In*: Peshin, R. and A. Dhawan (eds.). Natural Resource Management: Ecological Perspectives. Sustainability in Plant and Crop Protection. Springer, Cham, Switzerland.

Chen, K., Y. Wang, R. Zhang, H. Zhang and C. Gao. 2019. CRISPR/Cas genome editing and precision plant breeding in agriculture. Annu. Rev. Plant Biol. 70: 667–697.

Chen, Y., X.R. Zhou, Z.J. Zhang, P. Dribnenki, S. Singh and A. Green. 2015. Development of high oleic oil crop platform in flax through RNAi-mediated multiple FAD2 gene silencing. Plant Cell Rep. 34: 643–653.

Chikkaputtaiah, C., J. Debbarma, I. Baruah, L. Havlickova, H.P.D. Boruah and V. Curn. 2017. Molecular genetics and functional genomics of abiotic stress-responsive genes in oilseed rape (Brassica napus L.): a review of recent advances and future. Plant Biotechnol. Rep. 11: 365–384.

Chu, H., Q. Qian, W. Liang, C. Yin, H. Tan, X. Yao et al. 2006. The FLORAL ORGAN NUMBER4 gene encoding a putative ortholog of Arabidopsis CLAVATA3 regulates apical meristem size in rice. Plant Physiol. 142: 1039–1052.

Consonni, C., P. Bednarek, M. Humphry, F. Francocci, S. Ferrari, A. Harzen et al. 2010. Tryptophan-derived metabolites are required for antifungal defense in the Arabidopsis mlo2 mutant. Plant Physiol. 152: 1544–1561.

Daszkowska-Golec, A., A. Skubacz, K. Sitko, M. Słota, M. Kurowska and I. Szarejko. 2018. Mutation in barley ERA1 (Enhanced Response to ABA1) gene confers better photosynthesis efficiency in response to drought as revealed by transcriptomic and physiological analysis. Environ. Exp. Bot. 148: 12–26.

Demorest, Z.L., A. Coffman, N.J. Baltes, T.J. Stoddard, B.M. Clasen, S. Luo et al. 2016. Direct stacking of sequence-specific nuclease-induced mutations to produce high oleic and low linolenic soybean oil. BMC Plant Biol. 16: 225.

Derbyshire, M.C. and M. Denton-Giles. 2016. The control of Sclerotinia stem rot on oilseed rape (Brassica napus): current practices and future opportunities. Plant Pathol. 65: 859–877.

Dhanker, O.P. and C.H. Foyer. 2018. Climate resilient crops for improving global food security and safety. Plant Cell Environ. 41: 877–884.

Do, P.T., C.X. Nguyen, H.T. Bui, L.T.N. Tran, G. Stacey, J.D. Gillman et al. 2019. Demonstration of highly efficient dual gRNA CRISPR/Cas9 editing of the homeologous GmFAD2-1A and GmFAD2-1B genes to yield a high oleic, low linoleic and α-linolenic acid phenotype in soybean. BMC Plant Biol. 19: 311.

Dolowy, M. and A. Pyka. 2015. Chromatographic methods in the separation of long-chain mono- and polyunsaturated fatty acids. J. Chem. 2015: 20.

Dou, S., T. Zhang, J. Tu, J. Shen, B. Yi, J. Wen et al. 2021. Generation of novel self-incompatible Brassica napus by CRISPR/Cas9. Plant Biotechnol. J. 19: 875–877.

Du, H., X. Zeng, M. Zhao, X. Cui, Q. Wang, H. Yang et al. 2016. Efficient targeted mutagenesis in soybean by TALENs and CRISPR/Cas9. J. Biotechnol. 217: 90–97.

Dunn, M.W. and L.G. Gaynor. 2020. Impact and control of powdery mildew on irrigated soybean varieties grown in southeast Australia. Agronomy 10: 514.

Elferjani, R. and R. Soolanayakanahally. 2018. Canola responses to drought, heat, and combined stress: shared and specific effects on carbon assimilation, seed yield, and oil composition. Front. Plant Sci. 9: 1224.

Ercolano, M.R., W. Sanseverino, P. Carli, F. Ferriello and L. Frusciante. 2012. Genetic and genomic approaches for R-gene mediated disease resistance in tomato: retrospects and prospects. Plant Cell Rep. 31: 973–985.

Fahad, M., M.T. Altaf, A. Jamil, A. Basit, M.M. Aslam, W. Liaqat et al. 2023. Functional characterization of transcriptional activator gene SIARRI in tomato reveals its role in fruit growth and ripening. Transgenic Res. 32(1): 77–93.

Fan, C., Y. Wu, Q. Yang, Y. Yang, Q. Meng, K. Zhang et al. 2014. A novel single-nucleotide mutation in a CLAVATA3 gene homolog controls a multilocular silique trait in Brassica rapa L. Mol. Plant 7: 1788–1792.

Fan, S., L. Zhang, M. Tang, Y. Cai, J. Liu, H. Liu et al. 2021. CRISPR/Cas9-targeted mutagenesis of the BnaA03.BP gene confers semi-dwarf and compact architecture to rapeseed (Brassica napus L.). Plant Biotechnol. J. 19: 2383–2385.

Fang, W., Z. Wang, R. Cui, J. Li and Y. Li. 2012. Maternal control of seed size by EOD3/CYP78A6 in *Arabidopsis thaliana*. Plant J. 70: 929–939.

Farooq, R., K. Hussain, S. Nazir, M.R. Javed and N. Masood. 2018. CRISPR/Cas9: A robust technology for producing genetically engineered plants. Cell. Mol. Biol. 64: 31–38.

Farooq, S. and Z. Ngaini. 2024. Introduction to vegetable oils. pp. 1–20. *In*: Vegetable Oil-Based Composites: Processing, Properties and Applications. Singapore: Springer Nature Singapore.

Feng, Y., R. Cui, S. Wang, M. He, Y. Hua, L. Shi et al. 2020. Transcription factor BnaA9.WRKY47 contributes to the adaptation of Brassica napus to low boron stress by up-regulating the boric acid channel gene BnaA3.NIP5;1. Plant Biotechnol. J. 18: 1241–1254.

Food and Agriculture Organization of the United Nations. Retrieved from http://www.fao.org/3/a-ap106e.pdf.

Frye, C.A., D. Tang and R.W. Innes. 2001. Negative regulation of defense responses in plants by a conserved MAPKK kinase. Proc. Natl. Acad. Sci. USA 98: 373–378.

Gaj, T., C.A. Gersbach and C.F. Barbas III. 2013. ZFN, TALEN, and CRISPR/Cas-based methods for genome engineering. Trends Biotechnol. 31: 397–405.

Goritschnig, S., T. Weihmann, Y. Zhang, P. Fobert, P. McCourt and X. Li. 2008. A novel role for protein farnesylation in plant innate immunity. Plant Physiol. 148: 348–357.

Guo, J., J. Wang, L. Xi, W.-D. Huang, J. Liang and J.-G. Chen. 2009. RACK1 is a negative regulator of ABA responses in Arabidopsis. J. Exp. Bot. 60: 3819–3833.

Gupta, S., A. Kumar, R. Patel and V. Kumar. 2021. Genetically modified crop regulations: scope and opportunity using the CRISPR/Cas9 genome editing approach. Mol. Biol. Rep. 48(5): 4851–4863.

Gupta, D., O. Bhattacharjee, D. Mandal, M.K. Sen, D. Dey, A. Dasgupta et al. 2019. CRISPR-Cas9 system: a new-fangled dawn in gene editing. Life Sci. 232: 116636.

Hao, Y.-J., W. Wei, Q.-X. Song, H.-W. Chen, Y.-Q. Zhang, F. Wang et al. 2011. Soybean NAC transcription factors promote abiotic stress tolerance and lateral root formation in transgenic plants. Plant J. 68: 302–313.

Haun, W., A. Coffman, B.M. Clasen, Z.L. Demorest, A. Lowy, E. Ray et al. 2014. Improved soybean oil quality by targeted mutagenesis of the fatty acid desaturase 2 gene family. Plant Biotechnol. J. 12: 934–940.

Ho, T., H. Pak, C. Ryom and M. Han. 2019. Overexpression of OsmiR393a gene confers drought tolerance in creeping bentgrass. Plant Biotechnol. Rep. 13: 85–93.

Hu, J.H., K.M. Davis and D.R. Liu. 2016. Chemical biology approaches to genome editing: understanding, controlling, and delivering programmable nucleases. Cell Chem. Biol. 23: 57–73.

Hu, L., H. Zhang, Q. Yang, Q. Meng, S. Han, C.C. Nwafor et al. 2018. Promoter variations in a homeobox gene, BnA10. LMI1, determine lobed leaves in rapeseed (Brassica napus L.). Theor. Appl. Genet. 131: 2699–2708.

Hu, Q., W. Hua, Y. Yin, X.K. Zhang, L.J. Liu, J.Q. Shi et al. 2017. Rapeseed research and production in China. Crop J. 5: 127–135.

Hu, Z.Y., H.L. Yang, L. Zhang, X.F. Wang, G.H. Liu, H.Z. Wang et al. 2015. A large replum-valve joint area is associated with increased resistance to pod shattering in rapeseed. J. Plant Res. 128: 813–819.

Huang, H., T. Cui, L. Zhang, Q. Yang, Y. Yang, K. Xie et al. 2020. Modifications of fatty acid profile through targeted mutation at BnaFAD2 gene with CRISPR/Cas9-mediated gene editing in Brassica napus. Theor. Appl. Genet. 133: 2401–2411.

Jaradat, A.A. 2016. Breeding oilseed crops for climate change. pp. 421–472. *In*: Gupta, S.K. (ed.). Breeding Oilseed Crops for Sustainable Production. Academic Press, San Diego, CA.

Jedličková, V., K. Mácová, M. Štefková, J. Butula, J. Staveníková, M. Sedláček et al. 2022. Hairy root transformation system as a tool for CRISPR/Cas9-directed genome editing in oilseed rape (Brassica napus). Front. Plant Sci. 13: 919290.

Jiang, C.-H., Z.-Y. Huang, P. Xie, C. Gu, K. Li, D.-C. Wang et al. 2016. Transcription factors WRKY70 and WRKY11 served as regulators in Bacillus cereus AR156-induced systemic resistance to Pseudomonas syringae pv. tomato DC3000 in Arabidopsis. J. Exp. Bot. 67: 157–174.

Jiang, L., D. Li, L. Jin, Y. Ruan, W.H. Shen and C. Liu. 2018. Histone lysine methyltransferases Bna SDG 8.A and Bna SDG 8.C are involved in the floral transition in Brassica napus. Plant J. 95: 672–685.

Jiang, W.Z., I.M. Henry, P.G. Lynagh, L. Comai, E.B. Cahoon and D.P. Weeks. 2017. Significant enhancement of fatty acid composition in seeds of the allohexaploid, Camelina sativa, using CRISPR/Cas9 gene editing. Plant Biotechnol. J. 15: 648–657.

Joung, J.K. and J.D. Sander. 2013. TALENs: a widely applicable technology for targeted genome editing. Nat. Rev. Mol. Cell Biol. 14: 49–55.

Journot-Catalino, N., I.E. Somssich, D. Roby and T. Kroj. 2006. The transcription factors WRKY11 and WRKY17 act as negative regulators of basal resistance in *Arabidopsis thaliana*. Plant Cell 18: 3289–3302.

Kanazashi, Y., A. Hirose, I. Takahashi, M. Mikami, M. Endo, S. Hirose et al. 2018. Simultaneous site-directed mutagenesis of duplicated loci in soybean using a single guide RNA. Plant Cell Rep. 37: 553–563.

Khan, M.H.U., L. Hu, M. Zhu, Y. Zhai, S.U. Khan, S. Ahmar et al. 2021. Targeted mutagenesis of EOD3 gene in Brassica napus L. regulates seed production. J. Cell. Physiol. 236: 1996–2007.

Kim, R.J., H.U. Kim and M.C. Suh. 2019. Development of Camelina enhanced with drought stress resistance and seed oil production by co-overexpression of MYB96A and DGAT1C. Ind. Crop Prod. 138: 111475.

Kim, T.H. and S.W. Lee. 2022. Therapeutic application of genome editing technologies in viral diseases. Int. J. Mol. Sci. 23(10): 5399.

Kumar, N., A. Chaudhary, D. Singh and S. Teotia. 2020a. Transcriptional regulation of seed oil accumulation in *Arabidopsis thaliana*: role of transcription factors and chromatin remodelers. J. Plant Biochem. Biotechnol. 29: 754–768.

Kumar, K., G. Gambhir, A. Dass, A.K. Tripathi, A. Singh, A.K. Jha et al. 2020b. Genetically modified crops: current status and future prospects. Planta 251(4): 91.

Lapointe, G., M.D. Luckevich, M. Cloutier and A. Seguin. 2001. 14-3-3 gene family in hybrid poplar and its involvement in tree defence against pathogens. J. Exp. Bot. 52: 1331–1338.

Li, J., G. Brader and E.T. Palva. 2004. The WRKY70 transcription factor: a node of convergence for jasmonate-mediated and salicylate-mediated signals in plant defense. Plant Cell 16: 319–331.

Li, X. and S. Dhaubhadel. 2011. Soybean 14-3-3 gene family: identification and molecular characterization. Planta 233: 569–582.

Li, X., S. Sandgrind, O. Moss, R. Guan, E. Ivarson, E.S. Wang et al. 2021. Efficient protoplast regeneration protocol and CRISPR/Cas9-mediated editing of glucosinolate transporter (GTR) genes in rapeseed (Brassica napus L.). Front. Plant Sci. 12: 680859.

Liaqat, W., M.T. Altaf, C. Barutçular, E.M. Zayed and T. Hussain. 2023a. Drought and sorghum: a bibliometric analysis using VOS viewer. J. Biomol. Struct. Dyn. 1–13.

Liaqat, W., M.T. Altaf, C. Barutçular and S. Yasmin. 2023b. Bioenergy crops in the perspective of climate change. pp. 1–27. *In*: Biotechnology and Omics Approaches for Bioenergy Crops. Singapore: Springer Nature Singapore.

Liaqat, W., S. Yasmin, C. Barutçular and M.T. Altaf. 2023c. Agronomic and molecular approaches: a solution to the water and environmental crisis. *In*: A Solution to the Crises of Soil, Water, and Climate in Plant Production, pp. 200.

Liaqat, W., M.T. Altaf, C. Barutçular, H.I. Mohamed, Z. Ali and M.O. Khan. 2024. Drought stress in sorghum: physiological tools, breeding technology, Omics approaches and genomic-assisted breeding—A review. J. Soil Sci. Plant Nutr. 1–27.

Lincoln, C., J. Long, J. Yamaguchi and S.S. Hake. 1994. A knotted1-like homeobox gene in Arabidopsis is expressed in the vegetative meristem and dramatically alters leaf morphology when overexpressed in transgenic plants. Plant Cell 6: 1859–1876.

Lino, C.A., J.C. Harper, J.P. Carney and J.A. Timlin. 2018. Delivering CRISPR: a review of the challenges and approaches. Drug Deliv. 25: 1234–1257.

Liu, J.Z., Y. Fang and H. Pang. 2016. The current status of the soybean–soybean mosaic virus (SMV) pathosystem. Front. Microbiol. 7: 1906.

Liu, H., B. Lin, Y. Ren, P. Hao, L. Huang, B. Xue et al. 2022. CRISPR/Cas9-mediated editing of double loci of BnFAD2 increased the seed oleic acid content of rapeseed (Brassica napus L.). Front. Plant Sci. 13.

Liu, T., B. Tao, H. Wu, J. Wen, B. Yi, C. Ma et al. 2021. Bn.YCO affects chloroplast development in Brassica napus L. Crop J. https://doi.org/10.1016/j.cj.2020.10.015.

Liu, W., M.R. Rudis, M.H. Cheplick, R.J. Millwood, J.-P. Yang, C.A. Ondzighi-Assoume et al. 2020a. Lipofection-mediated genome editing using DNA-free delivery of the Cas9/gRNA ribonucleoprotein into plant cells. Plant Cell Rep. 39: 245–257.

Liu, Y., S. Ye, G. Yuan, X. Ma, S. Heng, B. Yi et al. 2020b. Gene silencing of BnaA09.ZEP and BnaC09.ZEP confers orange color in Brassica napus flowers. Plant J. 104: 932–949.

Lopisso, D.T., J. Knüfer, B. Koopmann and A. Von Tiedemann. 2017. The vascular pathogen Verticillium longisporum does not affect water relations and plant responses to drought stress of its host, Brassica napus. Phytopathology 107: 444–454.

Lu, Q.S.M. and L. Tian. 2022. An efficient and specific CRISPR-Cas9 genome editing system targeting soybean phytoene desaturase genes. BMC Biotechnol. 22(1): 7.

Luo, Y., R. Na, J.S. Nowak, Y. Qiu, Q.S. Lu, C. Yang et al. 2021. Development of a Csy4-processed guide RNA delivery system with soybean-infecting virus ALSV for genome editing. BMC Plant Biol. 21(1): 419.

Ma, J., S. Sun, J. Whelan and H. Shou. 2021. CRISPR/Cas9-mediated knockout of GmFATB1 significantly reduced the amount of saturated fatty acids in soybean seeds. Int. J. Mol. Sci. 22: 3877.

Ma, M., Q. Wang, Z.J. Li, H.H. Cheng, X. Liu, W. Song et al. 2015a. Expression of TaCYP78A3, a gene encoding cytochrome P450 CYP78A3 protein in wheat (Triticum aestivum L.), affects seed size. Plant J. 83: 312–325.

Ma, X., Q. Zhang, Q. Zhu, W. Liu, Y. Chen, R. Qiu et al. 2015b. A robust CRISPR/Cas9 system for convenient, high-efficiency multiplex genome editing in monocot and dicot plants. Mol. Plant 8: 1274–1284.

Mabuza, L.M., N.P. Mchunu, B.G. Crampton and D.Z. Swanevelder. 2023. Accelerated breeding for Helianthus annuus (sunflower) through doubled haploidy: an insight on past and future prospects in the era of genome editing. Plants 12(3): 485.

Mall, T., L. Han, L. Tagliani and C. Christensen. 2018. Transgenic crops: status, potential, and challenges. pp. 451–485. *In*: Biotechnologies of Crop Improvement, Volume 2: Transgenic Approaches. Cham: Springer International Publishing.

McVetty, P.B.E. and R. Scarth. 2002. Breeding for improved oil quality in Brassica oilseed species. J. Crop Prod. 5: 345–369.

Metje-Sprink, J., J. Menz, D. Modrzejewski and T. Sprink. 2018. DNA-free genome editing: past, present and future. Front. Plant Sci. 9: 1957.

Ming, M., Q. Ren, C. Pan, Y. He, Y. Zhang, S. Liu et al. 2020. CRISPR-Cas12b enables efficient plant genome engineering. Nat. Plants 6: 202–208.

Mohanta, T.K., T. Bashir, A. Hashem, E.F. Abd Allah and H. Bae. 2017. Genome editing tools in plants. Genes 8: 399.

Mohanta, T.K., T. Bashir, A. Hashem, E.F. Abd Allah, A.L. Khan and A.S. Al-Harrasi. 2018. Early events in plant abiotic stress signaling: interplay between calcium, reactive oxygen species and phytohormones. J. Plant Growth Regul. 37: 1033–1049.

Morineau, C., Y. Bellec, F. Tellier, L. Gissot, Z. Kelemen, F. Nogué et al. 2017. Selective gene dosage by CRISPR-Cas9 genome editing in hexaploid Camelina sativa. Plant Biotechnol. J. 15: 729–739.

Mushtaq, M., A. Ahmad Dar, M. Skalicky, A. Tyagi, N. Bhagat, U. Basu et al. 2021. CRISPR-based genome editing tools: insights into technological breakthroughs and future challenges. Genes 12(6): 797.

Nadeem, M.A., M.T. Altaf, M.A.B. Zia, T. Karaköy, F.S. Baloch and M. Aasim. 2022. Breeding advancements in fenugreek for environmental stresses. pp. 449–460. *In*: Environmental Challenges and Medicinal Plants: Sustainable Production Solutions under Adverse Conditions. Cham: Springer.

Nakashima, K. and K. Yamaguchi-Shinozaki. 2013. ABA signaling in stress-response and seed development. Plant Cell Rep. 32: 959–970.

Nekrasov, V., C. Wang, J. Win, C. Lanz, D. Weigel and S. Kamoun. 2017. Rapid generation of a transgene-free powdery mildew resistant tomato by genome deletion. Sci. Rep. 7: 482.

Nir, I., H. Shohat, I. Panizel, N. Olszewski, A. Aharoni and D. Weiss. 2017. The tomato DELLA protein PROCERA acts in guard cells to promote stomatal closure. Plant Cell 29: 3186–3197.

Okuzaki, A., T. Ogawa, C. Koizuka, K. Kaneko, M. Inaba, J. Imamura et al. 2018. CRISPR/Cas9-mediated genome editing of the fatty acid desaturase 2 gene in Brassica napus. Plant Physiol. Biochem. 131: 63–69.

Paul, A.-L., P.C. Sehnke and R.J. Ferl. 2005. Isoform-specific subcellular localization among 14-3-3 proteins in Arabidopsis seems to be driven by client interactions. Mol. Biol. Cell 16: 1735–1743.

Pessina, S., L. Lenzi, M. Perazzolli, M. Campa, L. Dalla Costa, S. Urso et al. 2016. Knockdown of MLO genes reduces susceptibility to powdery mildew in grapevine. Hortic. Res. 3: 16016.

Pröbsting, M., D. Schenke, R. Hossain, C. Häder, T. Thurau, L. Wighardt et al. 2020. Loss of function of CRT1a (calreticulin) reduces plant susceptibility to Verticillium longisporum in both Arabidopsis thaliana and oilseed rape (Brassica napus). Plant Biotechnol. J. 18: 2328–2344.

Purty, R.S., G. Kumar, S.L. Singla-Pareek and A. Pareek. 2008. Towards salinity tolerance in Brassica: an overview. Physiol. Mol. Biol. Plants 14: 39–49.

Qi, L.L., G.J. Ma, X.H. Li and G.J. Seiler. 2019. Diversification of the downy mildew resistance gene pool by introgression of a new gene, Pl35, from wild Helianthus argophyllus into oilseed and confection sunflowers (Helianthus annuus L.). Theor. Appl. Genet. 132: 2553–2565.

Rahman, M.U., S. Zulfiqar, M.A. Raza, N. Ahmad and B. Zhang. 2022. Engineering abiotic stress tolerance in crop plants through CRISPR genome editing. Cells 11(22): 3590.

Rahman, M. and M.M. de Jiménez. 2016. Designer oil crops. pp. 361–376. *In*: Gupta, S.K. (ed.). Breeding Oilseed Crops for Sustainable Production. Academic Press, Cambridge, MA.

Raitskin, O., C. Schudoma, A. West and N.J. Patron. 2019. Comparison of efficiency and specificity of CRISPR-associated (Cas) nucleases in plants: an expanded toolkit for precision genome engineering. PLoS One 14: e0211598.

Raman, H., R.K. Uppal and R. Raman. 2019. Genetic solutions to improve resilience of canola to climate change. pp. 75–131. *In*: Kole, C. (ed.). Genomic Designing of Climate-Smart Oilseed Crops. Springer, Cham, Switzerland.

Ramirez, C., J.E. Foley, D.A. Wright, F. Müller-Lerch, S.H. Rahman, T.I. Cornu et al. 2008. Unexpected failure rates for modular assembly of engineered zinc fingers. Nat. Methods 5: 374–375.

Running, M.P., M. Lavy, H. Sternberg, A. Galichet, W. Gruissem, S. Hake et al. 2004. Enlarged meristems and delayed growth in plp mutants result from lack of CaaX prenyltransferases. Proc. Natl. Acad. Sci. USA 101: 7810–7820.

Saharan, G.S. and N. Mehta. 2008. Sclerotinia Diseases of Crop Plants: Biology, Ecology and Disease Management. Springer, Berlin/Heidelberg, Germany.

Sahni, S., B.D. Prasad, Q. Liu, V. Grbic, A. Sharpe, S.P. Singh et al. 2016. Overexpression of the brassinosteroid biosynthetic gene DWF4 in Brassica napus simultaneously increases seed yield and stress tolerance. Sci. Rep. 6: 28298.

Sanjana, N.E., L. Cong, Y. Zhou, M.M. Cunniff, G.G. Feng and F. Zhang. 2012. A transcription activator-like effector toolbox for genome engineering. Nat. Protoc. 7: 171–192.

Sashidhar, N., H.J. Harloff, L. Potgieter and C. Jung. 2020. Gene editing of three BnITPK genes in tetraploid oilseed rape leads to significant reduction of phytic acid in seeds. Plant Biotechnol. J. 18: 2241–2250.

Saxena, S.C., P. Salvi, N.U. Kamble, P.K. Joshi, M. Majee and S. Arora. 2020. Ectopic overexpression of cytosolic ascorbate peroxidase gene (Apx1) improves salinity stress tolerance in Brassica juncea by strengthening antioxidative defense mechanism. Acta Physiol. Plant. 42: 45.

Seehaus, K. and R. Tenhaken. 1998. Cloning of genes by mRNA differential display induced during the hypersensitive reaction of soybean after inoculation with Pseudomonas syringae pv. glycinea. Plant Mol. Biol. 38: 1225–1234.

Shan, Q., Y. Wang, K. Chen, Z. Liang, J. Li, Y. Zhang et al. 2013. Rapid and efficient gene modification in rice and Brachypodium using TALENs. Mol. Plant 6: 1365–1368.

Shen, Q., J. Zhao, C. Du, Y. Xiang, J. Cao and X. Qin. 2012. Genome-scale identification of MLO domain-containing genes in soybean (Glycine max L. Merr.). Genes Genet. Syst. 87: 89–98.

Shi, J., H. Gao, H. Wang, H.R. Lafitte, R.L. Archibald, M. Yang et al. 2017. ARGOS8 variants generated by CRISPR-Cas9 improve maize grain yield under field drought stress conditions. Plant Biotechnol. J. 15: 207–216.

Shim, J.S., C. Jung, S. Lee, K. Min, Y.-W. Lee, Y. Choi et al. 2013. AtMYB44 regulates WRKY70 expression and modulates antagonistic interaction between salicylic acid and jasmonic acid signaling. Plant J. 73: 483–495.

Singer, S.D., J. Zou and R.J. Weselake. 2016. Abiotic factors influence plant storage lipid accumulation and composition. Plant Sci. 2.

Singer, S.D., R.J. Weselake and H. Rahman. 2014. Development and characterization of low α-linolenic acid Brassica oleracea lines bearing a novel mutation in a 'class a' fatty acid desaturase 3 gene. BMC Genet. 15(1): 94.

Sriboon, S., H. Li, C. Guo, T. Senkhamwong, C. Dai and K. Liu. 2020. Knock-out of TERMINAL FLOWER 1 genes altered flowering time and plant architecture in Brassica napus. BMC Genet. 21: 1–13.

Subedi, U., J.A. Ozga, G. Chen, N.A. Foroud and S.D. Singer. 2020. CRISPR/Cas-mediated genome editing for the improvement of oilseed crop productivity. Crit. Rev. Plant Sci. 39(3): 195–221.

Subedi, U., K.N. Jayawardhane, X. Pan, J. Ozga, G. Chen, N.A. Foroud et al. 2020b. The potential of genome editing for improving seed oil content and fatty acid composition in oilseed crops. Lipids 55(5): 495–512.

Sun, Q., L. Lin, D. Liu, Y. Wu, F. Fang, J. Wu et al. 2018. CRISPR/Cas9-mediated multiplex genome editing of the BnWRKY11 and BnWRKY70 genes in Brassica napus L. Int. J. Mol. Sci. 19: 2716.

Sun, Y., X. Liu, L. Fu, P. Qin, T. Li, X. Ma et al. 2019. Overexpression of TaBADH increases salt tolerance in Arabidopsis. Can. J. Plant Sci. 99: 546–555.

Suzaki, T. 2004. The gene FLORAL ORGAN NUMBER1 regulates floral meristem size in rice and encodes a leucine-rich repeat receptor kinase orthologous to Arabidopsis CLAVATA1. Development 131: 5649–5657.

Taguchi-Shiobara, F., Z. Yuan, S. Hake and D. Jackson. 2001. The fasciated ear2 gene encodes a leucine-rich repeat receptor-like protein that regulates shoot meristem proliferation in maize. Genes Dev. 15: 2755–2766.

Teichmann, T. and M. Muhr. 2015. Shaping plant architecture. Front. Plant Sci. 6: 233.

Trenberth, K.E., A. Dai, G. Van Der Schrier, P.D. Jones, J. Barichivich, K.R. Briffa et al. 2014. Global warming and changes in drought. Nat. Clim. Chang. 4(1): 17–22.

Uloth, M.B., M.P. You and M.J. Barbetti. 2018. Plant age and ambient temperature: significant drivers for powdery mildew (Erysiphe cruciferarum) epidemics on oilseed rape (Brassica napus). Plant Pathol. 67: 445–456.

Venglat, S.P., T. Dumonceaux, K. Rozwadowski, L. Parnell, V. Babic, W. Keller et al. 2002. The homeobox gene BREVIPEDICELLUS is a key regulator of inflorescence architecture in Arabidopsis. Proc. Natl. Acad. Sci. USA 99: 4730–4735.

Villanueva-Mejia, D. and J.C. Alvarez. 2017. Genetic improvement of oilseed crops using modern biotechnology. pp. 295–317. *In*: Jimenez-Lopez, J.C. (ed.). Advances in Seed Biology. InTech, London, UK.

Wang, F., C. Wang, P. Liu, C. Lei, W. Hao, Y. Gao et al. 2016. Enhanced rice blast resistance by CRISPR/Cas9-targeted mutagenesis of the ERF transcription factor gene OsERF922. PLoS One 11.

Wang, J., H. Kuang, Z. Zhang, Y. Yang, L. Yan, M. Zhang et al. 2020a. Generation of seed lipoxygenase-free soybean using CRISPR-Cas9. Crop J. 8: 432–439.

Wang, L., S. Sun, T. Wu, L. Liu, X. Sun, Y. Cai et al. 2020b. Natural variation and CRISPR/Cas9-mediated mutation in GmPRR37 affect photoperiodic flowering and contribute to regional adaptation of soybean. Plant Biotechnol. J. 18: 1869–1881.

Wang, X., M. Tu, D. Wang, J. Liu, Y. Li, Z. Li et al. 2018. CRISPR/Cas9-mediated efficient targeted mutagenesis in grape in the first generation. Plant Biotechnol. J. 16: 844–855.

Wang, Y., M. Beaith, M. Chalifoux, J. Ying, T. Uchacz, C. Sarvas et al. 2009. Shoot-specific down-regulation of protein farnesyltransferase (alpha-subunit) for yield protection against drought in canola. Mol. Plant 2: 191–200.

Wang, Y., X. Cheng, Q. Shan, Y. Zhang, J. Liu, C. Gao et al. 2014. Simultaneous editing of three homoeoalleles in hexaploid bread wheat confers heritable resistance to powdery mildew. Nat. Biotechnol. 32: 947–951.

Wang, Y., J. Ying, M. Kuzma, M. Chalifoux, A. Sample, C. McArthur et al. 2005. Molecular tailoring of farnesylation for plant drought tolerance and yield protection. Plant J. 43: 413–424.

Wang, Z., M. Yang, Y. Sun, Q. Yang, L. Wei, Y. Shao et al. 2019. Overexpressing Sesamum indicum L.'s DGAT1 increases the seed oil content of transgenic soybean. Mol. Breed. 39: 101.

Wang, Z., L. Wan, Q. Xin, X. Zhang, Y. Song, P. Wang et al. 2021. Optimizing glyphosate tolerance in rapeseed by CRISPR/Cas9-based geminiviral donor DNA replicon system with Csy4-based single-guide RNA processing. J. Exp. Bot. 72: 4796–4808.

Wei, Y., J. Jin, S. Jiang, S. Ning and L. Liu. 2018. Quantitative response of soybean development and yield to drought stress during different growth stages in the Huaibei Plain, China. Agronomy 8: 97.

Weselake, R.J., D.C. Taylor, M.H. Rahman, S. Shah, A. Laroche, P.B.E. McVetty et al. 2009. Increasing the flow of carbon into seed oil. Biotechnol. Adv. 27: 866–878.

Wu, E.-J., Y.-P. Wang, L. Yahuza, M.-H. He, D.-L. Sun, Y.-M. Huang et al. 2020a. Rapid adaptation of the Irish potato famine pathogen Phytophthora infestans to changing temperature. Evol. Appl. 13: 769–781.

Wu, J., C. Chen, G. Xian, D. Liu, L. Lin, S. Yin et al. 2020b. Engineering herbicide-resistant oilseed rape by CRISPR/Cas9-mediated cytosine base-editing. Plant Biotechnol. J. 18: 1857–1859.

Wu, J., G. Yan, Z. Duan, Z. Wang, C. Kang, L. Guo et al. 2020c. Roles of the Brassica napus DELLA protein BnaA6.RGA in modulating drought tolerance by interacting with the ABA signaling component BnaA10.ABF2. Front. Plant Sci. 11: 577.

Wu, J.-R., L.-C. Wang, Y.-R. Lin, C.-P. Weng, C.-H. Yeh and S.-J. Wu. 2017. The Arabidopsis heat-intolerant 5 (hit5)/enhanced response to aba1 (era1) mutant reveals the crucial role of protein farnesylation in plant responses to heat stress. New Phytol. 213: 1181–1193.

Xie, T., X. Chen, T. Guo, H. Rong, Z. Chen, Q. Sun et al. 2020. Targeted knockout of BnTT2 homologues for yellow-seeded *Brassica napus* with reduced flavonoids and improved fatty acid composition. J. Agric. Food Chem. 68: 5676–5690.

Xu, Z., J. Li, X. Guo, S. Jin and X. Zhang. 2016. Metabolic engineering of cottonseed oil biosynthesis pathway via RNA interference. Sci. Rep. 6(1): 33342.

Xu, C., K.L. Liberatore, C.A. MacAlister, Z. Huang, Y.-H. Chu, K. Jiang et al. 2015. A cascade of arabinosyltransferases controls shoot meristem size in tomato. Nat. Genet. 47: 784–792.

Xu, P., S.Q. Cao, K.N. Hu, X.H. Wang, W. Huang, G. Wang et al. 2017. Trilocular phenotype in Brassica juncea L. resulted from interruption of CLAVATA1 gene homologue (BjMc1) transcription. Sci. Rep. 7: 3498.

Xun, H., X. Yang, H. He, M. Wang, P. Guo, Y. Wang et al. 2019. Over-expression of GmKR3, a TIR-NBS-LRR type R gene, confers resistance to multiple viruses in soybean. Plant Mol. Biol. 99: 95–111.

Yalovsky, S., A. Kulukian, M. Rodriguez-Concepcion, C.A. Young and W. Gruissem. 2000. Functional requirement of plant farnesyltransferase during development in Arabidopsis. Plant Cell 12: 1267–1278.

Yan, G., P. Yu, X. Tian, L. Guo, J. Tu, J. Shen et al. 2021. DELLA proteins BnaA6.RGA and BnaC7.RGA negatively regulate fatty acid biosynthesis by interacting with BnaLEC1s in Brassica napus. Plant Biotechnol. J. 19(10): 2011–2026.

Yang, H., J.-J. Wu, T. Tang, K.-D. Liu and C. Dai. 2017. CRISPR/Cas9-mediated genome editing efficiently creates specific mutations at multiple loci using one sgRNA in Brassica napus. Sci. Rep. 7: 7489.

Yang, Y., K. Zhu, H. Li, S. Han, Q. Meng, S.U. Khan et al. 2018. Precise editing of CLAVATA genes in Brassica napus L. regulates multilocular silique development. Plant Biotechnol. J. 16: 1322–1335.

Yang, Y., Q. Kong, A.R. Lim, S. Lu, H. Zhao, L. Guo et al. 2022. Transcriptional regulation of oil biosynthesis in seed plants: Current understanding, applications, and perspectives. Plant Commun. 3(5).

Yildiz, M., M.T. Altaf, F.S. Baloch, M. Koçak, G. Sadık, C. Kuzğun et al. 2022. Assessment of genetic diversity among 131 safflower (Carthamus tinctorius L.) accessions using peroxidase gene polymorphism (POGP) markers. Mol. Biol. Rep. 49(7): 6531–6539.

Zafar, S., Y.L. Li, N.N. Li, K.M. Zhu and X.L. Tan. 2019. Recent advances in enhancement of oil content in oilseed crops. J. Biotechnol. 301: 35–44.

Zaman, Q., W. Chu, M. Hao, Y. Shi, M. Sun, S.-F. Sang et al. 2019. CRISPR/Cas9-mediated multiplex genome editing of JAGGED gene in Brassica napus L. Biomolecules 9: 725.

Zaman, Q.U., C. Wen, S. Yuqin, M. Mengyu, D. Mei, J. Bouchereau et al. 2021. Characterization of SHATTERPROOF homoeologs and CRISPR-Cas9-mediated genome editing enhances pod-shattering resistance in Brassica napus L. CRISPR J. 4: 360–370.

Zegeye, W.A., M. Tsegaw, Y. Zhang and L. Cao. 2022. CRISPR-based genome editing: advancements and opportunities for rice improvement. Int. J. Mol. Sci. 23(8): 4454.

Zhai, Y., S. Cai, L. Hu, Y. Yang, O. Amoo, C. Fan et al. 2019. CRISPR/Cas9-mediated genome editing reveals differences in the contribution of INDEHISCENT homologues to pod shatter resistance in Brassica napus L. Theor. Appl. Genet. 132: 2111–2123.

Zhai, Y., K. Yu, S. Cai, L. Hu, O. Amoo, L. Xu et al. 2020. Targeted mutagenesis of BnTT8 homologs controls yellow seed coat development for effective oil production in Brassica napus L. Plant Biotechnol. J. 18: 1153–1168.

Zhan, G.-M., J. Tong, H.-Z. Wang and W. Hua. 2010. Molecular analysis and expression patterns of four 14-3-3 genes from Brassica napus L. Agric. Sci. China 9: 942–950.

Zhang, D., C. Li, B. Lv and J. Liang. 2013. The scaffolding protein RACK1: a platform for diverse functions in the plant kingdom. J. Plant Biol. Soil Health 1: 7.

Zhang, K., L. Nie, Q. Cheng, Y. Yin, K. Chen, F. Qi et al. 2019. Effective editing for lysophosphatidic acid acyltransferase 2/5 in allotetraploid rapeseed (Brassica napus L.) using CRISPR-Cas9 system. Biotechnol. Biofuels 12: 1–18.

Zhang, Y., S. Huang, X. Wang, J. Liu, X. Guo, J. Mu et al. 2018a. Defective APETALA2 genes lead to sepal modification in Brassica crops. Front. Plant Sci. 9: 367.

Zhang, Y., X. Ma, X. Xie and Y.-G. Liu. 2017. CRISPR/Cas9-based genome editing in plants. Prog. Mol. Biol. Transl. Sci. 149: 133–150.

Zhang, Z., X. Ge, X. Luo, P. Wang, Q. Fan, G. Hu et al. 2018b. Simultaneous editing of two copies of Gh14-3-3d confers enhanced transgene-clean plant defense against Verticillium dahliae in allotetraploid upland cotton. Front. Plant Sci. 9: 842.

Zhao, Y., Y. Huang, Y. Wang, Y. Cui, Z. Liu and J. Hua. 2018. RNA interference of GhPEPC2 enhanced seed oil accumulation and salt tolerance in Upland cotton. Plant Sci. 271: 52–61.

Zheng, P., W.B. Allen, K. Roesler, M.E. Williams, S. Zhang, J. Li et al. 2008. A phenylalanine in DGAT is a key determinant of oil content and composition in maize. Nat. Genet. 40(3): 367–372.

Zheng, M., L. Zhang, M. Tang, J. Liu, H. Liu, H. Yang et al. 2020. Knockout of two BnaMAX1 homologs by CRISPR/Cas9-targeted mutagenesis improves plant architecture and increases yield in rapeseed (Brassica napus L.). Plant Biotechnol. J. 18: 644–654.

Zheng, W.-N., D.-H. Li, F.-J. Chen, X.-P. Liu and H.-Y. Li. 2019. Abiotic stress tolerance and ABA responses of transgenic Glycine max plants with modulated RACK1 expression. Can. J. Plant Sci. 99: 250–267.

Zhu, M., J.G. Monroe, Y. Suhail, F. Villiers, J. Mullen, D. Pater et al. 2016. Molecular and systems approaches towards drought-tolerant canola crops. New Phytol. 210: 1169–1189.

Zhu, X., Y. Wang, Y. Liu, W. Zhou, B. Yan, J. Yang et al. 2018. Overexpression of BcHsfA1 transcription factor from Brassica campestris improved heat tolerance of transgenic tobacco. PLoS One 13: e0207277.

CHAPTER 12

Improvement of Vegetable Crops Using CRISPR-Cas9 as A Gene Editing Tool

Dilawar Khan,[1,*] *Kumail Ali Rizvi,*[2] *Asim Gulzar*[1] and *Tayyaba Zainab*[2,3]

1. Introduction

By 2050 the global population forecasts a count of 9.7 billion people which poses major challenges to worldwide food security (Wudil et al. 2022). Food production must achieve exponentially higher returns because climate change, degraded soils, and depleted resources need to be resolved (Hemathilake and Gunathilake 2022). Vegetable crops constitute vital elements of human nutrition because they contain vital vitamins and minerals in addition to essential dietary fiber (Ebabhi and Adebayo 2022). The cultivation of vegetable crops encounters several obstacles, which combine pests and diseases with water shortage and exposure to environmental stresses that involve both drought and salt damage. Advanced, efficient agricultural practices must be developed because society requires better yields and nutrient quality in agriculture. Following limitations within traditional breeding approaches, researchers are forced to investigate new ways of crop improvement due to time constraints and imprecise adjustments, and a narrow application scope. The groundbreaking CRISPR-Cas9 represents a leading agricultural biotechnology innovation that allows scientists to perform exact genome transformations among living organisms, particularly plants (Demir and Erbaş 2023; Akram et al. 2023).

CRISPR-Cas9 technology uses bacterial defense mechanisms to provide efficient DNA sequence modification through precise edits of an organism's genes (Hossain 2021). Traditional genetic modification procedures differ from CRISPR-Cas9 because they select specific genes instead of randomly adding outside components as they modify DNA sequences. The exact gene editing capabilities of CRISPR-Cas9 technologies offer prospective benefits to vegetable crops because they enhance productivity alongside disease defense capability and nutritional quality, and environmental resilience (Devi et al. 2022).

CRISPR-Cas9 is a naturally occurring system found in bacteria and archaea, which they use to defend against viral infections (El-Ashry 2023). The name CRISPR stands for "Clustered Regularly Interspaced Short Palindromic Repeats," referring to the repetitive DNA sequences found in bacterial

[1] Department of Entomology, PMAS-Arid Agriculture University, Rawalpindi.
[2] National Center of Industrial Biotechnology, PMAS-Arid Agriculture University, Rawalpindi.
[3] University Institute of Biochemistry and Biotechnology, PMAS-Arid Agriculture University, Rawalpindi.
* Corresponding author: dilawarkhan.arid@gmail.com

DOI: 10.1201/9781003518204-12

genomes. Cas9, which stands for "CRISPR-associated protein 9," is an enzyme that can cut DNA at a specific location (Asmamaw and Zawdie 2021). In nature, CRISPR sequences store segments of viral DNA, allowing bacteria to "remember" previous viral infections and defend against them by targeting the same DNA sequences in subsequent infections.

Scientists have adapted the CRISPR-Cas9 system for biological technology applications, which enables gene editing across various organisms, including plants. Two essential parts of this procedure are the Cas9 enzyme functioning as DNA-cutting molecular scissors and the guide RNA that directs Cas9 enzyme activity toward its desired target gene. The plant maintains its DNA repair processes after DNA cuts to either destroy gene function by adding random errors or modify DNA through homologous recombination (Zhan et al. 2021; Singh et al. 2023). Research indicates that CRISPR-Cas9 stands superior to Agrobacterium-mediated transformation and zinc finger nucleases because of its precise DNA cutting capabilities and guided RNA direction (Zhou et al. 2023). High-accuracy editing of genes through research enables scientists to introduce and remove crop traits by causing minimal unintended side effects.

The disease resistance of vegetable crops are at risk because bacteria and fungi, together with viruses, cause devastating disease outbreaks. The diseases cause major yield reduction and quality deterioration, which endanger food supply systems and require increased usage of chemical pesticides (Singh et al. 2021). The traditional cultivation techniques for breeding disease-resistant crops face impediments when scientists attempt to combat new pathogens that lack resistance characteristics in current plant resources. CRISPR-Cas9 serves as a time-effective technology for disease resistance improvement through targeted modifications of plant immune response genes. Scientific studies demonstrated how CRISPR modifications enabled researchers to make tomato plants resistant to the Tomato Yellow Leaf Curl Virus (TYLCV), which negatively affects tomato production worldwide (Shawky et al. 2024; Pramanik et al. 2021). Researchers improve plant defense powers by modifying genetic elements that determine viral reactions so they remain virus-resilient while avoiding chemical protectants.

Through CRISPR-Cas9 technology, scientists can develop mechanisms to protect various plants from bacterial and fungal infections. Crop health benefits from enhanced resistance when scientists use CRISPR-Cas9 to edit immune receptor genes in pepper as well as lettuce plants against pathogens, including Xanthomonas and Fusarium (Bigini et al. 2021; Rato et al. 2021).

Climate change has exacerbated the occurrence of abiotic stresses such as drought, extreme temperatures, and soil salinity, which significantly impact vegetable crop production (Wang et al. 2020). Traditional breeding methods have struggled to produce crops with enhanced tolerance to these stresses, mainly due to the complexity of traits involved and the long breeding cycles required. CRISPR-Cas9 provides a targeted approach to improving stress tolerance by editing genes that regulate plant responses to environmental stresses. For instance, in drought-prone regions, CRISPR has been used to modify genes in crops like rice and tomato to improve their ability to conserve water and maintain productivity under dry conditions (Sami et al. 2021). Similarly, genes related to salt tolerance have been targeted in crops like lettuce, enhancing their ability to grow in saline soils (Riu et al. 2023). These improvements could significantly increase the resilience of vegetable crops to changing climate conditions, helping to ensure stable food production in the face of environmental uncertainties.

The growing worldwide human population leads to higher requirements for nutritious food. The nutritional value of vegetable crops is currently restricted due to genetic limits, according to research by Thakur et al. (2022). CRISPR-Cas9 technology demonstrates exceptional performance in improving vegetable crop nutrition when it increases the amounts of essential nutrients, including vitamins and antioxidants, and minerals. Through CRISPR-Cas9 technology, scientists accomplished higher levels of beta-carotene (a vitamin A precursor) in tomato and sweet potato crops (Berman et al. 2019). Plants with enhanced nutrition through biofortification help minimize nutrition deficits, primarily in vitamin A deficits. The CRISPR system enables researchers to boost the content levels of vital nutrients such as zinc and iron in dietary plants, including beans and spinach, for better human health benefits (Chatterjee et al. 2025).

Using CRISPR lets scientists' lower anti-nutritional factors, lectins, and oxalates in food plants, thus enabling better nutrient absorption and healthy digestion. The editing of genes responsible for creating these compounds by CRISPR-Cas9 enhances vegetable crops while making them safer for human consumption (Wan et al. 2021).

CRISPR-Cas9 demonstrates great promise for increasing crop yields because this technology directly addresses food security issues. The research community can boost vegetable crop harvest capacity through genetic manipulations of developmental control elements that manage plant development stages and flowering. The CRISPR system has successfully altered genes that manage tomato and cucumber aspects, including fruit dimensions, together with weight and ripening stages, thus leading to superior quality yields (Sharma et al. 2023). In addition to yield, CRISPR can also be used to enhance the post-harvest quality of vegetables. For example, modifications to genes involved in ethylene production can delay ripening in tomatoes, extending shelf life and reducing food waste (Nie et al. 2024). Similarly, editing genes involved in fruit texture and taste can improve the eating quality of crops such as peppers and cucumbers (Liu et al. 2024).

The implementation of CRISPR-Cas9 technology in vegetable cultivation systems meets multiple hurdles in its use. The main concern about the Cas9 enzyme includes its ability to perform accidental genomic cuts that target areas beyond the intended sequencing. Off-target mutations persist as a principal concern during CRISPR applications mainly because of complex genomic editing and multiple gene manipulation activities (Guo et al. 2023). Gene-editing technologies applied to crops require their associated ethical rules and regulatory issues to be resolved. Senior government officials in some nations apply strict controls to Genetically Modified Organisms (GMOs) and show similar reluctance toward CRISPR-edited crops. The public expresses concern about the environmental release of genetically modified crops since it could lead to unintended damage to non-target species or the development of resistance in pests and pathogens (Reddy and Karthik 2024).

CRISPR-Cas9 functions as an important tool in genetic crop development for vegetable cultivation purposes. CRISPR provides the capacity for accurate genetic modifications, which can resolve critical issues in contemporary agriculture, from disease protection to heightened nutritional levels and increased yield quality. The sustained development and improvement of CRISPR technology represents an empowering chance to achieve sustainable agriculture together with food security despite existing challenges and ethical issues. The success of CRISPR-Cas9 applications in vegetable crop improvement demands proper management of innovation and rules while ensuring public support until all its potential can be maximized. This chapter delivers an extensive study of CRISPR-Cas9 gene-editing technology alongside its various uses for boosting vegetable crop cultivation methods. The chapter demonstrates the capability of CRISPR-Cas9 to enhance crop production by combining benefits such as enhanced yield output with minimized pest and disease susceptibility and better nutritional value, and improved tolerance to drought and saline environments. The chapter discusses both the working challenges of CRISPR-Cas9 and diverse barriers that include off-target effects, together with regulatory constraints and ethical obstacles. This section analyzes existing innovations together with future possibilities to understand how CRISPR-Cas9 plays an essential part in delivering food security solutions and creating stronger agricultural systems.

2. Genome Editing Technology

The development of genome editing technologies has revolutionized molecular biology and genetic engineering. Among the most powerful and widely used methods is the CRISPR-Cas9 system, which is a bacterial adaptive immune mechanism that has been adapted for use in a number of organisms, including plants, animals, and humans. The CRISPR-Cas9 system, initially discovered in bacteria and archaea, allows organisms to protect themselves from invading viruses and plasmids by utilizing a unique RNA-based mechanism to recognize and cleave foreign DNA (Allemailem 2024). This system has now been repurposed to create precise genome-editing tools for genetic modification, allowing scientists to target specific genes with great accuracy and efficiency (Fig. 1).

CRISPR-Cas9 functions with three integral components that include an endonuclease (Cas9) and sgRNA and the PAM sequences (5' NGG-3') required for target site recognition (Gupta et al. 2021; Přibylová and Fischer 2024). The sgRNA-Cas9 complex transitions into being formed in cells through sgRNA and Cas9 transcription and subsequently binds target sites containing matching PAM sequences. The process leads to site specific Double-Strand Breaks (DSBs) that either error prone NHEJ or HDR can repair (Fig. 2).

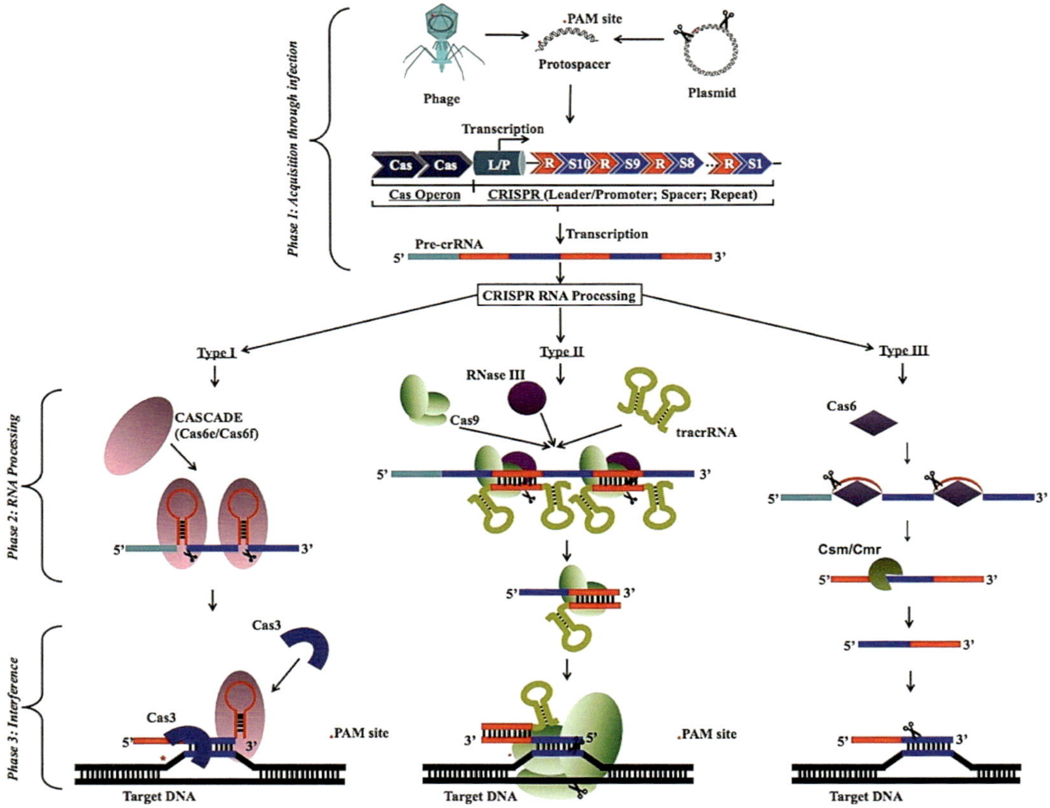

Figure 1. The CRISPR-Cas defense system involves acquisition, RNA processing, and interference (Types I, II, III), targeting foreign DNA through Cas proteins, with sequence-specific or non-specific degradation via ribonucleoprotein complexes (Singh et al. 2017).

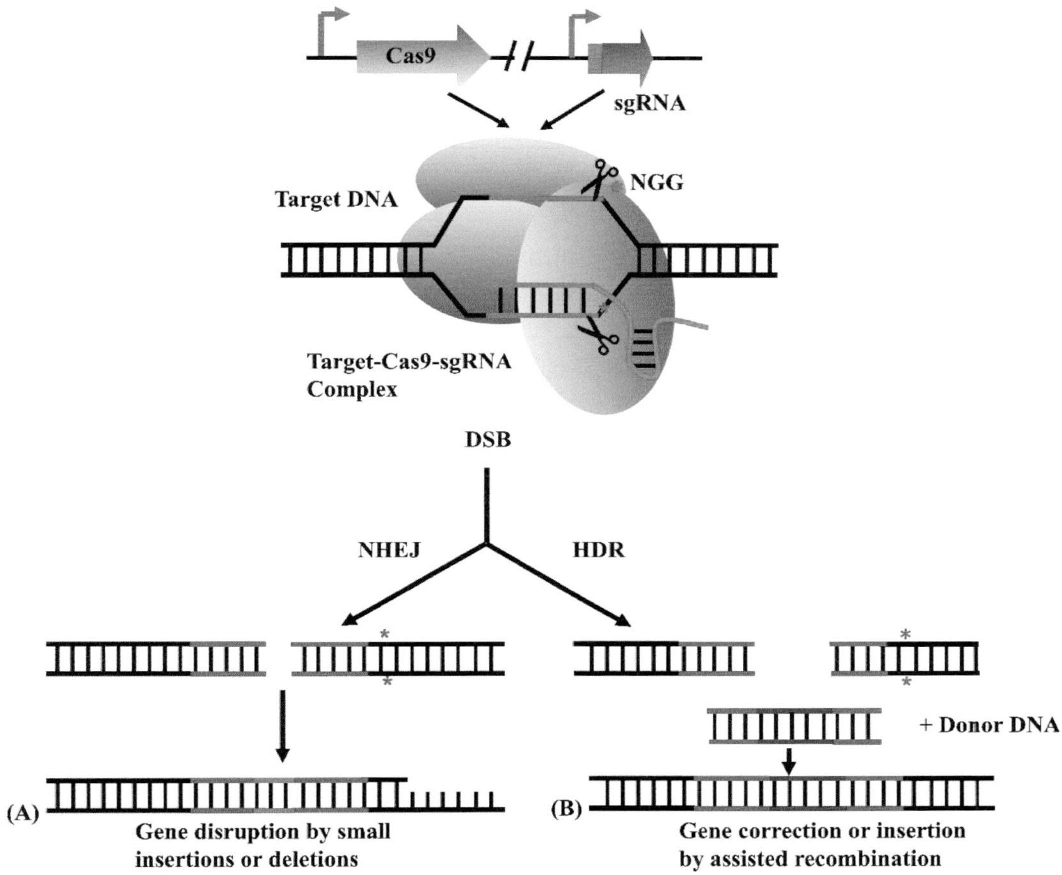

Figure 2. CRISPR-Cas9 genome editing process. Figure shows target DNA cleavage and repair via nNnon-Homologous End Joining (NHEJ) for gene disruption or Homology-Directed Repair (HDR) for precise gene correction or insertion (Singh et al. 2017).

3. Problems Faced in Vegetable Production

The global climate challenges, such as global warming and water availability, have exacerbated policymakers' concerns about attaining food security, especially for impoverished people and countries (Toromade et al. 2024). The environmental impact of increased food production concerns scientists because it could destroy natural vegetation and require expanded agricultural development (Yang et al. 2024). Food security represents a state where every person consistently maintains access to appropriate amounts of secure nutritious food for active living as per their food choices alongside both financial and physical capacity to obtain this food according to FAO (2023).

The body of human rights laws now introduces regulations for how people can acquire food together with rules about their access to it (Olson et al. 2021). Globalization along with rapid urbanization now drives changes throughout food systems across the world. Many interconnected variables (Fig. 3) determine how food security develops (Fig. 3). Among these factors is climate change. The combined consequences of short-term dramatic events such as drought events largely affect regions already facing continuous long-term stresses because available coping solutions either prove unaffordable to the population or completely unavailable. The combination of food production failures during drought leads to higher purchased food requirements that trigger poverty increases through additional factors including job loss and elevated food pricing (Munialo and Mellor 2024).

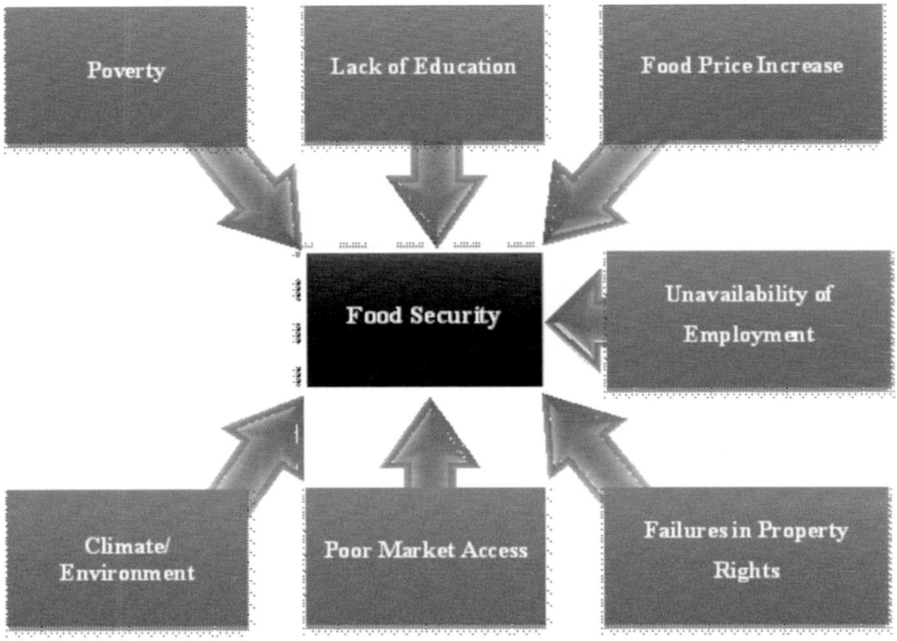

Figure 3. Factors affecting food security.

4. Need for Genome Editing Technology

Improving vegetable crops is crucial for several reasons, encompassing both societal and environmental aspects.

4.1 Food Security

Crucial nutrients, including vitamins, minerals, and dietary fiber, are found in vegetables, which are important parts of human diets. Increasing the resilience and production of vegetable crops contributes to food security, especially in areas where vegetables are the main source of dietary intake (Annepu et al. 2021).

4.2 Nutritional Health

A variety of vegetable crops are abundant in important nutrients, such as iron, calcium, and vitamins A, C, and K. Malnutrition and diet-related health problems, including vitamin deficiencies and micronutrient imbalances, can be addressed by increasing the nutritional content of vegetable crops (Kumar et al. 2022).

4.3 Dietary Diversity

A balanced diet and general health depend on dietary diversity, which is fostered by improved vegetable crops. Improved vegetable variety, flavor, and nutritional value encourage people to include more of them in their diets, which improves overall health results (Gupta et al. 2023).

4.4 Environmental Sustainability

Environmental sustainability depends largely on sustainable vegetable production since this practice reduces environmental degradation while securing natural resources alongside combating climate change. Vegetable cultivars showing improved resistance to pests and diseases, together with tolerance to drought and salinity, can guide farmers toward sustainable agriculture by reducing pesticide use (Giordano et al. 2021).

4.5 Water and Land Conservation

The preservation of water, together with land, represents a vital topic since numerous vegetable plants demand significant irrigation water. The improvement of vegetable crops combined with better water management practices serves to protect water reserves and reduce land requirements, particularly in regions facing land deterioration and water shortage challenges (Boutheina et al. 2022).

4.6 Adaptation to Climate Change

Climate change-driven obstacles to vegetable production include modified precipitation patterns as well as temperature increases and the rising occurrence of extreme weather events. By breeding for heat tolerance, disease resistance, and drought tolerance, vegetable crops can become more resilient to climate change and continue to produce even in the face of changing environmental conditions (Jain et al. 2024).

5. Advanced Genome Editing Technology in Vegetable Crops

Vegetables are high-value crops that have a positive influence on health and the environment. Recent reports on vegetables have shown that it is possible to use stable or transient transformation techniques to produce mutations that are heritable. Protoplast transfection was used to demonstrate the direct transfer of preassembled and purified Cas9-sgRNA ribonucleoprotein complexes (RNPs) in a number of species (Zhang et al. 2021). The variables, such as the available transformation/ regeneration methods, influence which approach is best for various crops.

6. Applications of CRISPR/Cas9 Technology in Vegetable Crops

6.1 Nutritional Value

The quality of vegetables is crucial for their overall appeal and is influenced by factors like taste, size, color, and nutrient content. Vegetables are essential sources of nutrients, and their quality can be improved through the precise editing of genes associated with desired characteristics using CRISPR/Cas9 genome editing technology (Shu et al. 2019; Das et al. 2023). The carotenoid metabolic pathway leads to the accumulation of antioxidant lycopene in fruits, after researchers targeted specific genes. The CRISPR/Cas9 increased lycopene content approximately 5.1 times in genome-edited tomato fruits (Shu et al. 2019). Researchers also employed CRISPR/Cas9 to enhance gamma-aminobutyric acid (GABA) accumulation in tomato fruits by removing the auto-inhibitory domain of SlGAD2 and SlGAD3 (Sakthivel et al. 2025). The quality of potato starch is crucial in various food applications. CRISPR-mediated genome editing improved starch quality by fully knocking out genes like Granule-Bound Starch Synthase (GBSS), starch synthase gene (SS6). Mutations have been induced in the StPPO2 gene, reducing polyphenol oxidase (PPO) activity in potato tubers (Tiwari et al. 2022). CRISPR/Cas technology has been applied in potatoes for traits like carotenoid biosynthesis and glycoalkaloids (Devi et al. 2022). In brinjal, CRISPR-Cas9-based mutagenesis eliminated the three polyphenol oxidase (PPO) genes (SmelPPO4, SmelPPO5, and SmelPPO6) linked to enzymatic browning (Table 1), preventing fruit flesh browning (Maioli et al. 2020).

Table 1. List of crops modified through the CRISPR/Cas9 technique for improved quality.

Crop	Trait	Targeted gene	Reference
Eggplant	Enzymatic browning	*SmelPPO1-10*	(Maioli et al. 2020)
Sweet potato	Starch quality	*GBSSI, SBEII*	(Wang et al. 2019)
Potato	Enzymatic browning	*StPPO2*	(González et al. 2020)
Tomato	Carotenoids	ANT1	(Čermák et al. 2015)
Kale	Carotenoids	*BoaCRTISO*	(Sun et al. 2020)
Rapeseed	Seed oil content	*SFAR*	(Karunarathna et al. 2020)

6.2 Post-harvest Shelf Life

Fruits and vegetables require advanced post-harvest technologies because they possess a short lifespan and high perishability (Palumbo et al. 2022). Scientists utilized HDR pathway techniques to switch the original allele in ALC with the alc gene, which led to producing T1 homozygous tomato plants with extended shelf life according to Das et al. (2023). Researchers utilized CRISPR to eliminate the RIN gene from tomato fruit, which expanded its shelf life according to Ito et al. (2015). The GBSS gene mutagenesis in Yukon gold potatoes produced with CRISPR/Cas9 technology led to improved shelf life through lowering amylose starch content, according to Toinga-Villafuerte et al. (2022). The vacuolar invertase gene mutations produced by CRISPR/Cas9 technology stop potatoes from developing symptoms of cold-induced sweetening (Yasmeen et al. 2022).

6.3 Commercial Value

Commercial varieties exhibit distinctions in size, texture, and fruit color compared to their wild counterparts. Pink tomatoes are favored by Chinese and Japanese consumers, while Europe and the United States prefer white and red varieties. The pink phenotype is attributed to a deficiency in yellow flavonoids, particularly naringenin chalcone, regulated by the transcription factor SlMYB12 linked to the monogenic recessive yellow (y) locus. By employing CRISPR/Cas9, the silencing of SlMYB12 successfully suppressed naringenin chalcone biosynthesis with a high editing efficiency of 90.9%. This method demonstrated quicker results (within 1 year) compared to traditional breeding. Additionally, CRISPR/Cas9 was used to cultivate purple and yellow tomatoes by targeting anthocyanin mutant gene 1 (ANT1) and phytoene synthase gene, respectively. Reducing unwanted compounds is another approach to enhance fruit and vegetable quality. In potatoes, silencing the St16DOX gene using CRISPR/Cas9 prevents the accumulation of bitter and toxic steroidal glycoalkaloids (α-solanine and α-chaconine) (Nakayasu et al. 2018). The CRISPR/Cas9 method was applied to delete polyphenol oxidase (PPO) genes in both potatoes and eggplants. PPO enzymes cause browning through polyphenol oxidation. Mutations induced in potato resulted in a substantial decrease in PPO activity (69% in tuber) and browning (73%) (González et al. 2020; Das et al. 2023).

6.4 Better Yield

Crop yields are increased by the collective effects of disease and pest resistance, improved stress tolerance, optimized nutrient utilization, induced parthenocarpy for consistent fruit set, modified plant architecture, expedited breeding processes, and enhanced photosynthetic efficiency. These applications contribute to the development of crops with improved traits, increased resilience, and higher overall yield (Table 2).

Table 2. Applications of CRISPR in improving the yield of vegetable crops.

Crop	Trait	Targeted gene	Reference
Tomato	Parthenocarpy	SlIAA9 Gene	(Ueta et al. 2017)
	Shortened flowering intervals, early harvest	SP5G Gene	(Soyk et al. 2017)
Cucumber	Gynoecious phenotype	CsWIP1 Gene	(Hu et al. 2017)

6.5 Abiotic Stress Improvement in Vegetables

Abiotic stress refers to the adverse environmental effects on plants. Heat, cold, drought, heavy metals, and salinity are some of the major abiotic stresses encountered by the plant. Abiotic stress can alter the growth and development of plants, resulting in low yield, leading to economic losses. To deal with abiotic stress, plants have specialized sensors that detect specific changes and generate signals inside the plant cell (Fig. 4). These signals stimulate different ion channels, Reactive Oxygen Species (ROS), and plant hormones that finally result in gene expression to fight the stress (Kesawat et al. 2023).

Elevated temperatures trigger the excessive production of reactive oxygen species (ROS), resulting in oxidative damage and disruption of normal plant cell functions. Mitogen-Activated Protein Kinases (MAPKs) play a crucial role in vegetable crop responses to heat stress. In tomatoes knockout of BRASSINAZOLE RESISTANT 1 (BZR1) impedes the induction of RESPIRATORY BURST OXIDASE HOMOLOG1 (RBOH1), leading to increased hydrogen peroxide (H_2O_2) production and enhanced heat tolerance (Table 3). The introduction of exogenous H_2O_2 restores heat tolerance in the bzr1 tomato mutant (Devi et al. 2022). CRISPR has enabled the development of novel germplasm with enhanced cold tolerance, exemplified by CBF1 (C-repeat Binding Factor 1), a regulator pivotal for plants to withstand cold conditions (Li et al. 2022). The presence of salinity stress led to elevated expression levels of RBOHD (Respiratory Burst Oxidase Homolog D), GRF12 (14-3-3 protein), and AHA1 (plasma membrane H+-ATPase) in pumpkin. These genes were knocked out via CRISPR/cas9 (Das et al. 2023).

While traditional breeding methods have proven effective to a certain degree in addressing environmental stresses, the advent of advanced CRISPR-Cas9 opens new avenues for creating highly resilient germplasms. This innovative approach holds the promise of significantly enhancing

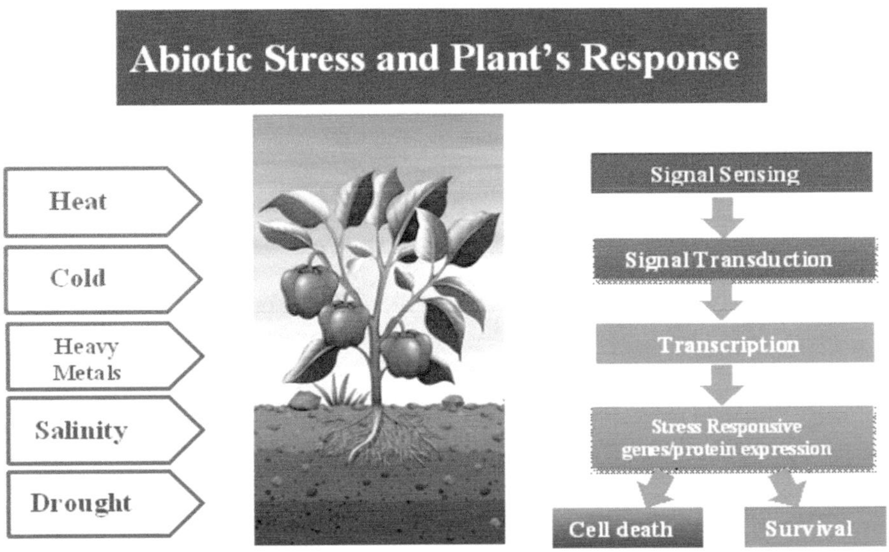

Figure 4. Plant response to abiotic stress.

Table 3. List of traits in vegetable crops for abiotic stress tolerance.

Vegetables	Type of stress	Target Gene	Reference
Tomato	Heat	slmapk3	(Bertier et al. 2018)
	Drought	MAPKs	(Li et al. 2018)
	UV-B	SlUVR8	(Liu et al. 2020)
	Salinity	HKT1 and 2 allele	(Das-Dangol et al. 2019)
Lettuce	Heat	LsNCED4	(Bertier et al. 2018)
Pumpkin	Salinity	RBOHD	(Das et al. 2023)

our capacity to confront and overcome various challenges posed by environmental stressors (Li et al. 2022). Table 3 provides a brief list of plants with enhanced abiotic stress tolerance induced through CRISPR.

6.6 *Insects, Pests, and Disease Resistance*

Biotic stress is the harmful effect of living organisms on plants. It can be due to pathogens, insects, weeds, or herbivores. Biotic stresses exert a significant influence on the growth and development of crops, leading to substantial losses in yield. The intricate signaling mechanisms in plants, influenced by stress, affect secondary metabolism through the generation of ROS, hormonal components, and enzymatic activity (Nicoletto et al. 2019). Conventional methods are commonly used to deal with biotic stress in plants, but these methods are laborious and time-consuming. CRISPR/Cas9, on the contrary, is a more efficient, cost-effective, and easy-to-design technique with a higher rate of success to generate disease-resistant plants (Munir et al. 2023).

Genetic modifications using CRISPR/Cas9 technology have been applied to enhance disease resistance in various vegetable crops. Susceptibility genes for diseases caused by *Pseudomonas, Phytophthora*, viruses, and bacteria have been successfully knocked out via CRISPR. Additionally, CRISPR-mediated mutations in vegetable crops such as tomato, potato, watermelon, cucumber, *Brassica napus,* etc., have provided resistance against powdery mildew, late blight, potato virus Y, late blight, *Fusarium oxysporum*, *Fusarium* wilt, Sclerotinia spp (Table 4).

Table 4. List of some of the modified genes for biotic stress tolerance in vegetable crops.

Crop	Disease/Pathogen	Targeted gene/sites	References
Tomato	Powdery mildew	Mlo allele SlMlo1	(Nekrasov et al. 2017)
	Pseudomonas syringae pv. tomato	SlDMR6-1 orthologue	(de Toledo Thomazella et al. 2016)
	Phytophthora capsici	Solyc03g080190.2	(de Toledo Thomazella et al. 2016)
	Bacterial speck resistance	SlJAZ2	(Ortigosa et al. 2019)
	Resistance to PtoDC3000	Jasmonate-Zim Domain (JAZ2) repressors (SlJAZ2jas)	(Ortigosa et al. 2019)
	Tomato yellow leaf curl virus	Coat protein and replication sites	(Zaidi et al. 2016)
	Fusarium Wilt Disease	Solyc08g075770	(Prihatna et al. 2018)
Potato	Potato Virus Y	P3, CI, Nib, CP (viral genes)	(Zhan et al. 2019)
	Late blight	StDND1, StCHL1, DMG400000582 (StDMR6-1)	(Kieu et al. 2021)
Cucumber	Virus Resistance:	eIF4E (eukaryotic translation initiation factor 4E)	(Chandrasekaran et al. 2016)
Brassica napus	Sclerotinia spp	BnWRKY11, BnWRKY70	(Sun et al. 2018)

7. Improvement in Important Vegetable Crops Using CRISPR/Cas9 System

7.1 Tomato (Solanum lycopersicum)

The CRISPR/Cas9 technique has received wide attention for the genetic improvement of tomato due to its status as a major agricultural study subject. Scientists use Tomato (*Solanum lycopersicum*) for research because it matters economically and reproduces fast and scientists have successfully modified genes to increase its yield and resistance to diseases, and improve its quality and tolerance toward stress (Zhu et al. 2024).

7.1.1 Enhancing Fruit Quality

The main objective of tomato breeding research involves improving fruit resistance, shelf life, and flavor quality. CRISPR techniques have been applied to alter multiple genes which resulted in enhanced improvement of target characteristics (Wang et al. 2024).

Through SlAGL6 regulation, the fruit maturation process is controlled. Researchers have achieved delayed ripening and longer shelf life, and improved firmness in tomato fruits after using CRISPR to knockout SlAGL6. Post-harvest losses occur less frequently in modified tomatoes, and the products have increased potential for both storage and transportation.

The SlNOR and SlRIN genes control the natural process of ethylene production, which results in fruit maturation. Gene alterations in these fundamental elements reduce ripening speed while stopping fruit from becoming overripe and enhance storage time without changing the original taste quality (Kaur et al. 2023).

7.1.2 Disease Resistance Improvement

Several disease-causing fungi and bacteria, alongside viruses, attack tomatoes, resulting in substantial yield loss. The CRISPR/Cas9 technology serves as an important tool for creating disease-resistant tomato plant types.

The SlMlo1 gene protects tomato plants from the destructive effects of powdery mildew fungi disease. The research team altered tomato plants by deleting the SlMlo1 gene to make them resistant to powdery mildew disease (Li et al. 2023).

The Bs5 gene controls the susceptibility of plants to bacterial spot disease. The resistance of Xanthomonas spp.-caused bacterial spot disease increases in CRISPR-modified tomatoes that have Bs5 variant mutations.

The Ty-5 gene confers resistance to Tomato Yellow Leaf Curl Virus (TYLCV), a devastating viral disease. The modification of this gene using CRISPR techniques improved viral resistance properties in tomato crops (Patel et al. 2023).

7.1.3 Yield Improvements

The disruption of the SlCLV3 gene expression enables researchers to obtain tomatoes with both enhanced fruit numbers along with greater yield production. The changed genetic material enables the production of high-yielding tomato cultivars. Examination of the SlWUS gene through modification led to higher fruit production alongside more flowers, thereby increasing total yield levels (Huang and Zhao 2022).

7.1.4 Abiotic Stress Tolerance

Studies of SlERF6 demonstrate its success in improving tomato resistance to drought and salt stress, which enables plants to grow under unfavorable environmental conditions.

7.2 Potatoes (Solanum tuberosum)

Together with other qualities, potatoes have also been edited using CRISPR/Cas9 with a targeted emphasis on disease control. Another constraint that affects potato production is enzymatic browning, mostly experienced in cut potatoes exposed to air. CRISPR/Cas9 was used to selectively delete the PPO gene that is accountable for the browning mechanism and awarded potatoes that seldom turn brown when sliced or cut, enhancing their storage duration and marketability (Andersson et al. 2017).

7.2.1 Disease Resistance

The disease resistance of potatoes remains poor against *Phytophthora infestans* and bacterial pathogens because they are extremely susceptible. Scientists have employed CRISPR technology to achieve disease resistance improvement as they have altered the StDMR6 gene through gene knockout experiments to boost resistance levels against *Phytophthora infestans*. The modified version minimizes dependence on fungal disease chemical fungicide treatments.

Similarly, Noureen et al. (2022) targeted eIF4E and developed lines showed enhanced resistance to Potato Virus Y (PVY) strains, are a major threat to potato production.

7.2.2 Nutritional and Health-Related Improvements

Glycoalkaloid biosynthesis genes occur naturally in wild potatoes because these plants contain the toxic solanine compound. Through CRISPR technology, scientists eliminated genes that generate glycoalkaloids, thus enhancing potato safety. The StAs1 gene has been edited to decrease the production of acrylamide, which is a dangerous compound created during frying (Rodriguez-Leal et al. 2019).

7.3 Eggplant (Solanum melongena)

7.3.1 Disease Resistance

Crop yield in eggplants decreases as pathogenic infections attack these plants. The introduction of SmPMR4 provides eggplants with elevated resistance against powdery mildew, so farmers can reduce their reliance on chemical fungicides.

Researchers used the CRISPR system to modify R-gene clusters to increase resistance against bacterial wilt diseases caused by *Ralstonia solanacearum* (Sun et al. 2021).

7.3.2 Improved Fruit Characteristics

The application of CRISPR technology resulted in seedless eggplants, which now have better marketability because they have a smoother surface texture.

The regulation of anthocyanin production enabled scientists to develop eggplants featuring enhanced pigmentation as well as better antioxidant properties (Sun et al. 2021).

7.4 Pepper (Capsicum annuum)

Editing technologies have been utilized for pepper cultivation to develop resistance against diseases, along with control of pungency and yield improvement.

7.4.1 Disease Resistance

Through CaERF gene editing, the resistance against bacterial spot disease has been enhanced, which serves as a critical defense against a common threat to pepper cultivation. The modification of Resistance (R) genes through CRISPR enhances the virus defense capability of peppers specifically against pepper mild mottle virus (PMMoV) (Zhang et al. 2018).

7.4.2 Capsaicin Biosynthesis Modification

Scientists have employed CRISPR to modify Pun1 so they could create pepper breeds with different capsaicin production levels from non-pungent to extra-spicy (Zhang et al. 2018).

7.5 Leafy Vegetables

Human dietary patterns largely depend on leafy vegetables, as these plants provide essential vitamins, minerals, and fiber. These crops tend to decay quickly due to various challenges posed by natural conditions and infectious diseases. The applications of CRISPR/Cas9 technology in leafy vegetables enhance storage duration, disease immunity, stress tolerance, and the quality of nutritional content. The following section details the precise implementation of CRISPR-based gene modifications in primary leafy vegetables.

7.6 Lettuce (Lactuca sativa)

The worldwide market favors lettuce as a main leafy vegetable product because it maintains its fresh appeal and usable diversity. The post-harvest deterioration, together with bacterial and fungal diseases, and susceptibility to abiotic stresses significantly minimize both market value and productivity of leafy vegetables. The CRISPR/Cas9 technology serves as a solution to resolve these problems (Ninama et al. 2024).

7.6.1 Delayed Senescence and Extended Shelf Life

The rapid degradation of lettuce after harvest becomes a critical production challenge because it produces massive food waste and economic losses.

Scientists edit LsRIN (Ripening Inhibitor) because it aids in ethylene synthesis. The CRISPR knockout mutation of this gene creates leaf aging delay and postpones standard yellowing and wilting. The change creates shelf stability that allows business-to-business transport of lettuce products.

Research utilizing CRISPR technology modified genes that control ROS pathways, thus decreasing stored lettuce tissue damage, together with browning deterioration (Lu 2023).

7.6.2 Disease Resistance

The plant lettuce develops severe problems from fungal infections and bacterial diseases, which include *Bremia lactucae* downy mildew and *Xanthomonas campestris* bacterial leaf spot. Specific applications of CRISPR technology enable the improvement of resistance properties in different crops. Research teams have targeted S genes through genetic editing, which led to *Bremia lactucae* resistance without needing fungicides.

The functionality of bacterium-fighting genes receives enhancement through CRISPR-based modifications, which enhances lettuce resistance to bacterial pathogen attacks (Ijaz et al. 2023).

7.6.3 Abiotic Stress Tolerance

DNA editing of stress response genes like ALS, HAS, and BZR1 has led to lettuce plants that demonstrate enhanced resistance against drought periods as well as salt stress conditions. Soil irrigation restrictions make this method especially beneficial (Noureen et al. 2022).

7.7 Spinach (Spinacia oleracea)

The leafy vegetable spinach holds dense nutritional content, which transports vital vitamins together with minerals to humans. The plant remains susceptible to fungal diseases, together with nutrient deficiencies. Scientists have applied CRISPR technology to spinach to enhance disease immunity, together with enhancing nutritional benefits and improving stress response capabilities.

7.7.1 Enhanced Disease Resistance

Spinach faces significant threats to its production from downy mildew caused by *Peronospora farinosa*. The application of CRISPR resulted in enhanced resistance against pathogenic organisms (Gao 2021).

Scientists have used CRISPR technology to modify genes in spinach that provide access to fungal pathogens, this modification work has established spinach with stronger defenses against downy mildew. The modification of plant immune receptors through CRISPR produces genetic changes in Resistance (R) genes, which strengthen spinach immune defenses against fungi and bacteria (Gao 2021).

7.7.2 Abiotic Stress Tolerance

The modification of genes tied to water retention and drought resistance through CRISPR enables spinach to survive in arid environments.

The modification of genes responsible for ion transport helped scientists to create spinach plants that survive well in environments with high salt content.

7.8 Cucurbitaceous Vegetables

The Cucurbitaceae family, which is also known as the gourd family, produces major vegetable plants including cucumber as well as melon, watermelon, pumpkin, and squash. The crops encounter various difficulties, including viral diseases together with fungal diseases, and decreased fruit production. Scientists have used the CRISPR/Cas9 system to increase disease resistance and improve fruit characteristics and modify growth behaviors, which have raised their economic value in farming (Fizikova et al. 2021).

7.9 Cucumber (Cucumis sativus)

Cucumber serves as a key cultivated vegetable because of its appetizing taste and its large water content. The susceptibility of cucumber crops remains high because they easily fall victim to viral infections in combination with powdery mildew and develop unnatural fruit forms. CRISPR/Cas9 technology has provided vital solutions to address the encountered difficulties.

7.9.1 Virus Resistance

Cucumber production suffers significantly from the attacks of Cucumber Mosaic Virus (CMV) and Zucchini Yellow Mosaic Virus (ZYMV). The target modification of *CsMLO* gene by CRISPR-based knockout resulted in plants that showed improved resistance against powdery mildew and viral pathogens. Scientists have utilized CRISPR to knock out cucumber genes which viruses rely on for their life cycle so infections can be prevented (Wan et al. 2021).

7.9.2 Seedless (Parthenocarpic) Varieties

The development of seedless cucumbers for consumer preference has been possible through CRISPR technology applications. CRISPR allowed scientists to manipulate hormonal pathways and target genes like the *IAA9* gene through genetic knockout, thus producing parthenocarpic (seedless) cucumbers (Gou et al. 2022).

7.9.3 Enhanced Shelf Life and Firmness

The gaseous plant hormone ethylene is a regulator of fruit shelf-life, one of the essential traits in fruits. Extending fruit shelf life reduces food loss, thereby contributing to food security. The CRISPR allowed scientists to target the enzyme 1-Aminocyclopropane-1-Carboxylic acid Oxidase (ACO), and its suppression has been shown to extend shelf-life (Nonaka et al. 2023).

7.10 Melon (Cucumis melo)

The preferred and common sweet fruit known as melon faces three major production challenges: short expiration time and fungal disease outbreaks, along with inconsistent sugar content throughout the fruit. The application of CRISPR technology has enabled scientists to enhance several traits.

7.10.1 Extended Shelf Life

They disabled the ethylene biosynthesis genes. Researchers have employed CRISPR to modify ethylene-related genes, which slows down fruit ripening, thus maintaining fresh melons that resist decay through shipping and storage procedures (Kumar et al. 2021).

7.10.2 Sugar Content Optimization

Scientists have used CRISPR techniques to adjust CMsuc1, a key sucrose accumulation regulator, thus producing more flavorful sweet melons that attract buyers in the market (Li et al. 2025).

7.10.3 Disease Resistance

The CRISPR tool succeeded in removing genes that make melons susceptible to *Fusarium oxysporum* until other major fungal pathogen damaging melon production becomes a problem (Jiménez Sánchez 2021).

7.11 Pumpkin (Cucurbita pepo)

The nutritious pumpkin plant suffers from multiple disease infections, as well as mildew problems, along with unpredictable yield consistency. CRISPR has proved essential for advancing pumpkin agronomic features through its applications (Hosen et al. 2021).

7.11.1 Disease Resistance

Physicians have utilized CRISPR editing to resist *Erysiphe cichoracearum,* which causes powdery mildew infections in plants (Aroge et al. 2024). CRISPR-mediated modifications block virus-binding proteins to decrease pumpkin exposure to the main infectious viruses.

7.11.2 Higher Yield and Improved Fruit Set

Plants have become larger, together with stable yield increases by using CRISPR-based deletion of growth-limiting genes that control plant hormone activity (Raza et al. 2023).

8. Implications

CRISPR/Cas9 technology creates substantial effects that affect vegetable farming, alongside global food safety and human wellness.

8.1 Increased Crop Yields and Productivity

The use of CRISPR has become vital for enhancing plant structural development as well as floral progression and fruiting ability, which generates greater vegetable harvests, specifically for tomatoes, potatoes, and cucumbers.

8.2 Disease Resistance and Reduced Pesticide Use

The modification of susceptibility genes through CRISPR technology has created bacterial, fungal and viral resistance, thus allowing plants to decrease their dependency on chemical pesticides.

8.3 Nutritional Enhancement and Biofortification

The application of CRISPR has elevated vitamin A, together with iron and antioxidants, in different vegetables, thereby combating nutritional deficiencies.

8.4 Extended Shelf Life and Food Waste Reduction

CRISPR has achieved the development of vegetables with longer shelf lives through genetic modifications in ripening-related genes, which minimize post-harvest losses in food (Zhu et al. 2024).

9. Limitations and Future Challenges

The implementation of CRISPR technology brings multiple significant difficulties.

9.1 Regulatory and Ethical Concerns

Most nations uphold strict guidelines that block the commercial opportunities for genetically edited crops.

9.2 Off-Target Effects and Unintended Mutations

CRISPR conducts unintended modifications to areas outside its target region, which produces unexpected genetic alterations that need additional development work.

9.3 Public Perception and Consumer Acceptance

The resistance to accept "genetically edited" foods, as well as false information about them, creates public refusal.

9.4 High Costs and Limited Access

Modern CRISPR research demands complex equipment together with specialized personnel who render the technology inaccessible to many developing nations (Zhu et al. 2024).

9.5 Future Research Directions

Research organizations need to develop exact gene-editing tools to stop the generation of undesired genetic changes. The scientific community should create affordable CRISPR systems for everyday farm use. Awareness campaigns in combination with public education initiatives will promote better consumer acceptability of CRISPR technologies (Huang and Zhao 2022).

10. Conclusion

Vegetables deliver important nutrients, together with vitamins and minerals, in large quantities to our bodies. Many nutritional choices and different farming systems for vegetable crops help establish global food security programs. Through CRISPR/Cas9 methodology, scientists have generated substantial changes in crop nutritional value while developing plants with stress tolerance to both abiotic and biotic factors. Plant breeding employing CRISPR/Cas9 technology produces elevated vegetable production standards, as shown by research studies implementing multiple plant breeding applications. The use of CRISPR technology for transgene-free genome editing has turned out to be vital for the commercial acceptance of vegetable crops because it fulfills consumer demands. Scientists must discover substantial applications of CRISPR after conducting thorough scientific research for crop development. Studying agricultural research about sustainable trait development alongside crop quality improvement indicates how future advancements can secure food production systems and increase sustainability in farming. Gene editing through a multiplex editing approach brings revolution by providing researchers the possibility to edit different gene sets simultaneously. Further investigations of CRISPR technology led to various breeding techniques for creating crop resistance against multiple environmental threats. This advancement enables farmers to improve production methods, leading to better food security systems. The scientific field requires a complete ethical evaluation of CRISPR technology to proceed. For sustainable agricultural practices to be achieved in the future, the scientific community should evaluate the ethics of CRISPR technology. Continuous adoption of CRISPR-based agricultural technology depends on how well innovation methods synchronize with ethical and legal requirements. The worldwide sustainable agricultural system benefits from CRISPR technology, which addresses nutritional needs by applying it to vegetable crops.

References

Andersson, M., H. Turesson, A. Nicolia, A.S. Fält, M. Samuelsson and P. Hofvander. 2017. Efficient targeted multiallelic mutagenesis in tetraploid potato (*Solanum tuberosum*) by transient CRISPR-Cas9 expression in protoplasts. Plant Cell Reports 36(1). https://doi.org/10.1007/s00299-016-2062-3.

Akram, F., S. Sahreen, F. Aamir, I.U. Haq, K. Malik, M. Imtiaz et al. 2023. An insight into modern targeted genome-editing technologies with a special focus on CRISPR/Cas9 and its applications. Mol. Biotechnol. 65(2): 227–242.

Allemailem, K.S. 2024. Recent advances in understanding the molecular mechanisms of multidrug resistance and novel approaches of CRISPR/Cas9-based genome-editing to combat this health emergency. Int. J. Nanomedicine 1125–1143.

Annepu, S.K., S.A. Nair, S. Thakur and V. Verma. 2021. Climate-resilient vegetable farming: Approaches for sustainable development. pp. 347–363. *In*: Climate Change and Resilient Food Systems: Issues, Challenges, and Way Forward. Singapore: Springer Singapore.

Anzalone, A.V., L.W. Koblan and D.R. Liu. 2020. Genome editing with CRISPR–Cas nucleases, base editors, transposases and prime editors. Nat. Biotechnol. 38(7): 824–844.

Aroge, T., Y. Zhu, D.N. Jin, M.Z.N. Dara, J. Feng, A.M. Olajuyin et al. 2024. Omics and CRISPR-Cas9 molecular perception: A progressive review approach for powdery mildew disease management. Physiol. Mol. Plant Pathol. 130: 102217.

Asmamaw, M. and B. Zawdie. 2021. Mechanism and applications of CRISPR/Cas-9-mediated genome editing. Biologics: Targets and Therapy 353–361.

Barman, A., B. Deb and S. Chakraborty. 2020. A glance at genome editing with CRISPR–Cas9 technology. Curr. Genet. 66: 447–462.

Berman, J., J. Meier and R. Verhoeven. 2019. CRISPR technology in plant breeding: Achievements and future directions. Agricultural Biotechnology Reviews 15(2): 45–57.

Bertier, L.D., M. Ron, H. Huo, K.J. Bradford, A.B. Britt and R.W. Michelmore. 2018. High-resolution analysis of the efficiency, heritability, and editing outcomes of CRISPR/Cas9-induced modifications of NCED4 in lettuce (Lactuca sativa). G3: Genes, Genomes, Genetics 8(5). https://doi.org/10.1534/g3.117.300396.

Bigini, V., F. Camerlengo, E. Botticella, F. Sestili and D.V. Savatin. 2021. Biotechnological resources to increase disease-resistance by improving plant immunity: A review.

Boutheina, D., M. Amel, S. Sami, B.S. Fatma and M. Bassem. 2022. Agricultural water management practices in MENA region facing climatic challenges and water scarcity. Water Conserv. Manage. 6(1): 39–44.

Burgos, G., T. Zum Felde, C. Andre and S. Kubow. 2020. The potato and its contribution to the human diet and health The Potato Crop: Its Agricultural, Nutritional and Social Contribution to Humankind 37–74.

Čermák, T., N.J. Baltes, R. Čegan, Y. Zhang and D.F. Voytas. 2015. High-frequency, precise modification of the tomato genome. Genome Biology 16(1). https://doi.org/10.1186/s13059-015-0796-9.

Chandrasekaran, J., M. Brumin, D. Wolf, D. Leibman, C. Klap, M. Pearlsman et al. 2016. Development of broad virus resistance in non-transgenic cucumber using CRISPR/Cas9 technology. Molecular Plant Pathology 17(7). https://doi.org/10.1111/mpp.12375.

Chatterjee, A., S. Purkaystha, S. Bhattacharyya and P. Sen. 2025. Genetic engineering and genome editing for enhancing nutritional quality of legumes. pp. 389–419. *In*: Recent Trends and Applications of Leguminous Microgreens as Functional Foods. Cham: Springer Nature Switzerland.

Dalastra, C., M. Teixeira Filho, M.R. da Silva, T.A. Nogueira and G.C. Fernandes. 2020. Head lettuce production and nutrition in relation to nutrient solution flow. Hortic. Bras. 38: 21–26.

Das, T., U. Anand, T. Pal, S. Mandal, M. Kumar, A.V. Radha et al. 2023. Exploring the potential of CRISPR/Cas genome editing for vegetable crop improvement: An overview of challenges and approaches. In Biotechnology and Bioengineering (Vol. 120, Issue 5).

Das-Dangol, S., A. Barakate, J. Stephens, M.E. Çalıskan and A. Bakhsh. 2019. Genome editing of potato using CRISPR technologies: current development and future prospective. Plant Cell Tiss Organ Cult. 139: 403–416. https://doi.org/10.1007/s11240-019-01662-y.

de Toledo Thomazella, D.P., Q. Brail, D. Dahlbeck and B.J. Staskawicz. 2016. CRISPR-Cas9 mediated mutagenesis of a DMR6 ortholog in tomato confers broad-spectrum disease resistance. In bioRxiv. https://doi.org/10.1101/064824.

Deb, S., A. Choudhury, B. Kharbyngar and R.R. Satyawada. 2022. Applications of CRISPR/Cas9 technology for modification of the plant genome. Genetica 1–12.

Demir, Ö. and O. Erbaş. 2023. Discovery and applications of CRISPR-Cas9 gene editing technology. Demiroğlu Bilim Üniversitesi Florence Nightingale Transplantasyon Dergisi 8(1): 056–067.

Devi, R., S. Chauhan and T.S. Dhillon. 2022. Genome editing for vegetable crop improvement: Challenges and future prospects. Front. Genet. 13: 1037091.

Dolničar, P. 2021. Importance of potato as a crop and practical approaches to potato breeding. *Solanum Tuberosum*: Methods and Protocols 3–20.

Ebabhi, A. and R. Adebayo. 2022. Nutritional values of vegetables. Vegetable Crops-Health Benefits and Cultivation.

El-Ashry, A. 2023. The CRISPR/Cas system: Gene editing by bacterial defense. Novel Research in Microbiology Journal 7(5): 2101–2115.

Fizikova, A., N. Tikhonova, Y. Ukhatova, R. Ivanov and E. Khlestkina. 2021. Applications of CRISPR/Cas9 system in vegetatively propagated fruit and berry crops. Agronomy 11(9): 1849.

Food and Agriculture Organization of the United Nations. 2023. The State of Food Security and Nutrition in the World 2023. FAO. Retrieved from https://www.fao.org/3/cc3017en/online/cc3017en.html.

Gan, W.C. and A.P. Ling. 2022. CRISPR/Cas9 in plant biotechnology: Applications and challenges. BioTechnologia 103(1): 81–93.

Gao, C. 2021. Genome engineering for crop improvement and future agriculture. Cell 184(6): 1621–1635.

Giordano, M., S.A. Petropoulos and Y. Rouphael. 2021. Response and defence mechanisms of vegetable crops against drought, heat and salinity stress. Agriculture 11(5): 463.

González, M.N., G.A. Massa, M. Andersson, H. Turesson, N. Olsson, A.S. Fält et al. 2020. Reduced enzymatic browning in potato tubers by specific editing of a polyphenol oxidase gene via ribonucleoprotein complexes delivery of the CRISPR/Cas9 system. Front. Plant Sci. 10.

Gou, C., P. Zhu, Y. Meng, F. Yang, Y. Xu, P. Xia et al. 2022. Evaluation and genetic analysis of parthenocarpic germplasms in cucumber. Genes 13(2): 225.

Guo, C., X. Ma, F. Gao and Y. Guo. 2023. Off-target effects in CRISPR/Cas9 gene editing. Front. Bioeng. Biotechnol. 11: 1143157.

Gupta, P., S.G. Karkute, P.K. Dash and A.K. Singh. 2023. Genome editing for improvement of nutrition and quality in vegetable crops. Applications of Genome Engineering in Plants 222.

Gupta, R., D. Gupta, K.T. Ahmed, D. Dey, R. Singh, S. Swarnakar et al. 2021. Modification of Cas9, gRNA and PAM: Key to further regulate genome editing and its applications. Prog. Mol. Biol. Transl. Sci. 178: 85–98.

Hemathilake, D.M.K.S. and D.M.C.C. Gunathilake. 2022. Agricultural productivity and food supply to meet increased demands. pp. 539–553. *In*: Future Foods. Academic Press.

Hosen, M., M.Y. Rafii, N. Mazlan, M. Jusoh, Y. Oladosu, M.F.N. Chowdhury et al. 2021. Pumpkin (Cucurbita spp.): a crop to mitigate food and nutritional challenges. Horticulturae 7(10): 352.

Hossain, M.A. 2021. CRISPR-Cas9: A fascinating journey from bacterial immune system to human gene editing. Prog. Mol. Biol. Transl. Sci. 178: 63–83.

Hu, B., D. Li, X. Liu, J. Qi, D. Gao, S. Zhao et al. 2017. Engineering non-transgenic gynoecious cucumber using an improved transformation protocol and optimized CRISPR/Cas9 system. Molecular Plant 10(12): 1575–1578. https://doi.org/10.1016/j.molp.2017.09.005.

Huang, Y. and Q. Zhao. 2022. CRISPR-Cas9 gene editing for fruit and vegetable crops: Advances and applications. Hortic. Res. 7(7): 193.

Ito, Y., A. Nishizawa-Yokoi, M. Endo, M. Mikami and S. Toki. 2015. CRISPR/Cas9-mediated mutagenesis of the RIN locus that regulates tomato fruit ripening. Biochem. Biophys. Res. Commun. 467(1). https://doi.org/10.1016/j.bbrc.2015.09.117.

Ijaz, M., F. Khan, H.E. Zaki, M.M. Khan, K.S. Radwan, Y. Jiang et al. 2023. Recent trends and advancements in CRISPR-based tools for enhancing resistance against plant pathogens. Plants 12(9): 1911.

Jain, S., K. Lamo, S. Walling, A. Imchen, J.F. Tirkey and A. Singh. 2024. Climate-resilient horticulture: Adapting to climate change through innovative practices and technologies. Int. J. Environ. Climate Change 14(11): 219–233.

Janik, E., M. Niemcewicz, M. Ceremuga, L. Krzowski, J. Saluk-Bijak and M. Bijak. 2020. Various aspects of a gene editing system—CRISPR–Cas9. Int. J. Mol. Sci. 21(24): 9604.

Jiménez Sánchez, D. 2021. Gene editing in melon (Cucumis melo L.) using CRISPR/Cas9 to functionally validate candidate genes for ripening and fruit shape (Bachelor's thesis, Universitat Politècnica de Catalunya).

Karunarathna, N.L., H. Wang, H.J. Harloff, L. Jiang and C. Jung. 2020. Elevating seed oil content in a polyploid crop by induced mutations in SEED FATTY ACID REDUCER genes. Plant Biotechnol. J. 18(11).

Kaur, H., R. Singh and P. Sharma. 2023. A critical review: Recent advancements in the use of CRISPR/Cas9 in plant breeding. Mol. Plant Biol. 15(2): 112–129.

Kesawat, M.S., N. Satheesh, B.S. Kherawat, A. Kumar, H.U. Kim, S.M. Chung et al. 2023. Regulation of reactive oxygen species during salt stress in plants and their crosstalk with other signaling molecules—Current perspectives and future directions. Plants 12(4): 864.

Kieu, N.P., M. Lenman, E.S. Wang, B.L. Petersen and E. Andreasson. 2021. Mutations introduced in susceptibility genes through CRISPR/Cas9 genome editing confer increased late blight resistance in potatoes. Sci. Rep. 11(1).

Kumar, D., A. Yadav, R. Ahmad, U.N. Dwivedi and K. Yadav. 2022. CRISPR-based genome editing for nutrient enrichment in crops: A promising approach toward global food security. Front. Genet. 13: 932859.

Kumar, M., B.N. Saha, W. Siddiqui, A.K. Singh, D. Srivastava, A. Nayyer et al. 2021. Molecular approaches for the postharvest losses in tomato by different biotic stresses. pp. 251–284. *In*: Biotic Stress Management in Tomato. Apple Academic Press.

Li, M., S. Chen and L. Zhou. 2023. Genome editing for vegetable crop improvement: Current status and future directions. Front. Genet. 14: 1037091.

Li, M., C. Sun and S. Wu. 2025. Unlocking sweetness: Gene editing of SlCDPKs to improve tomato flavor. Mol. Plant 18(2): 189-191.

Li, Y., X. Wu, Y. Zhang and Q. Zhang. 2022. CRISPR/Cas genome editing improves abiotic and biotic stress tolerance of crops. In Frontiers in Genome Editing (Vol. 4).

Li, R., L. Zhang, L. Wang, L. Chen, R. Zhao, J. Sheng et al. 2018. Reduction of tomato-plant chilling tolerance by CRISPR-Cas9-mediated SlCBF1 mutagenesis. Journal of Agricultural and Food Chemistry 66(34). https://doi.org/10.1021/acs.jafc.8b02177.

Liu, J., Y. Xu, P. Fang, Q. Guo, W. Huang, J. Hou et al. 2024. Genetic regulation of fruit shape in horticultural crops: A review. Horticulturae 10(11): 1151.

Liu, X., Q. Zhang, G. Yang, C. Zhang, H. Dong, Y. Liu et al. 2020. Pivotal roles of tomato photoreceptor SlUVR8 in seedling development and UV-B stress tolerance. Biochem. Biophys. Res. Commun. 522(1).

Lu, Y. 2023. Association Genetics Exploration in *Lactuca sativa* Pinking Discolouration Which is Independent of Browning (Doctoral dissertation, University of Warwick).

Maioli, A., S. Gianoglio, A. Moglia, A. Acquadro, D. Valentino, A.M. Milani et al. 2020. Simultaneous CRISPR/Cas9 editing of three PPO genes reduces fruit flesh browning in *Solanum melongena* L. Front. Plant Sci. 11.

Mali, P., L. Yang and K. Esvelt. 2013. RNA-guided human genome engineering via Cas9. Science 339(6121): 823–826.

Munialo, C.D. and D.D. Mellor. 2024. A review of the impact of social disruptions on food security and food choice. Food Sci. Nutr. 12(1): 13–23.

Munir, F., S. Azeem, R. Amir, M. Gillani, A. Gul, A. Mustafa et al. 2023. Genetic transformation methods in cereal crops. pp. 273–294. *In*: Cereal Crops. CRC Press.

Nakayasu, M., R. Akiyama, H.J. Lee, K. Osakabe, Y. Osakabe, B. Watanabe et al. 2018. Generation of α-solanine-free hairy roots of potato by CRISPR/Cas9 mediated genome editing of the St16DOX gene. Plant Physiology and Biochemistry 131. https://doi.org/10.1016/j.plaphy.2018.04.026.

Nekrasov, V., C. Wang, J. Win, C. Lanz, D. Weigel and S. Kamoun. 2017. Rapid generation of a transgene-free powdery mildew resistant tomato by genome deletion. Scientific Reports 7(1). https://doi.org/10.1038/s41598-017-00578-x.

Nguyen, T., J. Kim and H. Lee. 2022. Applications of CRISPR/Cas genome editing in economically important vegetable crops. Mol. Hortic. 4(3): 245–260.

Nicoletto, C., C. Maucieri, G. Zanin, F. Vianello and P. Sambo. 2019. Vegetables quality and biotic stress. *In*: Ansari, R. and I. Mahmood (eds.). Plant Health Under Biotic Stress. Springer, Singapore. https://doi.org/10.1007/978-981-13-6043-5_6.

Nie, H., X. Yang, S. Zheng and L. Hou. 2024. Gene-based developments in improving quality of tomato: Focus on firmness, shelf life, and pre- and post-harvest stress adaptations. Horticulturae 10(6): 641.

Ninama, N., L. Gangal, A. Khayum, H.B. SB, S.H.M. HM and A. Singh. 2024. Post-harvest biotechnology or genetic engineering solutions: Extending shelf life and reducing food waste. J. Adv. Biol. Biotechnol. 27(4): 1–26.

Nonaka, S., M. Ito and H. Ezura. 2023. Targeted modification of CmACO1 by CRISPR/Cas9 extends the shelf-life of *Cucumis melo* var. reticulatus melon. Front. Genome Editing 5: 1176125.

Noureen, A., M.Z. Khan, I. Amin, T. Zainab and S. Mansoor. 2022. CRISPR/Cas9-mediated targeting of susceptibility factor eIF4E-enhanced resistance against potato virus Y. Front. Genet. 13: 922019.

Olson, R., B. Gavin-Smith, C. Ferraboschi and K. Kraemer. 2021. Food fortification: The advantages, disadvantages and lessons from sight and life programs. Nutrients 13(4): 1118.

Ortigosa, A., S. Gimenez-Ibanez, N. Leonhardt and R. Solano. 2019. Design of a bacterial speck resistant tomato by CRISPR/Cas9-mediated editing of SlJAZ2. Plant Biotechnology Journal 17(3). https://doi.org/10.1111/pbi.13006.

Palumbo, M., G. Attolico, V. Capozzi, R. Cozzolino, A. Corvino, M.L.V. de Chiara et al. 2022. Emerging postharvest technologies to enhance the shelf-life of fruit and vegetables: An overview. Foods 11(23): 3925.

Patel, D., R. Gupta and S. Rao. 2023. Exploring the potential of CRISPR/Cas genome editing for vegetable improvement. Biotechnol. Bioeng. 27(4): 567–582.

Pramanik, D., R.M. Shelake, J. Park, M.J. Kim, I. Hwang, Y. Park et al. 2021. CRISPR/Cas9-mediated generation of pathogen-resistant tomato against tomato yellow leaf curl virus and powdery mildew. Int. J. Mol. Sci. 22(4): 1878.

Prihatna, C., M.J. Barbetti and S.J. Barker. 2018. A novel tomato fusarium wilt tolerance gene. Front. Microbiol. 9: 1226. doi: 10.3389/fmicb.2018.01226.

Přibylová, A. and L. Fischer. 2024. How to use CRISPR/Cas9 in plants: From target site selection to DNA repair. J. Exp. Bot. 75(17): 5325–5343.

Rato, C., M.F. Carvalho, C. Azevedo and P.R. Oblessuc. 2021. Genome editing for resistance against plant pests and pathogens. Transgenic Res. 30(4): 427–459.

Raza, A., J. Tabassum, A.Z. Fakhar, R. Sharif, H. Chen, C. Zhang et al. 2023. Smart reprogramming of plants against salinity stress using modern biotechnological tools. Crit. Rev. Biotechnol. 43(7): 1035–1062.

Reddy, N.V. and C.M. Karthik. 2024. Issues regarding genetically modified crops in the light of non-target species. Adv. 69: 55.

Riu, Y.S., G.H. Kim, K.W. Chung and S.G. Kong. 2023. Enhancement of the CRISPR/Cas9-based genome editing system in lettuce (*Lactuca sativa* L.) using the endogenous U6 promoter. Plants 12(4): 878.

Rodriguez-Leal, D., Z.H. Lemmon, J. Man, M.E.. Bartlett and Z.B. Lippman. 2019. CRISPR technology is revolutionizing the improvement of tomato. Horticulture Research 6: 15.

Sakthivel, K., R. Balasubramanian, V. Sampathrajan, R. Veerasamy, S.V. Appachi and K.K. KK. 2025. Transforming tomatoes into GABA-rich functional foods through genome editing: A modern biotechnological approach. Funct. Integr. Genomics 25(1): 27.

Sami, A., Z. Xue, S. Tazein, A. Arshad, Z. He Zhu, Y. Ping Chen et al. 2021. CRISPR–Cas9-based genetic engineering for crop improvement under drought stress. Bioengineered 12(1): 5814–5829.

Sharma, P., A. Pandey, R. Malviya, S. Dey, S. Karmakar and D. Gayen. 2023. Genome editing for improving nutritional quality, post-harvest shelf life and stress tolerance of fruits, vegetables, and ornamentals. Front. Genome Editing 5: 1094965.

Shawky, A., A. Hatawsh, N. Al-Saadi, R. Farzan, N. Eltawy, M. Francis et al. 2024. Revolutionizing tomato cultivation: CRISPR/Cas9 mediated biotic stress resistance. Plants 13(16): 2269.

Shu, P., X.X. Cui, F.J. Li, D.D. Min, X.H. Zhang, l.L. Dong et al. 2019. Application of CRISPR/Cas9 gene editing technology in fruit and vegetable crops. *In*: Zhiwu Shengli Xuebao/Plant Physiology Journal (Vol. 55, Issue 2). https://doi.org/10.13592/j.cnki.ppj.2018.0312.

Singh, S., R. Chaudhary, R. Deshmukh and S. Tiwari. 2023. Opportunities and challenges with CRISPR-Cas mediated homologous recombination based precise editing in plants and animals. Plant Mol. Biol. 111(1): 1–20.

Singh, V.K., R. Singh, A. Kumar and R. Bhadouria. 2021. Current status of plant diseases and food security. pp. 19–35. *In*: Food Security and Plant Disease Management. Woodhead Publishing.

Singh, V., D. Braddick and P.K. Dhar. 2017. Exploring the potential of genome editing CRISPR-Cas9 technology. Gene 599: 1–18. https://doi.org/10.1016/j.gene.2016.11.008.

Soyk, S., N.A. Müller, S.J. Park, I. Schmalenbach, K. Jiang, R. Hayama et al. 2017. Variation in the flowering gene SELF PRUNING 5G promotes day-neutrality and early yield in tomato. Nature Genetics 49(1). https://doi.org/10.1038/ng.3733.

Sun, B., M. Jiang, H. Zheng, Y. Jian, W.L. Huang, Q. Yuan et al. 2020. Color-related chlorophyll and carotenoid concentrations of Chinese kale can be altered through CRISPR/Cas9 targeted editing of the carotenoid isomerase gene BoaCRTISO. Hortic. Res. 7(1).

Sun, B., A. Zheng, M. Jiang, S. Xue, Q. Yuan, L. Jiang et al. 2018. CRISPR/Cas9-mediated mutagenesis of homologous genes in Chinese kale. Scientific Reports 8(1). https://doi.org/10.1038/s41598-018-34884-9.

Sun, Y., T. Zhang and Y. Wang. 2021. Application of CRISPR/Cas9 in crop quality improvement. Int. J. Mol. Sci. 22(8): 4206.

Thakur, V., A. Sharma, P. Sharma, P. Kumar and S. Shilpa. 2022. Biofortification of vegetable crops for vitamins, minerals and other quality traits. J. Hortic. Sci. Biotechnol. 97(4): 417–428.

Tiwari, J.K., T. Buckseth, C. Challam, R. Zinta, N. Bhatia, D. Dalamu et al. 2022. CRISPR/Cas genome editing in potato: Current status and future perspectives. In Frontiers in Genetics (Vol. 13).

Toinga-Villafuerte, S., M.I. Vales, J.M. Awika and K.S. Rathore. 2022. CRISPR/Cas9-mediated mutagenesis of the granule-bound starch synthase gene in the potato variety yukon gold to obtain amylose-free starch in tubers. Int. J. Mol. Sci. 23(9).

Toromade, A.S., D.A. Soyombo, E. Kupa and T.I. Ijomah. 2024. Reviewing the impact of climate change on global food security: Challenges and solutions. Int. J. Appl. Res. Soc. Sci. 6(7): 1403–1416.

Ueta, R., C. Abe, T. Watanabe, S.S. Sugano, R. Ishihara, H. Ezura et al. 2017. Rapid breeding of parthenocarpic tomato plants using CRISPR/Cas9. Scientific Reports 7(1). https://doi.org/10.1038/s41598-017-00501-4.

Wan, L., Z. Wang, M. Tang, D. Hong, Y. Sun, J. Ren et al. 2021. CRISPR-Cas9 gene editing for fruit and vegetable crops: strategies and prospects. Horticulturae 7(7): 193.

Wang, X., Z. Liu and P. Zhang. 2024. From sequencing to genome editing in Cucurbitaceae: Advances and future prospects. Agric. 14(1): 90.

Wang, Y., Z. Liu and J. Hu. 2020. Enhancing drought resistance in crops through CRISPR/Cas9 technology. Front. Plant Sci. 11: 545.

Wang, M., Y. Lu, J.R. Botella, Y. Mao, K. Hua and J.K. Zhu. 2017. Gene targeting by homology-directed repair in rice using a geminivirus-based CRISPR/Cas9 system. Molecular Plant 10(7): 1007–1010.

Wudil, A.H., M. Usman, J. Rosak-Szyrocka, L. Pilař and M. Boye. 2022. Reversing years for global food security: A review of the food security situation in Sub-Saharan Africa (SSA). Int. J. Environ. Res. Public Health 19(22): 14836.

Yang, Y., D. Tilman, Z. Jin, P. Smith, C.B. Barrett, Y.G. Zhu et al. 2024. Climate change exacerbates the environmental impacts of agriculture. Science 385(6713): eadn3747.

Yasmeen, A., S. Shakoor, S. Azam, A. Bakhsh, N. Shahid, A. Latif et al. 2022. CRISPR/Cas-mediated knockdown of vacuolar invertase gene expression lowers the cold-induced sweetening in potatoes. Planta 256(6).

Zaidi, S.S., M. Tashkandi, S. Mansoor and M.M. Mahfouz. 2016. Engineering plant immunity: Using CRISPR/Cas9 to generate virus resistance. Front. Plant Sci. 7: 1673. doi: 10.3389/fpls.2016.01673.

Zhan, X., F. Zhang, Z. Zhong, R. Chen, Y. Wang, L. Chang et al. 2019. Generation of virus-resistant potato plants by RNA genome targeting. Plant Biotechnology Journal 17(9). https://doi.org/10.1111/pbi.13102.

Zhan, X., Y. Lu, J.K. Zhu and J.R. Botella. 2021. Genome editing for plant research and crop improvement. J. Integr. Plant Biol. 63(1): 3–33.

Zhang, J., H. Zhang, J.R. Botella and J.K. Zhu. 2018. Generation of new glutinous rice by CRISPR/Cas9-targeted mutagenesis of the Waxy gene in elite rice varieties. J. Integr. Plant Biol. 60(5): 369–375.

Zhang, Y., B. Iaffaldano and Y. Qi. 2021. CRISPR ribonucleoprotein-mediated genetic engineering in plants. Plant Commun. 2(2).

Zhou, J., X. Luan, Y. Liu, L. Wang, J. Wang, S. Yang, et al. 2023. Strategies and methods for improving the efficiency of CRISPR/Cas9 gene editing in plant molecular breeding. Plants 12(7): 1478.

Zhu, H., Y. Li and J. Wang. 2024. CRISPR-Cas9 systems for the improvement of solanaceous vegetable crops. J. Plant Biotechnol. 28(3): 345–360.

Chapter 13

CRISPR-Cas Based Genome Editing Strategies Towards Adaptation of Rice Crop to Disease Resistance

Irfan Ullah[1] and *Tariq Mahmood*[2,*]

1. Introduction

Rice (*Oryza sativa* L.) provides the main carbohydrate source for more than two-thirds of the world population (Tagliapietra et al. 2024). Its productivity, however, faces major challenges due to some of the destructive diseases caused by pathogenic bacteria, fungi, or viruses, resulting in significant yield loss (FAO 2025). High-yielding and disease-resistant rice varieties are needed to support sustainable agricultural production and food security (Sabar et al. 2024). A combination of disease management strategies, including disease forecasting models, chemical and biological control agents, crop rotation, and breeding for resistance, have been utilized to control these diseases (Haq and Ijaz 2020). However, pathogens diversity, environmental toxicity of controlling chemicals, and inaccuracy in disease resistance assessment and prediction models have led to inefficient counter strategies.

Traditional and molecular markers-aided breeding have ensured the development of disease-resistant rice varieties and therefore contributed significantly toward global food security. Such approaches, however, are relatively slow, labor-intensive, and have limited genetic backgrounds (Lusser et al. 2012). On the other hand, random recombinant genetic transformation approaches are constrained by biosafety regulations, ethical issues, and public reluctance to accept genetically modified crops as their staple food (Lucht 2015). Hence, a deeper understanding of the molecular mechanisms underlying plant-pathogen interactions is required to develop effective and durable disease resistance strategies (Ijaz et al. 2024).

Over the millennia, plants have evolved complex protection mechanisms against pathogens that can be classified into different types. A preformed defense mechanism is the accumulation of antimicrobial secondary metabolites. Further, inducible resistance occurs when the immune response detects a pathogen through membrane-bound Pattern Recognition Receptors (PRRs) (Kumar et al. 2020; Iqbal et al. 2021). The PRRs detect Pathogen Associated Molecular Patterns (PAMPs), and initiate PAMP-Triggered Immunity (PTI) (Miller et al. 2017; Andersen et al. 2018). While the pathogens have evolved to secrete effector proteins to suppress PTI, the plants have developed intracellular

[1] Department of Biotechnology and Genetic Engineering, Hazara University Mansehra, Mansehra 21300, Pakistan.
[2] Department of Agriculture, Hazara University Mansehra, Mansehra 21300, Pakistan.
* Corresponding author: mahmoodt74@gmail.com; tmahmood74@hu.edu.pk

DOI: 10.1201/9781003518204-13

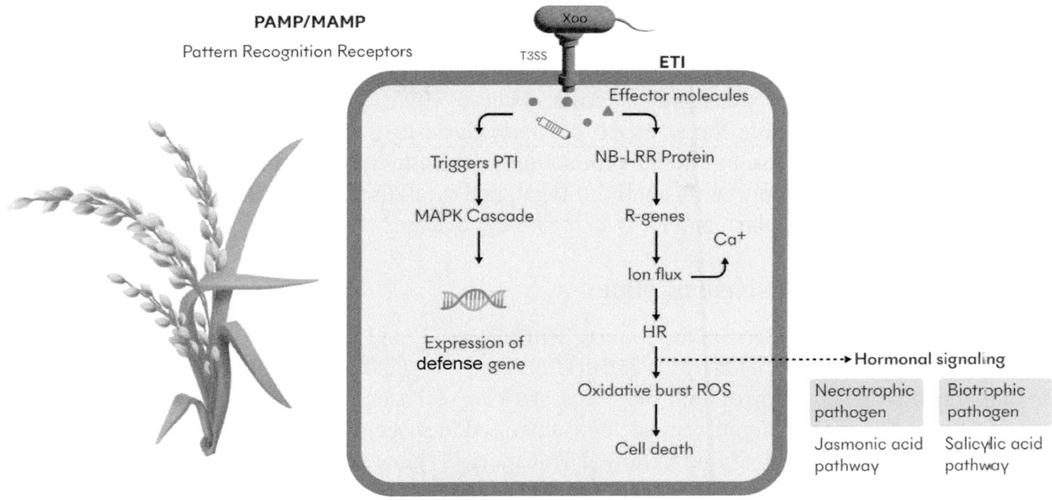

Figure 1. Overview of rice plant immune responses. Pathogen recognition triggers PTI through MAPK cascades and defense gene activation, while effector detection by *R*-genes induces ETI, involving ion flux, ROS production, HR, and localized cell death.

receptors to detect these effectors, resulting into Effectors-Triggered Immunity (ETI) (Thomma et al. 2011; Spoel and Dong 2012).

Xanthomonas oryzae pv. *oryzae* (Xoo), the Bacterial Blight (BB) pathogen of rice and probably the largely sophisticated bacterial plant pathogen, has evolved to suppress the ETI through specialized effectors, the Transcription-Activation-Like Effectors (TALEs). TALEs, after injection into the host cell by the Type III Secretion System (T3SS) of the bacterium, are guided toward the cell nucleus where these TALEs modulate the cognate host *susceptibility* (*S*) or *resistance* (*R*) genes (White and Yang 2009). Induction of *S* gene probably suppresses immunity, while that of *R* gene triggers resistance that is usually expressed as Hypersensitive Response (HR), due to rapid cell death and local necrosis of the underlying cells (Oh and Martin 2011). Traditional rice breeding for resistance mainly exploits *R* genes that directly or indirectly recognize pathogen effectors, and activate ETI (Zhao et al. 2022) (Fig. 1).

The advancement of conventional and speed breeding coupled with high-throughput genotyping, and genome editing, have transformed crop improvement (Li et al. 2020). Plant biotechnology has made use of several genome editing tools, including zinc finger nucleases, transcription activator-like effector nucleases and CRISPR-Cas nucleases. CRISPR-Cas nucleases, however, have earned great popularity because of their simplicity, specificity, and efficiency, and emerged as powerful tools for precise genetic modifications in plants (Voytas and Gao 2014; Zaidi et al. 2018). The double stranded breaks created in the target DNA by the CRISPR-Cas technology are either repaired by non-homologous end joining, an error prone mechanism or by homologous recombination, a precise mechanism free of any error in the target DNA (Sander and Joung 2014).

CRISPR-Cas technology has been effectively used to confer resistance against plant pathogenic bacteria, fungi, and viruses (Wang et al. 2016; Zaidi et al. 2018). The approach has also been successfully used in rice, modifying rice genomes to confer diseases-resistance (Li et al. 2019; Zhang et al. 2020). The BB pathogen remains the main limitation to Asian rice production because it leads to yield decreases between 40–80%, depending on environmental factors, infection stage, and rice susceptibility (Ullah et al. 2022). The fungal pathogen *Magnaporthe oryzae* induces rice blast which remains one of the major threats to agriculture. The pathogen creates annual global losses exceeding USD 7 billion, while averaging 10–30% yield reduction and leading to complete crop failure during severe outbreaks (Nalley et al. 2016). Another disease, The Rice Tungro Disease

(RTD) complex, caused by Rice Tungro Bacilliform Virus (RTBV) and Rice Tungro Spherical Virus (RTSV), leads to yield losses that range from 10% to 100%, and results in an annual economic damage of approximately USD 1.5 billion to Southeast Asia's rice production due to poor yield performance and increased management costs (Hibino 1996; Chancellor et al. 2006; Dai and Beachy 2009). The collective threat posed by these diseases demonstrates the critical need for effective management approaches in rice production. In the current chapter, the use of CRISPR-Cas technology in managing important bacterial, fungal, and viral diseases of rice has been highlighted with respect to BB, fungal blast, and RTD as the model diseases.

2. Bacterial Blight Disease of Rice

The BB, manifested by water-soaked lesions, leaf blighting, and grain discoloration, is especially damaging to rice, reducing yields by 20–80% (Ullah et al. 2022, 2024). CRISPR/Cas9 system allows for simultaneous modification of multiple genes, stacking resistance against the Xoo pathogen (Oliva et al. 2019). The approach targets critical genes involved in susceptibility pathways, particularly the *SWEET* (Sugars Will Eventually be Exported Transporters) gene family, exploited by the pathogen to access nutrients in the host plant (Chen et al. 2010; Oliva et al. 2019).

2.1 Mechanism of Susceptibility in Rice

Understanding rice susceptibility to Xoo is essential for developing effective strategies against BB (White and Yang 2009). Xoo utilizes type 3 effectors, known as TALEs, to manipulate the host genes and thereby, its transcriptome to promote the disease (Boch et al. 2014). Although varied in number, TALEs are present in all Xoo strains and induce host *S* genes for successful Xoo colonization (Yang et al. 2006). TALEs have a conserved modular structure, including N-terminal type III secretion signals and nuclear localization signals, a C-terminal acidic activation domain, and central tandem repeats of 33–35 amino acids. The central repeats are almost identical to each other, except for residues 12 and 13 which vary in amino acid composition from repeat to repeat, termed as Repeat Variable Di-residues (RVDs). The RVDs anchor to specific nucleotides in EBEs of *S* genes (Fig. 2) (Boch and Bonas 2010). Major TALEs, such as PthXo1 and

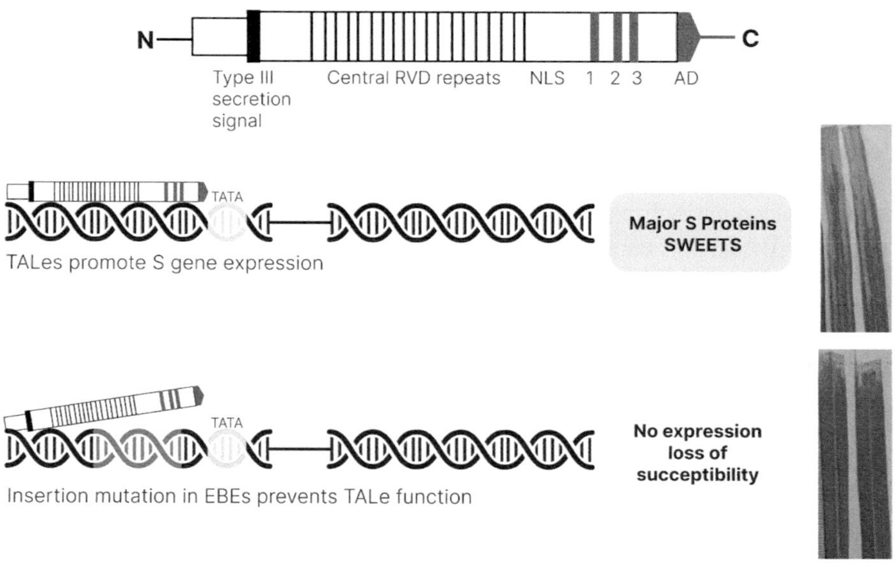

Figure 2. A Schematic representation of a typical TALE from Xoo and TALE-mediated susceptibility and resistance in rice. TALEs activate *S* gene expression by binding to the cognate EBEs, causing disease symptoms (top). Insertion/deletion mutations in EBEs block TALE binding, preventing the *S* gene expression, thereby conferring resistance (bottom).

PthXo2, are found in PXO99 and T7174, the Philippines and Japanese Xoo strains, respectively. Several *R* genes have been identified against Xoo. However, the continuous evolution of Xoo diminishes the effectiveness of gene-for-gene fashioned *R* genes in rice, complicating disease management. Hence targeting the *S* genes in the host seems to be a viable strategy to attain durable BB resistance (Mahmood and White 2019; Oliva et al. 2019).

2.2 *Major TALEs in Xoo and their Cognate Major S Genes in Rice*

Major *S* genes targeted by major Xoo TALEs in the field are members of the *OsSWEET* gene family and include *OsSWEET11*, *OsSWEET13*, and *OsSWEET14*. The former two are specifically induced by the TALEs PthXo1 and PthXo2, respectively, while the latter is activated by multiple TALEs, i.e., AvrXa7, PthXo3, TalC, and Tal5/TalF harbored by extant Xoo populations (Yang and White 2004; Zhou et al. 2015; Eom et al. 2019; Shah et al. 2025). The *OsSWEET*14-associated TALEs, however, either individually bind to a single EBE motif or overlap on EBE to induce the gene (Antony et al. 2010; Streubel et al. 2013). Further, other *OsSWEET* genes, such as *OsSWEET12* and *OsSWEET15*, have been implicated in rice susceptibility, though their cognate TALEs have not been identified yet (Streubel et al. 2013). The role of unidentified TALEs in Xoo strains from rice paddies is critical, as they may contribute to the evolution of novel TALEs and facilitate host infection (Mahmood and White 2019).

2.3 *Recessive Resistance in BB - Promoter Variations*

Recessive resistance of rice to BB can be termed as the non- or loose binding of a specific TALE to its cognate *S* gene EBE(s), thereby resulting in non-expression or lower than the susceptibility threshold expression of *S* gene(s). This is either because of variations in the promoters of otherwise *S* genes in a rice cultivar or due to the lack of any major virulent TALE in a Xoo strain (Zhou et al. 2015). For example, the recessive resistance conferred by the allele *xa13* is by the virtue of a 14.8 kb DNA fragment insertion in the promoter of the otherwise susceptible *OsSWEET11* allele, leading to non, or low expression of *OsSWEET11* gene by the TALE PthXo1 (Yang et al. 2006). Similarly, the TALE PthXo2 is unable to bind to the EBE of *xa25*, a recessive resistance allele of *OsSWEET13* due to single nucleotide polymorphisms in its EBE (Huang et al. 2016). Nevertheless, many Asian Xoo strains have defeated *xa13* by adapting to *OsSWEET* gene(s) alternate to *OsSWEET11* (Huang et al. 2016). For example, the strains PXO86 and PXO61 from the Philippines harbor TALEs AvrXa7 and PthXo3, respectively that target an alternate *SWEET* gene, the *OsSWEET14,* to cause the BB (Antony et al. 2010). RNAi-mediated silencing of *OsSWEET11* open reading frame in plants could also confer resistance to Xoo strains that rely on major TALE PthXo1 for its induction. However, the silenced plants showed low pollen viability, indicating a developmental compromise to attain resistance (Yang et al. 2006). Hence, a more viable resistance strategy is required.

2.4 *CRISPR-Cas for Recessive Resistance in BB*

The CRISPR-Cas9 system offers a target-oriented, highly efficient approach to achieve broad-spectrum BB resistance based on mutations achieved in the EBEs of *SWEET* gene promoters (Fig. 3; Table 1). For example, Oliva et al. (2019) and Shah et al. (2025) edited the promoters of three *SWEET* genes, *OsSWEET11, OsSWEET13,* and *OsSWEET14*, resulting in resistance to the corresponding Xoo strain. In addition, five promoter mutations were introduced into rice lines Kitake, IR64, and Ciherang-Sub1, all of which led to broad-spectrum BB-resistance (Oliva et al. 2019). CRISPR-Cas, technology was also used to design a BB diagnostic tool to predict the optimum set of BB-resistance genes for a particular geographic region (Eom et al. 2019). Similarly, Pakistani export brand Super Basmati rice was edited in *OsSWEET14* EBEs compatible with TALEs AvrXa7, PthXo3, and TalF. Resultantly, four edited lines from susceptible Super Basmati rice exhibited resistance against a local Xoo strain that harbored AvrXa7 to cause the disease (Zafar et al. 2020).

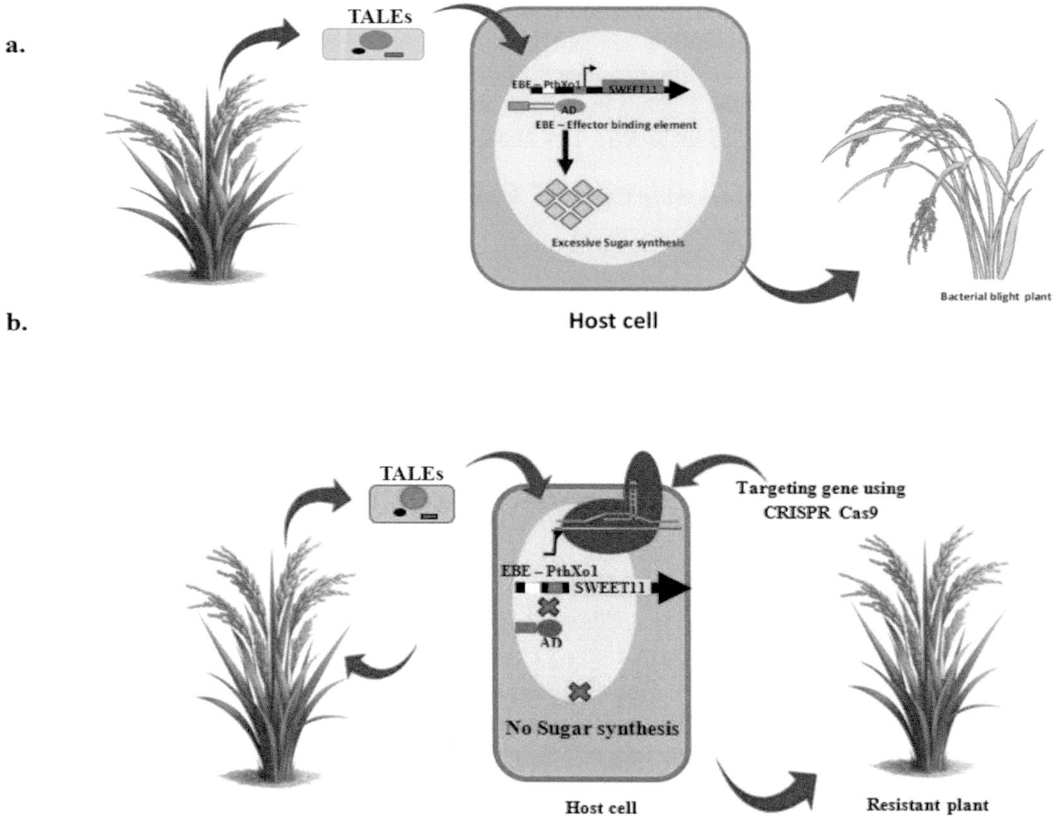

Figure 3. CRISPR-Cas based editing of TALE binding EBE motif in rice for resistance to Xoo. a. Susceptibility mode incurred by TALE PthXo1-dependent induction of *OsSWEET11* gene. b. CRISPR-associated mutation in *OsSWEET11* EBE conferred resistance to Xoo.

The Vietnamese commercial rice line, TBR225, also exhibited BB-resistance to two Vietnamese Xoo races when the AvrXa7, PthXo3, and TalF respective EBEs in the *OsSWEET14* promoter were edited by the CRISPR-Cas technique (Duy et al. 2021).

Similarly, a broad resistance against extant Xoo strains was achieved in transgenic rice lines by editing the EBEs in the promoter regions of *OsSWEET11, OsSWEET13*, and *OsSWEET14* using CRISPR/Cas9 technology (Xu et al. 2019). Targeted mutagenesis of these EBEs in various rice cultivars has led to resistance while maintaining essential agronomic traits (Oliva et al. 2019; Li et al. 2019; Blanvillain-Baufume et al. 2017; Zhou et al. 2015; Kumar et al. 2023). Zhu et al. (2024) further demonstrated that a single Cas9/gRNA construct targeting homologous sequences of *Xa13* and *Xa25* conferred effective resistance in rice varieties Yuxiang, Youzhan, the indica restorer line Shuhui 143, and a sterile indica line Zhinong, providing valuable germplasm for durable resistance breeding. Additionally, Zeng et al. (2020) generated nine *OsSWEET14* mutant alleles in the Zhonghua 11 cultivar by editing its protein coding sequence, resulting in strong resistance to both Asian and African Xoo strains, including AXO1947. These developments suggest a vital role for CRISPR-Cas genome editing with no, or least collateral damage to the normal physiology of rice plants, and hence can serve as a leaping off point for breeding for BB-resistance programs.

Table 1. Genes edited by CRISPR-Cas9 for diseases resistance in rice.

Diseases	Target gene	Function	Reference
Bacterial disease resistance	*OsSWEET11/OsSWEET11a, SWEET13, SWEET14, SWEET15, Xa23*	BB resistance	Eom et al. 2019; Kim et al. 2019; Oliva et al. 2019; Zafar et al. 2020; Duy et al. 2021; Ni et al. 2021; Wei et al. 2021; Li et al. 2022; Schepler-Luu et al. 2023; Kumar et al. 2023
Fungal disease resistance	*OsERF922*	Improved rice blast resistance	Wang et al. 2016
	OsSEC3A	//	Ma et al. 2018
	OsPi21	//	Nawaz et al. 2020; Li et al 2019
	OsPFT1	Sheath blight resistance in rice	Shah et al. 2019
Viral disease resistance	*eIF4G*	Improved resistance to the tungro disease	Macovei et al. 2018

3. Rice Blast as Fungal Disease and Its Control Through CRISPR-Cas

An important disease of rice, the rice blast, caused by *Magnaporthe oryzae*, leads to significant yield losses (Giraud et al. 2010; Sharma et al. 2012). The disease affects rice at all growth stages, with new pathogen races causing over 30% global production losses, enough to feed 60 million people (Nalley et al. 2016; Xiao et al. 2019). Several factors contribute to the susceptibility of rice varieties to the blast disease. The pathogen overcompensates the host defense mechanism by employing its effectors, the Avr proteins that antagonize the host immune responses. The crucial defense components of the host, including major chitin receptors, are impeded by the fungal Avr proteins which enable fungus' invasion and colonization of the host. In addition, the pathogen modulates host hormone resistance signaling pathways, including Jasmonic Acid (JA) and Salicylic Acid (SA), to inhibit necrosis-based Systemic Acquired Resistance (SAR) of the host, thus promoting disease progression. *M. oryzae* also induces the expression of microRNA, miR319 in rice which modulates the expression of *OsTCP21*, a transcription factor which positively regulates blast resistance, thus exacerbating rice susceptibility to the pathogen (Patkar and Naqvi 2017; Zhang et al. 2018).

Rice and *M. oryzae* undergo a constant evolutionary battle which determines the permanent need to create blast-resistant host varieties to maintain their durability (Skamnioti and Gurr 2009). Three fundamental *R* genes in rice against *M. oryzae* include *Pi54, Pi9* and *Pita* that detect pathogen effectors in a gene-for-gene interaction manner and trigger ETI. Most *R* genes contain NBS-LRR proteins that serve both as effectors recognizer and the subsequent activators of Hypersensitive Responses (HR), resulting in localized cell death restriction on pathogen spread. The developments, in one way or another, are the result of generation of Reactive Oxygen Species (ROS) or production of phytoalexins during the process. The *Pi54* gene releases defense mechanisms which simultaneously activate gene and microRNAs to provide enhanced protection against *M. oryzae* (Gupta et al. 2012; Arora et al. 2021).

Out of 100 identified blast *R* genes, 30 genes have been cloned, including *Pi-ta/Pi-ta2, Pi-z, Pi-b,* and *Pi-k/h/m/s* that have been used in breeding for resistance in rice programs (Xiao et al. 2019; Sharma et al. 2021). Allele surveying, both genotypically and phenotypically, is a feasible and trustworthy approach that has identified potent blast resistance genes. The Pigm locus, for example, encodes NBS-LRR receptors that confers broad-spectrum resistance with no penalty to yield. For instance, PigmR promotes resistance against the blast fungus, while PigmS balances the seed production (Deng et al. 2017). Further, the blast resistance is also enhanced by single-base modifications in regulatory regions, such as *bsr-d1* (Li et al. 2017).

CRISPR-Cas9 has been demonstrated to be useful for enhancing blast resistance through editing genes like *OsERF922* (Fig. 4; Table 1), and creating novel resistance alleles (Thakur et al. 2015; Wang et al. 2016; Pacher and Puchta 2017). About 42% mutation frequency was generated in the T_0 plants which was stably inherited to the following generations (Wang et al. 2016). Showing decreased lesion length and no cost to agronomic traits. Multiplex CRISPR targeting of *OsERF922* significantly enhanced mutation efficiency, with up to 90% of T_0 plants showing edits at all three target sites, and the resultant homozygous mutants showing improved blast resistance with no collateral damage to the agronomic traits (Wang et al. 2016). Disruption of *OsSEC3A* gene enhanced blast resistance, however, with undesirable agronomical traits that may include a small plant height or reduced yields, as mentioned by Ma et al. (2018). CRISPR-Cas has been successfully used to edit a single gene, as in case of *Pi21* for blast resistance, or multiple genes, as in case of *tms5/pi21/xa13*, which resulted in enhanced resistance as a breeding substrate for hybrid production (Li et al. 2019).

MicroRNAs (miRNAs) have become key players in plant-pathogen interactions as susceptibility targets manipulated by *M. oryzae* to induce rice blast disease (Zhan and Meyers 2023). Some

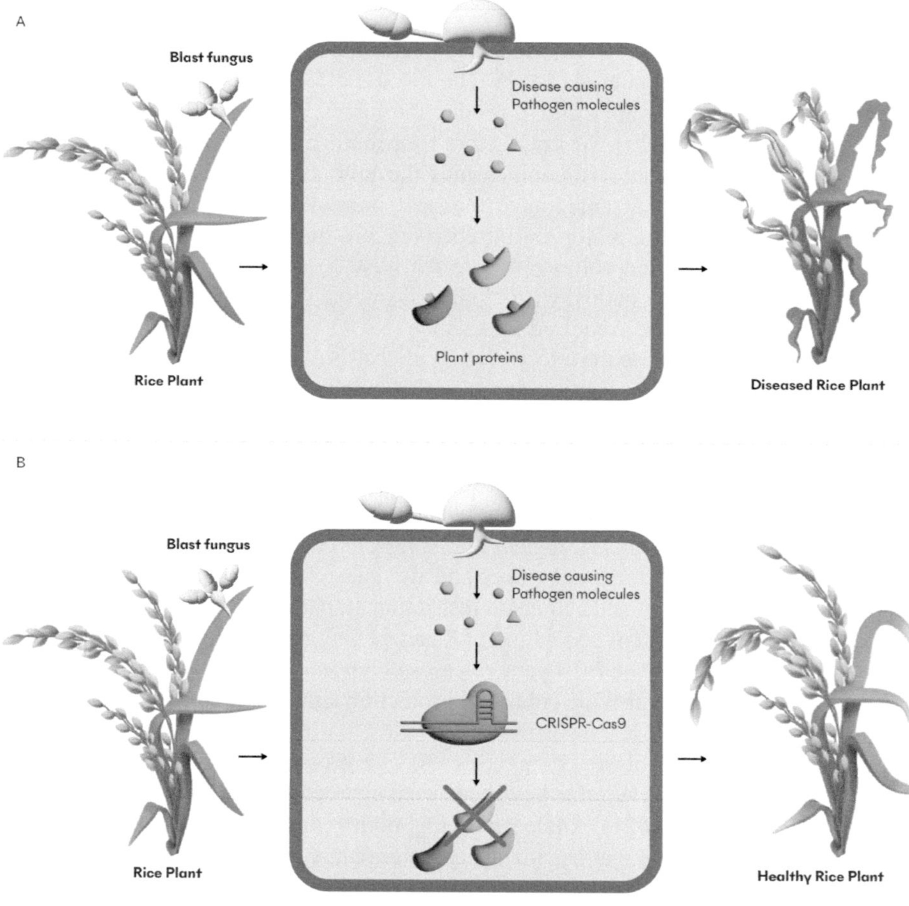

Figure 4. CRISPR-Cas9 based editing of plant proteins in rice for resistance to fungal blast. a. Susceptibility mode incurred by attachment of the pathogen molecules to the plant molecules. b. CRISPR-associated mutation in plant proteins conferred resistance to fungal blast.

miRNAs, modulate the expression of Jasmonic Acid (JA), a resistance signaling molecule, to induce susceptibility. For example, the enhanced accumulation of miRNA, *Osa-miR319* downregulate the JA expression by repressing *OsTCP21*, a transcription factor required for JA expression, thus promoting fungal invasion (Li et al. 2019). Similarly, the *Osa-miR827*, a microRNA associated with phosphate homeostasis, has been identified as a factor that makes plants more susceptible to pathogens, and whose overexpression led to decreased activation of defense genes and enhanced fungal colonization (Bundo et al. 2024). Yet there are miRNAs which repress the expression of Transcription Factors (TF) to cause susceptibility as is done by Osa-miR444b.2, which inhibits the expression of TFs MADS27b and MADS57 to facilitate *M. oryzae* proliferation (Zhang et al. 2020).

Conversely, some miRNAs act as resistance targets as is exemplified by miRNA *Osa-miR162a* enhances rice resistance to *M. oryzae* by modulating the expression of *OsDCL1* and vice versa (Li et al. 2020). *Osa-miR1432* brokers growth-defense trade-offs via *OsEFH1* culminating into both immunity and yield (Wang et al. 2021). These points emphasize miRNA-mediated regulation to be an imperative strategy employed by *M. oryzae* to modulate host immunity, and hence can prove potential genetic intervention targets to engineer blast resistance.

CRISPR/Cas9-mediated knockout of *Bsr-d1* has been shown to confer broad-spectrum resistance to rice blast in Northeast China, where japonica rice cultivars lack the resistance allele of *OsBsr-d1* (Zhang et al. 2024). Similarly, gene editing and overexpression of *OsOPR10* enhanced the host resistance to both blast and BB by positively regulating defense responses via the JA and SA-mediated pathways (Xu et al. 2024). Gene-editing technology was also employed to knock out the *pid3* and *pi21* genes in the rice variety Dalixiang, generating high-quality, blast-resistant rice lines without introducing foreign genetic material or altering grain quality (Wu et al. 2021; Chen et al. 2024).

4. Viral Rice Tungro Disease Complex and Its Control Through CRISPR-Cas

Viral diseases significantly hinder global rice productivity, with Rice Tungro Disease (RTD) being a major threat, affecting approximately 3.5 million hectares in Asia (Chancellor et al. 2006). RTD causes significant yield and economic losses, with yield reductions ranging from 10 to 100%, depending on the cultivar used and infection severity. Up to 50–100% and 10–30% losses have been reported in susceptible cultivars and moderately resistant rice cultivars, respectively (Hibino 1983; Chancellor et al. 2006). The infection severity also depends on weather conditions and abundance of vector populations, particularly in the case of viral diseases (Azzam and Chancellor 2002; Chancellor et al. 2006; Muralidharan et al. 2003; Macovei et al. 2018). Economically, RTD results in substantial losses, with estimates of USD 1.5 billion annually in Southeast Asia due to reduced yields and increased management costs (Azzam and Chancellor 2002; Chancellor et al. 2006; Muralidharan et al. 2003).

RTD symptoms include plant yellowing, growth reduction, decreased tillering, and sterile panicles which result in major losses in rice production (Hull 1996; Macovei et al. 2018). The RTD complex-forming RTSV is a single-stranded RNA virus, while RTBV contains double-stranded DNA genome (Hull 1996). The vectors, green leafhoppers (*Nephotettix virescens* and *N. nigropictus*), contract the RTSV after RTBV creates primary symptoms (Hibino 1989), which are worsened by RTSV that facilitates virus transmission toward further progression and disease severity (Macovei et al. 2018).

The host proteins, eukaryotic translation initiation factors (eIFs), including eIF4G, are essential for RNA viruses to translate viral RNAs, as they facilitate the recruitment of ribosomes to viral RNA (Sanfacon 2015; Shopan et al. 2020). Given that eIF4G and eIF(iso)4G serve as susceptibility factors, these proteins serve as targets to be modulated to build resistance in the host (Fig. 5). The eIF4G determines RTSV rice susceptibility as it controls viral RNA translation. RTSV utilizes the same host translation mechanism as other plant RNA viruses to create viral proteins and repress host genes (Kneller et al. 2006; Sonenberg and Hinnebusch 2009). IR64 rice, along with other susceptible rice varieties, carries wild-type alleles of *eIF4G* that promote functional viral replication and spread throughout the plant system. The vital part played by host factors during virus pathogenesis demonstrates their ability to resist infection (Dreher and Miller 2006; Macovei et al. 2018). Host resistance breeding programs have identified rice varieties exhibiting resistance to one or both viruses involved in the tungro disease (Khush et al. 2004). However, finding and deploying homozygous recessive mutations in *eIF4G* and *eIF(iso)4G* prohibits host's translational machinery to be hijacked by rice black-streaked dwarf virus (RBSDV) and RTSV (Lee et al. 2010; Macovei et al. 2018; Wang et al. 2022).

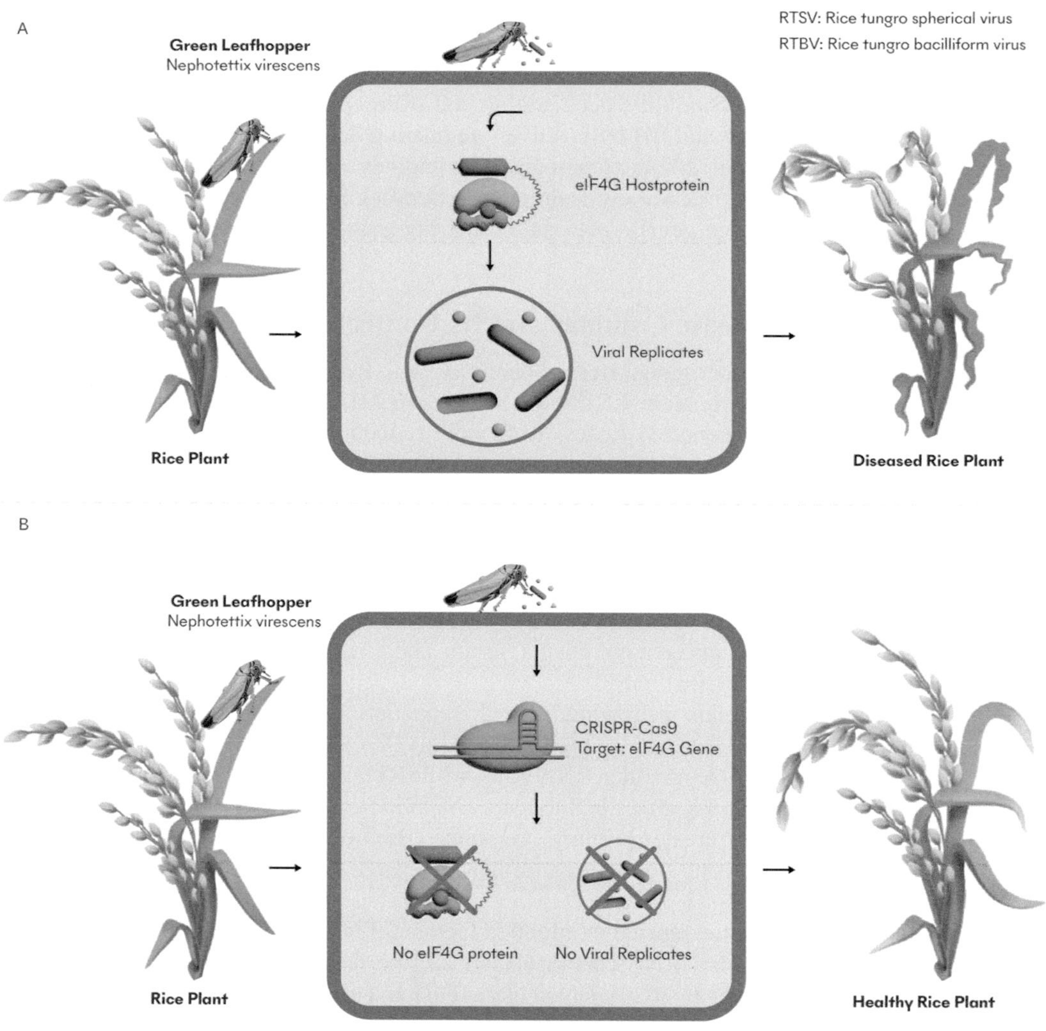

Figure 5. CRISPR-Cas9 based editing of *eIF4G* gene in rice for viral resistance. a. Susceptibility mode incurred by expression of *eIF4G* gene and the replication of the viruses increased. b. CRISPR-associated mutation in *eIF4G* gene conferred resistance to viruses.

For example, in the indica rice variety Utri Merah, RTSV resistance has been implicated to be related to two possibilities: a single-nucleotide polymorphism or a deletion in *eIF4G*; in either case, the YVV amino acid residues of the virus are affected (Lee et al. 2010). Further, mutations in the residues $Y^{1059}V^{1060}V^{1061}$ of eIF4G avert both viral protein binding and viral replication, accompanied by systemic infection (Lee et al. 2010). Latest advances in CRISPR/Cas9-mediated genome editing have made it feasible to acquire site-specific mutations in *eIF4G* at the SVLFPNLAGKS motif adjacent to the YVV residues that are critical for viral interaction (Macovei et al. 2018). These designed mutations cause in-frame deletions that do not affect normal plant development but confer a significant block during RTSV infection by preventing the assembly of viral ribonucleoprotein complexes required for translation initiation (Albar et al. 2006). In another example, CRISPR-Cas9-based editing in *eIF4G* in the susceptible rice variety IR64 resulted in 36–86% mutation frequencies, while inheritance was stable, bearing no apparent off-target implications (Macovei et al. 2018).

Co-infections of RTSV and RTBV, as seen in tungro disease, exacerbate yield losses and complicate disease management strategies due to the complex interactions between the two viruses (Bunawan et al. 2014). Correspondingly, and given the fast evolution of RNA viruses, CRISPR systems are designed in a way to provide extensive immunity to rice host, as was done against RBSDV and rice stripe mosaic virus (Zhang et al. 2019; Zhan et al. 2024).

5. Conclusion

CRISPR-Cas enables the development of rice with recessive resistance aligning with sustainable and eco-friendly agricultural practices. As the CRISPR-Cas technology progresses, it shows great promise in tackling challenges like BB, blast fungus, and RTD viral complex diseases, thus enhancing crop resistance, and boosting global food security. The technology, followed by classical breeding, and disease monitoring and detection system, provides a robust strategy to mitigate the effects of these diseases and ensure their sustainable production. It is plausible to state that the strategy, though essentially genetic engineering based, shall still be considered non-GMO while keeping the pathogens at bay.

References

Albar, L., M. Bangratz-Reyser, E. Hébrard, M.N. Ndjiondjop, M. Jones and A. Ghesquière. 2006. Mutations in the eif (iso) 4g translation initiation factor confer high resistance of rice to rice yellow mottle virus. Plant J. 47: 417–426.

Andersen, E.J., S. Ali, E. Byamukama, Y. Yen and M.P. Nepal. 2018. Disease resistance mechanisms in plants. Genes 9: 339.

Antony, G., J. Zhou, S. Huang, T. Li, B. Liu, F. White et al. 2010. Rice *xa13* recessive resistance to bacterial blight is defeated by induction of the disease susceptibility gene os-11n3. Plant Cell. 22: 3864–3876.

Arora, K., A.K. Rai, B. Devanna, H. Dubey, A. Narula and T.R. Sharma. 2021. Deciphering the role of micrornas during pi54 gene mediated magnaporthe oryzae resistance response in rice. Physiol. Mol. Biol. Plants. 27: 633–647.

Azzam, O. and T.C. Chancellor. 2002. The biology, epidemiology, and management of rice tungro disease in asia. Plant Dis. 86: 88–100.

Blanvillain-Baufume, S., M. Reschke, M. Sole, F. Auguy, H. Doucoure, B. Szurek et al. 2017. Targeted promoter editing for rice resistance to *Xanthomonas oryzae* pv. *oryzae* reveals differential activities for SWEET14-inducing TAL effectors. Plant Biotechnol. J. 15: 306–317.

Boch, J. and U. Bonas. 2010. Xanthomonas avrbs3 family-type iii effectors: Discovery and function. Annu. Rev. Phytopathol. 48: 419–436.

Boch, J., U. Bonas and T. Lahaye. 2014. Tal effectors–pathogen strategies and plant resistance engineering. New Phytol. 204: 823–832.

Bunawan, H., L. Dusik, S.N. Bunawan and N.M. Amin. 2014. Rice tungro disease: From identification to disease control. World Appl. Sci. J. 31: 1221–1226.

Bundo, M., B. Val-Torregrosa, H. Martín-Cardoso, M. Ribaya, L. Campos-Soriano, M. Bach-Pages et al. 2024. Silencing osa-mir827 via crispr/cas9 protects rice against the blast fungus magnaporthe oryzae. Plant Mol. Biol. 114: 105.

Chancellor, T., J. Holt, S. Villareal, E. Tiongco and J. Venn. 2006. Spread of plant virus disease to new plantings: A case study of rice tungro disease. Adv. Virus Res. 66: 1–29.

Chen, J., Z. Miao, D. Kong, A. Zhang, F. Wang, G. Liu et al. 2024. Application of crispr/cas9 technology in rice germplasm innovation and genetic improvement. Genes 15: 1492.

Chen, L.Q., B.H. Hou, S. Lalonde, H. Takanaga, M.L. Hartung, X.Q. Qu et al. 2010. Sugar transporters for intercellular exchange and nutrition of pathogens. Nature 468: 527–532.

Dai, S. and R.N. Beachy. 2009. Genetic engineering of rice to resist rice tungro disease. *In Vitro* Cell. Dev. Biol. Plant. 45: 517–524.

Deng, Y., K. Zhai, Z. Xie, D. Yang, X. Zhu, J. Liu et al. 2017. Epigenetic regulation of antagonistic receptors confers rice blast resistance with yield balance. Science 355: 962–965.

Dreher, T.W. and W.A. Miller. 2006. Translational control in positive strand RNA plant viruses. Virology 344: 185–197.

Duy, P.N., D.T. Lan, H. Pham Thu, H.P. Thi Thu, H. Nguyen Thanh, N.P. Pham et al. 2021. Improved bacterial leaf blight disease resistance in the major elite vietnamese rice cultivar tbr225 via editing of the ossweet14 promoter. PloS One 16: e0255470.

Eom, J.S., D. Luo, G. Atienza-Grande, J. Yang, C. Ji, V. Thi Luu et al. 2019. Diagnostic kit for rice blight resistance. Nat. Biotechnol. 37: 1372–1379.

FAO, Food and AUtUMN Nations. 2025. The Future of Food and Agriculture: Trends and Challenges.

Giraud, T., P. Gladieux and S. Gavrilets. 2010. Linking the emergence of fungal plant diseases with ecological speciation. Trends Ecol. Evol. 25: 387–395.

Gupta, S.K., A.K. Rai, S.S. Kanwar, D. Chand, N.K. Singh and T.R. Sharma. 2012. The single functional blast resistance gene pi54 activates a complex defence mechanism in rice. J. Exp. Bot. 63: 757–772.

Haq, I.U. and S. Ijaz. 2020. Plant disease management strategies for sustainable agriculture through traditional and modern approaches. Springer Nature. 13.

Hibino, H. 1983. Transmission of two rice tungro-associated viruses and rice waika virus from doubly or singly infected source plants by leafhopper vectors.774–777.

Hibino, H. 1989. Insect-borne viruses of rice. Adv. Dis. Vect. Res. 6: 209–241.

Hibino, H. 1996. Biology and epidemiology of rice viruses. Annu. Rev. Phytopathol. 34: 249–274.

Huang, S., G. Antony, T. Li, B. Liu, K. Obasa, B. Yang et al. 2016. The broadly effective recessive resistance gene xa5 of rice is a virulence effector-dependent quantitative trait for bacterial blight. Plant J. 86: 186–194.

Hull, R. 1996. Molecular biology of rice tungro viruses. Annu. Rev. Phytopathol. 34: 275–297.

Ijaz, U., C. Zhao, S. Shabala and M. Zhou. 2024. Molecular basis of plant–pathogen interactions in the agricultural context. Biology 13: 421.

Iqbal, Z., M.S. Iqbal, A. Hashem, E.F. Abd_Allah and M.I. Ansari. 2021. Plant defense responses to biotic stress and its interplay with fluctuating dark/light conditions. Front. Plant Sci. 12: 631810.

Khush, G., E. Angeles, P. Virk and D.S. Brar. 2004. Breeding rice for resistance to tungro virus at IRRI. Annu. Rev. Phytopathol. 42: 143–165.

Kim, Y.A., H. Moon and C.J. Park. 2019. Crispr/cas9-targeted mutagenesis of os8n3 in rice to confer resistance to xanthomonas oryzae pv. Oryzae. Rice. 12: 1–13.

Kneller, E.L.P., A.M. Rakotondrafara and W.A. Miller. 2006. Cap-independent translation of plant viral RNAs. Virus Res. 119: 63–75.

Kumar, A., C. Changwal, A. Hada and P.K. Singh. 2020. Molecular insight of plant–pathogen interaction. Sustainable Agriculture in the Era of Climate Change: 481–511.

Kumar, J., S.N. Char, T. Weiss, H. Liu, B. Liu, B. Yang et al. 2023. Efficient protein tagging and cis-regulatory element engineering via precise and directional oligonucleotide-based targeted insertion in plants. Plant Cell. 35: 2722–2735.

Lee, J.H., M. Muhsin, G.A. Atienza, D.Y. Kwak, S.M. Kim, T.B. De Leon et al. 2010. Single nucleotide polymorphisms in a gene for translation initiation factor (eif4g) of rice (*Oryza sativa*) associated with resistance to rice tungro spherical virus. Mol. Plant-Microbe Interact. 23: 29–38.

Li, C., L. Zhou, B. Wu, S. Li, W. Zha, W. Li et al. 2022. Improvement of bacterial blight resistance in two conventionally cultivated rice varieties by editing the noncoding region. Cells 11: 2535.

Li, S., L. Shen, P. Hu, Q. Liu, X. Zhu, Q. Qian et al. 2019. Developing disease-resistant thermosensitive male sterile rice by multiplex gene editing. J. Integr. Plant Biol. 61: 1201–1205.

Li, W., Z. Zhu, M. Chern, J. Yin, C. Yang, L. Ran et al. 2017. A natural allele of a transcription factor in rice confers broad-spectrum blast resistance. Cell 170: 114–126.

Li, X.P., X.C. Ma, H. Wang, Y. Zhu, X.X. Liu, T.T. Li et al. 2020. Osa-mir162a fine-tunes rice resistance to magnaporthe oryzae and yield. Rice 13: 1–13.

Liu, L., Y. Li, Q. Wang, X. Xu, J. Yan, Y. Wang et al. 2024. Constructed rice tracers identify the major virulent transcription activator-like effectors of the bacterial leaf blight pathogen. Rice 17: 30.

Lucht, J.M. 2015. Public acceptance of plant biotechnology and gm crops. Viruses 7: 4254–4281.

Lusser, M., C. Parisi, D. Plan and E. Rodríguez-Cerezo. 2012. Deployment of new biotechnologies in plant breeding. Nat. Biotechnol. 30: 231–239.

Ma, J., J. Chen, M. Wang, Y. Ren, S. Wang, C. Lei et al. 2018. Disruption of ossec3a increases the content of salicylic acid and induces plant defense responses in rice. J. Exp. Bot. 69: 1051–1064.

Macovei, A., N.R. Sevilla, C. Cantos, G.B. Jonson, I. Slamet-Loedin, T. Cermak et al. 2018. Novel alleles of rice eif4g generated by crispr/cas9-targeted mutagenesis confer resistance to rice tungro spherical virus. Plant Biotechnol. J. 16: 1918–1927.

Mahmood, T. and F.F. White. 2019. Disease resistance and susceptibility genes to bacterial blight of rice. *In*: Protecting Rice Grains in the Post-genomic Era. IntechOpen London. UK. 327.

Miller, R.N.G., G.S. Costa Alves and M.A. Van Sluys. 2017. Plant immunity: Unravelling the complexity of plant responses to biotic stresses. Ann. Bot. 119: 681–687.

Muralidharan, K., D. Krishnaveni, N. Rajarajeswari and A. Prasad. 2003. Tungro epidemics and yield losses in paddy fields in India. Curr. Sci. 85: 1143–1147.

Nalley, L., F. Tsiboe, A. Durand-Morat, A. Shew and G. Thoma. 2016. Economic and environmental impact of rice blast pathogen (magnaporthe oryzae) alleviation in the united states. PloS One 11: e0167295.

Nawaz, G., B. Usman, H. Peng, N. Zhao, R. Yuan, Y. Liu et al. 2020. Knockout of pi21 by CRISPR/Cas9 and iTRAQ-based proteomic analysis of mutants revealed new insights into *M. oryzae* resistance in elite rice line. Genes 11: 735.

Ni, Z., Y. Cao, X. Jin, Z. Fu, J. Li, X. Mo et al. 2021. Engineering resistance to bacterial blight and bacterial leaf streak in rice. Rice 14: 38.

Oh, C.S. and G.B. Martin. 2011. Effector-triggered immunity mediated by the pto kinase. Trends Plant Sci. 16: 132–140.

Oliva, R., C. Ji, G. Atienza-Grande, J.C. Huguet-Tapia, A. Perez-Quintero, T. Li et al. 2019. Broad-spectrum resistance to bacterial blight in rice using genome editing. Nat. Biotechnol. 37: 1344–1350.

Pacher, M. and H. Puchta. 2017. From classical mutagenesis to nuclease-based breeding–directing natural DNA repair for a natural end-product. Wiley Online Library 819–833.

Patkar, R.N. and N.I. Naqvi. 2017. Fungal manipulation of hormone-regulated plant defense. PLoS Pathog. 13: e1006334.

Sabar, M., S.E. Mustafa, M. Ijaz, R.A.R. Khan, F. Shahzadi, H. Saher et al. 2024. Rice breeding for yield improvement through traditional and modern genetic tools. Eur. J. Ecol. Biol. Agric. 1: 14–19.

Sander, J.D. and J.K. Joung. 2014. Crispr-cas systems for editing, regulating and targeting genomes. Nat. Biotechnol. 32: 347–355.

Sanfacon, H. 2015. Plant translation factors and virus resistance. Viruses 7: 3392–3419.

Schepler-Luu, V., C. Sciallano, M. Stiebner, C. Ji, G. Boulard, A. Diallo et al. 2023. Genome editing of an african elite rice variety confers resistance against endemic and emerging xanthomonas oryzae pv. oryzae strains. Elife. 12: e84864.

Shah, P.R., S. Varanavasiappan, E. Kokiladevi, A. Ramanathan and K.K. Kumar. 2019. Genome editing of rice PFT1 gene to study its role in rice sheath blight disease resistance. Int. J. Curr. Microbiol. Appl. Sci. 8: 2356–2364.

Shah, S.A., R. Ahsan, L. Liu, Y. Li, Q. Wang, Y. Wang et al. 2025. TALome and phenotypic analysis of Pakistani Xanthomonas oryzae pv. oryzae population revealed novel virulent TALEs contributing to bacterial blight of rice. Phytopathol. Res. https://doi.org/10.1186/s42483-024-00292-3.

Sharma, S.K., D. Sharma, R.P. Meena, M.K. Yadav, R. Hosahatti, A.K. Dubey et al. 2021. Recent insights in rice blast disease resistance. Blast Disease of Cereal Crops: Evolution and Adaptation in Context of Climate Change. 89–123.

Sharma, T., A. Rai, S. Gupta, J. Vijayan, B. Devanna and S. Ray. 2012. Rice blast management through host-plant resistance: Retrospect and prospects. Agric. Res. 1: 37–52.

Shopan, J., X. Lv, Z. Hu, M. Zhang and J. Yang. 2020. Eukaryotic translation initiation factors shape RNA viruses resistance in plants. Hortic. Plant J. 6: 81–88.

Skamnioti, P. and S.J. Gurr. 2009. Against the grain: Safeguarding rice from rice blast disease. Trends Biotechnol. 27: 141–150.

Sonenberg, N. and A.G. Hinnebusch. 2009. Regulation of translation initiation in eukaryotes: Mechanisms and biological targets. Cell. 136: 731–745.

Spoel, S.H. and X. Dong. 2012. How do plants achieve immunity? Defence without specialized immune cells. Nat. Rev. Immunol. 12: 89–100.

Streubel, J., C. Pesce, M. Hutin, R. Koebnik, J. Boch and B. Szurek. 2013. Five phylogenetically close rice sweet genes confer tal effector-mediated susceptibility to xanthomonas oryzae pv. oryzae. New Phytol. 200: 808–819.

Tagliapietra, B.L., C.F. Soares and M.T.P.S. Clerici. 2024. Rice (*Oryza sativa* L.) and its products for human consumption: General characteristics, nutritional properties, and types of processing. Food Sci. Technol. 44.

Thakur, S., P.K. Singh, A. Das, R. Rathour, M. Variar, S. Prashanthi et al. 2015. Extensive sequence variation in rice blast resistance gene pi54 makes it broad spectrum in nature. Front. Plant Sci. 6: 345.

Thomma, B.P., T. Nürnberger and M.H. Joosten. 2011. Of pamps and effectors: The blurred pti-eti dichotomy. Plant Cell. 23: 4–15.

Ullah, I., H. Ali, T. Mahmood, M.N. Khan, M. Haris, H. Shah et al. 2022. Pyramiding of four broad spectrum bacterial blight resistance genes in cross breeds of basmati rice. Plants 12: 46.

Ullah, I., T. Mahmood, H. Ali, M. Islam, A. Din, M. Haris et al. 2024. Molecular investigation of bacterial blight of rice in the foothills of the western himalayas, pakistan. Pak. J. Bot. 56: 1603–1609.

Voytas, D.F. and C. Gao. 2014. Precision genome engineering and agriculture: Opportunities and regulatory challenges. PLoS Biol. 12: e1001877.

Wang, F., C. Wang, P. Liu, C. Lei, W. Hao, Y. Gao et al. 2016. Enhanced rice blast resistance by crispr/cas9-targeted mutagenesis of the erf transcription factor gene oserf922. PloS One 11: e0154027.

Wang, P., J. Liu, Y. Lyu, Z. Huang, X. Zhang, B. Sun et al. 2022. A review of vector-borne rice viruses. Viruses 14: 2258.

Wang, W., S. Ma, P. Hu, Y. Ji and F. Sun. 2021. Genome editing of rice eif4g loci confers partial resistance to rice black-streaked dwarf virus. Viruses 13: 2100.

Wei, Z., M. Abdelrahman, Y. Gao, Z. Ji, R. Mishra, H. Sun et al. 2021. Engineering broad-spectrum resistance to bacterial blight by CRISPR-Cas9-mediated precise homology-directed repair in rice. Mol. Plant. 14: 1215–1218.

White, F.F. and B. Yang. 2009. Host and pathogen factors controlling the rice-xanthomonas oryzae interaction. Plant Physiol. 150: 1677–1686.

Wu, X., J. Li, Q. Zeng, X. Zhang, D. Zhang, H. Xu et al. 2021. Improvement of blast resistance of rice variety dalixiang by crispr/cas9 gene editing technology. Seed. 40: 50–55.

Xiao, W., Q. Yang, M. Huang, T. Guo, Y. Liu, J. Wang et al. 2019. Improvement of rice blast resistance by developing monogenic lines, two-gene pyramids and three-gene pyramid through mas. Rice 12: 1–11.

Xu, D., Q. Lin, Z. Li, X. Zhuang, Y. Ling, M. Lai et al. 2024. Osopr10 positively regulates rice blast and bacterial blight resistance. Chin. J. Rice Sci. 38: 364.

Xu, Z., X. Xu, Q. Gong, Z. Li, Y. Li, S. Wang et al. 2019. Engineering broad-spectrum bacterial blight resistance by simultaneously disrupting variable TALE-binding elements of multiple susceptibility genes in rice. Mol. Plant. 12: 1434–1446.

Yang, B. and F.F. White. 2004. Diverse members of the avrbs3/ptha family of type iii effectors are major virulence determinants in bacterial blight disease of rice. Mol. Plant-Microbe Interact. 17: 1192–1200.

Yang, B., A. Sugio and F.F. White. 2006. Os8n3 is a host disease-susceptibility gene for bacterial blight of rice. Proc. Natl. Acad. Sci. U.S.A. 103: 10503–10508.

Zafar, K., M.Z. Khan, I. Amin, Z. Mukhtar, S. Yasmin, M. Arif et al. 2020. Precise crispr-cas9 mediated genome editing in super basmati rice for resistance against bacterial blight by targeting the major susceptibility gene. Front. Plant Sci. 11: 575.

Zaidi, S.S.A., M.S. Mukhtar and S. Mansoor. 2018. Genome editing: Targeting susceptibility genes for plant disease resistance. Trends Biotechnol. 36: 898–906.

Zeng, X., Y. Luo, N.T.Q. Vu, S. Shen, K. Xia and M. Zhang. 2020. CRISPR/Cas9-mediated mutation of OsSWEET14 in rice cv. Zhonghua11 confers resistance to *Xanthomonas oryzae* pv. *oryzae* without yield penalty. BMC Plant Biol. 20: 313.

Zhan, J. and B.C. Meyers. 2023. Plant small rnas: Their biogenesis, regulatory roles, and functions. Annu. Rev. Plant Biol. 74: 21–51.

Zhan, X., F. Zhang, N. Li, K. Xu, X. Wang, S. Gao et al. 2024. Crispr/cas: An emerging toolbox for engineering virus resistance in plants. Plants 13: 3313.

Zhang, L.L., Y. Li, Y.P. Zheng, H. Wang, X. Yang, J.F. Chen et al. 2020. Expressing a target mimic of mir156fhl-3p enhances rice blast disease resistance without yield penalty by improving spl14 expression. Front. Genet. 11: 327.

Zhang, T., Y. Zhao, J. Ye, X. Cao, C. Xu, B. Chen et al. 2019. Establishing crispr/cas13a immune system conferring RNA virus resistance in both dicot and monocot plants. Plant Biotechnol. J. 17: 1185.

Zhang, X., Y. Bao, D. Shan, Z. Wang, X. Song, Z. Wang et al. 2018. Magnaporthe oryzae induces the expression of a microrna to suppress the immune response in rice. Plant Physiol. 177: 352–368.

Zhang, Y., X.F. Lin, L. Li, R.H. Piao, S. Wu, A. Song et al. 2024. Crispr/cas9-mediated knockout of bsr-d1 enhances the blast resistance of rice in northeast china. Plant Cell Rep. 43: 100.

Zhao, C., F. Yin, L. Chen, D. Li, S. Xiao, Q. Zhong et al. 2022. Identification of bacterial blight resistance genes in rice landraces from yunnan province, China. Australas. Plant Pathol. 1–11.

Zhou, J., Z. Peng, J. Long, D. Sosso, B. Liu, J.S. Eom et al. 2015. Gene targeting by the tal effector pthxo2 reveals cryptic resistance gene for bacterial blight of rice. Plant J. 82: 632–643.

Zhou, Y., S. Xu, N. Jiang, X. Zhao, Z. Bai, J. Liu et al. 2022. Engineering of rice varieties with enhanced resistances to both blast and bacterial blight diseases via crispr/cas9. Plant Biotechnol. J. 20: 876–885.

Zhu, Y., X. Yang, P. Luo, J. Yan, X. Cao, H. Qian et al. 2024. Creation of bacterial blight resistant rice by targeting homologous sequences of Xa13 and Xa25 genes. Agronomy 14: 800.

CHAPTER 14

Advancement in CRISPR/Cas9 Genome Editing in Field of Crop Improvement

Husnain Rauf

1. Introduction

By 2050, the world population is projected to reach more than 10 billion individuals, resulting in an unprecedented and requirement of high food demand all over the world. To address this need, food production needs to grow by 60 to 100% (Francisco Ribeiro and Camargo Rodriguez 2020). However, despite significant progress in agricultural technology, the existing global food system is struggling hard to provide sufficient food for an increasing population (Hickey et al. 2019). Challenges include climate change, limited arable land, and unstable weather conditions, which exacerbate the challenges in achieving food security.

Crop improvement is crucial for addressing these challenges by concentrating on yield improvements, nutritional quality, and stress tolerance. Genetic engineering has evolved as a method to improve agricultural production, addressing the nutritional needs of the growing population with consumer preferences and health concerns. Diverse genetic engineering approaches, including Zinc Finger Nucleases (ZFNs) and Transcription Activator-Like Effector Nucleases (TALENs), have been used to achieve these points (Akanmu et al. 2024). The CRISPR/Cas9 (Clustered Regularly Interspaced Short Palindromic Repeats) has emerged as a transformative tool due to its efficiency, accuracy, and accessibility. The CRISPR/Cas9 system utilizes guide RNAs (gRNAs) to direct the Cas9 protein to a specific DNA sequence, which leads to a DSB. CRISPR/Cas9-, Genome Editing (GE) technology that was originally discovered in bacteria as a component of their adaptive immune system, has demonstrated its effectiveness in the precise modification of genetic elements in plants (Cong et al. 2013; Mali et al. 2013). Therefore, comparing with GE technologies, such as ZFNs and TALENs, CRISPR/Cas9 systems are more easily engineered, exhibit greater efficiency, and reduced costs, which makes them more suitable for several crop enhancement applications (Ghoshal 2024).

The effectiveness and accuracy of CRISPR/Cas9 have made it especially preferred for plant breeding, with several important crops like wheat, maize, rice, and soybean produced from benefiting from this approach. These staple crops, which provide the primary caloric supply for significant portion of the world population, are increasingly endangered by pests, diseases, and climate change. Ensuring total yield, nutritional quality, and resilience of crops is more crucial for guaranteeing food

University of Georgia, Department of Horticulture, 1111 Miller Plant Sciences Building, Athens, GA 30602.
Email: hrauf@uga.edu

DOI: 10.1201/9781003518204-14

security in the coming few decades (Morrow et al. 2023). CRISPR/Cas9-Cas9 has been extensively employed to improve traits like disease resistance (Katoch et al. 2020), drought tolerance (Shi et al. 2017), and salt tolerance (Zhang et al. 2019a). In crops like wheat (Qi 2019), maize (Barman et al. 2019), and soybean (Chilcoat et al. 2017).

These developments demonstrate that the CRISPR/Cas9 technology has ability to increase food demand, quality, and ensure food safety for the entire globe. The CRISPR/Cas9 system was initially discovered as a component of bacterial immune system, aiding in the defense against viral infections. The system consists of CRISPR/Cas9 loci inside the genome and Cas proteins, with Cas9 being the most prevalent. The CRISPR/Cas9 loci are made up of tandem repeat sequences and spacers derived from foreign DNA sequences, including those found in bacteriophages (Kim and Kim 2014). Along with the Cas9 protein, CRISPR/CAS9 loci include non-coding RNAs such as CRISPR/Cas9 RNA (crRNA) and trans-activating CRISPR/Cas9 RNA (tracrRNA), which form the guide RNA (gRNA) complex. The gRNA is essential for directing Cas9 to a particular DNA sequence, where it causes a Double-Strand Breaks (DSBs) and activates cell's repair activities (Jinek et al. 2013). The adaptability of the CRISPR/Cas9 system to target DNA sequences with the use of gRNA, which may be tailored to attach to almost any intended target sequence. The capacity for precise genetic modifications has transformed both fundamental and practical plant research (Doudna and Charpentier 2014). Beyond its use in GE, CRISPR/Cas9 has been adapted to create RNA-Guided Engineered Nucleases (RGENs) that facilitate precise genetic modifications, hence expanding its utility in crop improvement (Jin 2017). The approach has shown significant efficiency by changing gene activities in plants, including applications such as gene deletion, gene activation, and gene silencing (Sander and Joung 2014).

CRISPR/Cas9 has enabled enhancement for numerous crops resilience to environmental stresses, which is essential in climate change. Several studies on CRISPR/Cas9-modified crops have concentrated on improving traits like insect resistance, drought tolerance, and salinity tolerance, all that are significantly influenced by crop yields and quality. The development of salt tolerance in rice, which is essential for production in saline soils, has been achieved using CRISPR/Cas9 (Zhang et al. 2019). CRISPR/Cas9 has shown effective outcomes in creating drought-resistant maize cultivars, presenting potential for enhanced maize yields in arid areas (Barman et al. 2019). The future of crop improvement is closely associated with the advancements in GE technologies. The accuracy and range of CRISPR/Cas9 are improving with advances in innovative techniques like Prime Editing (PE) and Base Editing (BE), which facilitate more precise and efficient genetic modifications (Swartjes et al. 2020). These technologies are expected to accelerate the development of crops that address the rising food demand and enhance agricultural sustainability among global challenges. The aim of this chapter is to investigate the applications of CRISPR/Cas9 genome editing in improving key agricultural traits such as yield, disease resistance, abiotic stress tolerance, and nutritional quality in field crops. By highlighting both the potential and limitations of CRISPR/Cas9, we aim to provide an overview of its role in transforming crop breeding, as well as the challenges and future directions for its integration into sustainable agricultural practices.

2. Mechanism of CRISPR/Cas9 GE

2.1 CRISPR/Cas9 System and Its Function

The CRISPR/Cas9 system has revolutionized biotechnology by providing a precise mechanism for genome alteration in several crops. CRISPR/Cas9 was first revealed in the adaptive immune system of bacteria, where it facilitates the guide RNA (gRNA) in directing the Cas9 nuclease to a particular DNA sequence inside the plant genome (Basu et al. 2023). While binding of the gRNA-targeted sequence, Cas9 induces DSBs in the DNA, then cell repair mechanisms address this breakdown by either introducing small insertions/deletions through the Non-Homologous End Joining (NHEJ) pathway or by precisely incorporating new DNA sequences via the Homology-Directed Repair (HDR) pathway, depending upon the availability of a repair template

(Globus and Qimron 2018). The CRISPR/Cas9 system enables precise DNA cleavage at specific locations, making it a powerful tool for gene editing. This ability allows researchers to interfere with gene function, introduce new genes, and modify existing genetic sequences. This adaptability has enabled rapid implementation for both research and practical applications of crop improvement. CRISPR/Cas9 presents a significant advantage compared to earlier genome-editing technologies such as ZFN and TALEN due to the ease of designing new single-guide RNAs (sgRNAs) for each target gene, eliminating the requirement to engineer proteins as necessary with ZFNs and TALENs (Basu et al. 2023). CRISPR/Cas9-induced NHEJ is frequently applied to create targeted gene knockouts in plants, whereas HDR, albeit with lower efficiency in plants, is used to introduce or replace genes for trait enhancement (Arora and Narula 2017). The CRISPR/Cas9 system allows targeted and efficient editing, significantly impacting agricultural biotechnology. This versatile "search-and-edit" tool has transformed plant breeding within a few years of its discovery (Basu et al. 2023).

2.2 Variants and Applications

Beyond this classic Cas9 protein from Streptococcus pyogenes, several other variants like Cas9 and other CRISPR/Cas9-associated proteins (Cas) have been developed, thus increasing the CRISPR/Cas9 GE toolkit. Cas12a (initially referred to as Cpf1) presents specific advantages for applications. Cas12a identifies distinct Protospacer Adjacent Motif (PAM), generally characterized by T-rich sequence, and produces staggered DNA cuts with some sticky ends, unlike blunted cuts produced by Cas9 (Basu et al. 2023). The capacity to generate sticky ends allows precise insertions or deletions within specific genomes. Cas12a demonstrates high editing efficiency in crops such as rice, facilitating targeted modifications in genes that are challenging to edit with Cas9 (Zhang et al. 2021). Cas13 is a significant variant of Cas that specifically targets RNA instead of DNA. Cas13 has a characteristic of both binding and cleaving to single-stranded RNA, directed by a CRISPR/Cas9 RNA. It has demonstrated utility in applications like transient silencing of viral RNA in plants, particularly in initiatives aimed at addressing RNA viruses in crops like potato (Kavuri et al. 2022). In contrast to DNA-targeting Cas proteins, Cas13 serves as a distinctive mechanism for the post-transcriptional regulation of gene expression, allowing for modifications without inducing permanent alterations to the genome. Furthermore, customized Cas9 variants have been established for improving specificity or add new capabilities. Mutants of "Dead" Cas9 (dCas9), which lack function of nuclease, could be combined with other proteins to control gene expression by CRISPR/Cas9 interference/activation. This enables reversible and flexible control of gene activity without affecting the genetic sequence. Smaller Cas nucleases, notably Cas12f, is being studied for their ability to enable transport into plant cells, addressing a key barrier in plant GE because of complexity of the cell wall in plants (Rasheed et al. 2021). Temperature-tolerant Cas variations, like as LbCas12a, are being extensively developed for application in plants grown in hot climes. These variations increase GE efficiency in crops like cotton and maize, where high temperatures typically reduce the activity of conventional Cas9 proteins (Li et al. 2020). The inclusion of these variations in the CRISPR/Cas9 toolset enables more accurate and successful GE across a several plant species. By expanding the CRISPR/Cas9 toolkit (Cas9, Cas12a, Cas13, and others), scientists may choose the best system for each crop and target, whether objectives are to cut DNA or RNA, create diverse mutation patterns, or control genes without causing irreversible changes. Table 1 outlines numerous CRISPR/Cas9 variations and their uses in plant genome editing, emphasizing their distinct properties and essential genes targeted across diverse crops.

Table 1. CRISPR/CAS9/Cas System variants and their applications.

Cas Protein Variant	Targeted DNA/RNA	Unique Features	Application in Crop Editing
Cas9	DNA (Double-Strand Breaks)	Most common and widely used, needs a PAM sequence	Gene knockouts, insertions, and precise mutations (Basu et al. 2023)
Cas12a (Cpf1)	DNA (Double-Strand Breaks)	T-rich PAM, staggered cuts, more efficient for editing certain genes	Targeting difficult-to-edit genes like *OsRR22* in rice (Zhang et al. 2021)
Cas13	RNA (Single-Stranded)	Targets RNA instead of DNA, no permanent genome edits	RNA silencing, transient gene regulation (e.g., viral RNA in potato) (Kavuri et al. 2022)
dCas9 (Dead Cas9)	DNA (No Double-Strand Breaks)	No nuclease activity, can be fused with other proteins	Gene activation or repression (Rasheed et al. 2021)

2.3 Precision and Efficiency in Editing

The primary challenge is to enhance precision while minimizing off-target effects. In the initial generations of the Cas9 protein, off-target activity frequently emerged, as slight variations between the sgRNA and target DNA could promote unintended cleavage in similar sequences. High-fidelity Cas9 variants, such as eSpCas9 and SpCas9-HF1, have been engineered with specific amino acid substitutions to reduce non-specific DNA binding and off-target activity, addressing this significant issue (Guo et al. 2019). CRISPR/Cas9 serves as a more reliable apparatus for precision genetic engineering, owing to the markedly enhanced specificity offered by these variants. The utilization of joined Cas9 nickases represents a sophisticated strategy aimed at improving precision. Two sgRNAs are utilized to guide the Cas9 nickase in creating nicks at distinct adjacent locations, with Cas9 being modified to cleave only one strand of DNA in this approach. These kinds of approaches markedly minimize risks associated with off-target effects, given that DSBs require the occurrence of two independent incidents (Kulcsár et al. 2020). Paired nickases provide enhanced specificity, especially when exact mutations are needed, as they require simultaneous binding at two neighboring sites. GE technologies like PE and BE have been advanced to enhance the accuracy of genetic modifications, especially concerning the Cas9 nuclease. A DNA modifying enzyme is combined with a catalytically inactive Cas9 or Cas9 nickase in base editors, which enable targeted conversion of exact base pairs without DSBs induction. For example, Cytosine Base Editors (CBEs) utilize a cytidine deaminase to convert C $\Rightarrow$ T at a target site, whereas Adenine Base Editors (ABEs) convert A $\Rightarrow$ G. These editors help in the precise substitution of individual nucleotides, which is particularly beneficial for trait enhancement in the development of allele variants, such as herbicide resistance, or the correction of point mutations (Basu et al. 2023).

PE develops this concept through integrating the reverse transcriptase enzyme with a Cas9 nickase. The prime editor can introduce precise nucleotide alterations, such as base substitutions, insertions, and deletions, with high accuracy when directed by a specialized PE gRNA (pegRNA). This method has the potential to be an invaluable instrument for genetic modifications in both fundamental research and agricultural breeding, as it can potentially enable the correction of any base substitution without requiring donor DNA (Butt et al. 2020; Li et al. 2023). Thus, PE is still in the process of being optimized, it has already demonstrated success in crops (maize, rice, and wheat). These crops have been primarily used for developing herbicide-resistant varieties by introducing specific mutations in key genes (Butt et al. 2020). Although these specialized tools, including base and prime editors, prevent the formation of DSBs, they are associated with trade-offs in terms of editing efficacy. Nevertheless, their capacity to implement highly precise alterations with minimal off-target effects makes them essential for the future of plant breeding and engineering (Vu et al. 2024). The efficacy and targeting range of these technologies in crops are expected to increase as research continues to refine them, enabling even greater control over the genetic modification of plants.

3. Applications of CRISPR/Cas9 in Field Crop Improvement

3.1 Yield Enhancement

Enhancing crop yield potential is crucial for complementary food security, especially as the population is expected to reach 10 billion by 2050 in the entire world. CRISPR/Cas9 technology is considered a real method for improving crop quantity and quality through modification of genes that regulate key factors in plant growth, including plant architecture, grain size, and seed quantity. A primary strategy for improving yield using CRISPR/Cas9 involves targeting and disrupting genes that function as negative regulators of yield. These genes frequently inhibit processes such as grain development or biomass accumulation, thereby constraining overall productivity. Eliminating yield-repressing genes enables plants to generate increased seed production or biomass, resulting in enhanced yields.

The *Gn1a* gene encodes the *OsCKX2* enzyme in rice, which is responsible for the degradation of cytokinin, a hormone essential for plant growth and grain development. Loss-of-function mutations in Gn1a lead to elevated cytokinin levels in inflorescences, resulting in an increased number of grains/panicles. Utilizing CRISPR/Cas9, researchers have been used to induce mutations in the *Gn1a* gene of elite rice varieties, resulting in larger panicles and a substantial increase in grain numbers (Li et al. 2016; Lu et al. 2022). This CRISPR/Cas9-based modification reflects the natural mechanism that enhances grain production in rice and illustrates how targeting negative regulators of yield can expedite crop development. CRISPR/CAS9 has been employed to transform additional rice yield-related genes, including both *GS3* and *DEP1*, which affect grain size and panicle, respectively. Editing these genes has enabled researchers to enhance both grain weight and number, thus improving the yield potential of rice (Thiruppathi et al. 2024). Wheat, a polyploid species, presents more difficulties for genetic alteration because of its many genomes (homeologs). It is often essential to modify all alleles of a gene that effect significant changes in characteristics. Researchers have successfully addressed this difficulty using CRISPR/Cas9. The *TaGW2* gene has a negative regulatory effect on grain size in wheat. Researchers have already increased grain weight by eliminating all three homeologs of TaGW2. In a separate study, researchers concurrently inactivated four genes, *TaDEP1*, *TaNAC2*, *TaGW2*, and *TaGASR7*, that govern different yield metrics. This deletion resulted in elongated spikes, enlarged grains, and a significant increase in total yield per plant, illustrating CRISPR/CAS9's capacity to augment wheat production (Basu et al. 2023). CRISPR/Cas9 may also be used to enhance gene expression that functions as a positive regulator of yield. An exemplary example is the gene *ARGOS8* in maize, which contributes to drought resistance and grain production in water-scarce environments. Researchers enhanced the expression of the *ARGOS8* gene by adding a robust promoter upstream of it using CRISPR/Cas9 technology. This led to maize plants exhibiting enhanced grain yields under drought conditions, underscoring the potential of CRISPR/Cas9 to augment gene expression and enhance complex characteristics such as yield in response to environmental stress (Shi et al. 2017). The modified maize plants had no exogenous DNA, demonstrating CRISPR/Cas9 capability for precise genetic alterations without the incorporation of foreign genes. This approach eliminates the requirement for extensive backcrossing, hence expediting the breeding process for enhanced crop characteristics.

3.2 Disease Resistance

Field crops are facing continuous threats from different pathogens, including fungi, bacteria, viruses, and pests, which can significantly reduce crop yields. Addressing these challenges and improving food security necessitates the development of crops that exhibit robust disease resistance.

Conventional breeding techniques are successful, typically necessitating multiple generations of crossing and selection to help create those varieties which are resistant to disease. GE technologies like CRISPR/Cas9 introduced a more rapid and accurate method for creating disease-resistant crops through direct targeting and modification genome. Enhancing disease resistance with CRISPR/Cas9 only focuses on the targeting and alteration of plant susceptibility genes, which are utilized to invade and induce disease. By disabling these susceptibility genes, plants can achieve resistance to the pathogen independently of incorporating foreign resistance genes. The powdery mildew fungus (*Blumeria*) is responsible for considerable diseases that affect both wheat and barley. It was found that loss-of-function mutations in the *TaMLO* gene, essential for successful fungus infections, provide lasting resistance against powdery mildew. By the application of CRISPR/Cas9 technology, breeders have effectively eliminated the *TaMLO* gene in wheat, leading to the development of mildew-resistant plants (Tripathi et al. 2020; Wang et al. 2014). CRISPR/Cas9 induced mutations across three wheat *TaMLO* homologs developed significant resistance against powdery mildew, with an editing efficiency of 28% observed in the first generation (Shan et al. 2014). This method emulates the inherent resistance system found in top-performing wheat varieties, greatly shortening the duration needed for developing disease-resistant crops.

A *TaEDR1* gene, which is considered a transcriptional repressor in wheat, has also been deleted using CRISPR/Cas9 to improve resistance to powdery mildew. Furthermore, removing this gene increased resistance to fungus (Char and Yang 2020). To produce broad-spectrum, persistent disease resistance without introducing foreign genes, these examples demonstrated the capability of CRISPR/Cas9 for specific target plant susceptibility genes. The application of CRISPR/Cas9 technology to treat bacterial illnesses has also shown promise. A significant hazard to rice productivity, for instance, in rice is bacterial blight, which is brought on by *Xanthomonas oryzae pv. oryzae* (*Xoo*). The pathogen activates rice susceptibility genes, including those in the *OsSWEET* family, using TALEs. To successfully stop pathogens from taking over the genes, researchers used CRISPR/Cas9 to adapt promoter regions of genes like *OsSWEET13* and *OsSWEET14* by removing TALE effector binding sites. They targeted four Effector Binding Elements (EBEs) in the *OsSWEET14* promoter region of superior basmati rice using CRISPR/Cas9. The modified lines that were produced demonstrated up to 9% editing efficiency and resistance to bacterial blight (Zafar et al. 2020). This research determines how CRISPR/Cas9 may be used for the protection of local rice varieties against common *Xoo* strains by fostering bacterial disease resistance in them. For example, eukaryotic translation initiation factors (*eIFs*), such as *eIF4E*, are encoded by several recessive resistance genes in plants that are implicated in viral infections. As susceptibility genes to different potyviruses, these variables are essential in viral infection. Tomato (*Solanum lycopersicum*) is resistant to potyviruses, an *eIF4E1* gene was targeted using CRISPR/Cas9. These results established the increased efficiency of CRISPR/Cas9 and induced mutations for viral resistance, as multiple lines of edited tomato showed resistance to pepper mottle virus (PepMoV) while no accumulation of Tobacco Etch Virus (TEV) was seen (Yoon et al. 2020).

Tomatoes are the most widely consumed vegetable in the world, which may be infected by a number of diseases, such as fungi and viruses. Two problematic diseases that significantly reduce production are powdery mildew and Tomato Yellow Leaf Curl Virus (TYLCV). TYLCV susceptibility is transmitted by the Ty-5 gene (*SlPelo*), while powdery mildew susceptibility is linked to the *SlMlo1* gene. CRISPR/CAS9 was applied to target these susceptibility genes in the superior BN-86 tomato cultivar to increase tomato resistance. Both *SlPelo* and *SlMlo1* mutant lines were successfully produced, yielding lines with high resistance to powdery mildew and TYLCV. Pramanik et al. (2021) investigated that *SlMlo1* knockout lines demonstrated resistance to powdery mildew, although *SlPelo* knockout lines demonstrated limited viral dissemination and decreased TYLCV accumulation. CRISPR/Cas9 introduces targeted mutations that give resistance to numerous diseases, as shown in this study, providing an effective approach to increase tomato crop resilience.

Despite the potential of CRISPR/Cas9 GE for developing disease-resistant crops, various challenges remain, particularly regarding viral resistance. The use of CRISPR/Cas9 interference

Table 2. CRISPR/Cas9 applications for disease resistance.

Crop	Disease	Genes Edited	Impact on Crop	Reference
Rice	Bacterial blight	*OsSWEET14*	Resistance to Xanthomonas oryzae (Xoo)	(Zafar et al. 2020)
Wheat	Powdery mildew	*TaMLO, TaEDR1*	Resistance to Blumeria graminis (powdery mildew)	(Shan et al. 2014; Tripathi et al. 2020)
Tomato	Viral diseases	*eIF4E1*	Resistance to potyviruses (PepMoV, TEV)	(Yoon et al. 2020)
Cassava	African cassava mosaic virus	*eIF4E*	Resistance to African cassava mosaic virus (ACMV)	(Mehta et al. 2019)

facilitated the development of resistance to African cassava mosaic virus in studies focused on cassava. However, the findings were not as good as expected, with some edited viral genomes developing single-nucleotide changes that enabled the virus to escape CRISPR/Cas9 cleavage (Mehta et al. 2019). The mutagenic nature of the CRISPR/Cas9 system could result in the emergence of virus strains that are resistant to targeted editing. Recent studies and technological progress, including high-fidelity Cas9 variants and enhanced guide RNAs, and the effectiveness of CRISPR/Cas9 in crop breeding. As these technologies advance, CRISPR/Cas9-based disease resistance strategies will emerge as a crucial asset in combating plant pathogens, presenting new possibilities for sustainable crop production amid global agricultural challenges. Table 2 presents a summary of CRISPR/Cas9 applications for disease resistance across various crops, identifying the target genes and the specific pathogens they protect against.

3.3 Abiotic Stress Tolerance

Crop production significantly affects food security globally, which is at risk due to abiotic stress, including cold, salinity, drought, and heat. It is crucial to recover tolerance of cereals to these stresses to sustain and increase yields, especially climate change. Traditional breeding methods for enhancing abiotic stress tolerance can be labor-intensive and slow. However, CRISPR/Cas9 GE technology provides a more precise and accelerated method for creating crops that help flourishing in challenging environmental surroundings. Researchers have effectively improved salinity tolerance by altering the gene *OsRR22* with CRISPR/Cas9 in rice. Different mutations in the gene *OsRR22* have been demonstrated to enhance tolerance to salinity, and it is implicated in regulating the salinity response. This gene was directed through researchers exploiting CRISPR/Cas9, identification development of nine mutant plants from 14 transgenic plants. These kinds of mutations were effectively transmitted. CRISPR/Cas9 improved rice salinity tolerance, already described by enhancing salinity tolerance of T2 homozygous mutation lines at the seedling stage, which did not show any significant difference in agronomic attributes compared with wild-type plants (Zhang et al. 2019a). Abiotic stress (drought, salinity, heat, and cold) are significant limiting factors to produce wheat, a staple product that provides substantial nutritional value for the world population. The hexaploid nature of wheat presents a challenge for genetic modification, as it requires the editing of multiple copies of genes to accomplish a discernible change. Thus, CRISPR/Cas9 has demonstrated its efficacy in the development of wheat that is resistant to herbicides and desertification. Recent research has demonstrated CRISPR/Cas9 potential to target genes that are largely involved in abiotic stress tolerance, thereby facilitating the improvement of wheat varieties that are more resilient to environmental stress. The traditional challenges are related with breeding, which can only be reduced with precision and adaptability of CRISPR/Cas9, and also contributes to problems associated with food security concerns worldwide (Afroz et al. 2023). The Abscisic Acid (ABA) signaling systems, control rice drought resistance. Numerous ABA receptors and regulatory factors, including the *OsPYL/RCAR* family and *OsABAR1*, show a role in regulating plant response to drought stress. CRISPR/Cas9 was applied to modify many genes, including *OsPYL9*, which improves drought stress by controlling ABA buildup and production of stress-related genes. Mutants changed with *OsPYL9* demonstrated drought tolerance and yield, which indicated that CRISPR/Cas9 improved drought resistance without sacrificing other agronomic factors in rice (Ma et al. 2024). Heat tolerance is also important for improving crop development, particularly in locations where temperatures are increasing because of climate change. For improving thermo-tolerance in tomatoes, CRISPR/Cas9 was applied and target the *SlMAPK3* gene, and associated with heat stress. Mutants with changes in this gene have better heat tolerance than wild-type plants, indicating that *SlMAPK3* is considered for negative controller of thermo-tolerance (Yu et al. 2019). In maize crops, CRISPR/Cas9 changes to the thermosensitive *TMS5* gene that are followed in enhancing thermosensitive male sterile plants, important for increasing yield under high temperatures (Zhou et al. 2016). Another key crop, lettuce, improved its heat tolerance with CRISPR/Cas9 (Bertier et al. 2018). Finally, CRISPR/Cas9 was applied and targeted to key genes that are involved in plant growth and stress response. In rice, researchers have used CRISPR/Cas9 to target the *OsTOR* and *OsSnRK1* genes, both of which are essential for the growth and response to environmental-related stress. While modifying these specific key genes showed difficulties due to their critical roles in plant life, CRISPR/Cas9 could be applied to induce knock-out mutations, allowing for additional investigation on their activities in rice (Pathak et al. 2022). The application of CRISPR/Cas9, which increased abiotic stress tolerance in crops such as wheat, rice, tomato, lettuce, and maize is considered a major advancement in agricultural biotechnology. CRISPR/CAS9 increases the growth of climate-resilient crops through enhancing tolerance to heat, salt, drought, and other stressors.

3.4 Herbicide Tolerance

Herbicide tolerance in crops is the focus of agricultural biotechnology, caused by the growing problem of herbicide-resistant weeds that substantially reduce crop yields. The use of herbicides in rice cultivation has proven successful for weed management. However, prolonged application has led in the emergence of herbicide resistance in weeds. In GE, CRISPR/Cas9 is used to create herbicide-tolerant crops by targeting and altering genes linked to herbicide resistance. The acetolactate synthase (*ALS*) gene is the primary target for herbicide tolerance, particularly for herbicides such as bispyribac sodium in rice. CRISPR/Cas9 employed for the insertion of point mutations in the *ALS* gene, like as P178S variant, resulting in herbicide-tolerant rice. Researchers have utilized homologous recombination to produce rice lines showing resistance to, and demonstrating CRISPR/Cas9 potential for enhancing herbicide tolerance (Luo and Liu 2025). In wheat, researchers have successfully created herbicide-tolerant wheat varieties by targeting the *ALS* gene and acetyl-CoA carboxylase (*ACCase*) gene by using BE. This method has shown effectiveness in transferring resistance to many kinds of herbicides, such as sulfonylureas, imidazolinones, and aryloxyphenoxy propionates. Editing the *TaALS-P174* codon by using BE created wheat plants with sufficient resistance to the nicosulfuron herbicide in wheat. This combination of various herbicide resistance features inside a single plant, only accomplished with the help of BE in CRISPR/Cas9, which enhanced control of weeds in wheat (Zhang et al. 2019b).

Herbicide resistance is obtained through mechanisms like target site mutations, which provide resistance to herbicides. The *ALS* gene is the target for herbicides that inhibit ALS, which are extensively utilized in the agricultural sector. Mutations in the *ALS* gene, such as amino acid residues, have been demonstrated to confer resistance to multiple *ALS*-inhibiting herbicides. Researchers have used initiate point-mutations in *ALS* gene with CRISPR/Cas9, resulting in the creation of herbicide-tolerant plants. This method has been applied to numerous crops such as maize, rice, and soybean, presenting successful accomplishment with CRISPR/Cas9 for the development of herbicide-resistant varieties (Dong et al. 2021). A few *ALS* inhibitors, herbicide tolerance in crops have been accomplished by editing genes. HPPD inhibitors, including mesotrione (MST), affect photosynthetic systems, leading to albinism and later, the death of entire plants. These herbicides demonstrate significant efficacy in controlling weeds within rice production systems. Researchers have been using CRISPR/Cas9 and CRISPR/Cas12a for editing 3′UTR of the gene *OsHPPD*, resulting in an improvement of herbicide-tolerant rice plants. Mutations in the 3′ UTR of *OsHPPD* resulted in augmented resistance to HPPD inhibitors for herbicides and offered a novel strategy for enhancing herbicide tolerance. This research emphasized the capability of modifying non-coding regions, including 3′ UTR, to provide herbicide resistance while preserving gene capacities (Wu et al. 2023). Improvement of herbicide-resistant crops through CRISPR/Cas9 is considered to address the increasing problem related with herbicide-resistant weeds. Both *ALS* and HPPD are considered as targeted genes which enable researchers to develop herbicide-resistant crops, thus enhancing weed management and increasing crop yields. CRISPR/Cas9-based herbicide tolerance strategies have the potential for enhancing crop performance and managing weeds.

4. Case Studies in Major Field Crops

4.1 Rice

Rice is considered a staple food and is consumed by about half the population in the world. Climate change has profoundly affected rice production, making it more vulnerable to different abiotic and biotic challenges, including drought, heat, salt, and bacterial infections. To address these challenges, CRISPR/Cas9 GE has emerged as an appropriate approach for creating climate-resilient rice cultivars. In the last 10 years, researchers have successfully utilized CRISPR/Cas9 to implement precise alterations in many rice genes that improve stress tolerance, disease resistance, and nutritional quality. These developments have created new opportunities for developing rice cultivars resilient to detrimental climate changes. Salinity tolerance could be enhanced with the *OsRR22* gene with CRISPR/Cas9 in rice. Lines that are mutants with targeted mutations in this specific gene exhibited improved salt tolerance while preserving agronomic characters, hence sustaining yield potential in the saline environment (Zhang et al. 2019a). A gene called *OsSWEET14*, implicated in bacterial disease caused by *Xanthomonas oryzae*, has been improved for resistance in rice. CRISPR/Cas9 enhances disease resistance and a significant reduction in bacterial blight symptoms that resulted from targeted mutations in the gene promoter region *OsSWEET14* in rice (Zafar et al. 2020). Moreover, CRISPR/Cas9 was applied and target genes like *OsVDE* for salt tolerance and *OsBadh2* for drought tolerance, illustrating the extensive application in enhancing rice resistance (Kumar et al. 2023).

In rice, CRISPR/Cas9 improved herbicide tolerance. *ALS* gene, which is essential for branched-chain amino acids production and is considered a target of numerous herbicides. Introducing specific changes in *ALS* gene allows rice plants which show resistance to herbicides such as bispyribac sodium and imidazolinones. A substitution of tryptophan with leucine at position 548 in the *ALS* gene produced herbicide-resistant basmati rice, providing a novel approach for weed control (Zafar et al. 2020). Additionally, in rice, auxin receptor from auxin, a gene *OsAFB4,* provided tolerance to herbicides such as picloram and 2,4-D, further as a representative of CRISPR/Cas9 in increasing resistance to herbicides (Guo et al. 2021).

CRISPR/Cas9 has enabled a meaningful amount of yield traits in rice. Researchers have successfully generated rice lines with improved grain size, weight, and number by editing multiple Quantitative Trait Loci (QTLs), including *OsGn1a, OsGS3,* and *OsGW2*. These modifications led to a remarkable enhancement in yield per panicle, with certain lines exhibiting yields almost 68% higher when compared with wild-type rice (Thiruppathi et al. 2024). It emphasizes the significance of CRISPR/Cas9 in enhancing yield, climate-resilient rice varieties. In rice, there are several efforts to enhance herbicide tolerance by editing the 3′ UTR of the *OsHPPD* gene that have yielded promising results. Researchers have conferred resistance to herbicides such as mesotrione (MST), which inhibit plant photosynthesis and induce albinism, by introducing mutations in this non-coding region. This strategy presents a novel method for improving herbicide resistance in rice, addressing the increasing problematic weeds for herbicide resistance in CRISPR/Cas9 transformed crop improvement by precise genetic modifications, which are associated with stress, disease, herbicide tolerance, and yield improvement in rice production. Despite obstacles in gene delivery and efficacy of specific editing methods, recent advancements demonstrate potential usage of CRISPR/Cas9 to address the urgent issues associated with climate change.

4.2 Wheat

Wheat is the major worldwide crop, contributing over 20% of the world's caloric consumption and 25% of protein needs. Enhancing wheat for increased yields and improved tolerance to diverse pressures is essential for global food security. CRISPR/Cas9 is used for promoting wheat breeding and overcoming the constraints of conventional breeding strategies. CRISPR/Cas9 applications have effectively developed cultivars of wheat having great resistance to diseases like powdery mildew and tolerance to abiotic stress such as drought, salt, and heat, and improved herbicide resistance. *TaMLO* gene has been employed with CRISPR/Cas9, which is essential for mildew susceptibility and enhanced disease resistance in wheat. This specific knockout technique yielded mildew-resistant wheat varieties having a mutation frequency around 28.5%, showing the benefits of CRISPR/Cas9 to create disease-resistant crops (Shan et al. 2014; Tripathi et al. 2020). In addition to *TaMLO* gene, other genes such as *TaEDR1* have been identified for enhancing resistance against powdery mildew. The CRISPR/Cas9 disrupted *TaEDR1* has shown an increase in resistance against powdery mildew in wheat cultivars, emphasizing the importance of CRISPR/CAS9 in controlling disease in the plants (Char and Yang 2020). Genes like *TaLox2* and *Tapx1* have been modified to improve disease resistance, demonstrating mutation rates of around 9 and 45%, respectively, showing the benefits of CRISPR/Cas9 in allowing targeted modification of genes (Shan et al. 2014).

Apart from disease resistance, it has been widely used to increase wheat tolerance against abiotic conditions (drought, heat, and salt). Utilizing CRISPR/Cas9, genes like *TaDEP1*, *TaNAC2*, *TaGASR9,* and *TaGW2* were transformed to improve wheat stress tolerance by increasing grain length, width, and total weight of grains. Furthermore, in rice, *OsRR22* gene was targeted to enhance salt tolerance, and similar strategies are being investigated for wheat to address challenges of soil salinity in various wheat-producing locations (Zhang et al. 2019a). Moreover, CRISPR/Cas9 supports knockout genes, including *TaEDR1* has enhanced disease resistance and stress tolerance, providing a dual advantage in wheat improvement (Zhang et al. 2017). CRISPR/Cas9 has been employed to modify genes for herbicide resistance. For example, in wheat, the BE *ALS* gene confers resistance to herbicides such as nicosulfuron, which block important metabolic processes (Zhang et al. 2019a). Wheat plants were made herbicide-tolerant by producing precise point mutations in the *TaALS* genes, indicating a feasible approach for weed management. Basically, this approach is used with the BE system, which usually provides a precise method to introduce herbicide-resistant characteristics without the use of transgenic elements, addressing environmental concerns and regulatory barriers associated with GMOs (Dong et al. 2021). CRISPR/Cas9 in wheat breeding has allowed quicker and more precise genetic modifications, enabling breeders to address complex characteristics like stress tolerance, herbicide and disease resistance more effectively. However, challenges remain in the regulatory aspects and the possibility of off-target consequences of gene editing. High-fidelity Cas9 variants and better delivery techniques are likely to improve the accuracy and efficiency of CRISPR/Cas9-based GE in wheat.

4.3 Maize

Maize is also an important staple crop, and plays a critical part in food security globally. GE technologies, including CRISPR/Cas9 used to improve maize characteristics, including stress tolerance, crop improvements, and disease resistance in response to the increasing needs for food and to enhance resilience against adverse environmental conditions. The accuracy of CRISPR/Cas9 has facilitated rapid advancement in maize breeding, providing an effective method for developing novel genetic variants to tackle problems with maize production, including drought stress, reduced yield, and disease susceptibility in maize. CRISPR/Cas9 was applied to the key genes associated with maize development and growth. The *ZmSMC3* gene, which is essential to sister chromatid cohesion and meiotic centromere pairing, was targeted, resulting in the emergence of slow-growing and dwarfed plants after knockout in maize (Zhang et al. 2020). Likewise, *ZmNRPC2*, which plays a role in RNA polymerase III function and kernel growth, has been altered, leading to substantially reduced kernel size in edited plants (Zhao et al. 2020). *ZmThx20* is one example of a gene that influences kernel size and accumulation of storage proteins. The increase in the growth of small kernels has been seen because of modifications made to the *ZmThx20* gene (Li et al. 2021). Because of targeted gene knockouts, researchers have developed molecular mechanisms that regulate crop growth and kernel formation, which has facilitated improvement in maize development. CRISPR/Cas9 has been effective in altering genes linked to yield-associated traits such as ear diameter, ear length, number of kernels, and various other agronomic characteristics. Editing the *ZmCEP1* gene, which manages transport of nitrate and sugar to the kernel, resulted in improvement in height of the plant, kernel size, ear length, and weight of the kernel (Xu et al. 2021). Different mutations in *ZmACO2*, a gene that influences ear length and kernel number, have been demonstrated to significantly increase ear weight and total yield of grains (Wang et al. 2022). GE of gene promoters such as *ZmCLE7* and *ZmFCP1* has resulted in enhanced size of meristem and total grain yield and illustrating the benefit of CRISPR/Cas9 to improve yield in maize via targeted genetic modifications (Liu et al. 2021).

CRISPR/Cas9 is employed in maize varieties that exhibit higher yields in drought conditions. Researchers have enhanced maize lines for improved performance under field drought stress by inserting a GOS2 promoter into the 5′UTR of the gene *ARGOS8*, a gene associated with drought tolerance (Shi et al. 2017). This application highlights the effectiveness of CRISPR/Cas9 in mitigating abiotic stresses, which are becoming more significant due to climate change. Maize breeders have employed CRISPR/Cas9 to target genes affecting maize's response to herbicides and environmental stresses. CRISPR/Cas9 modifications in the *ZmLIG1, ZmM26,* and *Zm45* genes have been utilized to improve herbicide resistance, an essential characteristic considering the increasing prevalence of weeds that are herbicide resistant (Ahmar et al. 2024). The identification and modifications of the *stiff1* gene have been employed to enhance maize resistance to stalk lodging, a significant issue in high-density planting scenarios. The introduction of a 2 bp omission in gene *stiff1* resulted in the production of plants with enhanced stalk strength, thereby decreasing lodging and improving yield stability (Zhang et al. 2019c). These progressions highlight the significance of CRISPR/Cas9 in maize breeding through precise modification of essential genes associated with growth, stress resilience, and yield improvement.

4.4 Cotton

Cotton serves as a significant crop for the textile industry and an extremely rich source of protein and oil. Cotton can withstand high temperatures, positioning it as an ideal candidate for application of heat-inducible genetic engineering. The AaCas12b enzyme was utilized in cotton genetic engineering, taking advantage of its ability to function efficiently at elevated temperatures. In this study (Wang et al. 2020), investigated that exposing cotton explants to 45°C for four days during callus formation resulted in the greatest GE efficiency around 17.4% with slight negative effects on cell culture of cotton. This research shows that AaCas12b may cause deletions of 1–16 base pairs, but with most changes ranging from 9 to 14 base pairs. Interestingly, mutations in *GhCLA,* a gene in the sub-genome, occurred more effectively than at the sub-genome, indicating that chromatin structure differences may impact GE efficiency in cotton. CRISPR/Cas9 was utilized for both quality and quantity of the cotton fibers by targeting several key genes associated with fiber production. The *GbPDF1* gene, expressed in cells of the fiber at the start of fiber maturity, shows a fundamental role in the regulation of hydrogen peroxide production and ethylene synthesis. Employing CRISPR/Cas9 to inhibit the expression could improve the initiation of cotton fiber related to this gene (Deng et al. 2012). Similarly, *GhGAI3a* gene, responsible for converting into DELLA proteins, regulates fiber growth by suppressing gibberellin (GA) signaling at the start of fiber initiation (Saleem et al. 2024). *GhFLA1* is a significant target gene that shows a decisive role and regulates elongation, fiber initiation, and composition of cell wall (Huang et al. 2013). The elongation and initiation of cotton fibers may be enhanced by employing CRISPR/Cas9 to knockout specific genes, thereby improving the overall quality and yield of the fiber. Furthermore, the signaling pathway involving Jasmonic Acid (JA) has been specifically regulated by the *GhJAZ2* gene, which encodes a protein that acts as a repressor during the initiation of fiber development. The elongation and initiation of cotton fibers may be improved by altering this pathway through the completed application of CRISPR/Cas9 (Hu et al. 2016). Similarly, the *GhVIN1* gene is essential for the initiation and development of fibers, as it is implicated in MYB transcription factors and auxin signaling components (Wang et al. 2014). Another Cas variant related to temperature sensitivity, the *LbCas12a* gene, which are used in hot climates. These variants will improve GE efficiency in crops, including cotton and maize, where raised temperatures reduced the standard Cas9 proteins (Li et al. 2020). By modifying the cotton genome, disruption of this gene may result from increased formation of fiber. The developments in GE are an important advance in improving the quality and amount of fiber that cotton produces, as well as addressing difficulties connected to the toxicity of cotton seeds and the use of cotton seeds for human consumption. Table 3 provides a consolidated overview of

Table 3. CRISPR/Cas9 applications of targeted genes in rice, wheat, maize, and cotton improvement.

Crop	CRISPR/Cas9 Application	Key Gene Targets/ Approach	Impact/Improvement	Reference
Rice	Salinity Tolerance	Editing of *OsRR22*	Improved salt tolerance while preserving agronomic traits	(Zhang et al. 2019a)
	Disease Resistance	Editing of *OsSWEET14*	Resistance to bacterial blight (*Xanthomonas oryzae*)	(Zafar et al. 2020)
	Herbicide Resistance	Editing of the *ALS* gene	Herbicide tolerance (bispyribac sodium, imidazolinones)	(Zafar et al. 2020)
	Yield Enhancement	Editing of *OsGS3, OsGW2, OsGn1a*	Enhanced grain size, weight, and yield per panicle	(Thiruppathi et al. 2024)
Wheat	Disease Resistance	Editing of *TaMLO*	Resistance to powdery mildew (Blumeria graminis)	(Shan et al. 2014; Thiruppathi et al. 2024)
	Abiotic Stress Tolerance	Editing of *TaDEP1, TaNAC2, TaGW2, TaGASR7*	Enhanced drought, heat, and salt tolerance	(Zhang et al. 2019a)
	Herbicide Resistance	Editing of *TaALS*	Herbicide tolerance (nicosulfuron, sulfonylureas)	(Dong et al. 2021)
Maize	Yield Enhancement	Editing of *ZmCEP1,* and *ZmACO2*	Enhanced ear length, kernel size, and total yield	(Wang et al. 2022; Xu et al. 2021)
	Abiotic Stress Tolerance	Enhancing *ARGOS8* gene expression	Enhanced drought tolerance and yield stability	(Shi et al. 2017)
	Herbicide Resistance	Editing of *ZmLIG1, ZmM26*	Herbicide tolerance (glyphosate, atrazine)	(Ahmar et al. 2024)
Cotton	Yield Enhancement	Editing of *GbPDF1*	Improved fiber quality and quantity	(Deng et al. 2012)
	Abiotic Stress Tolerance	Editing of *LbCas12a*	Enhanced tolerance under high-temperature conditions	(Li et al. 2020).
	Yield Enhancement	Editing of *GhGAI3a, GhFLA1*	Fiber elongation and initiation	(Huang et al. 2013)

CRISPR/Cas9 application in the improvement of rice, wheat, maize, and cotton, highlighting the targeted genes and their associated benefits for crop impact/improvement.

5. Advances in Tools and Techniques for CRISPR/CAS9/Cas in Crop Improvement

Effective application of CRISPR/Cas9 in crop improvement depends on both the molecular capabilities of the system and the developments in delivery methods and precise procedures. A primary issue in CRISPR/Cas9 about delivery, including CRISPR/Cas9 components into plant cells and the subsequent regeneration of whole plants. Different methods are used with their distinct benefits and limitations. Agrobacterium-mediated transformation is the predominant technique for crops such as rice and wheat that are vulnerable to Agrobacterium tumefaciens. The CRISPR/Cas9 gene and sgRNA expression cassettes are included in the T-DNA plasmid region, which is transported into plant cells through Agrobacterium, where it participates in the genome of plants. For crops such as maize, particle bombardment (biolistic) is more appropriate. Alternatively, ribonucleoprotein (RNP) delivery methods may directly introduce pre-assembled Cas9 proteins and sgRNAs into plant cells without requiring incorporation into the plant genome, making them appropriate for rapid and transient modifications (Basu et al. 2023). A significant development in CRISPR/Cas9 technology is the ability for multiplex GE. It involves multiple simultaneous targeting of many genes, significantly enhancing crop efficiency. In rice, recent studies revealed that up to 16 sites may be at the same time using Cas12a, exhibiting greater effectiveness than Cas9 in targeting AT-rich genomic regions. Utilizing multiplexed systems, researchers may simultaneously edit several genes, therefore enhancing the efficacy of breeding programs by directly targeting many traits, such as yield improvement and resilience to stress, in a single editing operation. This approach is especially beneficial for polyploid crops, such as wheat, where multiple homeologs must be addressed to induce a phenotypic change. The use of multiplexing allows for the concurrent editing of genes across various genomes within a single transformation (Zhang et al. 2021). Despite the precision of CRISPR/Cas9, unintended modifications of the genome, known as off-target effects, continue to be a significant concern. This concern could be particularly harmful in agricultural settings, as unintended alterations might affect plant performance or result in undesirable characteristics. Advancements in high-fidelity Cas9 proteins, like SpCas9-HF1, have been achieved to minimize off-target activity by improving the selectivity of the Cas9 protein. Furthermore, careful design of the sgRNAs is essential to minimize off-target effects. It was demonstrated that employing sgRNAs with high specificity for their target sequences, along with a minimum mismatch of three bases from other genomic regions, can significantly reduce off-target effects. Furthermore, advanced sequencing methods, such as Digenome-seq, are currently employed to identify and assess off-target sites in edited plants, facilitating more accurate editing and enhanced oversight. The advancements are crucial for guaranteeing that the CRISPR/Cas9-edited crops attain optimal precision, thereby readily satisfying regulatory standards (Young et al. 2019). The regulations surrounding crops modified with CRISPR/CAS9 vary considerably depending on the geographical context. In the United States and Japan, crops edited with CRISPR/CAS9 that do not contain foreign genetic material are frequently classified as non-GMOs, which aids in their acceptance. In regions such as the European Union, the legal framework is more stringent, and crops edited with CRISPR/CAS9 may encounter similar regulatory challenges as genetically modified organisms, which could impede their market introduction. The regulatory disparities present various challenges to the global application of CRISPR/Cas9 in plant breeding (Ewa et al. 2022). CRISPR/Cas9 technologies are advancing day by day, and they provide significant potential to expedite the creation of crops with increased yields, superior stress resilience, and enhanced nutritional values. However, different challenges for off-target consequences, regulatory problems, and public acceptability must be resolved to fully harness the promise of CRISPR/Cas9. Current advancements in CRISPR/Cas9 technology, together with clear communication and regulatory adjustments, will play a critical role

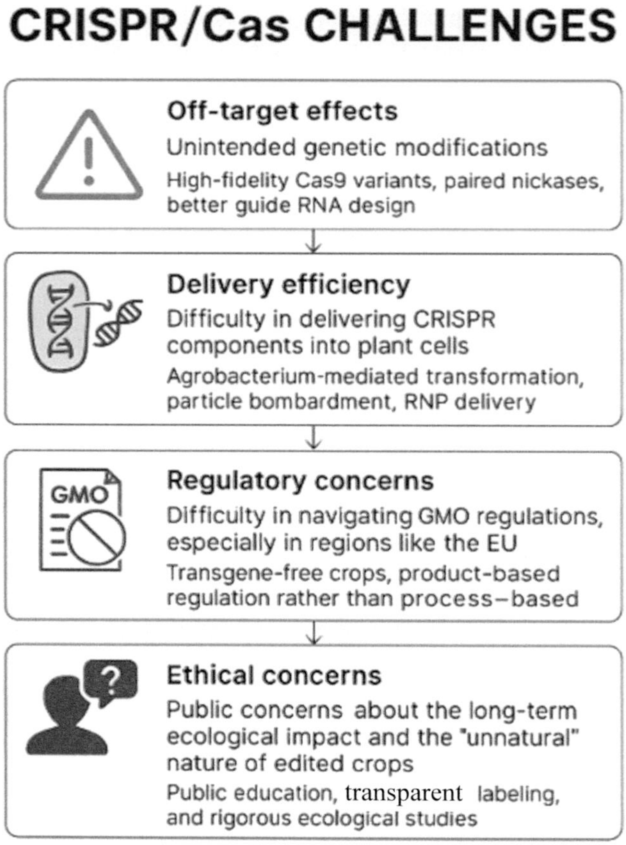

Figure 1. Challenges in CRISPR/Cas9 for crop improvement.

that GE is pivotal in sustainable crop production and ensuring global food security (Globus and Qimron 2018). Figure 1 illustrates the major challenges associated with CRISPR/Cas9 technology in crop breeding, including ethical concerns, off-target effects, and regulatory and ethical concerns.

6. Ethical, Environmental, and Social Considerations

Increasing usage of gene editing in plants has led to a lot of interest and debate about its ethical, environmental, and societal implications. This advancement in GE, particularly with CRISPR/Cas9, is rendering these issues more essential for the acceptance technology. Despite these inequalities, marketing challenges persist, mostly owing to unfamiliarity with technology, concerns about labeling, and public risk perceptions. These problems persist in shaping the future trajectory of gene-edited crops. The adoption of gene-edited crops is largely dependent on public views, which are often influenced by the extent of scientific understanding. One research accomplished in Japan where different perspectives of molecular biology professionals, specialists from other disciplines, and the public were examined about genetic alteration and gene editing. This research indicated that molecular biology specialists had more positive perspectives on gene editing, seeing it low-risk, high-benefit technique. Conversely, the general populace exhibited a preference for gene editing compared to traditional genetic alteration, albeit the distinctions were less pronounced than those between genetic modification and conventional breeding. Research found that scientific literacy significantly influenced public attitudes, particularly regarding the advantages of gene-edited crops

(Kato-Nitta et al. 2019). Public perception regarding gene-edited crops is perceived as totally unnatural and strange, which is frequently shaped by the perceived benefits and technology applications in Japan. Some consumers showed favorable attitudes toward gene-edited crops, especially those believed to offer direct advantages, including increased nutritional values and enhanced crop resilience. Concerns regarding the artificial nature of these foods were widespread, indicating a broader skepticism towards technological interventions in food production. A considerable segment of the population exhibited indifference towards gene-edited foods, neither endorsing nor rejecting them, indicating that societal acceptance is not uniform. These mixed perceptions highlighted the importance and complexity of public responses, influenced by factors including perceived benefits as well as risks, and naturalness of the food products (Yamaguchi et al. 2024). There are considerable concerns about the ecological impact of genetically modified crops when viewed from an environmental standpoint. An important concern is the unforeseen outcomes of gene flow between genetically modified crops and their wild counterparts, which could lead to the development of invasive species. The enhancement of plant requirements through herbicide resistance, insect resistance, and drought tolerance could allow GM crops to surpass wild relatives and non-GM crops in competition. Research suggests that genetically modified crops could transfer specific traits to wild populations, increasing their resilience and possibly resulting in invasiveness. The ecological impacts of gene flow present significant challenges in prediction, as they depend on the specific traits involved, particularly within the environmental context (Warwick et al. 2009). Current models used to assess gene flow risk frequently fail to predict long-term ecological consequences, highlighting the need for more thorough risk assessment strategies that take into account the complex relationships between genetically modified (GM) crops and their surrounding environments (Bauer-Panskus et al. 2020). In contrast, the European Union enforces a stringent and meticulous licensing system for GMOs, primarily overseen by the European Food Safety Authority (EFSA). The safety and ecological impacts of GMOs are evaluated by EFSA, whereas the final approval decision rests with the European Commission. Nonetheless, significant public opposition to GMOs is present in various member states, rendering the licensing and cultivation of GM crops a contentious issue. In the EU, even with the approval of GM crops, production could still face prohibition at the state level due to national restrictions. Furthermore, within the EU, legislation mandates that GM foods be labeled, a strategy aimed at ensuring consumers have transparent information regarding their food choices (Ewa et al. 2022). The complex nature of GMO regulation is exacerbated by the advent of novel technologies, such as CRISPR/CAS9, which provide precise gene alterations without incorporation of exogenous DNA into the plant genome. This method, which allows the development of crops free of transgenes, may bypass some regulatory challenges that are directly linked to GMOs. Consequently, regulatory agencies must identify and classify GMOs according to the methodologies used in their development or resulting products. Some believe that the regulation regarding GMOs should shift from a process-based approach to ⇒ product-based approach, emphasizing the evaluation of the completed product's safety rather than the processes of production used. This modification may accelerate the licensing process for gene-edited crops, allowing for individual assessments. The ethical issues associated with CRISPR/Cas9 and GMOs continue to be contentious, especially around intellectual property rights, technological accessibility, and the possibility of different effects on small-scale farmers. The advancement and commercialization of GM crops need substantial investment, and the patents linked to CRISPR/Cas9 technology have engendered a competitive market, alongside issues over the ownership of intellectual rights. Since CRISPR/Cas9 technology offers significant promise for enhancing agricultural productivity and pest and disease resistance, the large amount of expenses is associated with product development and market entry may benefit larger corporations, thus disadvantaging smaller farmers because they have limited resources. This highlights problems over the consolidation of power and money within a limited number of multinational agricultural corporations, so reducing anticipated advantages of CRISPR/Cas9-based agriculture for small-scale farmers (Globus and Qimron 2018).

7. Future Directions and Challenges

Global malnutrition necessitates long-term development, financial, community, and medical concerns for people, societies, and countries all over the world. Current biotechnological advancements have integrated biofortification into various stable crops to combat malnutrition. However, all methods were unsustainable and encountered numerous limitations. CRISPR/Cas9 system effectively mitigates numerous limitations by offering a precise method for gene editing, positioning it as a superior tool devoid of the ethical issues associated with transgenic crops. CRISPR/Cas9 is engaged in crop improvement initiatives due to its effectiveness, duplicability, and rapid turnaround time. This system is currently exploited in biofortification of cereal crops, including wheat, rice, maize, barley, as well as vegetable crops such as potatoes and tomatoes. CRISPR/Cas9 editing has been utilized to improve aroma, shelf life, sweetness, and various other traits, alongside enhancing gamma aminobutyric acid (GABA), starch, anthocyanin, protein, oleic acid, phytic acid, and steroidal glycoalkaloids in crops. Certain varieties have been genetically altered to enhance resistance to diseases and environmental stressors. CRISPR/Cas9 GE is demonstrating nutraceutical properties of crops (Kumar et al. 2022). This rapid increase in population, alongside climate change, has intensified the necessity for the advancement of crop plants that exhibit high yield and enhance nutritional value. Hence, conventional techniques of breeding are not adequate to meet the necessities of the entire world. Increasing agricultural production and meeting the demands of the global food supply may be accomplished only with the use of cutting-edge technology, CRISPR/Cas9. Based on the findings of (Mubarik et al. 2021), CRISPR/Cas9 is recognized to enhance a wide range of agricultural characteristics, making it an efficient instrument for addressing concerns about global food security. CRISPR/Cas9 systems represent an innovative advancement, facilitating precise genetic modifications in crops that enhance resistance, yield, and nutritional quality. Recent advancements have been evidenced in cereal crops. Recent advances, including PE and BE, along with the creation of new CRISPR/Cas9-related proteins, have considerably enhanced the precision, scope, and productivity of genetic engineering. CRISPR/Cas9 systems are employed to improve resistance to various abiotic factors like drought, as well as biotic threats including diseases and pests. The CRISPR/Cas9 technology enhances crop quality and production by increasing photosynthetic efficiency, nutrient absorption, and lodging resistance, while improving texture and flavor, shelf life, and nutritional value through biofortification. Considering these significant advances, challenges persist, including off-target effects, the necessity for further effective delivery systems, and ethical and regulatory considerations. The highlighted issues emphasize the significance of systematic investigation and the amalgamation of CRISPR/Cas9 technology with other cutting-edge fields, including synthetic biology, Machine Learning (ML), and nanotechnology for reaching the potential of crop-resilient, production, and sustainable systems (Chen et al. 2024).

7.1 Machine Learning

Machine Learning (ML) serves as an innovative technology by enhancing the precision of CRISPR/Cas9 GE. A notable obstacle in the CRISPR/Cas9 technology is the need to attain both specificity and competence in targeting of correct genomic regions, while simultaneously diminishing off-target effects. ML addresses these challenges by analyzing extensive data sets from CRISPR/Cas9 experiments for the prediction of the most effective guide RNA sequences, therefore improving the accuracy of the Cas9 enzyme and minimizing unintended consequences.

Furthermore, ML can be utilized to predict the phenotypic outcomes of gene modifications, which adds complexity to polygenic traits like yield and stress resilience. The integration of CRISPR/Cas9 and ML holds promise for enhancing crop performance under field conditions, thereby promoting more sustainable agricultural practices (Chen et al. 2024; Das et al. 2022). Future inquiries should concentrate on enhancing the complexity of ML procedures, which can address complexities linked with polygenic traits, alongside the creation of wide-ranging datasets for efficiently training those models. The incorporation of CRISPR/Cas9 and machine learning into precision agriculture systems facilitates the delivery of tailored recommendations designed to improve crop performance and tackle global food security confronts (Chen et al. 2024).

This area of study has great potential to develop the efficiency and sustainability of agricultural systems, especially for those regions where traditional breeding methods have struggled to meet rising food demands. The combination of CRISPR/CAS9-Cas systems and machine learning is set to transform agricultural biotechnology, resulting in developments that are increasingly precise, efficient, and sustainable.

8. Conclusion

The rising global demand for food, due to growing populations and climate change, require the advancement of greater durability, high-yield, and crops rich in nutrients. Conventional breeding techniques are still insufficient, highlighting the importance of CRISPR/CAS9 GE is an essential tool for enhancing crop characteristics like crop production, disease resistance, abiotic stress-tolerance, and nutritional value. This technology shows significant potential for enhancing staple crops such as wheat, rice, maize, and soybean, addressing issues like drought, heat, and salinity. Furthermore, CRISPR/Cas9 applied for enhancing crop quality via biofortification, which could play a significant role in addressing global malnutrition. Yet, challenges such as unintended consequences and regulations are still unclear. Innovations in high-fidelity Cas9 variants, BE, and ML are considered to enhance accuracy, effectiveness, and scope of CRISPR/Cas9. Broadening its consumption to lesser-known crops may also tackle regional food security challenges, particularly in regions at risk from climate change. The combination of CRISPR/Cas9 with additional breeding technologies enhanced significantly improve its efficacy in the creation of sustainable, high-yield crops.

References

Afroz, N., M.W.R. Ansary and T.I.A.S.i. Wheat. 2023. CRISPR-cas genome editing for the development of abiotic stress-tolerant wheat. Elsevier. https://www.sciencedirect.com/science/article/pii/B978032395368900014X.

Ahmar, S., B. Usman, G. Hensel, K.H. Jung and D. Gruszka. 2024. CRISPR enables sustainable cereal production for a greener future. Trends in Plant Science 29(2): 179–195. https://doi.org/10.1016/j.tplants.2023.10.016.

Akanmu, A.O., M.D. Asemoloye, M.A. Marchisio and O.O. Babalola. 2024. Adoption of CRISPR-Cas for crop production: present status and future prospects. PeerJ 12: e17402. https://pmc.ncbi.nlm.nih.gov/articles/PMC11164064/pdf/peerj-12-17402.pdf.

Arora, L. and A. Narula. 2017. Gene editing and crop improvement using CRISPR-Cas9 system. Frontiers in Plant Science 8: 1932.

Barman, H.N., Z. Sheng, S. Fiaz, M. Zhong, Y. Wu, Y. Cai et al. 2019. Generation of a new thermo-sensitive genic male sterile rice line by targeted mutagenesis of TMS5 gene through CRISPR/Cas9 system. BMC Plant Biology 19: 1–9.

Basu, U., S. Riaz Ahmed, B.A. Bhat, Z. Anwar, A. Ali, A. Ijaz et al. 2023. A CRISPR way for accelerating cereal crop improvement: Progress and challenges. Frontiers in Genetics 13: 866976.

Bauer-Panskus, A., J. Miyazaki, K. Kawall and C. Then. 2020. Risk assessment of genetically engineered plants that can persist and propagate in the environment. Environmental Sciences Europe 32(1): 32. https://doi.org/10.1186/s12302-020-00301-0.

Bertier, L.D., M. Ron, H. Huo, K.J. Bradford, A.B. Britt and R.W. Michelmore. 2018. High-resolution analysis of the efficiency, heritability, and editing outcomes of CRISPR/Cas9-induced modifications of NCED4 in Lettuce (Lactuca sativa). G3 (Bethesda) 8(5): 1513–1521. https://doi.org/10.1534/g3.117.300396.

Butt, H., G.S. Rao, K. Sedeek, R. Aman, R. Kamel and M. Mahfouz. 2020. Engineering herbicide resistance via prime editing in rice. Plant Biotechnology Journal 18(12): 2370.

Char, S.N. and B. Yang. 2020. Genome editing in grass plants. Abiotech, 1(1): 41–57.

Chen, F., L. Chen, Z. Yan, J. Xu, L. Feng, N. He et al. 2024. Recent advances of CRISPR-based genome editing for enhancing staple crops [Review]. Frontiers in Plant Science 15. https://www.frontiersin.org/journals/plant-science/articles/10.3389/fpls.2024.1478398.

Chen, L., G. Liu and T. Zhang. 2024. Integrating machine learning and genome editing for crop improvement. Abiotech, 5(2): 262–277. https://doi.org/10.1007/s42994-023-00133-5.

Chilcoat, D., Z.-B. Liu and J. Sander. 2017. Use of CRISPR/Cas9 for crop improvement in maize and soybean. Progress in Molecular Biology and Translational Science 149: 27–46.

Cong, L., F.A. Ran, D. Cox, S. Lin, R. Barretto, N. Habib et al. 2013. Multiplex genome engineering using CRISPR/Cas systems. Science 339(6121): 819–823.

Das, J., S. Kumar, D.C. Mishra, K.K. Chaturvedi, R.K. Paul and A. Kairi. 2022. Machine learning in the estimation of CRISPR-Cas9 cleavage sites for plant system. Front Genet. 13: 1085332. https://doi.org/10.3389/fgene.2022.1085332.

Deng, F., L. Tu, J. Tan, Y. Li, Y. Nie and X. Zhang. 2012. GbPDF1 is involved in cotton fiber initiation via the core cis-element HDZIP2ATATHB2. Plant Physiol. 158(2): 890–904. https://doi.org/10.1104/pp.111.186742.

Dong, H., Y. Huang and K. Wang. 2021. The development of herbicide resistance crop plants using CRISPR/Cas9-mediated gene editing. Genes (Basel) 12(6). https://doi.org/10.3390/genes12060912.

Doudna, J.A. and E. Charpentier. 2014. The new frontier of genome engineering with CRISPR-Cas9. Science 346(6213): 1258096.

Ewa, W.G., T. Agata, P. Milica, B. Anna, E. Dennis, V. Nick et al. 2022. Public perception of plant gene technologies worldwide in the light of food security. GM Crops Food 13(1): 218–241. https://doi.org/10.1080/21645698.2022.2111946.

Francisco Ribeiro, P. and A.V. Camargo Rodriguez. 2020. Emerging advanced technologies to mitigate the impact of climate change in Africa. Plants 9(3): 381.

Ghoshal, B. 2024. A birds-eye-view on CRISPR-Cas system in agriculture. The Nucleus 67(1): 89–96.

Globus, R. and U. Qimron. 2018. A technological and regulatory outlook on CRISPR crop editing. Journal of Cellular Biochemistry 119(2): 1291–1298.

Guo, F., Y. Huang, P. Qi, G. Lian, X. Hu, N. Han et al. 2021. Functional analysis of auxin receptor OsTIR1/OsAFB family members in rice grain yield, tillering, plant height, root system, germination, and auxinic herbicide resistance. New Phytologist 229(5): 2676–2692.

Guo, M., K. Ren, Y. Zhu, Z. Tang, Y. Wang, B. Zhang et al. 2019. Structural insights into a high fidelity variant of SpCas9. Cell Research 29(3): 183–192.

Hickey, L.T.N., A. Hafeez, H. Robinson, S.A. Jackson, S.C.M. Leal-Bertioli, M. Tester et al. 2019. Breeding crops to feed 10 billion. Nature Biotechnology 37(7): 744–754.

Hu, H., X. He, L. Tu, L. Zhu, S. Zhu, Z. Ge et al. 2016. GhJAZ2 negatively regulates cotton fiber initiation by interacting with the R2R3-MYB transcription factor GhMYB25-like. Plant J. 88(6): 921–935. https://doi.org/10.1111/tpj.13273.

Huang, G.Q., S.Y. Gong, W.L. Xu, W. Li, P. Li, C.J. Zhang et al. 2013. A fasciclin-like arabinogalactan protein, GhFLA1, is involved in fiber initiation and elongation of cotton. Plant Physiol, 161(3): 1278–1290. https://doi.org/10.1104/pp.112.203760.

Jin, Z. 2017. The molecular mechanism of CRISPR/Cas9 system and its application in gene therapy of human diseases. Microreviews in Cell and Molecular Biology 3.

Jinek, M., A. East, A. Cheng, S. Lin, E. Ma and J. Doudna. 2013. RNA-programmed genome editing in human cells. elife 2: e00471.

Kato-Nitta, N., T. Maeda, Y. Inagaki and M. Tachikawa. 2019. Expert and public perceptions of gene-edited crops: Attitude changes in relation to scientific knowledge. Palgrave Communications 5: 137. https://doi.org/10.1057/s41599-019-0328-4.

Katoch, S., N. Kumari, R. Salwan, V. Sharma and P.N. Sharma. 2020. Recent developments in social network disruption approaches to manage bacterial plant diseases. Biological Control 150: 104376.

Kavuri, N.R., M. Ramasamy, Y. Qi and K. Mandadi. 2022. Applications of CRISPR/Cas13-based RNA editing in plants. Cells 11(17): 2665.

Kim, H. and J.-S. Kim. 2014. A guide to genome engineering with programmable nucleases. Nature Reviews Genetics 15(5): 321–334.

Kulcsár, P.I., A. Tálas, E. Tóth, A. Nyeste, Z. Ligeti, Z. Welker et al. 2020. Blackjack mutations improve the on-target activities of increased fidelity variants of SpCas9 with 5′ G-extended sgRNAs. Nature Communications 11(1): 1223.

Kumar, D., A. Yadav, R. Ahmad, U.N. Dwivedi and K. Yadav. 2022. CRISPR-based genome editing for nutrient enrichment in crops: a promising approach toward global food security. Front Genet. 13: 932859. https://doi.org/10.3389/fgene.2022.932859.

Kumar, M., M.R. Prusty, M.K. Pandey, P.K. Singh, A. Bohra, B. Guo et al. 2023. Application of CRISPR/Cas9-mediated gene editing for abiotic stress management in crop plants. Front Plant Sci. 14: 1157678. https://doi.org/10.3389/fpls.2023.1157678.

Li, B., S. Liang, M. Alariqi, F. Wang, G. Wang, Q. Wang et al. 2020. The application of temperature sensitivity CRISPR/LbCpf1 (LbCas12a) mediated genome editing in allotetraploid cotton (G. hirsutum) and creation of nontransgenic, gossypol-free cotton. Plant Biotechnology Journal 19(2): 221.

Li, J., C. Zhang, Y. He, S. Li, L. Yan, Y. Li et al. 2023. Plant base editing and prime editing: The current status and future perspectives. J. Integr. Plant Biol. 65(2): 444–467. https://doi.org/10.1111/jipb.13425.

Li, M., X. Li, Z. Zhou, P. Wu, M. Fang, X. Pan et al. 2016. Reassessment of the four yield-related genes Gn1a, DEP1, GS3, and IPA1 in rice using a CRISPR/Cas9 system. Frontiers in Plant Science 7: 377.

Li, P., Z. Li, G. Xie and J. Zhang. 2021. Trihelix transcription factor ZmThx20 is required for kernel development in maize. Int. J. Mol. Sci. 22(22). https://doi.org/10.3390/ijms222212137.

Liu, L., J. Gallagher, E.D. Arevalo, R. Chen, T. Skopelitis, Q. Wu et al. 2021. Enhancing grain-yield-related traits by CRISPR–Cas9 promoter editing of maize CLE genes. Nature Plants 7(3): 287–294.

Lu, Y., M. Chuan, H. Wang, R. Chen, T. Tao, Y. Zhou et al. 2022. Genetic and molecular factors in determining grain number per panicle of rice. Frontiers in Plant Science 13: 964246.

Luo, Q. and Y.G. Liu. 2025. Breeding herbicide-resistant rice using CRISPR-Cas gene editing and other technologies. Plant Commun. 6(2): 101172. https://doi.org/10.1016/j.xplc.2024.101172.

Ma, Y., M. Tang, M. Wang, Y. Yu and B. Ruan. 2024. Advances in understanding drought stress responses in rice: molecular mechanisms of ABA signaling and breeding prospects. Genes (Basel) 15(12). https://doi.org/10.3390/genes15121529.

Mali, P., J. Aach, P.B. Stranges, K.M. Esvelt, M. Moosburner, S. Kosuri et al. 2013. CAS9 transcriptional activators for target specificity screening and paired nickases for cooperative genome engineering. Nature Biotechnology 31(9): 833–838.

Mehta, D., A. Stürchler, R.B. Anjanappa, S.S. Zaidi, M. Hirsch-Hoffmann, W. Gruissem et al. 2019. Linking CRISPR-Cas9 interference in cassava to the evolution of editing-resistant geminiviruses. Genome Biol. 20(1): 80. https://doi.org/10.1186/s13059-019-1678-3.

Morrow, N., J.S. Borrell, N.B. Mock, L. Büchi, A. Gatto and E. Lulekal. 2023. Measure of indigenous perennial staple crop, Ensete ventricosum, associated with positive food security outcomes in southern Ethiopian highlands. Food Policy 117: 102451.

Mubarik, M.S., S.H. Khan and M. Sajjad. 2021. Key applications of CRISPR/Cas for yield and nutritional improvement. CRISPR Crops: The Future of Food Security 213–230. https://doi.org/10.1007/978-981-15-7142-8_7.

Pathak, B., C. Maurya, M.C. Faria, Z. Alizada, S. Nandy, S. Zhao et al. 2022. Targeting TOR and SnRK1 genes in rice with CRISPR/Cas9. Plants (Basel) 11(11). https://doi.org/10.3390/plants11111453.

Pramanik, D., R.M. Shelake, J. Park, M.J. Kim, I. Hwang, Y. Park et al. 2021. CRISPR/Cas9-mediated generation of pathogen-resistant tomato against tomato yellow leaf curl virus and powdery mildew. Int. J. Mol. Sci. 22(4). https://doi.org/10.3390/ijms22041878.

Qi, Y. 2019. Plant Genome Editing with CRISPR Systems. Springer.

Rasheed, A., R.A. Gill, M.U. Hassan, A. Mahmood, S. Qari, Q.U. Zaman et al. 2021. A critical review: recent advancements in the use of CRISPR/Cas9 technology to enhance crops and alleviate global food crises. Current Issues in Molecular Biology 43(3): 1950–1976.

Saleem, M.S., S.H. Khan, A. Ahmad, I.A. Rana, Z.A. Naveed and A.I. Khan. 2024. The 4Fs of cotton: genome editing of cotton for fiber, food, feed, and fuel to achieve zero hunger. Front Genome Ed. 6: 1401088. https://doi.org/10.3389/fgeed.2024.1401088.

Sander, J.D. and J.K. Joung. 2014. CRISPR-Cas systems for editing, regulating and targeting genomes. Nature Biotechnology 32(4): 347–355.

Shan, Q., Y. Wang, J. Li and C. Gao. 2014. Genome editing in rice and wheat using the CRISPR/Cas system. Nature Protocols 9(10): 2395–2410.

Shi, J., H. Gao, H. Wang, H.R. Lafitte, R.L. Archibald, M. Yang et al. 2017. ARGOS 8 variants generated by CRISPR-Cas9 improve maize grain yield under field drought stress conditions. Plant Biotechnology Journal 15(2): 207–216.

Swartjes, T., R.H.J. Staals and J. van der Oost. 2020. Editor's cut: DNA cleavage by CRISPR RNA-guided nucleases Cas9 and Cas12a. Biochemical Society Transactions 48(1): 207–219.

Thiruppathi, A., S.R. Salunkhe, S.P. Ramasamy, R. Palaniswamy, V.R. Rajagopalan, S.A. Rathnasamy et al. 2024. Unleashing the potential of CRISPR/Cas9 genome editing for yield-related traits in rice. Plants 13(21): 2972. https://doi.org/10.3390/plants13212972.

Tripathi, L., V.O. Ntui and J.N. Tripathi. 2020. CRISPR/Cas9-based genome editing of banana for disease resistance. Current Opinion in Plant Biology 56: 118–126.

Vu, T.V., N.T. Nguyen, J. Kim, J.C. Hong and J.Y. Kim. 2024. Prime editing: Mechanism insight and recent applications in plants. Plant Biotechnol J. 22(1): 19–36. https://doi.org/10.1111/pbi.14188.

Wang, L., A. Cook, J.W. Patrick, X.Y. Chen and Y.L. Ruan. 2014. Silencing the vacuolar invertase gene GhVIN1 blocks cotton fiber initiation from the ovule epidermis, probably by suppressing a cohort of regulatory genes via sugar signaling. Plant J. 78(4): 686–696. https://doi.org/10.1111/tpj.12512.

Wang, Q., M. Alariqi, F. Wang, B. Li, X. Ding, H. Rui et al. 2020. The application of a heat-inducible CRISPR/Cas12b (C2c1) genome editing system in tetraploid cotton (G. hirsutum) plants. Plant Biotechnol J. 18(12): 2436–2443. https://doi.org/10.1111/pbi.13417.

Wang, Y., X. Cheng, Q. Shan, Y. Zhang, J. Liu, C. Gao et al. 2014. Simultaneous editing of three homoeoalleles in hexaploid bread wheat confers heritable resistance to powdery mildew. Nature Biotechnology 32(9): 947–951.

Wang, Y., Q. Tang, L. Pu, H. Zhang and X. Li. 2022. CRISPR-Cas technology opens a new era for the creation of novel maize germplasms [Review]. Frontiers in Plant Science 13. https://www.frontiersin.org/journals/plant-science/articles/10.3389/fpls.2022.1049803.

Warwick, S.I., H.J. Beckie and L.M. Hall. 2009. Gene flow, invasiveness, and ecological impact of genetically modified crops. Ann. N Y Acad. Sci. 1168: 72–99. https://doi.org/10.1111/j.1749-6632.2009.04576.x.

Wu, Y., N. Xiao, Y. Cai, Q. Yang, L. Yu, Z. Chen et al. 2023. CRISPR-Cas9-mediated editing of the OsHPPD 3' UTR confers enhanced resistance to HPPD-inhibiting herbicides in rice. Plant Commun. 4(5): 100605. https://doi.org/10.1016/j.xplc.2023.100605.

Xu, R., Y. Li, Z. Sui, T. Lan, W. Song, M. Zhang et al. 2021. A C-terminal encoded peptide, ZmCEP1, is essential for kernel development in maize. J. Exp. Bot. 72(15): 5390–5406. https://doi.org/10.1093/jxb/erab224.

Yamaguchi, T., K. Ezaki and K. Ito. 2024. Exploring the landscape of public attitudes towards gene-edited foods in Japan. Breed Sci. 74(1): 11–21. https://doi.org/10.1270/jsbbs.23047.

Yoon, Y.-J., J. Venkatesh, J.-H. Lee, J. Kim, H.-E. Lee, D.-S. Kim et al. 2020. Genome editing of eIF4E1 in tomato confers resistance to pepper mottle virus. Frontiers in Plant Science 11: 1098.

Young, J., G. Zastrow-Hayes, S. Deschamps, S. Svitashev, M. Zaremba, A. Acharya et al. 2019. CRISPR-Cas9 editing in maize: systematic evaluation of off-target activity and its relevance in crop improvement. Scientific Reports 9(1): 1–11. https://doi.org/10.1038/s41598-019-43141-6.

Yu, W., L. Wang, R. Zhao, J. Sheng, S. Zhang, R. Li et al. 2019. Knockout of SlMAPK3 enhances tolerance to heat stress involving ROS homeostasis in tomato plants. BMC Plant Biol. 19(1): 354. https://doi.org/10.1186/s12870-019-1939-z.

Zafar, K., M.Z. Khan, I. Amin, Z. Mukhtar, S. Yasmin, M. Arif et al. 2020. Precise CRISPR-Cas9 mediated genome editing in super basmati rice for resistance against bacterial blight by targeting the major susceptibility gene. Frontiers in Plant Science 11. https://doi.org/10.3389/FPLS.2020.00575/FULL.

Zhang, A., Y. Liu, F. Wang, T. Li, Z. Chen, D. Kong et al. 2019a. Enhanced rice salinity tolerance via CRISPR/Cas9-targeted mutagenesis of the OsRR22 gene. Molecular Breeding 39: 1–10.

Zhang, R., J. Liu, Z. Chai, S. Chen, Y. Bai, Y. Zong et al. 2019b. Generation of herbicide tolerance traits and a new selectable marker in wheat using base editing. Nat. Plants 5(5): 480–485. https://doi.org/10.1038/s41477-019-0405-0.

Zhang, Y., Y. Bai, G. Wu, S. Zou, Y. Chen, C. Gao et al. 2017. Simultaneous modification of three homoeologs of Ta EDR 1 by genome editing enhances powdery mildew resistance in wheat. The Plant Journal 91(4): 714–724.

Zhang, Y., Q. Ren, X. Tang, S. Liu, A.A. Malzahn, J. Zhou et al. 2021. Expanding the scope of plant genome engineering with Cas12a orthologs and highly multiplexable editing systems. Nature Communications 12(1): 1944.

Zhang, Z., X. Zhang, Z. Lin, J. Wang, H. Liu, L. Zhou et al. 2019c. A large transposon insertion in the stiff1 promoter increases stalk strength in maize [OPEN]. The Plant Cell 32(1): 152–165. https://doi.org/10.1105/tpc.19.00486.

Zhang, J., C. Feng, H. Su, Y. Liu, Y. Liu and F. Han. 2020. The Cohesin complex subunit ZmSMC3 participates in meiotic centromere pairing in maize. Plant Cell. 2020 Apr; 32(4): 1323–1336. doi: 10.1105/tpc.19.00834. Epub 2020 Jan 29. PMID: 31996400; PMCID: PMC7145474.

Zhao, H., Y. Qin, Z. Xiao, Q. Li, N. Yang, Z. Pan et al. 2020. Loss of function of an RNA polymerase III subunit leads to impaired maize kernel development. Plant Physiol. 184(1): 359–373. https://doi.org/10.1104/pp.20.00502.

Zhou, H., M. He, J. Li, L. Chen, Z. Huang, S. Zheng et al. 2016. Development of commercial thermo-sensitive genic male sterile rice accelerates hybrid rice breeding using the CRISPR/Cas9-mediated TMS5 editing system. Scientific Reports 6(1): 37395.

CHAPTER 15

Linking miRNAs and Nano-technology with CRISPR/Cas System for Improvement of Agricultural Crops

Dilawar Khan, Asim Gulzar* and *Sobia Khaliq*

1. Introduction

Agricultural biotechnology development remains essential to fulfill the escalating world food requirements because the planet confronts complex environmental issues while both land availability and human numbers expand (Kumar et al. 2025). Genetic engineering combined with nanotechnology enables entirely new possibilities to boost agricultural outputs by enhancing disease resistance as well as elevating nutritional content in crops (Squire et al. 2023). Research has launched a cooperative framework combining microRNA (miRNA) technology with nanotechnology and CRISPR/Cas9 (Clustered Regularly Interspaced Short Palindromic Repeats) genome-editing for agricultural crop improvements. Through precise plant genome modification, these technological methods develop better crop types that increase production while remaining resilient against environmental stressors. The chapter studies the combined effects of these three contemporary technologies on crop enhancement as they work together.

The gene expression functions of plants are organized by short non-coding RNA molecules known as microRNAs (miRNAs) (Bhogireddy et al. 2021). These Southern RNAs operate in multiple biological processes to facilitate growth-related and stress response functions and metabolic pathway activities. Plants improve under miRNA technology because it generates new crop traits for protecting against both biological and environmental stressors. The control mechanism of important plant growth genes by miRNA molecules makes them valuable traits for genetic modifications, according to research (Jones et al. 2020). Plants gain better agronomic feature regulation capability through miRNA integration into their genetic makeup, thus achieving optimized performance across different environments.

Scientists now use the CRISPR/Cas9 genome-editing system to make exact modifications in plant genomes due to its groundbreaking transformation in genetic engineering (Wang and Doudna 2023). The original purpose behind bacteria developing this system centered around protecting themselves through exact genome modifications at specific target genes. Plant genetic engineering implements CRISPR/Cas9 system technology because it functions efficiently while maintaining its

Department of Entomology, Faculty of Agriculture, PMAS-Arid Agriculture University, Rawalpindi, Pakistan.
* Corresponding author: dilawarkhan.arid@gmail.com

DOI: 10.1201/9781003518204-15

basic operation point and provides flexible functionality. Scientists Zhang et al. (2019) exploited CRISPR/Cas9 to develop crops resistant to pests and display enhanced ability to tolerate drought conditions as well as a better nutritional profile. Through multiplexing CRISPR/Cas9 enables plant breeding of crops featuring multiple desirable characteristics which were not possible through natural breeding techniques.

Nanotechnology functions as a core base of this innovative method because scientists can use it to develop unique tools from small materials that demonstrate specific properties (Malik et al. 2023). Plant cell genome editing benefits from CRISPR/Cas9 technologies that are enhanced by nanotechnology applications in the agricultural sectors. Plants shoulder their targeted cellular destinations due to nanotechnological materials such as nanoparticles and nanotubes and nanocarrier systems that transfer CRISPR/Cas9 elements along with miRNAs. Researchers have designed nanoparticles that deliver growth regulators together with nutrients and genetic materials and antimicrobial agents, and therefore enhance both crop health and production levels (Khan et al. 2021). Nanotechnology's combination with CRISPR/Cas9 and miRNAs solves the main delivery method problems of suboptimal efficiency and imprecision by targeting gene delivery research.

Agricultural biotechnology receives significant benefits from the collective power of miRNAs with CRISPR/Cas9 technology and nanotechnology-based delivery systems that help solve crop production problems. Scientific teams can create stress-resistant and nutritious high-yielding crops by using CRISPR/Cas9 genome editing in combination with miRNA gene regulation and improved delivery achieved through nanotechnology principles. Scientists investigate using the CRISPR/Cas9 system to modify genes underlying miRNA synthesis as this generates plants with modified miRNA activity for better stress defense and nutrient absorption (Wang et al. 2020). Researchers combine nanomaterials with techniques to transport the CRISPR/Cas9 components and miRNAs into plant cells, which results in accurate genetic transformations of the plant genome.

The combined method functions as a vital solution for sustainable agriculture because it decreases farmers' dependence on chemical pesticides and fertilizers. Scientific attempts to create pest- and disease-resistant crops allow farmers to reduce their reliance on chemical pest control products, which makes agricultural production more environmentally friendly. Scientists utilize modified crop genetics to develop foods which endure drought conditions and soils with poor nutrition to safeguard vulnerable food production areas from climate-related environmental stresses.

The substantial advantages of using miRNAs with CRISPR/Cas9 and nanotechnology in agriculture require additional challenges to overcome. The application of CRISPR/Cas9 technology creates weaknesses because off-target mutations in the plant genome threaten to occur unintentionally. Scientists actively work on creating high-fidelity Cas9 variants to enhance CRISPR/Cas9 specificity because this will reduce potential risks (Liu et al. 2020). Research about the environmental safety and ecosystem changes caused by nanomaterial introduction needs expanded study because scientists have not fully understood these effects.

A promising development in agricultural biotechnology emerges from uniting miRNAs with CRISPR/Cas9 along with nanotechnology. Scientists leverage specific capabilities of these technologies to generate improved crops that deliver high yields while resisting environmental challenges, including climate change. Strategic research advancement in the field indicates we may soon reach a new era of genetically improved crops because they will supply food requirements and decrease agricultural environmental effects. The combination of miRNAs with CRISPR/Cas9 and nanotechnology directs our path toward maximizing precision agriculture, which secures food availability for forthcoming generations.

2. CRISPR as an Advanced Genome Editing Technology

The genome editing field received a revolutionary breakthrough through CRISPR technology. CRISPR emerged through bacterial immune processes and functions together with Cas9 protein to deliver precise genome alterations in organisms (Hossian 2021). Since scientists developed

CRISPR/Cas9 for genome editing, it brought complete transformations to genetic engineering through its superior performance against older techniques ZFNs (Zinc-Finger Nucleases) and TALENs (Transcription Activator-Like Effector Nucleases) (Jinek et al. 2012). Scientists can precisely modify the DNA of plants and animals, and microorganisms due to CRISPR/Cas9, which opens opportunities to better crops and advanced medical research, and create gene-based therapeutic solutions.

The main components in CRISPR/Cas9 function include the Cas9 protein together with a guide RNA (gRNA). The Cas9 protein functions as molecular scissors that make DNA cuts at defined positions but requires a gRNA to direct it exactly to genomic target sequences. The combination of Cas9 protein with gRNA provides researchers the ability to make double-strand DNA breaks at chosen genomic positions where cellular repair processes will fix the damage. The cell repair mechanisms choose between inserting or deleting genetic material at the cut site, or they can use homology-directed repair to perform specific genetic alterations (Ran et al. 2013). The DNA repair processes enabled by CRISPR open many possibilities for genetic engineering that span from gene knockout to gene insertions, thus making this technology widely applicable in biotechnology.

Plant genetic engineering experienced a historic development because of CRISPR/Cas9 technology's basic operational concept, together with its high precision abilities. The implementation of CRISPR for plant genome modifications occurs much faster when compared to conventional breeding methods and chemical mutagenesis procedures, since it takes only two months (Li et al. 2016). Research on genetically modified crops now permits scientists to develop strains with improved resistance toward diseases, pests, and environmental stress using this method. The CRISPR technique empowers researchers to enhance rice and maize crops by altering genes that control water retention and stress response (Li et al. 2018). The application of CRISPR technology within scientific research has led to disease-resistant improvements in both tomato and wheat crops against viral and fungal infections (Zhou et al. 2023).

The CRISPR system functions as a two-in-one technology since scientists can edit numerous genes at once under the multiplexing method. Multiple guide RNAs allow scientists to edit multiple genes at once for developing crops that show improved yield performance alongside better nutrition content and multi-factor resistance factors (Wang et al. 2017). CRISPR delivers exceptional power in crop development through its multiplexing function because this capability enables scientists to create multitailed crops that traditional breeding approaches would struggle to generate.

Through CRISPR technology, scientists obtain the capability to change plant genome genes and to insert new genetic sequences from outside sources. The technology provides significant benefits when scientists use it to incorporate attributes missing from natural species because it enables gene transfers through various living organisms. Scientists enhance crops using CRISPR through battlefield gene insertions that provide pest defense while enhancing nutritional elements, such as the Bt gene addition to cotton and corn for insect resistance (Bertier et al. 2020).

CRISPR/Cas9 operates as a powerful scientific tool, but scientists need to resolve several problems before it can be fully optimized. Researchers consider off-target effects as the primary significant concern with CRISPR, since this causes genes to be accidentally altered across regions different from their designated targets. The technology has achieved better precision, but off-target editing presents a critical medical demand that demands urgent fixing in agricultural production and therapeutic medicine. The CRISPR detection system continues to become more precise because researchers combine efforts to discover enhanced Cas9 versions while refining guide RNA sequences (Chen et al. 2018). Plant cell protection due to the thick walls results in challenges for researchers who try to transfer CRISPR components during genetic modification experiments. Tzfira and Vainstein (2019) analyzed Agrobacterium-mediated transformation along with biolistics and nanoparticle delivery systems to resolve the difficulties encountered during plant genetic modification, according to their research (Tzfira and Vainstein 2019).

2.1 Mechanisms and Genetic Transformation

A DNA transformation system integrates two fundamental elements, which include the Cas9 protein together with guide RNA (Zhang et al. 2021). The guide RNA contains sequences that match target DNA coding within the genome structure to direct Cas9 to its destined destination. After Cas9-gRNA recognizes the target DNA molecule, the Cas9 protein generates a Double-Strand Break (DSB) at the target site. The double-strand breaks activate cellular DNA repair processes for either NHEJ-Non-Homologous End Joining or HDR-Homology-Directed Repair that enables donor DNA templates to make accurate genomic edits (Doudna and Charpentier 2014).

This system makes accessible cuts at precise DNA locations so scientists can add changes and delete targeted genetic sequences. The modifications from plant genetic transformation result in enhanced traits which include disease resistance in addition to stress tolerance and improved yield potential. Cells aim to repair double stranded breaks that arise in the genome. The DNA repair process of NHEJ can cause small variations known as insertions or deletions (indels) that interrupt suitable gene expression. Science experts utilize HDR to achieve specific edits because they can swap out faulty genes with functional ones or insert beneficial genetic sequences (Jinek et al. 2012).

The CRISPR/Cas9 gene-editing system works through its core mechanism as shown in Fig. 1, which displays its structural parts alongside DNA repair routes. Scientists adapted this schematic from CRISPR/Cas9 System: A Breakthrough in Genome Editing, published # by Razzaq and Masood (2018). The illustration describes CRISPR/Cas9 operations at the molecular scale in a straightforward manner, appearing as an essential guide to understand its operational mechanisms.

The diagram illustrates Cas9 as molecular scissors connecting to single-guide RNA, which finds precise target DNA sequences to bind. The Cas9 protein makes a double-strand break (DSB) following its binding to the target location. After creating the break in the DNA, the cell activates its DNA repair systems, which use NHEJ and HDR to fix the damage. Through NHEJ, cells tend to create insertions or deletions (indels) that harm gene functionality, whereas HDR allows researchers to generate exact genetic modifications with donor DNA templates.

This diagram serves both as a visual explanation of CRISPR/Cas9 technology as well as an essential basis for discussions about genetic transformation and crop improvement.

Genome location targeting capabilities of CRISPR/Cas9 have transformed plant genetic transformation through precise locus manipulation. CRISPR provides direct genetic editing that produces minimal genetic alterations beyond the target site unlike traditional random mutagenesis and chemical treatment methods (Fu et al. 2013). The precise nature of CRISPR/Cas9 technology benefits agricultural applications by enabling researchers to safely add specific traits to crops without inheriting unneeded traits common in traditional breeding approaches.

The process of gene transformation in plants requires successful delivery of CRISPR/Cas9 components inside plant cells. The plant cell wall poses challenges that scientists have tackled through various developed techniques. Agrobacterium tumefaciens serves as the key bacterium in Agrobacterium-mediated transformation since this natural DNA-transfer organism delivers the CRISPR/Cas9 construct to plant cells (Zhang et al. 2016). Researchers successfully use this method to modify DNA in Arabidopsis together with rice and maize crops and various other plant species.

Two alternative techniques with CRISPR/Cas9 components delivery to plant cells exist, which include biolistics (gene gun) and electroporation. Scientists fire DNA-bound particles at plant cells under high pressure during gene gun procedures, and plant cells accept CRISPR/Cas9 via electroporation through membrane opening by electric fields. These experimental procedures find effective use in various plant species that do not accept Agrobacterium infection (Kwon et al. 2017).

The successful delivery of the CRISPR/Cas9 system into plant cells enables modification of target DNA, which results in various desirable traits. The CRISPR/Cas9 system generates enhanced disease-resistant plants that also show improved drought tolerance and increased nutritional value. Rice plants have benefited from CRISPR applications that modify drought-responsive gene sequences to develop drought-resistant crops (Shou et al. 2018). Scientists used CRISPR/Cas9

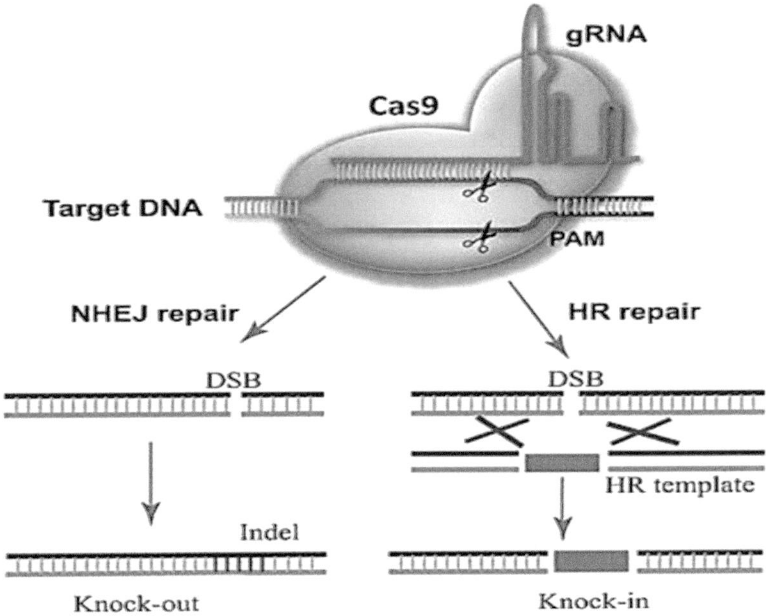

Figure 1. Working of the CRISPR/Cas9 system (Razzaq and Masood 2018).

technology to generate "Golden Rice" as a genetically modified crop that contains beta-carotene as its key ingredient for preventing vitamin A deficiency (Zhang et al. 2019).

CRISPR/Cas9 technology has brought new opportunities to plant genetic transformation, but many obstacles stand in its way. Scientists are concerned that the modification process could affect regions of the DNA genome except the intended target. The technological improvements have reduced off-target editing prevalence, yet this occurrence continues at minimal rates and remains imperative to track, particularly when used in agricultural settings (Fu et al. 2013). Researchers enhance Cas9 protein specificity through research that includes high-fidelity Cas9 variants as well as guide RNA design optimization to reduce off-target activity (Kleinstiver et al. 2016).

Research has made delivery of CRISPR/Cas9 components more efficient into plant cells, but some plant species remain difficult to penetrate for this process. Research teams investigate new gene-editing delivery approaches using nanoparticles and viral vectors to enhance both efficiency and precision in genome modifications, according to Shan et al. (2018). The development of guide RNA design and new delivery strategies will strengthen the agricultural applications of CRISPR/Cas9 technology for crop enhancement.

2.2 Knock-in and Knock-out of Gene

The field of genetic manipulation achieved significant advances through CRISPR/Cas9 because this tool provides scientists with an effective and precise way to edit genes (Barazesh et al. 2021). CRISPR/Cas9 enables scientists to perform two main techniques that serve as gene knock-in and gene knock-out. Gene modification through these two processes alters genetic material differently since they have separate goals and utilization protocols. When scientists perform the "knock-out" technique, they eliminate all functions of a gene (Ratner et al. 2021), but during "knock-in" science, they introduce new genetic material into the genome at designated locations (Rozov et al. 2022).

The genetic modification of organisms for agricultural use requires these fundamental mechanisms to develop GMOs with specific desirable characteristics.

Gene Knock-out

Scientists use the technique of gene knock-out to disable specific genes so they can study their functions and eliminate unneeded characteristics. Targeted double-strand breaks (DSBs) in the desired gene occur when the Cas9 protein acts as an enzyme in the genome. The break induces the cell's natural repair mechanisms to intervene. The most essential repair process is non-homologous end joining (NHEJ) that provides error-prone repair services, which usually create insertions or deletions (indels) at the break point. Points of genetic deletion and insertion introduced by Cas9 protein activity led to frameshift mutations which ruin the normal functional nature of genes (Jinek et al. 2012).

Plant genetic engineers have implemented the knock-out technique to disabled genes that control disease resistance and undesirable traits. Scientists use gene knockout techniques to mutable disease and pest resistance in plants, thereby increasing crop production while reducing the need for chemical pesticides (Zhou et al. 2023). Rice and maize crops have received knock-out modifications that eliminate fungal infection sensitivity and alter plant development patterns (Shou et al. 2018).

The knock-out technique helps scientists explore exactly how particular genes work through studies conducted with biological models, including Arabidopsis thaliana and rice. Researchers who destroy genes under observation gain insight about gene functions by analyzing resulting changes in observable traits. Researchers used this technique to unravel difficult plant features and their resistance mechanisms, and their capacity to withstand stresses and absorb nutrients according to Li et al. (2016).

Gene Knock-in

The gene knock-in methodology adds foreign genetic materials precisely to genome locations, thus differing from knock-out procedures. Scientists benefit from the precise DNA repair process, HDR, to achieve precise new genetic material integration at desired genomic locations. Scientists can deliver targeted sequences through the donor DNA template to specific genomic locations after breaking the DNA using Cas9. The repair process enables the mechanism to employ the template for genetic insertion at the break site, where it accurately integrates foreign DNA sequences into plant DNA (Ran et al. 2013).

Gene knock-in is vital in agricultural biotechnology since it delivers researchers the capacity to integrate advantageous characteristics into organisms with deficient features. The current genetic engineering instruments use knock-in techniques to support crops by integrating drought tolerance genes alongside pest protection genes while updating nutritional gene features. Rice crops developing more stress tolerance acquire specific genes from genetic engineering because these genes provide them better protection against drought and heatwaves (Li et al. 2018). The insertion of selected genes that enhance vitamin A content in Golden Rice has proven to be effective better crops by boosting their nutritional value (Zhang et al. 2019).

Gaining beneficial genes from different species becomes possible through this process. Scientists can obtain resistance against specific pests by transferring insect-resistance genes from *Bacillus thuringiensis* (Bt) bacteria into cotton and maize crops (Bertier et al. 2020).

The following image explains how CRISPR/Cas9 genome editing achieves both gene knock-in and knock-out through its operating principles. Figure 2 demonstrates how Cas9 protein makes a double-strand break at precise DNA locations and after which cells pursue either non-homologous end joining (NHEJ) repair for knock-out or homology-directed repair (HDR) for knock-in.

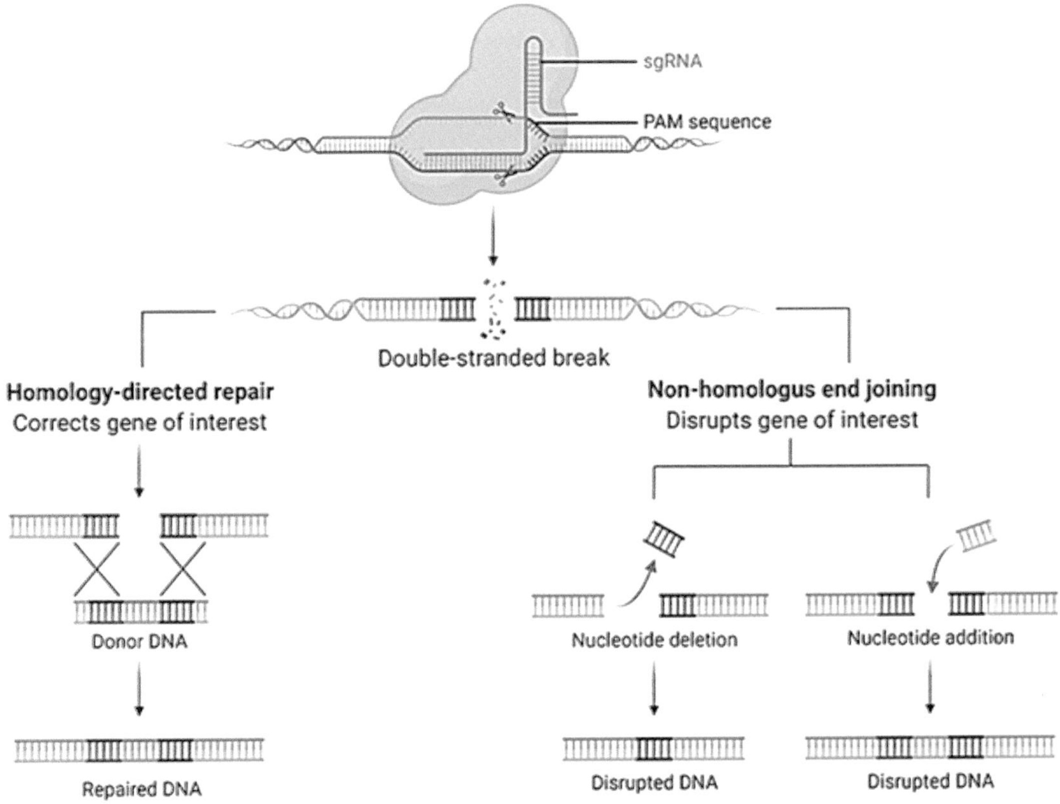

Figure 2. General methodology of CRISPR-Cas9-mediated gene knockout and gene knock-in (Wani et al. 2023).

3. miRNAs (Pathway and Approach to Link Genome Editing)

MicroRNAs operate by regulating gene expression through their role at the post-transcriptional level (Fontemaggi 2023). Plants use miRNA molecules in vital biological functions, including growth and development, as well as their reaction to environmental stress and their defense against diseases (Pagano et al. 2021). The mechanism involves specific bindings of miRNAs to messenger RNA (mRNA) for degradation or translation inhibition, enabling target gene expression control (Naeli et al. 2023). The exact control mechanisms of miRNAs during gene expression play a fundamental role in perfecting plant traits and environmental responses. The ability to control miRNAs along with CRISPR/Cas9 genome editing technologies creates opportunities within plant biotechnology to produce crops with superior qualities, including stress tolerance and disease resistance, and nutritional advancements.

Plant genetic engineering has been revolutionized through CRISPR/Cas9 because this tool enables scientists to modify selected genome positions specifically. When CRISPR/Cas9 pairs with miRNA pathways, scientists receive the ability to conduct plant gene modifications and the exact control of target gene expression regulation. The miRNA genes receive direct editing through CRISPR/Cas9 modifications to control the expression rates of specific miRNAs in the plant system. The modified expression controls which downstream genes become active for vital plant processes. Scientists can improve drought tolerance through the modification of ABA (abscisic acid) pathway miRNAs, which enhances plant response to water (Zhou et al. 2023).

CRISPR/Cas9 technology targets regulatory miRNA biogenesis genes DCL1 (Dicer-like 1) and HYL1 (Hyponastic Leaves 1) as well as editing miRNA genes directly. Plant cells obtain their mature miRNA production from these genes so modifications to them would lead to elevated miRNA production. Through miRNA regulation, scientists can use this strategy to modify gene expression levels for improving disease resistance along with enhanced nutrient uptake (Li et al. 2016).

Scientists use genome editing methods to combine miRNA pathways, which allows them to precisely modify plant attributes. Scientists introduce stress-resistant miRNAs to specific genes, which enables crops to become more resilient to drought, heat, and salinity stress (Fig. 3). DNA editors can modify nutrient-related miRNAs to enhance crop nutritive value through increased dietary vitamins while also enhancing the zinc and iron absorption of rice (Zhang et al. 2019). Through an integration of miRNA-based regulation with genome editing technology, scientists have the potential to create environmentally friendly agricultural methods which eliminate the need for chemical-based fertilizers and pesticides.

A number of obstacles limit the practical utilization of miRNA-based CRISPR editing. Problems arise because the gene regulation techniques of miRNAs target their effects specifically. When miRNAs possess multiple target genes, modifications to their expression could produce unintended off-target effects on multiple cellular pathways. Research teams work to advance miRNA editing technology to enable researchers to target cells precisely without producing unintended consequences (Xie et al. 2018). Plant cell transformation creates the main obstacle because scientists need to successfully transmit CRISPR/Cas9 components together with miRNA constructs, specifically when working with plant species that have challenging genome structures. Unprecedented delivery systems that apply nanoparticles or viral vectors might create more precise and effective editing procedures (Xu et al. 2020).

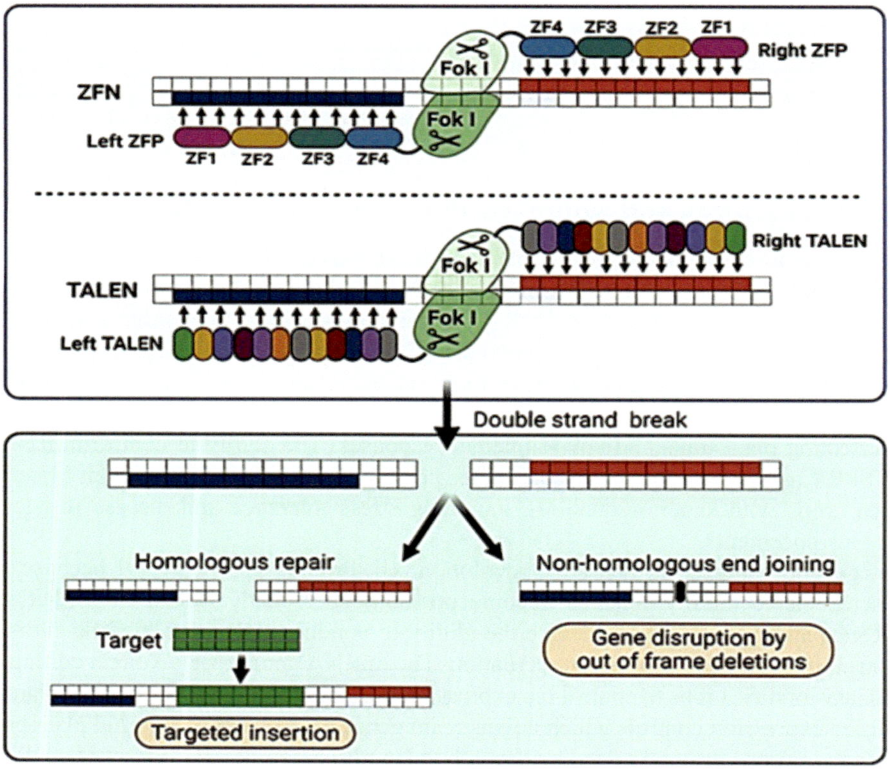

Figure 3. Illustration showing the miRNA pathway in plants and how CRISPR/Cas9 can be used to modify miRNA genes to regulate gene expression (Wani et al. 2023).

Farmers now possess an advanced tool through the union between CRISPR/Cas9 genome editing techniques and miRNA regulation systems for farm crop advancement. By controlling gene expression with miRNA modifications, the agricultural sector acquires stress-resistant crops that yield better results, including improved nutritional values. Under the present climate transformation and growing human population, this dual approach achieves the best possible results for agricultural sustainability as well as food protection.

4. Biological Function and Applications of miRNA in Plants

The biological processes of plants largely rely on miRNAs for their development and stress response mechanisms, and disease resistance capabilities (Naeli et al. 2023). The molecules bind to matching mRNA sequences to regulate genetic expression in two ways, which results in either degrading the mRNA or preventing its translation (Nalbant and Akkaya-Ulum 2024). Plant homeostasis regulation, along with environmental adaptation, depends specifically on miRNA regulation capacities, which makes these molecules indispensable (Pagano et al. 2021).

Plants rely on miRNA molecules to direct the development processes, including leaf patterning and stem cell maintenance, together with root development (Dong et al. 2022). The miR165/166 family controls the expression of HD-ZIP III transcription factor through mRNA binding that controls leaf and shoot apical meristem specification (Yadav et al. 2021). The flowering time and vegetative phase change depend on miR156 because it targets SPL (SQUAMOSA PROMOTER BINDING PROTEIN-LIKE) genes (Wang et al. 2009). Plant growth and development strictly depend on miRNAs since they control the expression of essential genes.

The plant needs miRNAs for survival under different abiotic and biotic stresses that it encounters. Plant miRNA expression patterns shift during stress events such as drought and salinity, and pathogen attack to help the plant manage these stress conditions (Ding et al. 2024). Plants maintain protection through miR398-triggered expression after oxidative stress because this miRNA targets the Cu/Zn superoxide dismutase gene (Zheng et al. 2023). The plant uses miR169 to manage drought stress challenges by influencing the expression of NF-YA transcription factors that function as stress response signaling elements (Chen et al. 2024).

The agricultural field benefits significantly from miRNA applications. MiRNAs benefit agriculture through genetic engineering and CRISPR-based methods that enhance crop yield capabilities as well as disease resistance, while providing better environmental stress tolerance. Research shows that elevating select miRNA expressions in crops generates drought resistance capacity in rice and maize (Singroha et al. 2021). Research supports the utilization of miRNAs in producing transgenic plants with improved pathogen resistance capabilities. MicroRNA silencing of genes involved in plant-pathogen interactions has been developed into a sustainable pest management strategy through its application for boosting plant immunity (Yang et al. 2021).

5. Nanotechnology in Agriculture

In recent years nanotechnology has established itself as a revolutionary agricultural tool which provides advanced solutions to improve crop yields while fighting pests and sustaining healthy soils and protecting the environment. Nanotechnology enables scientists to use nanoscale (1 to 100 nanometer range) manipulation for producing new substances and devices exhibiting unique properties that differ from bulk materials (Malik et al. 2023). The specific characteristics of nanomaterials such as large surface contact and active substance behavior together with enhanced dissolving ability result in their superior performance across agricultural applications (An et al. 2022).

Agricultural nanomaterials form the basis of major nanotechnology applications because they improve the delivery efficiency of agrochemicals including fertilizers along with side pesticides and herbicides. Numerous existing approaches to agrochemical applications lead to poor usage patterns that generate environmental pollution and extra financial expenditure (Mitra et al. 2021). Smart nanotechnology delivery systems developed through this method control agrochemical distribution so they directly reach plant target areas thus decreasing waste along with environmental impact (Singh et al. 2021). Nanosilver and nanoclay nanoparticles are used to develop pesticide-dosing capsules that deliver chemicals over prolonged periods to maintain pest protection without excessive chemical applications (Rico et al. 2011).

Soil health management largely depends on nanotechnology systems. Soil structure improves significantly when nanomaterials are applied because they enhance nutrient conservation and help beneficial microorganisms thrive (Wahab et al. 2024). The integration of biochar and nanoclay nanoparticles into soil leads to better water storage, together with enhanced nutritional substance accessibility, which creates fertile and productive conditions (Pavlicevic et al. 2022). Nanomaterials demonstrate their value for soil remediation because they adsorb contaminating pollutants from the soil to reduce its pollution level and thereby enhance quality (Mukhopadhyay et al. 2022).

Nanotechnology brings exciting opportunities to boost plant nutrient intake and strengthen plant tolerance against environmental stresses in plant development systems (Ijaz et al. 2023). Nanoscale fertilizers improve nutrient absorption by plant roots, which results in better growth and higher plant yield (Abdel-Hakim et al. 2023). Research proves the superior absorption of plants for the nanomaterials nanosulfur and nanozinc compared to standard fertilization approaches under conditions of nutrient shortage (Semenova et al. 2024). Nanomaterials show potential as agents that boost plant systems against abiotic stresses, which include drought and salinity. The utilization of nanoparticles regulates genetic stress responses, thus improving plant resistance against environmental threats (Thabet and Alqudah 2024).

The field of nanotechnology has developed new pest management solutions together with disease control methods (Hajji-Hedfi and Chhipa 2021). The delivery of plant protection agents, including fungicides and herbicides, and insecticides, through nanomaterials as nanocapsules or nanoparticles allows effective targeted and controlled release (Jiang et al. 2024). The precise delivery of chemicals to target organisms becomes possible through nanotechnology, which cuts down pesticide usage requirements and protects unfamiliar species (Punniyakotti et al. 2024). The antimicrobial features of nanomaterials prevent plant diseases caused by fungi and bacteria, and viruses (Sharmin et al. 2021). Scientists study nanosilver antimicrobial properties for plant disease management because scientists view it as a healthier alternative to conventional chemical plant sprays (Karuvelan et al. 2024).

Modern agriculture benefits from nanotechnology by producing enhanced crop yields through efficient water management and efficient chemical control as well as enhancing environmental soil conditions for creating lasting agricultural approaches (Ashraf et al. 2021). The incorporation of nanomaterials within food packaging systems enables product preservation and maintains food safety quality to provide global benefits toward food security.

The deployment of nanotechnology in agriculture industry generates uncertain environmental consequences and obscure health impacts on humans despite showing potential agricultural benefits. The wide-scale deployment of nanomaterials needs thorough safety assessments to establish clear knowledge of their impacts on both ecosystems and soil organisms and human beings. Data regarding nanoparticle accumulation throughout the food chain and non-target species must be carefully managed because of our incomplete knowledge about this topic. The implementation of diverse regulatory standards forms a fundamental basis for the secure use of nanotechnology-based agriculture methods.

6. CRISPR/Cas System and Nanotechnology for Crop Improvement

The DNA editing method depends on the Cas9 protein to generate precise double-strand breaks at specific DNA positions to achieve precise gene editing with minimal off-target activities (Guo et al. 2023). Crop enhancement through CRISPR/Cas9 system delivery improvements has become possible with nanotechnology because it facilitates more precise delivery systems and reduces conventional genetic engineering challenges (Ahmar et al. 2021). CRISPR and nanotechnology systems can enhance crop production while providing stronger disease protection and stronger resistance against stress factors.

Nanotechnology develops modern delivery systems for CRISPR/Cas9 components that enable better plant cell penetration of the CRISPR materials (Demirer et al. 2021). Nanoparticles contain physical features, including size and surface charge, along with membrane permeability, which makes them excellent carriers to transfer CRISPR genetic materials. Recent research studies using nanoparticles, including lipid nanoparticles and carbon-based nanomaterials, and gold nanoparticles, show promise for delivering both CRISPR/Cas9 elements and guide RNA and Cas9 protein into plant cells (Kumar et al. 2024). Nanoparticles improved the stability and uptake of CRISPR components by making genetic editing more efficient and reducing the risks of cell death in plant cells (Yan et al. 2022).

6.1 CRISPR Delivery Through Nano-Tech Approach

For plant genome editing the main obstacle involves appropriate delivery methods for CRISPR/Cas9 components. Again, the transformation methods, including Agrobacterium-mediated transformation or particle bombardment, demonstrate inefficient results when attempting to change plant genomes, particularly with non-model plant species (Mosey et al. 2021). The utilization of nanomaterials offers a beneficial delivery approach for CRISPR because it enables streamlined genetic material transportation into plant cells (Demirer et al. 2021).

Lipid Nanoparticles (LNPs) represent the leading delivery system for CRISPR/Cas9 in plant-related research. Plants delivered by these nanoparticles receive protected Cas9 protein and guide RNA, which effectively disengage inside the target cells after delivery. Research has demonstrated that LNPs boost genetic editing performance in plants since they serve as non-viral delivery methods, preventing the hazards of viral vectors (Bardhan et al. 2019). These engineered nanoparticles become more specific for plants, which increases their internal absorption level while minimizing unwanted side effects.

Gold nanoparticles serve as effective materials for transporting CRISPR delivery into target cells. Plant cells receive CRISPR/Cas9 components through AuNPs because these nanoparticles demonstrate exceptional DNA and RNA molecule binding capability and simple functionalization, and biocompatible characteristics. Gupta et al. (2019) proved that AuNPs function effectively to transport guide RNA and Cas9 protein into plant tissue, where they cause genome editing with reduced toxicity. Engineered nanotechnology enables precise CRISPR component delivery, which effectively enhances the genome-editing precision.

Multiple CRISPR delivery applications are developed using carbon-based nanomaterials, which include carbon nanotubes (CNTs) and Graphene Oxide (GO) (Safarkhani et al. 2024). The special characteristics of these materials, containing a large surface area and electrically conductive capability, enable effective interactions with CRISPR components (Huang et al. 2023). Plant cell genome editing through CRISPR/Cas9 components delivery uses carbon nanotubes, especially for their ability to anchor CRISPR systems on their surfaces, thus enabling efficient genetic modification (Pattnaik et al. 2020).

6.2 Applications and Limitations of CRISPR/Cas9 and Nanotechnology Systems in Modern Biology

The partnership between CRISPR/Cas9 technology and nanotechnology provides multiple uses in contemporary biological research, particularly within agricultural settings. The joint application of CRISPR/Cas9 and nanotechnology systems shows promise in accelerating the development of genetically modified crops that exhibit advantageous characteristics for enhanced production capabilities and better pest immunity as well as improved tolerance against drought and salinity conditions. CRISPR/Cas9-based editing utilizes genome precision to modify plant DNA but nanotechnology optimizes component transportation, which boosts the total precision of genome-editing processes.

CRISPR and nanotechnology together produce resistant crops through their agricultural applications. The CRISPR/Cas9 system enables researchers to edit plant genes responsible for viral susceptibility when used to modify genes affecting susceptibility to Tomato Yellow Leaf Curl Virus (TYLCV) in tomato plants (Wang et al. 2016). The use of nanomaterials enhances CRISPR delivery efficiency so that plant cells receive the system, which enables both rapid and precise genetic advancements.

Nanotechnology, applied together with CRISPR/Cas9, enables scientists to enhance nutritional content in crops. Scientists utilize gene editing methods to improve crops through alterations of genes controlling nutrient production which helps generate biofortified plants with vital nutrients at enhanced levels. Scientists utilize CRISPR to improve the provitamin A production in rice which leads to the development of Golden Rice with enhanced vitamin A levels (Amna et al. 2020).

The promising plant genome editing applications from CRISPR/Cas9 technology and nanotechnology encounter specific barriers during their implementation. Unexpected genomic changes cause the main drawback because they lead to unneeded features appearing in transformed plants during this approach. Research teams who implement the CRISPR/Cas9 system delivery using nanoparticles need to consider possible unwanted mutations that occur along with high system precision (Zhang et al. 2018). The field of research works to develop CRISPR/Cas9 systems that demonstrate enhanced precision and produce better delivery nanoparticles for precise transduction methods.

Nanomaterials display toxic behavior as the main obstacle when used extensively in various applications. Scientists need to address outstanding scientific questions about the prolonged impacts of nanoparticles that are suitable for biological use on plants, together with soil systems and human health. Future scientific research needs to investigate what safety level that nanomaterials employed in agriculture pose to environmental systems. New regulatory frameworks need to establish necessary conditions that maintain safety alongside sustainable performance standards for the agricultural usage of CRISPR/Cas9 and nanotechnology.

7. CRISPR/Cas System and miRNA Technology

CRISPR/Cas system technology united with miRNA regulation creates an improved genetic engineering method that helps control plant and animal genes specifically. MicroRNA regulators (miRNAs) work as essential controllers that destroy messenger RNA (mRNA) molecules or block their translation according to Sebastian-delaCruz et al. (2021). The advancement of development and immune response, together with stress adaptation, occurs through these small RNA molecules that execute various biological functions (Chen et al. 2022). CRISPR/Cas technology allows scientists to track miRNA pathways by activating or silencing target miRNAs due to its precise editing capabilities (Doudna and Charpentier 2014). These combined genome editing tools serve as powerful research instruments for miRNA functional examination and they open opportunities to determine their impact on crop development and disease response and therapeutic advancement (Zhang and Unver 2018).

The CRISPR/Cas systems enable precise research on biological systems by enabling activation or inactivation of miRNA genes (Gao et al. 2015). The combined technical competence allows researchers to make transgenic models where they can track miRNA expression for investigating the control system of miRNA-regulated gene networks while studying their relation to important biological traits in plants and animals. Researchers can use the precise molecular recognition of CRISPR/Cas9 to edit DNA sequences, which lets them specify guide RNAs and adjust the workings of the miRNA-processing machinery for enhanced miRNA research precision (Deng et al. 2022).

7.1 The Potential Application of CRISPR/Cas Systems in miRNA Research

CRISPR/Cas9 technology generates numerous potential applications in miRNA research to permit scientists to study miRNA while facilitating the creation of transgenic organisms with altered miRNA patterns, according to Saiyed et al. (2022). The tool called CRISPR/Cas9 helps research gene regulation and organism development through its knockout capabilities that target miRNAs and their corresponding genes as part of its miRNA research applications (Abkhooie and Saberianpour 2023). The analysis of biological context changes during miRNA depletion and overexpression is possible for scientists who employ genomic loci editing of miRNA genes and their regulatory elements, according to Gao et al. 2015.

Using CRISPR/Cas9, researchers can delete specific plant microRNA sequences for investigating their roles in stress response mechanisms and developmental sequences along with disease resistance tasks (Erdoğan et al. 2023). The CRISPR/Cas9 system has been applied to Arabidopsis plants to remove the complete major miR156 family because these microRNAs regulate essential vegetative-to-reproductive phase changes (Wang et al. 2009). Research into developmental and stress response pathways allows scientists to enhance crop improvement in agricultural biotechnology through miRNA-based regulatory pathfinding (Zhang et al. 2022).

Research laboratories that study animals apply CRISPR/Cas9 technology to eliminate miRNAs responsible for conditions such as cancer development, cardiovascular issues, and neurological diseases. Scientific research teams apply CRISPR/Cas9 to miRNA mouse genes to quantify disease-related behavior of these molecules for discovering new treatment targets that could use genetic approaches (Zhou et al. 2025). The study of cancer-related management activity by scientists became possible after using CRISPR/Cas9 technology to knock out miR-34a expression in mice.

Scientists can use CRISPR/Cas9 technology as an important equipment for investigating and modifying miRNA genes through genomic functional studies (Chen et al. 2018). This methodology shows large potential for researching complete biological actions of miRNAs, leading to the discovery of useful therapeutic targets, and generating improved knowledge about complex gene regulatory mechanisms.

7.2 Transcriptional Regulation and Precise Modifications of MIR Genes with CRISPR

Research in plants and animals has been transformed through precise editing of MIR genes that control the production of microRNA (miRNA). MIR gene transcription holds the primary function in regulating miRNA expression while also directing their effects on gene regulatory networks (Hajieghrari and Farrokhi 2022). CRISPR/Cas9 genome editing systems are used to change DNA sequences of MIR genes, thus gaining control over specific miRNA expression levels. The exact control of MIR genes unlocks significant potential solutions to address pathogen and pest problems as well as drought stress and extreme temperature and salt issues that currently reduce crop production and survival rates.

CRISPR/Cas9 technology modified MIR genes into stress-resistant organisms, including plants and animals (Ghosh and Dey 2022). CRISPR technology achieves plant stress response modulation by turning MIR genes either on or off, thus improving plant tolerance to drought and salinity pressure. Scientific research indicates that modifying miR156 and miR164 plant regulatory genes brings about superior stress resilience in abiotic conditions (Wang et al. 2009; Samynathan et al. 2023). Scientific studies using CRISPR to knockout specific miRNAs in laboratory mice have demonstrated decreased oxidative stress together with superior disease resistance results (Zhao et al. 2018).

Genetic material delivery and CRISPR/Cas9 component precision are both improved through nanotechnology implementation which enables precise genome editing control (Hasanzadeh et al. 2022). Stable delivery of CRISPR components becomes possible with the use of nanoparticles, which include lipid nanoparticles (LNPs), carbon-based materials, and gold nanoparticles for reaching plant cells or tissues that are difficult to transform through conventional methods (Bardhan et al. 2019). The cooperation of CRISPR/Cas9 technology with nanotechnology advances both agricultural crop performance and decreases the environmental burden of agricultural chemical support.

When CRISPR/Cas9 technology merges with miRNA technology, scientists achieve the ability to develop crops that resist multiple environmental challenges at once (Table 1). Scientists have used CRISPR-based systems to control miRNAs that manage both abiotic stress defense and pathogen resistance which led to more resilient plants (Wang et al. 2017). The delivery of CRISPR is enhanced through nanomaterial applications, which also increase both efficiency and precision during modifications.

Table 1. Summary of studies on using CRISPR/Cas9 in editing miRNAs in different plants.

Plant Species	Edited miRNA(s)	CRISPR Technique Used	Gene Function	Main Finding	Reference
Rice (*Oryza sativa*)	OsMIR390	Knockout (Cas9)	Regulates auxin signaling and root development	Knockout of OsMIR390 resulted in altered root morphology and improved drought tolerance.	Xu et al. 2022
Rice (*Oryza sativa*)	OsMIR168a	Knockout (Cas9)	Involved in stress response	Disruption of OsMIR168a led to pleiotropic effects, including altered stress responses and changes in gene expression.	Zhan et al. 2021
Rice (*Oryza sativa*)	OsMIR528	Knockout (Cas9)	Positive regulator in salt stress	Knockout mutants exhibited enhanced salt tolerance, confirming its regulatory role.	Zhang et al. 2020
Rice (*Oryza sativa*)	OsMIR408	Knockout (Cas9)	Involved in stress response	Disruption led to increased salt tolerance, indicating its role in abiotic stress response.	Zhang et al. 2019
Potato (*Solanum tuberosum*)	miR160a-5p, miR160b-5p, miR390a-5p	Fine-tuning (dCas9)	Modulates gene expression via RISC	Fine-tuning expression improved resistance to biotic and abiotic stresses in tetraploid potato.	Lukan et al. 2022
Rice (*Oryza sativa*)	OsMIRNA genes (9 loci)	Knockout (Cas12a)	Various regulatory roles in plant development and stress response	Cas12a-mediated knockout resulted in up to 100% genome editing efficiency, leading to altered gene expression and phenotypic changes.	Xu et al. 2022
Rice (*Oryza sativa*)	miR169a, miR827a	Knockout (Cas9)	Involved in stress response	Knockout of miR169a and miR827a revealed their roles in stress responses and gene regulation.	Ghosh and Chattopadhyay 2020
Soybean (*Glycine max*)	miR1514, miR1509	Knockout (Cas9)	Involved in stress response	Targeted knockout of miR1514 and miR1509 provided insights into their roles in stress responses.	Li et al. 2019
Arabidopsis (*Arabidopsis thaliana*)	miR169a, miR827a	Knockout (Cas9)	Involved in stress response	Knockout of miR169a and miR827a revealed their roles in stress responses and gene regulation.	Yoon et al. 2018

8. Advanced Breeding Methods through CRISPR Involving microRNAs and Nanotechnology

Today's breeding programs of crops and livestock develop through CRISPR/Cas9 technology, connecting with miRNA modulation while using nanotechnology applications. The methods of natural hybridization, together with traditional genetic selection, present technical limits for their precise operation and their degrees of effectiveness, and time requirements. Gene editing through CRISPR allows scientists to precisely modify targeted genes, thus accelerating the process to enhance vital characteristics, including yield quantity and disease resistance, drought and nutritional resistance. Using nanotechnology to deliver miRNA regulations serves as a scientific breakthrough that enables rapid genetic progress throughout plants and animals (Rao and Wang 2021; Schiml et al. 2014).

Research scientists currently use CRISPR/Cas9 technology as their main tool to modify specific genetic characteristics inside the DNA. Original research using this technology enables scientists to modify genetic material related to essential agronomic traits that include plant height and flowering time, and both stress tolerance and fruit size characteristics (Zhang et al. 2020; Borrelli et al. 2018). Researchers applied CRISPR genome editing to rice miR156 genes to produce plants with better shape and larger harvest quantities according to Wang et al. (2017). CRISPR enables scientific manipulation of genes regulating protein lipid and vitamin synthesis so researchers can improve crop nutritional value (Li et al. 2018).

The regulatory actions of miRNAs fundamentally control biological operations while directing development and managing growth plus responding to stressful conditions. Research scientists apply CRISPR to modify miRNA expression levels which enables them to obtain enhanced control over biological features of plants and animals. They specifically modify miR156 and miR164 expression for managing plant growth patterns as well as adjusting drought tolerance and salinity resistance abilities (Kantar et al. 2017; Wang et al. 2023). Scientists can achieve extra control of gene expression patterns through miRNA-based regulation that operates separately from genomic modifications (Shen et al. 2020).

Nanotechnology plays a vital role in enhancing the success rates of CRISPR-based breeding technologies. The main limiting factor in CRISPR-based gene editing is that researchers have trouble placing essential molecules like Cas9 proteins and guide RNA into target cells properly. Agrobacterium-mediated transformation, along with particle bombardment delivery methods, show low efficiency rates and are expensive when applied to genetically difficult crops, according to Rico et al. (2013). Nanotechnology has established promising delivery system solutions that direct their delivery to exact target locations. Efficient delivery of CRISPR components for plant cells depends on stable nanoparticles based on lipids (LNPs), carbon materials and gold nanoparticles (AuNPs) since these achieve stable delivery with low cytotoxic effects (Bardhan et al. 2019; Zhang et al. 2020). When engineered nanoparticles serve as tools for gene editing precision targets, they cause minor unwanted effects on adjacent targets (Sarkar et al. 2018).

Breeding programs achieve accelerated development of target-specific crops and livestock by combining CRISPR technology with miRNAs and the applications of nanotechnology. Through their integration with CRISPR/Cas9 and miRNA regulatory mechanisms, breeders acquire an effective tool to optimize gene networks involving growth regulation and stress tolerance, and disease resistance for creating resilient crops exposed to difficult environmental situations (Zhang et al. 2020). The combination of nanotechnology with CRISPR systems allows for better system delivery and specific delivery mechanics that leads to improved advanced breeding method accessibility and effectiveness (Zhang et al. 2020; Bardhan et al. 2019).

Scientists use the CRISPR system to change miRNA genes in rice and maize agricultural plants to boost resistance to drought and diseases according to Wang et al. (2017). Scientists operate CRISPR technology alongside miRNA adjustments to develop better livestock breeds that produce superior milk outputs and increase their feeder strengths particularly in cattle and pigs (Zhao et al.

2018; Borrelli et al. 2018). Through these advances in technology scientists speed up agricultural system development because these systems become durable and operate efficiently (Rico et al. 2013).

9. Prospects and Future Challenges

The combination of CRISPR/Cas9 technology with miRNA modulation and nanotechnology leads to revolutionary changes in modern breeding methods, which enables great potential for agricultural productivity while simultaneously improving sustainability and disease resistance. Modern research into these innovative technologies creates numerous future application opportunities for crop and livestock breeding endeavors. Quantifiable progress depends on overcoming various hurdles to complete the realization of their full potential.

Prospects for Future Applications

The greatest value of CRISPR/Cas9, miRNAs, and nanotechnology for agricultural applications lies in their ability to enhance both crop production along nutritional value, and resistance against biological and environmental hazards. These modern technologies will enable the creation of crops with tolerance to harsh environmental stress and enhanced nutrition security against global issues affecting food quality (Borrelli et al. 2018). Breeders can speed up trait development by using CRISPR and miRNA-based technology because these techniques provide efficient and precise breeding methods that eliminate the lengthy time constraints of conventional breeding practices.

Scientists have identified the development of climate-resilient crops as one of the significant upcoming applications because these plants will thrive under climate change conditions. The coming decades require plants that resist extreme weather such as droughts and temperature changes to provide global food security (Wang et al. 2017). CRISPR delivery systems support improvement from nanotechnology applications to enhance precision while boosting efficiency. This leads to better accessibility of crop breeding technology for farmers across the world and including areas without resources (Bardhan et al. 2019).

By applying CRISPR/Cas9 together with miRNA technology researchers can develop animals which display greater disease immunity alongside increased growth rates and enhanced environmental suitability. Through gene-editing innovations farmers will reduce their antibiotic dependency along with chemical treatment use so they can maintain better animal welfare along with sustainable farming practices (Zhao et al. 2018).

Future Challenges

The successful implementation of these technologies in agriculture and animal breeding demands that various barriers need to be resolved for achieving widespread adoption.

1. **Regulatory and Ethical Concerns**

   CRISPR/Cas9, together with other genome-editing tools in agricultural and livestock breeding, generates major ethical and regulatory concerns. The essential obstacle when it comes to controlling Genetically Modified Organisms (GMOs) through regulations exists because the world lacks a common governance framework. Many parts of the world oppose genetically modified crops because people are concerned about GMOs causing unintended ecological damage and food safety risks, as well as fears about company monopolies in biotechnology. Gene-editing technology applications need strict governmental oversight to create uniform procedures and prove their environmental safety as well as their suitability for human food consumption (Liu et al. 2020).

2. **Off-Target Effects and Precision**

   The CRISPR/Cas9 technology shows high precision, but scientists still need to address off-target effects, which happen when genetic alterations unintentionally modify DNA, particularly through nanotechnology-based CRISPR delivery. The improvement of CRISPR/Cas9 specificity requires additional research to minimize harmful mutations that can come from gene editing activities (Zhang et al. 2018). The development of advanced CRISPR delivery methods needs improvement because they struggle to modify agricultural plants through traditional genetic manipulation approaches.

3. **Public Perception and Consumer Acceptance**

   Public approval represents a major obstacle that stands in the way of gene-edited crop and animal acceptance. The population's belief about genetically modified organisms distorts market acceptance because they perceive these biological constructs as risky to nature. The public requires scientists and farmers, and regulatory bodies to work together so they can present detailed, transparent information about how CRISPR/Cas9-edited organisms deliver their safety benefits (Gura 2015).

4. **Environmental Impact**

   Scientists are unsure about the complete environmental consequences that GMO introduction into ecological systems would produce. Researchers need to conduct extensive research to determine how genetically modified plants impact ecosystem diversity and create lasting effects on soil structures, together with natural habitats. The environmental destiny of nanoparticles over time remains uncertain because researchers must conduct additional studies to determine how these deliverers affect the environment and soil and non-target organisms, and water systems (Sarkar et al. 2018).

Conclusion

CRISPR/Cas9 technology, together with miRNA manipulation and nanotechnology, powers revolutionary breakthroughs in agricultural breeding, which allows better crop and livestock development. New precision gene editing methods and delivery systems created through these technologies hold the promise of resolving major modern agricultural issues that involve climate change and food security, and sustainable farming systems. CRISPR and miRNA-based technologies show potential to change agricultural practices by developing characteristics that boost drought protection and disease immunity, together with better nutritional elements, thus optimizing agricultural production efficiency and reducing environmental harm.

The promising nature of their applications meets several technical obstacles that need to be addressed. Regulatory barriers, coupled with public perception issues and unintended effects, and environmental concerns, must undergo careful examination before these technologies transition from their research state to practical implementations. Scientific organizations must join forces with regulatory bodies and industry stakeholders to develop safe and sustainable approaches for implementing CRISPR/Cas9 along with nanotechnology applications.

Improved breeding techniques will create an innovative crop and animal generation through enhanced delivery systems research and development which ensures global food security through enhanced nutrition and productivity and improved resilience. CRISPR/Cas9, miRNA technology, and nanotechnology work together to create a powerful means for agricultural advancement, which delivers hope for a sustainable food-secure future.

References

Abdel-Hakim, S.G., A.S. Shehata, S.A. Moghannem, M. Qadri, M.F.A. El-Ghany, E.A. Abdeldaym et al. 2023. Nanoparticulate fertilizers increase nutrient absorption efficiency and agro-physiological properties of lettuce plant. Agronomy 13(3): 691.

Abkhooie, L. and S. Saberianpour. 2023. CRISPR/Cas9 tool for MicroRNAs editing in cardiac development, function, and disease. MicroRNA 12(1): 13–21.

Ahmar, S., T. Mahmood, S. Fiaz, F. Mora-Poblete, M.S. Shafique, M.S. Chattha et al. 2021. Advantage of nanotechnology-based genome editing system and its application in crop improvement. Front. Plant Sci. 12: 663849.

Amna, S. Qamar, A.Y. Tantray, S.S. Bashir, A. Zaid and S.H. Wani. 2020. Golden rice: genetic engineering, promises, present status and future prospects. Rice Res. Qual. Improv.: Genomics Genet. Eng. Vol. 2: Nutrient Biofortification and Herbicide Biotic Stress Resistance in Rice 581–604.

An, C., C. Sun, N. Li, B. Huang, J. Jiang, Y. Shen et al. 2022. Nanomaterials and nanotechnology for the delivery of agrochemicals: strategies towards sustainable agriculture. J. Nanobiotechnol. 20(1): 11.

Ashraf, S.A., A.J. Siddiqui, O.E. Abd Elmoneim, M.I. Khan, M. Patel, M. Alreshidi et al. 2021. Innovations in nanoscience for the sustainable development of food and agriculture with implications on health and environment. Sci. Total Environ. 768: 144990.

Barazesh, M., S. Mohammadi, Y. Bahrami, P. Mokarram, M.H. Morowvat, M. Saidijam et al. 2021. CRISPR/Cas9 technology as a modern genetic manipulation tool for recapitulating neurodegenerative disorders in large animal models. Curr. Gene Ther. 21(2): 130–148.

Bardhan, N.M., A. Routh and S. Soni. 2019. Lipid nanoparticles for CRISPR delivery in plants. ACS Appl. Nano Mater. 2(12): 7632–7642.

Bertier, L.D., Y. Zhang and D. Xie. 2020. CRISPR/Cas9-based editing of the Bt gene in maize for resistance to insect pests. Nat. Biotechnol. 38(5): 589–595.

Bhogireddy, S., S.K. Mangrauthia, R. Kumar, A.K. Pandey, S. Singh, A. Jain et al. 2021. Regulatory non-coding RNAs: a new frontier in regulation of plant biology. Funct. Integr. Genomics 21(3): 313–330.

Borrelli, V.M., D. Wolff and H. Zhang. 2018. CRISPR/Cas9-mediated editing of miRNA genes in plants for stress resistance. Front. Plant Sci. 9: 228.

Chen, X. and O. Rechavi. 2022. Plant and animal small RNA communications between cells and organisms. Nat. Rev. Mol. Cell Biol. 23(3): 185–203.

Chen, X., Z. Chen, A. Fiorentino, M. Kuess, N. Tharayil, R. Kumar et al. 2024. MicroRNA169 integrates multiple factors to modulate plant growth and abiotic stress responses. Plant Biotechnol. J. 22(9): 2541–2557.

Chen, Y., J. Zhang and Z. Lu. 2018. Optimizing guide RNA design for CRISPR-based genome editing. Front. Plant Sci. 9: 1017.

Demirer, G.S., T.N. Silva, C.T. Jackson, J.B. Thomas, D. Ehrhardt, S.Y. Rhee et al. 2021. Nanotechnology to advance CRISPR-Cas genetic engineering of plants. Nat. Nanotechnol. 16(3): 243–250.

Deng, F., F. Zeng, Q. Shen, A. Abbas, J. Cheng, W. Jiang et al. 2022. Molecular evolution and functional modification of plant miRNAs with CRISPR. Trends Plant Sci. 27(9): 890–907.

Ding, T., W. Li, F. Li, M. Ren and W. Wang. 2024. microRNAs: Key regulators in plant responses to abiotic and biotic stresses via endogenous and cross-kingdom mechanisms. Int. J. Mol. Sci. 25(2): 1154.

Dong, Q., B. Hu and C. Zhang. 2022. microRNAs and their roles in plant development. Front. Plant Sci. 13: 824240.

Doudna, J.A. and E. Charpentier. 2014. The new frontier of genome engineering with CRISPR-Cas9. Science 346(6213): 1258096.

Erdoğan, İ., B. Cevher-Keskin, Ö. Bilir, Y. Hong and M. Tör. 2023. Recent developments in CRISPR/Cas9 genome-editing technology related to plant disease resistance and abiotic stress tolerance. Biology 12(7): 1037.

Fontemaggi, G. 2023. Non-coding RNA regulatory networks in post-transcriptional regulation of VEGFA in cancer. IUBMB Life 75(1): 30–39.

Fu, Y., J.A. Foden and C. Khayter. 2013. High-frequency off-target mutagenesis induced by CRISPR-Cas nucleases in human cells. Nat. Biotechnol. 31(9): 822–826.

Gao, J., X. Xu and X. Cao. 2015. Applications of CRISPR/Cas9 in miRNA research. J. Mol. Biol. 427(2): 355–366.

Ghosh, S. and G. Dey. 2022. Biotic and abiotic stress tolerance through CRISPR-Cas mediated genome editing. J. Plant Biochem. Biotechnol. 31(2): 227–238.

Ghosh, S. and S. Chattopadhyay. 2020. Functional dissection of miR169a and miR827a in rice using CRISPR-Cas9. Sci. Rep. 10(1): 12299.

Guo, C., X. Ma, F. Gao and Y. Guo. 2023. Off-target effects in CRISPR/Cas9 gene editing. Front. Bioeng. Biotechnol. 11: 1143157.

Gupta, M., M. Kaur and P. Srivastava. 2019. Gold nanoparticles for delivery of CRISPR/Cas9 system in plants. Plant Biotechnol. Rep. 13(4): 497–507.

Gura, T. 2015. Public perception and regulatory challenges of CRISPR in agriculture. Nature 531(7593): 232–234.

Hajieghrari, B. and N. Farrokhi. 2022. Plant RNA-mediated gene regulatory network. Genomics 114(1): 409–442.

Hajji-Hedfi, L. and H. Chhipa. 2021. Nano-based pesticides: challenges for pest and disease management. Euro-Mediterr. J. Environ. Integr. 6(3): 69.

Hasanzadeh, A., H. Noori, A. Jahandideh, N. Haeri Moghaddam, S.M. Kamrani Mousavi, H. Nourizadeh et al. 2022. Smart strategies for precise delivery of CRISPR/Cas9 in genome editing. ACS Appl. Bio Mater. 5(2): 413–437.

Hossain, M.A. 2021. CRISPR-Cas9: A fascinating journey from bacterial immune system to human gene editing. Prog. Mol. Biol. Transl. Sci. 178: 63–83.

Hu, J., S. Huang, X. Liu, Y. Zhang, S. Wei and X. Hu. 2022. miR-155: an important role in inflammation response. J. Immunol. Res. 2022(1): 7437281.

Huang, Y., Q. Wen, Y. Xiong, Y. Chen, W. Li, J. Ren et al. 2023. Nanomaterials driven CRISPR/Cas-based biosensing strategies. Chem. Eng. J. 474: 145615.

Ijaz, M., F. Khan, T. Ahmed, M. Noman, F. Zulfiqar, M. Rizwan et al. 2023. Nanobiotechnology to advance stress resilience in plants: Current opportunities and challenges. Mater. Today Bio 22: 100759.

Jiang, X., F. Yang, W. Jia, Y. Jiang, X. Wu, S. Song et al. 2024. Nanomaterials and nanotechnology in agricultural pesticide delivery: A review. Langmuir 40(36): 18806–18820.

Jinek, M., K. Chylinski and I. Fonfara. 2012. A programmable dual-RNA–guided DNA endonuclease in adaptive bacterial immunity. Science 337(6096): 816–821.

Jones, J.D., J. Monroe and K.S. Koutmou. 2020. A molecular-level perspective on the frequency, distribution, and consequences of messenger RNA modifications. Wiley Interdisciplinary Reviews: RNA 11(4): e1586.

Kantar, M., M. Lucas and S. Jaramillo. 2017. miRNA regulation and plant abiotic stress: Applications of CRISPR technology. Front. Plant Sci. 8: 808.

Karuvelan, M., S.S. Sumukhi, G. Odukkathil, R. Chelliah and D.H. Oh. 2024. Use of silver nanoparticles as a nano-weapon in plant disease management. pp. 265–278. *In*: Plant Response to Silver Nanoparticles: Plant Growth, Development, Production, and Protection. Singapore: Springer Nature Singapore.

Khan, M.A., S.M. Lee and Y.J. Lee. 2021. Nanotechnology in agriculture: Opportunities and challenges. J. Nanoscience Nanotechnol. 21(9): 6112–6124.

Kleinstiver, B.P., S.Q. Tsai and M.S. Prew. 2016. High-fidelity CRISPR-Cas9 nucleases with no detectable genome-wide off-target effects. Nature 523(7561): 716–719.

Kumar, M., M.S. Akhtar, H. Bhardwaj, B. Bhoi and P. Kumar. 2025. Futuristic and prospective role of agricultural biotechnology in sustainable environment. pp. 1–23. *In*: Agricultural Biotechnology. Apple Academic Press.

Kumar, P., V.D. Rajput, A.K. Singh, S. Agrawal, R. Das, T. Minkina et al. 2024. Nano-assisted delivery tools for plant genetic engineering: A review on recent developments. Environ. Sci. Pollut. Res. 1–16.

Kwon, D.Y., Y.T. Zhao, J.M. Lamonica and Z. Zhou. 2017. Locus-specific histone deacetylation using a synthetic CRISPR-Cas9-based HDAC. Nature Communications 8(1): 15315.

Li, C., X. Hu and W. Yang. 2018. CRISPR/Cas9-mediated genome editing to improve drought resistance in crops. Nature Plants 4(7): 433–439.

Li, J., J. Liu and Z. Xie. 2018. CRISPR/Cas9 technology for plant breeding and crop improvement. Trends Biotechnol. 36(2): 141–154.

Li, S., Z. Liu and Y. Wang. 2016. CRISPR/Cas9-mediated regulation of miRNA expression in plants: Applications and challenges. Trends Biotechnol. 34(12): 962–975.

Li, Z., Y. Yang and Y. Zhang. 2019. CRISPR-Cas9 mediated knockout of miR1514 and miR1509 in soybean reveals their role in abiotic stress tolerance. BMC Plant Biol. 19: 89.

Liu, Q., Y. Li and F. Zhang. 2020. Global regulatory standards for CRISPR-based biotechnology in agriculture. Trends Biotechnol. 38(4): 430–433.

Lukan, T., F. Veillet, M. Križnik, A. Coll, T. Mahkovec Povalej, K. Pogačar et al. 2022. CRISPR/Cas9-mediated fine-tuning of miRNA expression in tetraploid potato. Horticulture Research 9: uhac147.

Malik, S., K. Muhammad and Y. Waheed. 2023. Nanotechnology: a revolution in modern industry. Molecules 28(2): 661.

Mitra, B., A.R. Chowdhury, P. Dey, K.K. Hazra, A.K. Sinha, A. Hossain et al. 2021. Use of agrochemicals in agriculture: alarming issues and solutions. Input Use Efficiency for Food and Environmental Security 85–122.

Mosey, M., D. Douchi, E.P. Knoshaug and L.M. Laurens. 2021. Methodological review of genetic engineering approaches for non-model algae. Algal Res. 54: 102221.

Mukhopadhyay, R., B. Sarkar, E. Khan, D.S. Alessi, J.K. Biswas, K.M. Manjaiah et al. 2022. Nanomaterials for sustainable remediation of chemical contaminants in water and soil. Crit. Rev. Environ. Sci. Technol. 52(15): 2611–2660.

Naeli, P., T. Winter, A.P. Hackett, L. Alboushi and S.M. Jafarnejad. 2023. The intricate balance between microRNA-induced mRNA decay and translational repression. FEBS J. 290(10): 2508–2524.

Nalbant, E. and Y.Z. Akkaya-Ulum. 2024. Exploring regulatory mechanisms on miRNAs and their implications in inflammation-related diseases. Clin. Exp. Med. 24(1): 142.

Pagano, L., R. Rossi, L. Paesano, N. Marmiroli and M. Marmiroli. 2021. miRNA regulation and stress adaptation in plants. Environ. Exp. Bot. 184: 104369.

Pattnaik, S., P. Kumari and P. Kachroo. 2020. Carbon nanotubes for plant gene editing using CRISPR/Cas9 technology. Front. Plant Sci. 11: 577222.

Pavlicevic, M., W. Abdelraheem, N. Zuverza-Mena, T. O'Keefe, S. Mukhtar, G. Ridge et al. 2022. Engineered nanoparticles, natural nanoclay and biochar, as carriers of plant-growth promoting bacteria. Nanomaterials 12(24): 4474.

Punniyakotti, P., S. Vinayagam, R. Rajamohan, S.D. Priya, M. Moovendhan and T. Sundaram. 2024. Environmental fate and ecotoxicological behaviour of pesticides and insecticides in non-target environments: Nanotechnology-based mitigation strategies. J. Environ. Chem. Eng. 113349.

Ran, F.A., P.D. Hsu and J. Wright. 2013. Genome engineering using the CRISPR-Cas9 system. Nat. Protoc. 8(11): 2281–2308.

Rao, M.J. and L. Wang. 2021. CRISPR/Cas9 technology for improving agronomic traits and future prospective in agriculture. Planta 254(4): 68.

Ratner, L.D., G.E. La Motta, O. Briski, D.F. Salamone and R. Fernandez-Martin. 2021. Practical approaches for knock-out gene editing in pigs. Front. Genet. 11: 617850.

Razzaq, A. and A. Masood. 2018. CRISPR/Cas9 System: A Breakthrough in Genome Editing. J. Agric. Biotechnol. 15(2): 81–92.

Rico, C.M., M. Manzano and S. Diez. 2013. Environmental impact of nanomaterials in agriculture. Nanotoxicology 7(1): 76–80.

Rico, C.M., S. Majumdar and S. Shankar. 2011. Nanotechnology applications for food production and processing. Food Res. Int. 44(5): 1037–1046.

Rozov, S.M., N.V. Permyakova, Y.V. Sidorchuk and E.V. Deineko. 2022. Optimization of genome knock-in method: Search for the most efficient genome regions for transgene expression in plants. Int. J. Mol. Sci. 23(8): 4416.

Safarkhani, M., M.R. Saeb, J.H. Lee, Y.S. Huh and N. Rabiee. 2024. Carbon-based nanomaterials for CRISPR/Cas delivery: A perspective on the design approach. Carbon Lett. 34(1): 387–397.

Saiyed, A.N., A.R. Vasavada and S.K. Johar. 2022. Recent trends in miRNA therapeutics and the application of plant miRNA for prevention and treatment of human diseases. Future J. Pharm. Sci. 8(1): 24.

Samynathan, R., B. Venkidasamy, A. Shanmugam, S. Ramalingam and M. Thiruvengadam. 2023. Functional role of microRNA in the regulation of biotic and abiotic stress in agronomic plants. Frontiers in Genetics 14: 1272446.

Sarkar, S., P. Das and P. Mukherjee. 2018. Carbon nanomaterials for targeted CRISPR delivery in plants: Progress and prospects. Sci. Rep. 8(1): 987.

Schiml, S., H. Puchta and O. Puchta. 2014. The CRISPR/Cas system in plants: Insights into its function and applications. J. Plant Physiol. 171(6): 726–736.

Sebastian-delaCruz, M., I. Gonzalez-Moro, A. Olazagoitia-Garmendia, A. Castellanos-Rubio and I. Santin. 2021. The role of lncRNAs in gene expression regulation through mRNA stabilization. Non-coding RNA 7(1): 3.

Semenova, N.A., D.E. Burmistrov, S.A. Shumeyko and S.V. Gudkov. 2024. Fertilizers based on nanoparticles as sources of macro-and microelements for plant crop growth: A review. Agronomy 14(8): 1646.

Shan, Q., Y. Wang and J. Li. 2018. Genome editing in rice: Past, present, and future. Plant Biotechnol. J. 16(1): 87–95.

Sharmin, S., M.M. Rahaman, C. Sarkar, O. Atolani, M.T. Islam and O.S. Adeyemi. 2021. Nanoparticles as antimicrobial and antiviral agents: A literature-based perspective study. Heliyon 7(3).

Shen, N.N., C. Zhang, Z. Li, L.C. Kong, X.H. Wang, Z.C. Gu et al. 2020. MicroRNA expression signatures of atrial fibrillation: The critical systematic review and bioinformatics analysis. Experimental Biology and Medicine 245(1): 42–53.

Shou, H., Q. Liu and Y. Zhang. 2018. CRISPR/Cas9-mediated genome editing for improved disease resistance in crops. Front. Plant Sci. 9: 1574.

Singh, H., A. Sharma, S.K. Bhardwaj, S.K. Arya, N. Bhardwaj and M. Khatri. 2021. Recent advances in the applications of nano-agrochemicals for sustainable agricultural development. Environ. Sci. Processes Impacts 23(2): 213–239.

Singroha, G., P. Sharma and R. Sunkur. 2021. Current status of microRNA-mediated regulation of drought stress responses in cereals. Physiologia Plantarum 172(3): 1808–1821.

Squire, H.J., S. Tomatz, E. Voke, E. González-Grandío and M. Landry. 2023. The emerging role of nanotechnology in plant genetic engineering. Nat. Rev. Bioengineering 1(5): 314–328.

Thabet, S.G. and A.M. Alqudah. 2024. Unraveling the role of nanoparticles in improving plant resilience under environmental stress conditions. Plant Soil 1–18.

Tzfira, T. and A. Vainstein. 2019. Advances in plant gene editing technologies for crop improvement. Crit. Rev. Plant Sci. 38(5): 330–355.

Wahab, A., M. Muhammad, S. Ullah, G. Abdi, G.M. Shah, W. Zaman et al. 2024. Agriculture and environmental management through nanotechnology: Eco-friendly nanomaterial synthesis for soil-plant systems, food safety, and sustainability. Sci. Total Environ. 171862.

Wang, J.W., B. Czech and D. Weigel. 2009. miR156-regulated SPL transcription factors define an endogenous flowering pathway in *Arabidopsis thaliana*. Cell 138(4): 738–749.

Wang, J.Y. and J.A. Doudna. 2023. CRISPR technology: A decade of genome editing is only the beginning. Science 379(6629): eadd8643.

Wang, L., J. Li and F. Zhang. 2016. CRISPR/Cas9-mediated editing of plant genomes and its application in agriculture. Front. Plant Sci. 7: 410.

Wang, Y., Z. Li and X. Chen. 2017. Multiplex CRISPR/Cas9: Efficient editing of plant genomes. Front. Plant Sci. 8: 1604.

Wang, Y., Z. Luo, X. Zhao, H. Cao, L. Wang, S. Liu et al. 2023. Superstar microRNA, miR156, involved in plant biological processes and stress response: A review. Scientia Horticulturae 316: 112010.

Wani, A.K., N. Akhtar, R. Singh, A. Prakash, S.H.A. Raza, S. Cavalu et al. 2023. Genome centric engineering using ZFNs, TALENs and CRISPR-Cas9 systems for trait improvement and disease control in animals. Vet. Res. Commun. 47(1): 1–16.

Xie, Z. and S. Guo. 2018. CRISPR/Cas9: An efficient tool for regulating miRNA expression in plants. Front. Plant Sci. 9: 1421.

Xu, Y., R. Liu and Z. Dai. 2020. Key considerations in designing CRISPR/Cas9-carrying nanoparticles for therapeutic genome editing. Nanoscale 12(41): 21001–21014.

Xu, P., X. Li and L. Zhang. 2022. CRISPR-Cas12a-based efficient MicroRNA knockout for plant functional genomics. Nat. Commun. 13: 1194.

Yadav, A., S. Kumar, R. Verma, C. Lata, I. Sanyal and S.P. Rai. 2021. microRNA 166: An evolutionarily conserved stress biomarker in land plants targeting HD-ZIP family. Physiol. Mol. Biol. Plants 1–15.

Yan, Y., X. Zhu, Y. Yu, C. Li, Z. Zhang and F. Wang. 2022. Nanotechnology strategies for plant genetic engineering. Adv. Mater. 34(7): 2106945.

Yang, X., L. Zhang, Y. Yang, M. Schmid and Y. Wang. 2021. miRNA-mediated regulation and interaction between plants and pathogens. Int. J. Mol. Sci. 22(6): 2913.

Yoon, J. and S. Lee. 2018. CRISPR-Cas9 mediated disruption of miR169a and miR827a in Arabidopsis results in altered stress responses. J. Exp. Bot. 69(19): 4529–4537.

Zhan, X., Z. Xu and C. Liu. 2021. CRISPR-Cas9 mediated OsMIR168a knockout reveals its pleiotropy in regulating stress responses. Plant J. 108(3): 751–765.

Zhang, F., J. Yang, N. Zhang, J. Wu and H. Si. 2022. Roles of microRNAs in abiotic stress response and characteristics regulation of plant. Front. Plant Sci. 13: 919243.

Zhang, B. and T. Unver. 2018. A critical and speculative review on microRNA technology in crop improvement: Current challenges and future directions. Plant Science 274: 193–200.

Zhang, H. and X. Ma. 2019. CRISPR-Cas9 mediated gene editing of miR408 for improved tolerance to abiotic stresses in rice. Front. Plant Sci. 10: 1379.

Zhang, L., X. Liu and J. Zhang. 2018. Nanotechnology in agriculture and food safety: A review. Environ. Toxicol. Pharmacol. 73: 103280.

Zhang, X., Y. Zhang and J. Wang. 2020. CRISPR-Cas9 Based Genome Editing Reveals New Insights into Plant miRNA Function in rice. Front. Plant Sci. 11: 1256.

Zhang, Y., H. Zhang and Y. Liu. 2016. Efficient and stable genome editing in maize plants using the CRISPR/Cas9 system. J. Agric. Sci. 154(5): 1074–1080.

Zhang, Y., Q. Wang, J. Wang and X. Tang. 2021. Chemical modification and transformation strategies of guide RNAs in CRISPR-Cas9 gene editing systems. ChemPlusChem 86(4): 587–600.

Zhang, Z., J. Liu and Y. Zhao. 2019. Advances in CRISPR/Cas9-mediated genome editing for crop improvement. Front. Plant Sci. 10: 1158.

Zheng, Z., J. Yang, X. Wang, N. Zhang and H. Si. 2023. Potato Stu-miR398b-3p negatively regulates Cu/Zn-SOD response to drought tolerance. International Journal of Molecular Sciences 24(3): 2525.

Zhou, J., G. Liu, Y. Zhao, R. Zhang, X. Tang, L. Li et al. 2023. An efficient CRISPR–Cas12a promoter editing system for crop improvement. Nature Plants 9(4): 588–604.

Zhou, W., H. Han, X. Sun, X. Guo, J. Wen and X. Zhao. 2025. CRISPR/Cas9 technology in tumor research and drug development application progress and future prospects. Frontiers in Pharmacology 16: 1552741.

CHAPTER 16

Legal, Ethical, and Social Issues Remained in CRISPR/Cas Genome Editing Technology

Sana Ullah

1. Introduction

The discovery of CRISPR/Cas9 technology has transformed genetic engineering, providing unmatched accuracy and efficacy in genome modification across several species with profound consequences for health and agricultural sciences. Originally derived from the adaptive immune systems of bacteria and archaea, CRISPR/Cas9 has been transformed into a potent instrument that allows researchers to effortlessly add, delete, or modify specific DNA sequences (Doudna and Charpentier 2014). In a time of increasing environmental concerns, resource depletion, and worldwide population increase, sustainable food systems are critically required. Agriculture must adjust to climatic change, insect pressures, soil degradation, and nutritional requirements. CRISPR/Cas9 provides means to expedite the enhancement of crops and livestock with superior precision compared to traditional breeding techniques, hence offering prospects for improved food security, resilience, and biodiversity preservation (Chen et al. 2019).

However, in addition to its transformational capabilities, CRISPR technology has a range of legal, ethical, and social implications. Concerns regarding intellectual property rights, ecological safety, socio-economic justice, and global governance must be carefully considered to guarantee that the implementation of CRISPR technologies promotes rather than hinders sustainable development objectives. This chapter examines the utilization of CRISPR/Cas9 technology in various disciplines, for instance, health and agriculture, assesses the related legal and ethical dilemmas, and identifies avenues for integrating innovation with ethical accountability. As the new era of biotechnology begins with CRISPR technology that influences modern society in diverse manners because of this outstanding discovery of the 21st century (Ansori et al. 2023), the utilization of these technologies in specific domains of human activity is being examined from an ethical perspective. Consequently, effective oversight and control at both national and international levels are essential (Ardekani 2009).

Department of Agricultural Leadership, Education & Communication, College of Agricultural & Environmental Sciences, University of Georgia, Athens, Georgia 30602, USA.
Email: sana.ullah@uga.edu

DOI: 10.1201/9781003518204-16

1.1 Overview of CRISPR/Cas9 Technology and Its Potential in Agriculture

CRISPR (Clustered Regularly Interspaced Short Palindromic Repeats) and the accompanying protein Cas9 operate as molecular scissors directed by an RNA sequence, facilitating precise alterations to an organism's DNA. The Nobel Prize in Chemistry in 2020 was awarded for the invention of a genome editing method utilizing CRISPR–Cas9 technology, occurring less than ten years after the identification of the key chemical components of the system. For the first time in history, a Nobel Prize was conferred on two women, Emmanuelle Charpentier and Jennifer Doudna, for their pivotal discoveries in DNA editing via the CRISPR–Cas9 system, referred described as "genetic scissors." In comparison to traditional genetic alteration techniques, CRISPR is more rapid, cost-effective, precise, and adaptable (Adli 2018). Agricultural applications encompass gene editing for drought tolerance, disease resistance, pest resistance, higher nutritional value, and extended shelf life, all of which are crucial goals for global food security and climate resilience (Mulvihill et al. 2017). In contrast to transgenic genetic modifications, CRISPR can effectuate alterations without the introduction of exogenous DNA, potentially mitigating regulatory challenges and public resistance in certain regions.

CRISPR has already enabled the development of:

- Tomatoes exhibiting extended shelf life and enhanced flavor profiles (Li et al. 2018),
- Soybeans with superior oil quality (Haun et al. 2014),
- Heat-resistant cattle adapted to climate change (Van Eenennaam 2019).

These breakthroughs illustrate the significant and powerful impact of CRISPR in building climate-resilient, nutritionally rich, and sustainable agricultural systems. The progression of these techniques has provided a revolutionary tool in agricultural biotechnology, with extensive implications for ecosystems and living beings overall (Mulvihill et al. 2017).

1.2 Legal and Intellectual Property Issues Related to Gene Editing

The legal framework of CRISPR/Cas9 is constantly changing, driven by its transformational potential and economic significance. In agriculture, Intellectual Property Rights (IPRs) dictate the permissible use, conditions, and pricing of CRISPR-edited crops. Multinational businesses possess numerous essential patents, prompting concerns that farmers, particularly in low- and middle-income nations, may encounter obstacles in getting these technologies inexpensively or may grow reliant on proprietary seed systems (Jasanoff et al. 2015). The monopolization of CRISPR technologies may jeopardize local knowledge systems and traditional breeding techniques, while also prompting inquiries regarding benefit-sharing and biopiracy, particularly when genetic resources or indigenous landraces are utilized without sufficient remuneration to the indigenous populations. It is essential to develop more equitable licensing arrangements and investigate open-access or public-sector options that promote food sovereignty and fair access to biotechnology.

2. Ethical Considerations of CRISPR Applications

2.1 In Agricultural Innovation

Ethical problems in CRISPR agriculture arise from ecological, socio-economic, and cultural considerations. Modifying genomes for increased yield or resistance may provide immediate advantages but also pose significant hazards, including the disruption of ecological balances or the creation of new dependencies. The potential for off-target mutations, however few, necessitates thorough risk evaluation, particularly for species introduced into the environment (Wolt et al. 2016). Moreover, gene drive technologies that can propagate altered features throughout populations create significant concerns regarding permanent alterations to ecosystems (Esvelt et al. 2014). Culturally, several groups see gene editing with skepticism, stemming from apprehensions about "interfering with nature." Ethical implementation necessitates substantial public involvement, especially with indigenous and smallholder agricultural communities, to guarantee that innovations correspond with local values and goals. CRISPR/Cas9 has developed an easy, precise, and quick genome editing tool beneficial to the agricultural and pharmaceutical industries (Almeida and Ranisch 2022; Ansori et al. 2023), but it must be evaluated alongside a multifaceted spectrum of ethical, legal, and ecological issues (Mulvihill et al. 2017).

2.2 In Food Security and Biodiversity Protection

CRISPR provides tools to enhance food production, improve quality, and minimize losses. However, its implementation must be evaluated regarding overarching sustainability objectives, such as preserving crop diversity, soil vitality, and ecological integrity. Dependence on a limited array of genetically modified crops may diminish agro-biodiversity, enhancing susceptibility to pests and environmental disturbances. Consequently, the ethical implementation of CRISPR should enhance, rather than supplant, agroecological practices and innovations spearheaded by farmers (Montenegro de Wit 2022).

3. Agricultural Applications of CRISPR/Cas9

3.1 Improving Crop Resilience and Yields

The need for global food is increasing continuously due to the overwhelming population and lack of agricultural land. To address this growing need, it is necessary to discover seed varieties that are high-yielding as well as have strong resistance to adverse climatic conditions. Diverse strategies have been established to enhance crop quality, including mutagenesis, intergeneric crosses, and translocation breeding. At this stage, active collaboration between developers and the public is essential for researchers to articulate their findings and effectively communicate, to dispel myths around genome editing technology in plants. Research communication should occur through the entire process, encompassing various developmental stages (Ishii and Araki 2016), particularly about the many applications of these approaches and their effects on public and environmental health. These achievements will undoubtedly facilitate consumer acceptance of genome-edited crops (Gupta et al. 2021).

A primary commitment of CRISPR/Cas9 in agriculture is the creation of stress-resistant and high-yield crop types. Researchers have genetically changed rice to enhance its resistance to salinity and drought, while wheat has been altered to withstand fungal diseases such as powdery mildew (Wang et al. 2014). These developments are particularly vital in the context of climate change, which is modifying precipitation patterns, rising temperatures, and exacerbating the extreme weather phenomenon. Genome editing can diminish the necessity for chemical inputs like pesticides and herbicides by bolstering plant defenses, therefore reducing environmental impact and production expenses (Chen et al. 2019). Genetic scientists have utilized CRISPR to engineer tomato plants (*Solanum lycopersicum*) against bacterial attack by precisely mutating susceptibility genes. Agriculture should benefit from the tremendous potential and power of this genomic technology to enhance food security and avoid world famines. The application of CRISPR Cas necessitates adherence to legal and bioethical norms, which should primarily preserve human dignity, ensure patient integrity, and secure the confidentiality of genetic information to prevent misuse (Gonzalez-Avila et al. 2021).

3.2 Enhancing Nutritional Value and Food Quality

CRISPR is utilized to create nutritionally enhanced crops that address worldwide micronutrient deficits. Gene editing for biofortification can enhance the concentrations of vital elements, including iron, zinc, folate, and vitamin A, in staple crops such as rice, cassava, and maize (Gao

2021). Researchers have engineered rice types with enhanced iron content by focusing on vacuolar iron transporters (Zafar et al. 2020). Tomatoes with elevated gamma-aminobutyric acid (GABA) concentrations, potentially beneficial for blood pressure regulation, have been developed through meticulous genome editing (Nonaka et al. 2017). These examples illustrate how CRISPR can enhance both food quantity and quality, providing innovative solutions for nutritional security.

3.3 Livestock Breeding and Disease Resistance

CRISPR applications encompass not only crops but also livestock. Researchers have employed gene editing to give illness resistance, improve reproductive qualities, and enhance heat tolerance in animals (Rodriguez 2016; Wani et al. 2022). A notable instance is pigs genetically modified to withstand Porcine Reproductive and Respiratory Syndrome Virus (PRRSV), which incurs substantial economic losses worldwide. By inactivating the CD163 gene, which encodes the PRRSV receptor, researchers have created pigs that are resistant to infection while maintaining their health. Researchers have developed heat-tolerant cattle by focusing on the prolactin receptor gene, an advancement relevant to livestock management in warm areas (Van Eenennaam 2019). Although intriguing, these applications also present ethical problems regarding animal welfare, biodiversity, and the long-term ecological consequences of gene-edited animals, requiring robust governmental monitoring.

4. Food Security and Socio-Economic Implications

4.1 Benefits for Global Food Systems

CRISPR technology possesses the capacity to revolutionize global food systems by increasing productivity, minimizing losses, and bolstering resilience. It facilitates accelerated and more accurate crop enhancement compared to traditional breeding, providing a potent tool for addressing the food requirements of a burgeoning worldwide population projected to surpass 9.7 billion by 2050 (FAO 2021). Gene-edited crops with prolonged shelf life and less spoilage can mitigate post-harvest losses, which contribute substantially to food waste, especially in low-income nations. Furthermore, climate-resilient varieties can stabilize food production amid increasingly changeable environmental conditions, so enhancing national and regional food security (Marsh et al. 2021). The application of CRISPR/Cas9 technology across diverse crops has facilitated precise gene modifications, thereby allowing for the targeted resolution of specific agronomic challenges. The improvement of rice cultivars through CRISPR-mediated gene modifications aimed at enhancing tolerance to drought and salinity serves as a prominent example presented by Wani et al. (2022).

4.2 Risks for Smallholder Farmers and Equity Concerns

Despite its promise, CRISPR-based agriculture could intensify disparities if not implemented fairly. Smallholder farmers, essential to food production in numerous developing nations, frequently lack access to new technologies, financial resources, and legal support (Scheben and Edwards 2018). Restrictions on intellectual property, high development costs, and restricted access to technical expertise may impede adoption and exacerbate reliance on multinational seed corporations. This may result in the marginalization of traditional knowledge systems and jeopardize food sovereignty. Ethical innovation must prioritize equitable access, transparent licensing frameworks, and public investment in CRISPR research and development targeted to the needs of small-scale farmers. In the same vein, the legal and ethical landscape surrounding CRISPR is extremely complicated. Misuse of CRISPR could result in the weaponization of the technology or its unethical application in ways that undermine public trust and put the safety of the entire world in jeopardy (Wani et al. 2022). It is consequently necessary for researchers and governments to take a forward-thinking posture, putting in place protections that anticipate misuse without limiting innovation.

5. Human Health and Biomedical Applications

5.1 Genetic Diseases

The incorporation of CRISPR/Cas9 technology into biomedical research has markedly enhanced methodologies for drug discovery and the simulation of human diseases (Brokowski and Adli 2019; Lorenzo et al. 2022; Wani et al. 2022). One of its most promising applications pertains to the editing of patient-derived induced Pluripotent Stem Cells (iPSCs) to replicate genetic mutations specific to certain diseases and to modify animals, insects, plants, and microorganisms (Javed et al. 2018; Piergentili et al. 2021). This enables researchers to develop accurate cellular models that replicate human disease states, facilitating more reliable assessments of therapeutic compounds in preclinical studies (Ayanoğlu et al. 2020). Presently, CRISPR–Cas9 and associated technologies have effectively been employed to treat life-threatening diseases (Rasul et al. 2022), develop coronavirus detection assays, and edit human embryonic cells, resulting in the birth of infants with the introduced mutations (Gostimskaya 2022), but the application of these genetic engineering technologies presents questions surrounding ethical issues that influence people and societies (Lorenzo et al. 2022; Piergentili et al. 2021; Rasul et al. 2022).

CRISPR technology has demonstrated potential in addressing genetic disorders, including sickle cell anemia and beta-thalassemia, via *ex vivo* modification of hematopoietic stem cells (Frangoul et al. 2021). Clinical trials are in progress for the modification of somatic cells to address inherited retinal illnesses, hematological disorders, and some malignancies. These interventions entail alterations in somatic cells and do not impact the germline, rendering them ethically less controversial. By focusing on specific DNA sequences, CRISPR/Cas9 facilitates the precise correction of harmful mutations, providing the possibility of curative treatments instead of solely addressing symptoms. One of the most significant applications is the therapeutic targeting of monogenic blood disorders, including sickle cell disease, β-thalassemia, other diseases such as cystic fibrosis, muscular dystrophy, malaria, HIV, Zika virus, dengue, or the recent SARS-CoV-2 (Ansori et al. 2023; Rasul et al. 2022; Wani et al. 2022). The conditions resulting from single-point mutations in genes that encode hemoglobin subunits have traditionally necessitated continuous supportive care throughout a patient's life. In addition to the application of CRISPR/Cas9 in treating several diseases, its utilization in editing the genome for treating cancer signifies a swiftly evolving field of research, especially about cancer immunotherapy (Ayanoğlu et al. 2020).

5.2 Ethical Boundaries in Human Genome Editing

The use of techniques related to genetically engineered genes on humans has been one of the most controversial and ethically significant and challenging topics in the field of genetic engineering (Baumann 2016). It is crucial to evaluate the social implications of wider use of the technology, particularly concerning costs, accessibility, potential disparities, and the socio-demographic circumstances of marginalized communities (Lorenzo et al. 2022). The 2018 incident involving gene-edited twins in China provoked global concern and prompted renewed demands for international regulation and ethical standards (Cyranoski 2019). The global scientific community largely concurs that germline editing should not advance without extensive public agreement, strong ethical guidelines, and effective regulatory structures (Jasanoff et al. 2019). Although CRISPR's medical promise is extensive, the ethical complexities, particularly about equity, consent, and long-term consequences, necessitate considerable caution. Nonetheless, due to its potential and benefits, the application of CRISPR in the medical field brings forth significant ethical and legal difficulties, particularly considering the complex nature of psychiatric disorders and the historical mistreatment of vulnerable groups in mental health studies (Foulkes et al. 2019). The dual-use nature of this technology raises concerns regarding its potential exploitation for bioterrorism and issues surrounding genetic privacy (D'Souza et al. 2023; Priyanka 2020; Wani et al. 2022). Such concerns underscore the necessity of establishing effective regulatory frameworks to ensure that the benefits of CRISPR/Cas9 are harnessed responsibly and ethically (Wani et al. 2022).

The ethical discourse can be encapsulated into four basic questions (Lehmann 2017) for the moral evaluation of any genetic change, irrespective of whether it pertains to the germ line. The author designated these criteria as the '4-S': (1) Safety, (2) the importance of the harm to be prevented, (3) successive generations, and (4) social implications. Hildt (2016) argued that attempts to genetically change the germline must satisfy three criteria that are now often unmet.

1. Prior robust and secure expertise in gene therapy utilizing somatic cells.
2. Animal tests that ensure safety and reproducibility, indicating that following human therapies would likely be successful.
3. Societal and political approval of the technique.

A critical ethical consideration in research is that the benefits must outweigh the risks. Many individuals have advocated for a public discourse regarding the social, ethical, and legal implications of employing genome-editing techniques in the human germline; nonetheless, additional pertinent difficulties and issues need discussion in connection with this technology. The application of CRISPR/Cas9 raises numerous social and ethical concerns, extending beyond human subjects to encompass other organisms and environmental considerations, including adherence to the principle of non-maleficence in risk assessment, safety measures to prevent ecological degradation, and the potential utilization of the technology for genetic enhancement (Rodriguez 2016).

6. Regulatory Frameworks and Governance Challenges

6.1 International Conventions and Agreements

The worldwide implications of CRISPR/Cas9 necessitate synchronized international regulation. Current accords, like the Cartagena Protocol on Biosafety (part of the Convention on Biological Diversity), govern the transboundary movement of genetically modified organisms but do not adequately encompass genome editing technology. The Nagoya Protocol underscores the importance of access to genetic resources and the equitable distribution of benefits; nonetheless, uncertainties persist over the applicability of benefit-sharing frameworks to gene-edited crops originating from indigenous landraces or local seed systems (Morgera et al. 2012). Emerging technologies such as CRISPR challenge the sufficiency of these frameworks, especially when gene-edited living beings lack foreign DNA and are not categorized as genetically modified in certain countries (e.g., the U.S.). This regulatory ambiguity has resulted in inconsistencies among nations, possibly hindering international trade and biosafety norms. The United Nations must be equipped to address diverse developing biological dangers and establish a cohesive regulatory framework to mitigate future risks (Wani et al. 2022).

6.2 National Regulatory Variations

Countries have implemented varying regulatory frameworks for CRISPR in agriculture. The United States Department of Agriculture (USDA) presently does not regulate CRISPR-edited crops that lack foreign genes, however, the European Court of Justice determined in 2018 that gene-edited species are subject to the same stringent restrictions as Genetically Modified Organisms (GMOs) (Callaway 2018). These disparities significantly impact international cooperation, market entry, and avenues for innovation. Regulatory harmonization, or at a minimum, mutual recognition systems, are essential to reduce barriers while ensuring environmental and public safety. The bioethical concerns that CRISPR-Cas9 technology may cause to the environment, agriculture, and animals must not be overlooked. Recent agreements between the scientific community and the pharmaceutical and biotechnology industries on the therapeutic application of CRISPR-Cas9 have raised public safety concerns (Shinwari et al. 2018). The regulations and statutes governing the content and implementation of these agreements should be formulated expeditiously and shared with the public (Ayanoğlu et al. 2020).

6.3 Need for Inclusive and Participatory Governance

Effective governance of CRISPR necessitates the inclusive engagement of several stakeholders, including farmers, indigenous people, scientists, ethicists, policymakers, and civil society organizations. Public engagement is crucial for formulating socially acceptable policies and fostering trust in gene-editing technology (Jasanoff et al. 2015). Clear communication, public engagement, and investment in scientific education can mitigate misconceptions and facilitate informed decision-making. Participatory governance ensures that technological advancement aligns with social values and supports collective sustainability goals.

7. Future Prospects and Responsible Innovation Pathways

The future of CRISPR/Cas9 in agriculture presents significant potential, accompanied by substantial responsibility. The technology provides a swift means for enhancing crops and livestock, although its implementation must adhere to inclusive ethics, transparent governance, and ecological sustainability. Moral decision-making must adapt alongside advancements in genomic engineering and contend that it is reasonable for national and supranational legislatures to contemplate evidence-based regulation of specific CRISPR uses to enhance human health and progress (Brokowski and Adli 2019).

7.1 Scaling Equitably and Responsibly

To ensure that CRISPR enhances sustainable food systems, its advantages must be available at all levels of production. Public-sector investments are essential to assist local breeding projects, especially in the Global South, utilizing gene-editing technologies to create crops adapted to local agroecological circumstances. Innovation should not be motivated exclusively by economic interests; instead, it ought to tackle genuine global issues such as climate resilience, nutrition, and the reduction of agricultural inputs like fertilizers and pesticides (Gao 2021). CRISPR should be included with agroecological techniques rather than being regarded as a technical silver bullet.

7.2 Building Public Trust and Policy Infrastructure

A robust policy framework is essential. Governments and international organizations must endorse regulatory frameworks that are grounded in evidence, precautionary in nature, and sufficiently adaptable to accommodate scientific progress. Public confidence in CRISPR technologies relies on transparency, ethical evaluation, and accountability. Frameworks such as Responsible Research and Innovation (RRI), which advocate for anticipation, reflection, inclusiveness, and responsiveness, can guide the deployment of CRISPR in agriculture (Stilgoe et al. 2020). The success of CRISPR will ultimately hinge on both technical proficiency and the integrity of decision-making processes, governance of technologies, and equitable distribution of advantages. "There needs to be transparency, you can't hide information; you shouldn't avoid telling people what you're doing until you secretly release a product. Every report that's come out since 2015 about gene editing has stressed the need for this widespread… engagement, [involving] not just global scientific stakeholders, but… the public… as well" (Lee 2021).

If CRISPR/Cas9 continues to thrive globally, we can anticipate a broader market for genetically modified species, which will raise questions regarding their regulation (Shinwari et al. 2018).

There is a significant necessity to adopt CRISPR technology across multiple platforms, with contributions from specialists in ethical, social, religious, legislative, and technological domains to formulate a sustainable strategy addressing the advantages and concerns associated with CRISPR technology. CRISPR/Cas9 engenders numerous ethical and societal dilemmas, including both human considerations and environmental concerns. Ecological and environmental risk assessments should also be conducted (Shinwari et al. 2018).

8. Conclusion

CRISPR/Cas9 technology represents a groundbreaking innovation with the potential to reshape agriculture, improve global food security, and contribute to sustainable development. Its ability to enhance crop traits, fortify nutritional quality, and improve livestock productivity makes it a powerful tool in responding to the global challenges of hunger, climate change, and ecological degradation. However, with such a transformative power comes equally significant ethical, legal, and socio-political responsibilities. Equitable access, ecological safeguards, and inclusive governance must be prioritized to ensure that CRISPR serves public interest and environmental sustainability, rather than corporate dominance or techno-centric solutions. By aligning CRISPR research and deployment with values of justice, transparency, and precaution, societies can harness their full potential to support resilient food systems and shared global wellbeing.

CRISPR/Cas9 technology signifies a revolutionary advancement capable of treating diseases, transforming agriculture, enhancing global food security, and fostering sustainable development. Its capacity to promote the pharmaceutical industry, increase agricultural characteristics, strengthen nutritional quality, and raise livestock productivity renders it a formidable tool in addressing the worldwide issues of health, hunger, climate change, and ecological degradation. Nevertheless, this transformative power entails equally substantial ethical, legal, and socio-political obligations. Prioritizing equitable access, ecological protections, and participatory governance is essential to guarantee that CRISPR advances public interest and environmental sustainability, rather than corporate hegemony or technology-driven solutions. By aligning CRISPR research and implementation with principles of fairness, openness, and caution, societies can fully leverage their promise to enhance resilient agricultural systems and promote collective global well-being.

Shortly after its introduction in 2012, CRISPR was quickly adopted due to its user-friendliness and simplicity. It is under consideration for many uses, ranging from agriculture to medicinal therapies, including human germline modifications to rectify genetic disorders. Rigorous laws and guidelines, along with global discourse and awareness, are essential to guarantee the responsible and judicious application of CRISPR-mediated genome editing technology. A broad dialogue among scientists, ethicists, industrialists, and policymakers is necessary regarding its societal ramifications. The perspectives of various societal aspects, including the general populace and religious experts, are also crucial (Shinwari et al. 2018).

References

Adli, M. 2018. The CRISPR tool kit for genome editing and beyond. Nat. Commun. 9: 1911.

Almeida, M. and R. Ranisch. 2022. Beyond safety: Mapping the ethical debate on heritable genome editing interventions. Humanit. Soc. Sci. Commun. 9: 1–14.

Ansori, A.N.M., Y. Antonius, R.J. Susilo, S. Hayaza, V.D. Kharisma, A.A. Parikesit et al. 2023. Application of CRISPR-Cas9 genome editing technology in various fields: A review. Narra J. 3(2): e184.

Ardekani, A.M. 2009. Genetic technologies and ethics. J. Med. Ethics Hist. Med. 2: 11.

Ayanoğlu, F.B., A.E. Elçin and Y.M. Elçin. 2020. Bioethical issues in genome editing by CRISPR-Cas9 technology. Turk. J. Biol. 44(2): 110–120.

Baumann, M. 2016. CRISPR/Cas9 genome editing–New and old ethical issues arising from a revolutionary technology. NanoEthics. 10: 139–159.

Brokowski, C. and M. Adli. 2019. CRISPR ethics: Moral considerations for applications of a powerful tool. J. Mol. Biol. 431: 88–101.

Callaway, E. 2018. CRISPR plants now subject to tough GM laws in European Union. Nature 560: 16–17.

Chen, K., Y. Wang, R. Zhang, H. Zhang and C. Gao. 2019. CRISPR/Cas genome editing and precision plant breeding in agriculture. Annu. Rev. Plant Biol. 70: 667–697.

Cyranoski, D. 2019. What CRISPR-baby prison sentences mean for research. Nature 577: 154–155.

D'Souza, R., M. Mathew and K.M. Surapaneni. 2023. A scoping review on the ethical issues in the use of CRISPR-Cas9 in the creation of human disease models. J. Clin. Diagn. Res. 17(12). 10.7860/JCDR/2023/68275.18809.

Doudna, J.A. and E. Charpentier. 2014. The new frontier of genome engineering with CRISPR-Cas9. Science 346: 1258096.

Esvelt, K.M., A.L. Smidler, F. Catteruccia and G.M. Church. 2014. Concerning RNA-guided gene drives for the alteration of wild populations. eLife 3: e03401.

FAO. 2021. The State of Food Security and Nutrition in the World 2021. https://www.fao.org/documents/card/en/c/cc0640en.

Foulkes, A.L., T. Soda, M. Farrell, P. Giusti-Rodríguez and G. Lázaro-Muñoz. 2019. Legal and ethical implications of CRISPR applications in psychiatry. N.C. Law Rev. 97(5): 1359.

Frangoul, H., D. Altshuler, M.D. Cappellini, Y.S. Chen, J. Domm, B.K. Eustace et al. 2021. CRISPR-Cas9 gene editing for sickle cell disease and β-thalassemia. N. Engl. J. Med. 384(3): 252–260.

Gao, C. 2021. Genome engineering for crop improvement and future agriculture. Cell 184(6): 1621–1635.

Gonzalez-Avila, L.U., J.M. Vega-López, L.I. Pelcastre-Rodríguez, O.A. Cabrero-Martínez, C. Hernández-Cortez and G. Castro-Escarpulli. 2021. The challenge of CRISPR-Cas toward bioethics. Front. Microbiol. 12: 657981.

Gostimskaya, I. 2022. CRISPR–Cas9: A history of its discovery and ethical considerations of its use in genome editing. Biochemistry (Moscow) 87(8): 777–788.

Gupta, S., A. Kumar, R. Patel and V. Kumar. 2021. Genetically modified crop regulations: Scope and opportunity using the CRISPR-Cas9 genome editing approach. Mol. Biol. Rep. 48: 4851–4863.

Haun, W., A. Coffman, B.M. Clasen, Z.L. Demorest, A. Lowy, E. Ray et al. 2014. Improved soybean oil quality by targeted mutagenesis of the fatty acid desaturase 2 gene family. Plant Biotechnol. J. 12(7): 934–940.

Hildt, E. 2016. Human germline interventions–Think first. Front. Genet. 7: 81.

Ishii, T. and M. Araki. 2016. Consumer acceptance of food crops developed by genome editing. Plant Cell Rep. 35: 1507–1518.

Jasanoff, S., J.B. Hurlbut and K. Saha. 2015. CRISPR democracy: Gene editing and the need for inclusive deliberation. Issues Sci. Technol. 32(1): 25–32.

Jasanoff, S., J.B. Hurlbut and K. Saha. 2019. Democratic governance of human germline genome editing. CRISPR J. 2(5): 266–271.

Javed, M.R., M. Sadaf, T. Ahmed, A. Jamil, M. Nawaz, H. Abbas et al. 2018. CRISPR-Cas system: History and prospects as a genome editing tool in microorganisms. Curr. Microbiol. 75: 1675–1683.

Lee, V.S. 2021. The ethics of CRISPR-Cas9 technology and its socioeconomic implications.

Lehmann, L.S. 2017. Is editing the genome for climate change adaptation ethically justifiable? AMA J. Ethics. 19(12): 1186–1192.

Li, T., B. Liu, M.H. Spalding, D.P. Weeks and B. Yang. 2018. High-efficiency TALEN-based gene editing produces disease-resistant rice. Nat. Biotechnol. 30(5): 390–392.

Lorenzo, D., M. Esquerda, F. Palau and F.J. Cambra. 2022. Ethics and genomic editing using the CRISPR-Cas9 technique: Challenges and conflicts. Nanoethics. 16(3): 313–321.

Marsh, J.I., H. Hu, M. Gill, J. Batley and D. Edwards. 2021. Crop breeding for a changing climate: Integrating phenomics and genomics with bioinformatics. Theor. Appl. Genet. 134: 1677–1690.

Montenegro de Wit, M. 2022. Can agroecology and CRISPR mix? The politics of complementarity and moving toward technology sovereignty. Agric. Hum. Values. 39(2): 733–755.

Morgera, E., M. Buck and E. Tsioumani. 2012. The 2010 Nagoya Protocol on Access and Benefit-sharing in Perspective: Implications for International Law and Implementation Challenges. Martinus Nijhoff.

Mulvihill, J.J., B. Capps, Y. Joly, T. Lysaght, H.A.E. Zwart and R. Chadwick. 2017. Ethical issues of CRISPR technology and gene editing through the lens of solidarity. Br. Med. Bull. 122(1): 17–29.

Nonaka, S., C. Arai, M. Takayama, C. Matsukura and H. Ezura. 2017. Efficient increase of ɣ-aminobutyric acid (GABA) content in tomato fruits by targeted mutagenesis. Sci. Rep. 7: 7057.

Piergentili, R., A. Del Rio, F. Signore, F. Umani Ronchi, E. Marinelli and S. Zaami. 2021. CRISPR-Cas and its wide-ranging applications: From human genome editing to environmental implications, technical limitations, hazards and bioethical issues. Cells 10(5): 969.

Priyanka, P. 2020. Patenting CRISPR-Cas9 therapeutic applications: Legal framework for human welfare-based applications in USA and EU. https://www.diva-portal.org/smash/get/diva2:1439274/FULLTEXT01.pdf.

Rasul, M.F., B.M. Hussen, A. Salihi, B.S. Ismael, P.J. Jalal, A. Zanichelli et al. 2022. Strategies to overcome the main challenges of the use of CRISPR/Cas9 as a replacement for cancer therapy. Mol. Cancer. 21(1): 1–30.

Rodriguez, E. 2016. Ethical issues in genome editing using CRISPR/Cas9 system. J. Clin. Res. Bioeth. 7(2): 1000266.

Scheben, A. and D. Edwards. 2018. Bottlenecks for genome-edited crops on the road from lab to farm. Genome Biol. 19: 1–7.

Shinwari, Z.K., F. Tanveer and A.T. Khalil. 2018. Ethical issues regarding CRISPR mediated genome editing. Curr. Issues Mol. Biol. 26(1): 103–110.

Stilgoe, J., R. Owen and P. Macnaghten. 2020. Developing a framework for responsible innovation. pp. 347–359. *In*: The Ethics of Nanotechnology, Geoengineering, and Clean Energy. Routledge.

Van Eenennaam, A.L. 2019. Genetic modification of food animals. Curr. Opin. Biotechnol. 61: 1–7.
Wang, Y., X. Cheng, Q. Shan, Y. Zhang, J. Liu, C. Gao et al. 2014. Simultaneous editing of three homoeoalleles in hexaploid bread wheat confers heritable resistance to powdery mildew. Nat. Biotechnol. 32(9): 947–951.
Wani, A.K., N. Akhtar and S. Shukla. 2022. CRISPR/Cas9: Regulations and challenges for law enforcement to combat its dual-use. Forensic Sci. Int. 334: 111274.
Wolt, J.D., K. Wang and B. Yang. 2016. The regulatory status of genome-edited crops. Plant Biotechnol. J. 14(2): 510–518.
Zafar, K., K.E. Sedeek, G.S. Rao, M.Z. Khan, I. Amin, R. Kamel et al. 2020. Genome editing technologies for rice improvement: Progress, prospects, and safety concerns. Front. Genome Ed. 2: 5.

Index

R

S

T

V

About the Editors

Abdul Basit is a Ph.D. researcher at the Department of Horticultural Science, Kyungpook National University, South Korea. He previously worked at the National Agricultural Research Center, Islamabad, and the Climate Change Center, The University of Agriculture, Peshawar. Supported by Pakistani and Korean government scholarships, he has published 67+ articles and 17+ book chapters. He has participated in over 35 scientific events. His research focuses on molecular cytogenetics, plant genomics, breeding, stress physiology, and horticultural crop production.

Heba I. Mohamed is a Professor of Plant Physiology at Ain Shams University. Her research focuses on plant stress physiology, biochemistry, eco-friendly stress alleviation, and genetic diversity. She has edited nine books published by international publishers and has also published 40 book chapters, eight reviews, and over 140 research articles. She obtained Encouragement State Award in Basic Sciences (Biological Sciences) 2018 and has been listed among Stanford–Elsevier's top 2% most influential scientists (2021–2024).

Syed Tanveer Shah is an Assistant Professor at the Department of Agriculture, Hazara University, and previously served over a decade in Horticulture at The University of Agriculture, Peshawar. He holds B.Sc (Hons), M.Sc (Hons), and Ph.D. degrees in Horticulture. Listed among H-index–based scientists (2022–23), he has 60+ research articles, and 8+ book chapters. He has participated in 20+ scientific events. His expertise includes pomology, postharvest science, stress physiology, plant genomics, and horticultural crop production.

Mona FA Dawood is an Associate Professor at the Botany and Microbiology Department, Assiut University, Egypt. She earned her M.Sc and Ph.D. in abiotic stress and has published 50+ research articles, 5 reviews, 9 book chapters, and 10 conference abstracts. With an H-index of 21, she serves as editor and guest editor for reputed journals and has reviewed over 210 articles. In 2023, she was recognized among Stanford University's top 2% scientists worldwide.